AF411367

Peptides for Health Benefits 2019

Peptides for Health Benefits 2019

Volume 2

Special Issue Editors

Blanca Hernández-Ledesma
Cristina Martínez-Villaluenga

MDPI • Basel • Beijing • Wuhan • Barcelona • Belgrade • Manchester • Tokyo • Cluj • Tianjin

Special Issue Editors
Blanca Hernández-Ledesma
Institute of Food Science Research (CIAL, CSIC-UAM, CEI UAM+CSIC)
Spain

Cristina Martínez-Villaluenga
Institute of Food Science, Technology and Nutrition (ICTAN, CSIC)
Spain

Editorial Office
MDPI
St. Alban-Anlage 66
4052 Basel, Switzerland

This is a reprint of articles from the Special Issue published online in the open access journal *International Journal of Molecular Sciences* (ISSN 1422-0067) (available at: https://www.mdpi.com/journal/ijms/special_issues/Peptides_2019).

For citation purposes, cite each article independently as indicated on the article page online and as indicated below:

LastName, A.A.; LastName, B.B.; LastName, C.C. Article Title. *Journal Name* **Year**, *Article Number*, Page Range.

Volume 2
ISBN 978-3-03936-084-0 (Hbk)
ISBN 978-3-03936-085-7 (PDF)

Volume 1-2
ISBN 978-3-03936-082-6 (Hbk)
ISBN 978-3-03936-083-3 (PDF)

Contents

About the Special Issue Editors

Blanca Hernández-Ledesma, Ph.D. earned a B.S. in Pharmacy in 1998, and defended her Ph.D. thesis in Pharmacy in 2002. Her research career has focused on the biological activity of food proteins/peptides, aiming to better understand their health implications and the development of novel food ingredients. She is author of 76 JCR articles, 9 popular science articles, and 30 book chapters, with an h-index 33 (WoS). Her results have been presented in 78 international and national conferences. She has supervised 4 Doctoral theses and 14 Master theses. She has participated in more than 30 international and national research projects. She has participated as a member of the Selection Board for Tenured Scientists, Ph.D. and Masters' thesis dissertation committees, reviewer of international Ph.D. theses, and member of national and international projects evaluation panels. She is member of the Editorial Committees of 3 books and 8 journals, and collaborates as a reviewer for more than 90 journals.

Cristina Martínez-Villaluenga (Ph.D.), B.S. in Biology by University Complutense of Madrid in 2001, Ph.D. in Food Science from the University Autonoma of Madrid in 2006. She joined the Spanish Research Council (CSIC) in 2009. The long-term goal of Dr. Martinez's research program is to enhance the health of individuals by identifying and determining the benefits of the bioactive components of plant foods with special focus on bioactive peptides. Dr. Martínez's research on legumes, cereals, and pseudocereals has led to increased understanding of the anti-inflammatory, anti-hypertensive, anti-diabetic, and other physiological properties of these foods. She is the author of 94 JCR articles and 9 book chapters with an h-index 31 (WoS). Her results have been disseminated in 84 international and national conferences and social media. In the last 10 years, she has supervised a total of 9 Ph.D. theses, 5 Masters' theses, and more than 20 undergraduate students. She has participated in a total of 35 international and national R&D projects and contracts with the agri-food sector. She is the member of the Editorial Committees of 3 books and 3 journals.

International Journal of
Molecular Sciences

Article

In Silico and In Vitro Assessment of Portuguese Oyster (*Crassostrea angulata*) Proteins as Precursor of Bioactive Peptides

Honey Lyn R. Gomez [1], Jose P. Peralta [1], Lhumen A. Tejano [1] and Yu-Wei Chang [2,*]

[1] Institute of Fish Processing Technology, College of Fisheries and Ocean Sciences, University of the Philippines Visayas, Miagao 5023, Iloilo, Philippines; honeylyngomez23@gmail.com (H.L.R.G.); f153mentor@yahoo.com (J.P.P.); lhumentejano@gmail.com (L.A.T.)

[2] Department of Food Science, National Taiwan Ocean University, Keelung 202, Taiwan

* Correspondence: bweichang@mail.ntou.edu.tw; Tel.: +886-2-2462-2192 (ext. 5152)

Received: 26 September 2019; Accepted: 17 October 2019; Published: 20 October 2019

Abstract: In this study, the potential bioactivities of Portuguese oyster (*Crassostrea angulata*) proteins were predicted through in silico analyses and confirmed by in vitro tests. *C. angulata* proteins were characterized by sodium dodecyl sulphate polyacrylamide gel electrophoresis (SDS-PAGE) and identified by proteomics techniques. Hydrolysis simulation by BIOPEP-UWM database revealed that pepsin (pH > 2) can theoretically release greatest amount of bioactive peptides from *C. angulata* proteins, predominantly angiotensin I-converting enzyme (ACE) and dipeptidyl peptidase IV (DPP-IV) inhibitory peptides, followed by stem bromelain and papain. Hydrolysates produced by pepsin, bromelain and papain have shown ACE and DPP-IV inhibitory activities in vitro, with pepsin hydrolysate (PEH) having the strongest activity of 78.18% and 44.34% at 2 mg/mL, respectively. Bioactivity assays of PEH fractions showed that low molecular weight (MW) fractions possessed stronger inhibitory activity than crude hydrolysate. Overall, in vitro analysis results corresponded with in silico predictions. Current findings suggest that in silico analysis is a rapid method to predict bioactive peptides in food proteins and determine suitable enzymes for hydrolysis. Moreover, *C. angulata* proteins can be a potential source of peptides with pharmaceutical and nutraceutical application.

Keywords: *Crassostrea angulata*; in silico; BIOPEP-UWM database; bioactive peptides; proteomics

---

1. Introduction

Oysters are the most popular and abundantly cultured shellfish in Taiwan [1]. They are considered as one of the major species of marine bivalves, comprising 33% of the total global production [2]. In most countries, these marine bivalves are consumed as food due to its health benefits, versatility, and easy-to-prepare characteristics. Furthermore, oysters have been used as raw materials for canning, bottling, and for the production of condiments like oyster sauce and powders. Despite being highly nutritious, oysters have not gained much attention because of its inherent characteristic flavor [3]. There are also some issues associated with post-harvest processing and handling of oysters, causing its low market value. Moreover, unprocessed oyster meat has a very short shelf life and are known to pose risk to public health [4,5]. Thus, many researchers are putting much effort into the search for new post-harvest application and development of high value products from oysters.

Hypertension and diabetes mellitus type-II are two of the most common chronic diseases affecting millions of people nowadays [6]. The development of these diseases is caused mainly by certain factors. Blood pressure is regulated by the renin-angiotensin system in the body. Renin catalyzes angiotensinogen to produce a vasodilator angiotensin I. Angiotensin-I converting enzyme (ACE) is an enzyme responsible for cleavage of angiotensin I, converting it to a potent vasoconstrictor angiotensin II [7,8]. Type 2 diabetes mellitus (T2DM) is characterized by hyperglycemia due to impaired insulin secretion, as a result of degradation of incretin hormones. During meals, endocrine cells release incretin hormones such as glucagon-like peptide-1 (GLP-1) and glucose-dependent insulinotropic polypeptide (GIP) [9]. These hormones stimulate pancreatic β-cell to boost glucose-dependent insulin secretion and suppress glucagon secretion, resulting to normal blood glucose levels [10,11]. The dipeptidyl peptidase IV (DPP-IV) is a ubiquitously expressed enzyme mainly involved in the modulation of biological activity of circulating peptide hormones by breaking down the two *N*-terminal amino acids X-Pro and X-Ala. Consequently, it could result to degradation and inactivation of numerous incretin hormones with Ala as the second *N*-terminal residue, such as GLP-1 and GIP [12,13]. These mechanisms have been the basis for the formulation of therapeutic drugs targeting those enzymes. Through the years, synthetic ACE and DPP-IV inhibitors are being tried in the management of these diseases [14–17]. However, these drugs are believed to cause negative effects to human health. Daily consumption of food containing ACE and DPP-IV inhibitory peptides are known to help lower blood pressure and blood sugar to healthy levels without exhibiting undesirable effects [18]. Thus, food proteins from natural sources are now being studied as alternative therapeutic agents.

Oyster is a rich source of proteins which generally ranges from 37%–81% on a dry weight basis [19–21]. In general, proteins contain peptides and essential amino acids which possess specific biological activity. Biologically active peptides are short sequences of amino acids that can be released from protein precursors through gastrointestinal digestion and food processing. They provide physiological effects in the body and function as regulatory compounds with hormone-like activity [22]. Peptides need to be released from the parent protein, be ingested, be bioaccessible, and reach the target site in sufficient quantities to exhibit biological bioactivity. Recently, several studies have been focused on the generation of bioactive peptides from food proteins and their utilization as functional ingredients [23]. Previous studies revealed that oyster is a good source of biologically active peptides with antioxidant, anti-cancer, ACE inhibitory, and anti-microbial activities in vitro and show antihypertensive activity in vivo [24–27]. Having these therapeutic potential, oysters can be considered as an alternative source of peptides that can be used as an ingredient for functional foods and nutraceuticals.

One of the most common methods used for the production of bioactive peptides from food proteins is by enzymatic hydrolysis. Traditionally, the selection of enzymes suitable for liberating potent peptides are based only on literature surveys, and in vitro analyses [28]. However, this approach is costly and time-consuming. Therefore, to overcome the drawbacks of this approach, in silico technique has been proposed and utilized. This technique is useful in predicting the release of bioactive peptides from known protein sequences and selecting suitable enzyme for hydrolysis [28,29]. Furthermore, the use of this technique for screening and identification of novel bioactive peptides had shown to be much more economical and time-saving [30].

Therefore, the objectives of this study were to assess the usefulness of in silico techniques in identifying the bioactive peptides encrypted in the *C. angulata* proteins and screening for the most suitable enzyme capable of releasing these peptides. Furthermore, it aimed to evaluate the bioactivities of *C. angulata* protein hydrolysate through in vitro analysis.

2. Results and Discussion

2.1. Identified Proteins From C. angulata

Freeze-dried Portuguese oyster (*C. angulata*) was subjected to SDS-PAGE to separate the proteins according to their molecular weights (MWs). Among all the bands observed in the SDS-PAGE gel, the eight most distinct bands were selected for further protein identification (Figure 1). These bands were subjected to in-gel digestion and nanoLC-nanoESI-MS/MS analysis and results obtained were then matched with the information from different protein databases through Mascot database search.

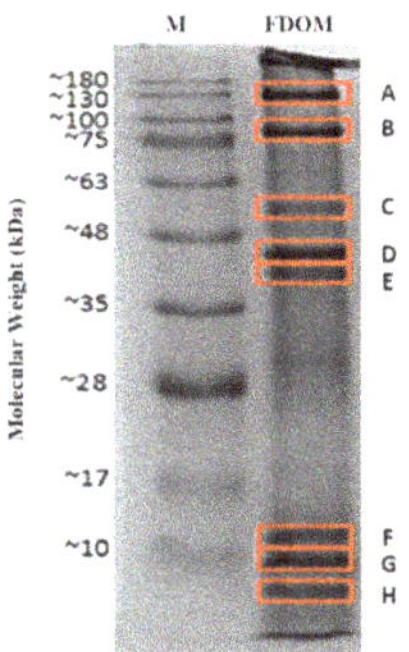

Figure 1. Protein patterns of Portuguese Oyster (*C. angulata*) by 12% sodium dodecyl sulphate polyacrylamide gel electrophoresis (SDS-PAGE). M: Protein marker; FDOM: freeze-dried oyster (*C. angulata*) meat.

The identification of *C. angulata* proteins is based mainly on the occurrence of matched tryptic peptides from oyster (resulting from trypsin digestion) within the sequences of known proteins from the database. Based on the result from Mascot MS/MS ion search, all the identified tryptic peptides from oyster were listed as doubly or triply charged peptides. Figure 2 illustrates how the doubly charged peptide IDSLEGSVSR (MW = 1061.53 Da) and triply charged peptide LTQENFDLQHQVQELDAANAGLAK (MW = 2652.32 Da) from paramyosin (band B) were identified using the mass spectrum from nanoLC-nanoESI-MS/MS analysis. The doubly charged peptide has an observed signal *m/z* of 531.78 (Figure 2A). Insert (a) displays the 0.5 difference between the adjacent signals while insert (b) shows the fragmentation spectra of the identified peptide. On the other hand, triply charged peptide with an observed signal *m/z* 885.45 was distinguished by a 0.3 difference between the adjacent signals, as shown in insert (a) (Figure 2B). The final result and the fragmentation of this peptide was illustrated in insert (b).

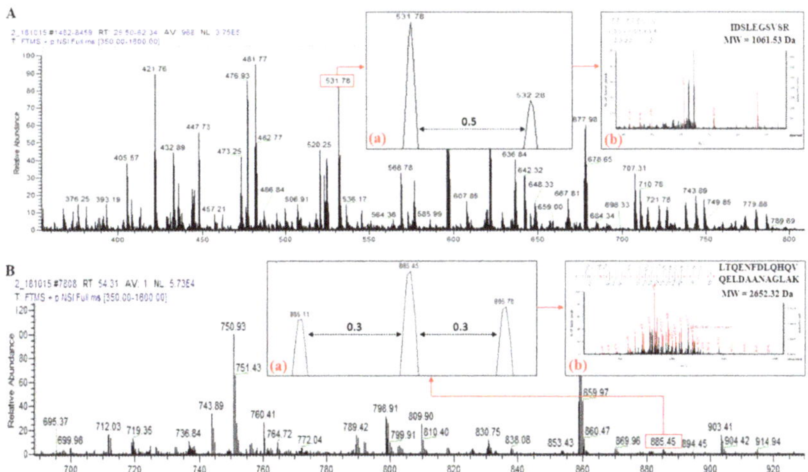

Figure 2. NanoLC-nanoESI-MS/MS spectra (*m/z* region 350 to 800 Da and 690 to 930 Da) of oyster protein band B with representative spectra of identified tryptic peptides in doubly (**A**) and triply (**B**) charged signal.

Out of 16,598,945 protein sequences discovered through Mascot database search, 352 proteins under the genus *Crassostrea* were identified. In each band, one protein (highest scoring *Crassostrea* sp. protein) was selected for further screening and evaluation. From that, five proteins belonging to *Crassostrea gigas* were chosen based on their high protein scores and sequence coverage. These proteins are the myosin heavy chain from striated muscle isoform X1, paramyosin isoform X2, tropomyosin isoform X1, myosin regulatory light chain B from smooth adductor muscle isoform X2, and actin. The accession numbers, protein lengths, scores, sequence coverage, and MWs of the selected proteins were listed in Table 1.

Table 1. Identified proteins from Portuguese oyster (*C. angulata*) meat and their characteristics.

Protein Name	Accession Number (NCBI)	Protein Score	Sequence Coverage (%)	Amino Acid Length	Molecular Weights from NCBI Database (kDa)	Molecular Weights Estimated from SDS-PAGE (kDa)
Myosin heavy chain, striated muscle isoform X1 *	XP_011442515.1	4670	48%	1901	222.66	130.00 (A) 82.50 (B)
Paramyosin isoform X2 *	XP_011429256.1	2736	65%	851	102.21	82.50 (B)
	EKC37566.1	789	20%	2001	229.67	55.14 (C)
						44.55 (D)
Myosin heavy chain, striated muscle						40.75 (E)
						11.66 (F)
						9.15 (G)
						5.76 (H)
Actin *	EKC30049.1	1272	58%	351	41.80	44.55 (D)
Tropomyosin isoform X1 *	XP_019925727.1	579	70%	251	33.02	40.75 (E)
Hypothetical protein CGI_10010027	EKC40031.1	220	27%	151	20.58	11.66 (F)
Myosin regulatory light chain B, smooth adductor muscle isoform X2 *	XP_011417566.1	112	50%	151	18.63	9.15 (G) 11.66 (F) 5.76 (H)

* Selected proteins for in silico analysis (based on protein score and sequence coverage).

As observed, myosin heavy chain is the largest among the five selected proteins with a theoretical MW of 220 kDa while myosin regulatory light chain B is the shortest (18.63 kDa). These proteins are both found in the muscle of oysters and other mollusks. Myosin is a contractile protein which plays a big role in muscle contraction. It is composed of six subunits: Two heavy chains and 4 light chains [31]. Paramyosin, on the other hand, is one unique protein that can be found in invertebrates such as oyster. It forms the core protein of the thick filaments of oysters which generally ranges from 3% to 9% (*w/w*) and constitute 38% to 48% (*w/w*) of the total myofibril [32]. Similarly, actin and tropomyosin are known as important proteins in oyster specifically for muscle contraction. To validate the representation of these proteins and its utilization in the subsequent enzymatic hydrolysis simulation, BLAST analysis was performed to compare the level of homology of these *C. gigas* proteins towards its *C. angulata* counterparts. BLAST analysis of myosin essential light chain protein from both oyster species resulted in 157/157 (100%) identities, 157/157 (100%) positives and 0/157 (0%) gaps, indicating total similarity of these two *Crassostrea* species in terms of their protein sequences. Thus, the identified *C. gigas* proteins were used to represent the proteins of *C. angulata* in the subsequent in silico analysis.

2.2. In Silico Prediction of Potential Bioactivities

In silico analysis of oyster proteins by BIOPEP-UWM database revealed that all five selected proteins are good precursors of biologically active peptides, predominantly with DPP-IV and ACE inhibitory activities, with a total of 2179 and 1391 peptides, respectively (Table 2). Most of the reported DPP-IV inhibitory peptides contain P (proline) and/or hydrophobic amino acids in their sequences. GP and PG sequences are observed to be most frequently found in meat and fish [33]. On the other hand, proteins with hydrophobic amino acid residues (W, F, Y, or P) and positively-charged group (R or K) at the C-terminal positions or branched aliphatic side chains (V and I) at the N-terminal positions are known to possess strong ACE inhibitory activity and most of these peptides contain 2–12 amino acid residues [28].

Table 2. Total number of potential bioactive peptides from oyster proteins predicted in silico using BIOPEP-UWM database (accessed on 6 March 2019).

Protein Name	Number of Potential Bioactive Peptides		
	ACE Inhibitor	DPP-IV Inhibitor	Other Activities *
myosin heavy chain, striated muscle isoform X1	721	1147	395
paramyosin isoform X2	294	517	147
actin	196	246	69
tropomyosin isoform X1	97	165	56
myosin regulatory light chain B, smooth adductor muscle isoform X2	83	104	39
Total	1391	2179	706

* Other activities include antiamnestic, antibacterial, antioxidative, antithrombotic, neuropeptide, renin inhibitor, immunomodulating, stimulating, regulating, alpha glucosidase inhibitor, activating ubiquitin-mediated proteolysis, etc.

Most of the bioactive peptides discovered in the protein sequences of *C. angulata* are di and tripeptides (Table S1). The dipeptide AE is the most frequently occurring DPP-IV inhibitory peptide in *C. angulata* proteins, especially in myosin heavy chain, paramyosin, and tropomyosin. A (alanine) belongs to the hydrophobic group of amino acids and is also occurring in considerable amount in oysters. E (glutamic acid), on the other hand, is observed to be abundant in oyster species [19,34]. However, it was observed that the activity of peptides containing the same amino acid residues but occurring in different position could exhibit different biological property. For example, the dipeptide AE which was characterized with DPP-IV inhibitory activity was also observed to demonstrate inhibitory effect against ACE when occurred in reversed form. This only means that the activity of a peptide could vary depending on the type of amino acid forming the peptide and its position in the protein sequence.

Result of hydrolysis simulation using commonly used commercial enzymes is presented in Figure 3. Among the 9 enzymes used, pepsin (pH > 2) (EC 3.4.23.1) exhibited most of the DPP-IV and ACE inhibitory peptides theoretically, followed by stem bromelain (EC 3.4.22.32) and papain (EC 3.4.22.2). The effectiveness of these enzymes to release peptides with bioactivity depends mainly on its cleavage specificity. Pepsin has broad specificity and preferentially cleaves peptides with aromatic or carboxylic L-amino acid linkages, F and L at C-terminal location and to a lesser extent E linkages. However, it does not cleave at V, A, or G [35]. Stem bromelain exhibits strong cleavage preference for Z-R-R-I-NHMec among small molecule substrates while papain cleaves peptide bonds containing basic amino acids like arginine, lysine, and residues following phenylalanine [36,37]. The bioactive peptides released by different enzymes are listed in Table S2. With reference to in silico predictions, these three enzymes were chosen for use in the subsequent in vitro enzymatic hydrolysis.

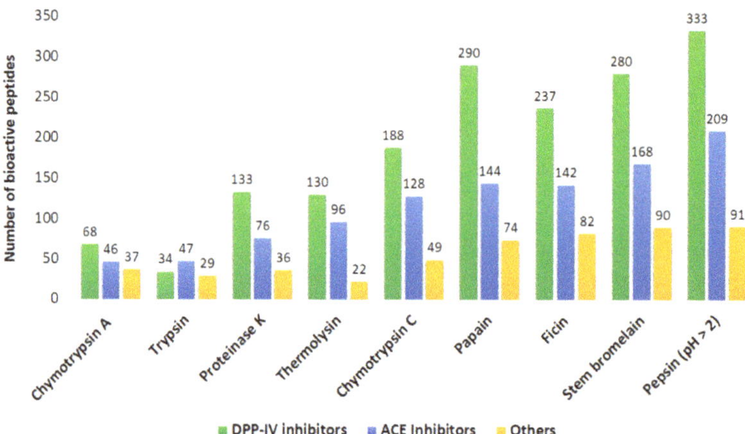

Figure 3. Total number of bioactive peptides released in silico by commercial enzymes through BIOPEP-UWM's "Enzyme Action" tool (accessed on 6 March 2019). Other activities include antiamnestic, antibacterial, antioxidative, antithrombotic, neuropeptide, renin inhibitor, immunomodulating, stimulating, regulating, alpha glucosidase inhibitor, and activating ubiquitin-mediated proteolysis. DPP-IV: dipeptidyl peptidase IV.

2.3. In Vitro Hydrolysis of Oyster Proteins

Oyster protein isolate (OPI) was used as raw material for in vitro enzymatic hydrolysis. Among the three enzymes, pepsin gave the highest DH after 4 h of hydrolysis with a maximum value of 22.20 ± 0.97%, followed by papain and bromelain with 18.57 ± 0.61% and 17.86 ± 0.08%, respectively. The DH values of the three reactions increased rapidly from time 0 to 0.5 followed by a slower linear effect as hydrolysis time progresses (Figure S1). Generally, the rate of hydrolysis is faster during the initial stages of the reaction followed by a more static state and becomes steady when the highest DH is reached. Apparently, in this study, the reaction rate displayed an increasing trend even after 4 h which means that the highest DH for the three enzyme-catalyzed reactions was not yet achieved. One of the factors causing these slow reaction rates and low DH values is the low E/S ratio used in this study since a higher enzyme concentration would develop more cleavage activity. Moreover, the rate and extent of hydrolysis can also be affected by the secondary and tertiary structures of proteins. Some protein tertiary structures are sensitive to environmental conditions like acidic pH, making it unsusceptible to proteases and difficult to hydrolyze [38]. Hydrolysis condition, yield, degree of hydrolysis, and peptide content of the hydrolysates produced by different enzymes are summarized in Table 3.

Table 3. Hydrolysis conditions, yield, and peptide content of *C. angulata* protein hydrolysate. PEH: pepsin hydrolysate; BRH: bromelain hydrolysate; PAH: papain hydrolysate.

Hydrolysate	Hydrolysis Conditions				Maximum DH (%)	Yield * (%)	Peptide Content (mg/mL)
	E/S Ratio	Time (h)	pH	Temp. (°C)			
PEH	1:100	4	2	37	22.20 ± 0.97 [a]	84.69	2.42 ± 0.06 [a]
BRH	1:100	4	7	50	17.86 ± 0.08 [b]	35.14	1.53 ± 0.03 [b]
PAH	1:100	4	7	65	18.57 ± 0.61 [b]	27.79	2.28 ± 0.03 [a]

* The yield was calculated based on the dry weight of the lyophilized hydrolysate over the dry weight of the protein isolate used during hydrolysis. Different superscript letters have significantly different ($p < 0.05$) mean values.

PEH, which demonstrated the highest DH, also obtained the highest yield (84.69%) among the three hydrolysate samples. However, in terms of peptide content, the value obtained by PEH was observed to be very comparable to that of PAH, despite of the differences in their DH values. This might be due to the unequal volume of solution at 4 h of hydrolysis wherein the collection of sample aliquots was done. The high temperature used during papain hydrolysis could have also led to evaporation which causes the reduction of sample volume and increase in concentration of peptides in the sample solution. Nevertheless, results showed that the increase in DH can lead to the production of more small peptides and free amino acids.

Based on the protein/peptide patterns of *C. angulata* hydrolysates and fractions (Figure 4), all hydrolysates (PEH, BRH, and PAH) showed dispersion around and below 10 kDa, which were not observed in the OPI. Among the three hydrolysates, PEH has the highest concentration of low molecular weight peptides which is attributed to its high DH. The degradation of actin (previously identified by compiled proteomics techniques) to different extent are also evident in all hydrolyzed samples.

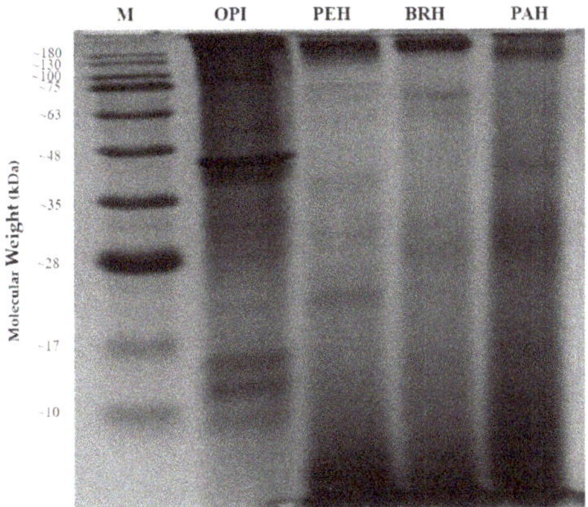

Figure 4. Peptide patterns of *C. angulata* protein hydrolysates by 15% SDS-PAGE. M: protein marker; OPI: oyster protein isolate; PEH: pepsin hydrolysate; BRH: bromelain hydrolysate; PAH: papain hydrolysate.

However, a band with the highest MW is observed to be visible even after hydrolysis, but appeared lighter in PAH than in PEH and BRH. This protein may be sensitive to high temperature applied during papain digestion. Moreover, the presence of light bands between 17 to 75 kDa indicates that there are still more protein substrates that were not cleaved even after 4 h of hydrolysis. This could be related to the low DH exhibited by the three enzymes. Overall, the electrophoretic pattern of *C. angulata* hydrolysate clearly supports the DH results, suggesting that pepsin's ability to break down oyster proteins into smaller peptides is better than bromelain and papain.

2.4. Confirmation of Bioactivities Through In Vitro Tests

2.4.1. ACE Inhibitory Activity

Angiotensin-I converting enzyme (ACE) is an enzyme responsible for the regulation of blood pressure. It converts angiotensin I into a potent vasoconstrictor angiotensin II and degrades the vasodilator, bradykinin, thus leading to an increase in blood pressure [39]. In this study, the potency of the three hydrolysates (PEH, BRH, and PAH) as inhibitors of ACE was evaluated. As shown in Figure 5A, all hydrolyzed samples exhibited an inhibitory activity against ACE which means that hydrolysis of proteins with pepsin, bromelain and papain were able to generate potent ACE inhibitory peptides. PEH displayed higher ACE inhibitory activity in all concentrations than BRH and PAH. The inhibition rates of the hydrolysate samples were observed to be dose-dependent except for BRH wherein a slight deviation was noticed at 1 mg/mL. Furthermore, the highest ACE inhibitory activity was noted in PEH prepared at 2 mg/mL with a value of 78.18 ± 2.19%, followed by BRH and PAH with 52.97 ± 1.01% and 42.65 ± 4.73%, respectively. It can be seen that PEH which have shown higher DH and peptide content than the other two hydrolysate samples also gave stronger inhibitory effect against ACE. One of the reasons for this is the high levels of free amino acids and smaller peptides liberated during hydrolysis that have ACE inhibitory properties. Basically, the biological activity of hydrolysates is influenced by the size, amount, composition of free amino acids and peptides, and the amino acid sequence [40]. This could also be associated to the cleavage specificity of pepsin, targeting the most bulky hydrophobic residues. The liberation of hydrophobic residues during pepsin hydrolysis results to their exposure to aqueous environment and susceptibility to reaction with different biomolecules, which leads to subsequent biological activities [41]. Overall, the results predicted in silico coincided with the results obtained in vitro with regards to the effectiveness of pepsin in releasing peptides with ACE inhibitory activity.

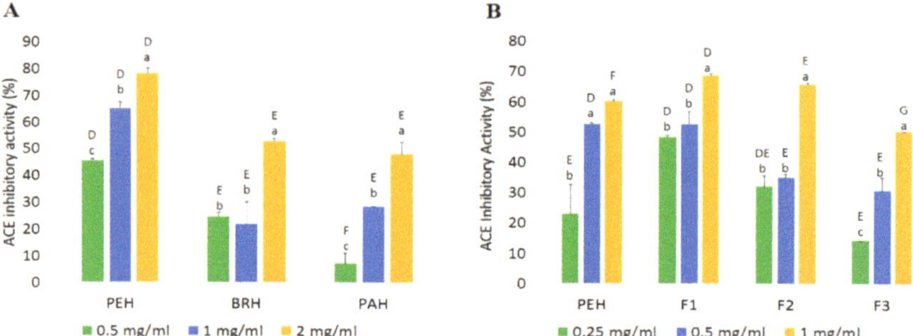

Figure 5. In vitro angiotensin I-converting enzyme (ACE) inhibitory activity of *C angulata* protein hydrolysates (**A**) and PEH fractions (**B**). Capital letters represent the significant difference ($p < 0.05$) among samples at specific concentrations; and small letters among concentrations within each sample. Each value (in percentage) represents the mean ± standard deviation ($n = 3$).

PEH was further separated into <1 kDa (F1), 1–5 kDa (F2), and >5 kDa (F3) fractions and their abilities to inhibit ACE activity were measured. The inhibition properties of PEH and peptide fractions against ACE followed a dose-dependent pattern in which an increase in concentration of peptides resulted in an increased inhibitory effect (Figure 5B). Result of ACE inhibitory activity assay revealed that F1 and F2 exhibited higher inhibitory activities (68.69 ± 0.82% and 65.95 ± 0.53%, respectively) compared to F3 (50.28 ± 0.09%) and PEH (60.32 ± 0.53%). Generally, peptides with very low MW are known to be most suitable for the formulation of therapeutic agents since these peptides can resist gastrointestinal digestion, thereby can be absorbed into the blood circulatory system in an intact form [42].

To test the inhibitory efficiency of crude PEH and fractions with respect to their MWs, the inhibitory efficiency ratio was calculated. Table 4 shows the peptide content, yield, and inhibition efficiency ratio of PEH, F1, F2, and F3. Result shows that F1 exhibited greatest efficiency in inhibiting ACE activity with an IER value of 217.05%/mg/mL compared to PEH and high MW fractions (F2 and F3), despite its low peptide content. This value is comparable to the IER of hard clam peptide fraction with a MW of 1360–1180 Da [43]. Analysis result indicates that products containing small peptides possess stronger ACE inhibitory activity. In addition, several studies have reported that those peptides with strong ACE inhibition are generally short peptides [44]. In most cases, peptides which contain 3–20 amino acids have greater potency as bioactive peptides than parent proteins [45].

The same with the unfractionated hydrolysate samples, the ACE inhibition properties of peptide fractions against ACE were observed to be dose-dependent in which an increase in the concentration of peptides resulted in increased inhibitory effect. Moreover, the inhibitory activity presented by PEH and its fractions is about half of the inhibitory activity of Captopril analyzed in this study (93.04%). Overall, results suggest that *C. angulata* proteins could be an important source of peptides that are capable of ACE inhibition.

Table 4. Inhibitory activity, peptide content, yield, and inhibitory efficiency ratio of pepsin hydrolysate and peptide fractions.

Fractions	Peptide Content (mg/mL)	Yield * (%)	Inhibition Efficiency Ratio [a] (%/mg/mL)	
			DPP-IV Inhibitory Peptides	ACE Inhibitory Peptides
PEH	2.42 ± 0.06 [a]	84.69	16.86	24.92
F1	0.32 ± 0.06 [c]	10.95	153.00	217.05
F2	0.45 ± 0.02 [c]	2.05	123.26	147.60
F3	1.73 ± 0.00 [b]	54.74	27.11	29.04

* Yield was calculated based on the dry weight of the lyophilized hydrolysate and fractions over the dry weight of the protein isolate and hydrolysate used during hydrolysis. [a] IER (inhibitory efficiency ratio) = % inhibition/peptide content. Different superscript letters represent significant difference between mean values ($n = 3$) at $p < 0.05$.

2.4.2. DPP-IV Inhibitory Activity

Dipeptidyl peptidase-IV (DPP-IV) is a postproline-cleaving enzyme that causes the degradation of incretins GLP-1 and GIP, leading to an increase in the blood glucose level. In this study, the ability of PEH, BRH, and PAH to inhibit DPP-IV was measured in vitro. Figure 6A shows that all the hydrolysate samples produced by different enzymes were able to inhibit DPP-IV activity. The strongest inhibition was observed in PEH prepared at 2 mg/mL (44.37 ± 0.09%), followed by BRH (23.98 ± 0.07%) and PAH (23.44 ± 1.44%). These results are in agreement with the in silico predictions. This strong inhibitory activity of PEH can be related to the ability of pepsin to cleave peptides with aromatic amino acid linkages. Previous in silico studies have shown that DPP-IV inhibitory peptides usually have a branched-chain amino acid or an aromatic residue containing a polar group in the side chain (primarily W) at their N-terminal position and/or P residue located at their P_1 [46]. In addition, the inhibitory activities of the hydrolysates were observed to be dose-dependent.

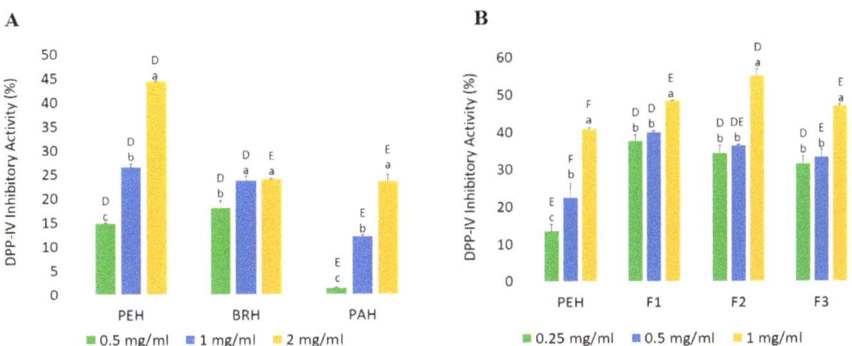

Figure 6. In vitro DPP-IV inhibitory activity of *C angulata* protein hydrolysates (**A**) and PEH fractions (**B**). Capital letters represent the significant difference ($p < 0.05$) among samples at specific concentrations; and small letters among concentrations within each sample. Each value (in percentage) represents the average of three samples ± standard deviation ($n = 3$).

PEH, the hydrolysate with highest inhibitory activity, was subjected to fractionation and the DPP-IV inhibitory activities of the fractions were also examined. Results show that the ability of all fractions to inhibit DPP-IV activity was observed to increase with increasing concentration. Among the samples, F1 presented the strongest inhibitory activity at different concentrations. However, for 1 mg/mL, F2 (55.08 ± 1.98%) displayed higher inhibition value than F1 (48.42 ± 0.06%) (Figure 6B). Bioactivity of peptides does not only depend on molecular weight, but also on other factors like amino acid composition and sequences in their chemical structure [47]. Results revealed that low MW fractions demonstrated higher efficiency as inhibitors of DPP-IV than high MW fractions with reference to their IER values (Table 4). Moreover, the inhibitory activity of Diprotin A (98.83%) obtained in this study was about 50% higher compared to that of the PEH fractions. Nevertheless, the above findings suggest that peptides from *C. angulata* proteins can be a good alternative for bioactive peptides suitable for DPP-IV inhibition.

3. Materials and Methods

3.1. Materials

Portuguese oysters (*Crassostrea angulata*) were purchased from Penghu Island, Taiwan. They were packed in a box with ice and sent to the laboratory by freight transport. Pepsin (from porcine gastric mucosa), bromelain (from pineapple stem), and papain (from papaya) were obtained from Sigma-Aldrich (St. Louis, MO, USA). The angiotensin I converting enzyme (ACE) from rabbit lung (≥ 2 units/mg), *N*-(3-[2-furyl]-acryloyl)-phenylalanyl glycyl glycine (FAPGG), Dipeptidyl Peptidase IV (DPP-IV) from human recombinant (≥ 1 unit/mg), and Gly-Pro p-nitroanilide hydrochloride ($\geq 99\%$) were also acquired from Sigma-Aldrich, USA. All chemical reagents used were of analytical grade.

3.2. Oyster Meat Preparation

Oysters (*C. angulata*) were manually shucked and the collected meat was washed with tap water and homogenized for 10 s using a food blender. The homogenized oyster meat was lyophilized for 48 to 72 h. It was then grounded into a fine powder (100 mesh). The resulting oyster powder were stored at $-20\ °C$ until further analysis.

3.3. Proteomics Techniques and In Silico Analysis

3.3.1. Sodium Dodecyl Sulfate Polyacrylamide Gel Electrophoresis (SDS-Page) Analysis

C. angulata proteins were separated through SDS-PAGE as described by Laemmli [48]. Firstly, 1 mg of sample (dry weight, protein basis) was diluted in 1 mL sample buffer [0.5 M Tris–HCl (pH 6.8), glycerol, 10% (*w/v*) SDS, 0.5% (*w/v*) bromophenol blue, and β-mercaptoethanol]. The solution was then heated at 95 °C for 4 min and centrifuged at $4000\times g$ for 15 min prior to loading. Electrophoresis was performed in a 12% running gel (ddH$_2$O, 30% Acrylamide/Bis (37.5:1), 1.5 M Tris-HCl (pH 8.8), 10% (*w/v*) SDS, 10% (*w/v*) ammonium persulfate and TEMED) and 4% stacking gel (ddH$_2$O, 30% Acrylamide/Bis (37.5:1), 0.5 M Tris-HCl (pH 6.8), 10% (*w/v*) SDS, 10% (*w/v*) ammonium persulfate and TEMED). Ten (10) μL of sample and 5 μL of standard (AccuRuler RGB prestained protein ladder, MaestronGen Inc., Taiwan) were loaded into each well of the gel. The voltage of power supply was set at 70 V for stacking gel and 110 V for running gel. After electrophoresis, the gel was stained with Coomassie Brilliant Blue for 30 min and subsequently destained with water/methanol/acetic acid (7/2/1, *v/v/v*) solution for 15 min with continuous shaking. The gel was then scanned using a gel image scanner and the MW of the visible bands was determined using the VisionCapt software (V16.08a, Vilber Lourmat, Paris, France).

3.3.2. Gel Slice and In-Gel Digestion

Distinct protein bands from SDS-PAGE gel were sliced and subjected to in-gel digestion following the method of Shevchenko and group [49] with some modifications. The sliced bands were cut into small cubes measuring around 1 mm^3. The gel sample was then placed in a siliconized eppendorf tube and spun down. Complete destaining was performed by subsequent addition of 50% and 25% acetonitrile/25 mM ammonium bicarbonate solution. Afterward, the gel sample was added with 100 μL DTE solution (50 mM dithioerythreitol/25 mM ammonium bicarbonate) and soaked for 1 h at 37 °C to break the disulfide bonds. After incubation, the mixture was spun down and the DTE solution was removed completely. The reduced gel sample was subjected to alkylation by incubating it with 100 μL of IAM solution (100 mM iodoacetamide/25 mM ammonium carbonate) for 1 h in the dark. After which, the gel sample was washed with 200 µL of 50% acetonitrile/25 mM ammonium bicarbonate for 15 min. The mixture was spun down and the buffer was pipetted out. Washing step was repeated 4 times to ensure that all buffers were removed completely. The gel sample was then soaked with 100 µL of 100% acetonitrile until it hardened and turned white. After removing all the acetonitrile, the remaining gel sample was dried for about 5 min in a SpeedVac concentrator. Digestion was performed by adding Lys-C/25 mM ammonium bicarbonate into the gel sample (1:50, enzyme:protein) followed by 3 h incubation at 37 °C. After that, the mixture was added with the same amount of trypsin and incubated at the same temperature for at least 16 h to complete the digestion. Prior to extraction, the enzyme was deactivated first by adding 50 µL of 50% acetonitrile/5% TFA to the mixture and sonicating it ten times (with 10 s interval). The mixture was centrifuged and the supernatant containing peptides was aspirated and transferred into a new tube. The remaining gel sample was extracted again following the above steps and the supernatant collected from the first and second extraction was combined and dried in a SpeedVac concentrator. The dried gel sample was then subjected to zip-tip purification prior to LC-ESI-MS/MS analysis.

3.3.3. NanoLC–NanoESI–MS/MS Analysis

The MS/MS data of the peptide mixtures were acquired from the Institute of Biological Chemistry, Academia Sinica, Nangang District, Taipei, Taiwan. The liquid chromatographic separation was done using C_{18} column and separated using a segmented gradient in 60 min from 5% to 35% solvent B (acetonitrile with 0.1% formic acid) at a flow rate of 300 nl/min and a column temperature of 35 °C. Solvent A was prepared with 0.1% formic acid in water. The mass spectrometry analysis was performed in a data-dependent mode and full scan MS spectra was acquired in the orbitrap (m/z 350–1600) with the resolution set to 60K at m/z 400 and automatic gain control (AGC) target at 10^6. The 20 most intense ions were sequentially isolated for CID MS/MS fragmentation and detection in the linear ion trap (AGC target at 10,000) with previously selected ions dynamically excluded for 60 s. Ions with singly and unrecognized charged state were also excluded.

3.3.4. Mascot Database Search

The collected MS/MS raw data were converted into MGF files and subjected to Mascot database search to identify the proteins and/or peptides detected by MS. The data were searched against National Center for Biotechnological Information (NCBI) Database for metazoa (animals). Search parameters used were carbamidomethyl (C) and oxidation (M) for variable modifications, ±10 ppm for peptide mass tolerance, 2+, 3+, and 4+ for peptide charge, ±0.6 Da for fragment mass tolerance, 2 for maximum missed cleavages, ESI-trap for the instrument used, and trypsin for the enzyme applied. All peptide masses were obtained as monoisotopic masses.

The Mascot ion score was −10*Log (P), where P is the probability that the observed match is a random event. Protein scores were derived from ion scores as a non-probabilistic basis for ranking protein hits. Mascot search results presented the protein sequence coverage in percentage (%) indicating the sequence homology of identified tryptic peptides from oyster fractions to corresponding protein hits.

3.3.5. BLAST Analysis of Oyster Proteins

Protein sequence of myosin essential light chain protein from *C. angulata* was compared to its *C. gigas* counterpart using BLAST (https://blast.ncbi.nlm.nih.gov/Blast.cgi), as described by Huang et al. [50], to determine the homology between the two proteins based on score, identities (%), positives (%), and gaps (%).3.3.6. BIOPEP-UWM analysis of bioactive peptides and enzyme cleavages.

The protein sequences obtained from the Mascot database search were analyzed using the BIOPEP database (http://www.uwm.edu.pl/biochemia/index.php/pl/biopep) to predict their bioactive peptide profile and enzyme cleavages. The activities, sequences, numbers and locations of bioactive peptides in protein sequences were determined using "profiles of potential biological activity" tool. The identified protein sequences were then examined through "enzyme action" tool wherein hydrolysis was simulated using different commercial enzymes available in the database. The theoretical peptides released by each enzyme were then directed to "search for active fragments" tool and the peptides with the highest number of potential bioactivities were selected for further analysis.

3.4. In Vitro Analyses

3.4.1. Protein Isolation

C. angulata proteins were isolated using alkaline solubilization/ isoelectric precipitation method described by Huang et al. [50]. In brief, the lyophilized oyster meat was mixed with 0.1 M NaOH at a ratio of 1:20 (powder: NaOH, *w/v*) and stirred for 2 h at room temperature. The mixture was centrifuged for 10 min at a speed of 8000× *g* at 4 °C. The supernatant was collected and adjusted to pH 5.5 with 0.1 N HCl to precipitate the myofibrillar proteins. The pH modified mixture was centrifuged at 8000× *g* speed for 10 min at 4 °C and the collected precipitate was lyophilized and stored at −20 °C until further analysis. The yield of the protein isolate was calculated based on the dry weight of oyster protein isolate over the dry weight of oyster meat used in isolation multiplied by 100.

3.4.2. Enzymatic Hydrolysis

Enzymatic hydrolysis was done following the combined methods of Dong et al. [19] and Jun et al. [51] with modifications. The oyster protein isolate (1 g, protein basis) was suspended in 100 mL deionized water. The homogenate was adjusted to 37 °C, pH 2 for pepsin; 50 °C, pH 7 for bromelain; and 65 °C, pH 7 for papain. The reaction started after adding the enzyme (1:100, *E/S*) and continued until 4 h. During hydrolysis, 1 mL aliquot of samples was taken at different time intervals (0, 0.5, 1, 1.5, 2, 2.5, 3, 3.5, and 4) for degree of hydrolysis determination. The enzyme activity was terminated by raising the temperature to 95 °C for 10 min. The sample was then cooled and centrifuged at 8000× *g* for 10 min. The recovered supernatant was neutralized to pH 7, lyophilized, and then stored at −20 °C for further analysis. The resulting dried hydrolysate was labeled as PEH (pepsin hydrolysate), BRH (bromelain hydrolysate), and PAH (papain hydrolysate).

3.4.3. Degree of Hydrolysis

The degree of hydrolysis was calculated based on the amount of free amino groups using modified o-phthalaldehyde method by Charoenphun et al. [52]. The OPA reagent was prepared fresh by mixing 12.5 mL of 100 mM sodium tetraborate, 1.25 mL of 20% sodium dodecyl sulfate (SDS), 20 mg of OPA reagent dissolved in 0.5 mL methanol and 50 μL of β-mercaptoethanol. The final volume of the solution was adjusted to 25 mL by adding deionized water. In a 96-well microplate, 10 μL of hydrolysate sample/blank/standard was combined with 200 μL of OPA reagent and incubated at 37 °C for 100 s. The absorbance was read at 340 nm using a UV/Vis microplate spectrometer (SPECTROstar Nano, BMG LABTECH) and the amount of amino groups was quantified using Gly-Gly-Gly as standard. The total number of primary amino groups in the sample was determined by performing a complete acid hydrolysis using 6N HCl for 24 h at 110 °C. The % DH was calculated as follows:

$$\text{DH } (\%) = [(NH_2)_{tx} - (NH_2)_{t0}]/[(NH_2)_{total} - (NH_2)_{t0}] \times 100\% \tag{1}$$

where, $(NH_2)_{tx}$ is the amount of free amino groups at X min; and $(NH_2)_{total}$ is the amount of total amino groups by total acid hydrolysis. $(NH_2)_{t0}$ represents the amount of free amino groups at 0 min of hydrolysis.

3.4.4. Fractionation

PEH was subjected to fractionation using Lefo Science-Spectrum Labs MAP-TFF Systems with a molecular weight cut-off of 5 and 1 kDa. The sample was dissolved in distilled water at 1% (*w/v*) concentration and placed in the ultrafiltration hollow fiber membrane. The recovered permeates, classified as F1 (<1 kDa), F2 (1–5 kDa), and F3 (>5 kDa) were then lyophilized and stored at −20 °C until further analysis. The yield of each fraction was calculated based on the dry weight of the fractions over the dry weight of the hydrolysate used multiplied by 100.

3.4.5. Peptide Content Determination

The peptide contents of the hydrolysates and fractions were measured using the modified OPA method by Charoenphun et al. [52], previously used in measuring the degree of hydrolysis. The peptide content was quantified using Gly-Gly-Gly as standard.

3.4.6. Angiotensin-I Converting Enzyme (ACE) Inhibitory Activity Assay

The ACE inhibitory activities of the protein hydrolysates and fractions were determined following the combined method of Raghavan and Kristinsson [53] and Udenigwe et al. [54]. In this method, *N*-[3-(2-furyl) acryloyl]-ʟ-phenylalanyl glycyl glycine (FAPGG) was used as synthetic substrate for ACE. Each assay sample was dissolved in 50 mM Tris-HCl buffer (pH 7.5) containing 0.3 NaCl at a final assay concentration of 0.5, 1.0, and 2.0 mg protein/mL (for hydrolysates) and 0.25, 0.5, and 1 mg protein/mL (for fractions). In a 96-well microplate, 20 μL of sample was combined with 170 μL of 0.5 mM FAPGG solution and pre-incubated at 37 °C for 10 min. Buffer solution was used as control. Thereafter, 10 μL of 0.5 U/mL ACE (pre-heated at 37 °C for 10 min) was added to each well and the rate of decrease in absorbance at 345 nm was checked and recorded for 30 min at 1 min interval using a UV/Vis microplate spectrometer (SPECTROstar Nano, BMG LABTECH) preset to 37 °C. Captopril (1 mg/mL) was used as a reference inhibitor for the assay. The ACE inhibitory activity was calculated as:

$$\text{ACE inhibitory activity (\%)} = [\Delta A \min^{-1}{}_{(control)} - \Delta A \min^{-1}{}_{(sample)}/\Delta A \min^{-1}{}_{(control)}] \times 100\% \quad (2)$$

where $\Delta A \min^{-1}{}_{(sample)}$ is the ACE activity in the presence of peptides while $\Delta A \min^{-1}{}_{(control)}$ is the ACE activity in the absence of peptides.

3.4.7. Dipeptidyl Peptidase IV (DPP-IV) Inhibitory Activity Assay

The inhibitory activities of the protein hydrolysates and fractions against the enzyme dipeptidyl peptidase-IV (DPP-IV) were determined following the combined methods of Lacroix and Li-Chan [55] and Zhang et al. [56]. Hydrolysate samples and fractions were dissolved in 100 mM Tris buffer (pH 8) to obtain a final assay concentration of 0.5, 1, and 2 mg protein/mL and 0.25, 0.5, and 1 mg protein/mL, respectively. In a 96-well microplate, 25 μL of assay sample was added with 25 μL of 1.6 mM Gly-Pro-p-nitroanilide and pre-incubated at 37 °C for 10 min. The mixture was then added with 50 μL of 0.008 U/mL DPP-IV (diluted with the same Tris-HCl buffer). Diprotin A (Ile-Pro-Ile) was used as reference inhibitor. Each sample was analyzed in triplicate and Tris-HCl buffer was used as blank. The positive control (DPP-IV activity in the absence of inhibitor) and negative control (no DPP-IV activity) was prepared using the same buffer solution in place of the sample and DPP-IV solution, respectively. The increase in absorbance per min was read in the UV/Vis microplate spectrometer (SPECTROstar Nano, BMG LABTECH) for 30 min at 37 °C and the rate of DPP-IV inhibition was calculated using the following equation:

$$\text{DPP-IV inhibitory activity (\%)} = [1 - (A_s - A_b)/(A_{pc} - A_{nc})] \times 100 \quad (3)$$

where A_s, A_b, A_{pc}, and A_{nc} are the absorbance of the sample, blank, positive control, and negative control, respectively.

3.4.8. Calculation of Inhibition Efficiency Ratio

The inhibition efficiency ratio of PEH and fractions was calculated based on the inhibitory activity (%) over the peptide content (mg/mL) of each sample.

3.5. Statistical Analysis

Data were presented as mean ± SD (standard deviation) for three replications for each sample. Data were analyzed using IBM SPSS (Statistical Package for Social Science,) version 20.0 (New York, NY, USA) for One-way Analysis of Variance (ANOVA) followed by Tukey's post-hoc test to estimate the significance among the main effects at the 5% probability level.

4. Conclusions

The application of in silico technique provided a rapid and reliable information on the identification of bioactive peptides from *C. angulata* proteins, and in the determination of suitable enzyme for the generation of these peptides. The results have shown the correspondence between in silico prediction and in vitro confirmation. Based on the above findings, *C. angulata* protein hydrolysates can be a good source of peptides with ACE and DPP-IV inhibitory activities. Moreover, pepsin (pH > 2) demonstrated most promise in releasing bioactive peptides form *C. angulata* proteins both in silico and in vitro. Furthermore, fractionation enhanced the ability of the hydrolysate to inhibit ACE and DPP-IV activities. Overall, peptides from *C. angulata* proteins can be an alternative source of bioactive peptides capable of ACE and DPP-IV inhibition and can be used as a functional ingredient with pharmaceutical and nutraceutical applications. However, in vivo testing is highly suggested to ensure safety and stability of these peptides during gastrointestinal digestion.

Supplementary Materials: The following are available online at http://www.mdpi.com/1422-0067/20/20/5191/s1.

Author Contributions: Conceptualization, H.L.R.G., J.P.P., and Y.-W.C.; Methodology, H.L.R.G. and L.A.T.; Validation, H.L.R.G.; Formal Analysis, H.L.R.G.; Investigation, H.L.R.G.; Resources, H.L.R.G. and Y.-W.C.; Data Curation, H.L.R.G., J.P.P. and Y.-W.C.; Writing—Original Draft Preparation, H.L.R.G.; Writing—Review and Editing, L.A.T.; Visualization, H.L.R.G.; Supervision, J.P.P. and Y.-W.C.; Funding acquisition, J.P.P. and Y.-W.C.

Funding: This research was funded by Ministry of Science and Technology, Taiwan (MOST: 106-2311-B-019-001) and Department of Science and Technology, Philippines.

Acknowledgments: We thank Academia Sinica Common Mass Spectrometry Facilities, Institute of Biological Chemistry, Academia Sinica, supported by Academia Sinica Core Facility and Innovative Instrument Project (AS-CFII-108-107) for the LTQ-Orbitrap data acquired.

Conflicts of Interest: The authors declare no conflict of interest. The funders had no role in the design of the study; in the collection, analyses, or interpretation of data; in the writing of the manuscript, or in the decision to publish the results.

References

1. Liu, C.W.; Liang, C.P.; Huang, F.M.; Hsueh, Y.M. Assessing the human health risks from exposure of inorganic arsenic through oyster (*Crassostrea gigas*) consumption in Taiwan. *Sci. Total Env.* **2006**, *361*, 57–66. [CrossRef] [PubMed]
2. Wijsman, J.W.M.; Troost, K.; Fang, J.; Roncarati, A. Global Production of Marine Bivalves. Trends and Challenges. *Goods Serv. Mar. Bivalves* **2019**, 7–26. [CrossRef]
3. Peralta, E.M.; Tumbokon, B.L.M.; Serrano Jr, A.E. Use of Oyster Processing Byproduct to Replace Fish Meal and Minerals in the Diet of Nile Tilapia Oreochromis niloticus Fry. *Isr. J. Aquac. Bamidgeh* **2016**.
4. Pawiro, S. Bivalves: Global production and trade trends. In *Safe Management of Shellfish and Harvest Waters*; IWA: London, UK, 2010; pp. 11–19.

5. Sikorski, Z.E.; Pan, B.S. The involvement of proteins and nonprotein nitrogen in postmortem changes in seafoods. *Seaf. Proteins* **1994**, 71–83.
6. Schutta, M.H. Diabetes and hypertension: Epidemiology of the relationship and pathophysiology of factors associated with these comorbid conditions. *J. Cardiometabolic Syndr.* **2007**, *2*, 124–130. [CrossRef]
7. Meisel, H.; Walsh, D.J.; Murray, B.; FitzGerald, R.J. ACE inhibitory peptides. *Nutraceutical proteins Pept. Health Dis.* **2005**, 273–319.
8. Ahn, C.B.; Jeon, Y.J.; Kim, Y.T.; Je, J.-Y. Angiotensin I converting enzyme (ACE) inhibitory peptides from salmon byproduct protein hydrolysate by Alcalase hydrolysis. *Process. Biochem.* **2012**, *47*, 2240–2245. [CrossRef]
9. McIntosh, C.H. Incretin-based therapies for type 2 diabetes. *Can. J. Diabetes* **2008**, *32*, 131–139. [CrossRef]
10. Holst, J.J.; Deacon, C.F. Glucagon-like peptide 1 and inhibitors of dipeptidyl peptidase IV in the treatment of type 2 diabetes mellitus. *Curr. Opin. Pharmacol.* **2004**, *4*, 589–596. [CrossRef]
11. Patel, K.; Pate, G., IV. Role of DPP-IV inhibitors in treatment of type II diabetes. *IRJP* **2010**, *1*, 19–28.
12. Hatanaka, T.; Inoue, Y.; Arima, J.; Kumagai, Y.; Usuki, H.; Kawakami, K.; Kimura, M.; Mukaihara, T. Production of dipeptidyl peptidase IV inhibitory peptides from defatted rice bran. *Food Chem.* **2012**, *134*, 797–802. [CrossRef] [PubMed]
13. Hsieh, C.H.; Wang, T.Y.; Hung, C.C.; Jao, C.L.; Hsieh, Y.L.; Wu, S.X.; Hsu, K.C. In silico, in vitro and in vivo analyses of dipeptidyl peptidase IV inhibitory activity and the antidiabetic effect of sodium caseinate hydrolysate. *Food Funct.* **2016**, *7*, 1122–1128. [CrossRef] [PubMed]
14. Deacon, C.F. Dipeptidyl peptidase-4 inhibitors in the treatment of type 2 diabetes: A comparative review. *DiabetesObes. Metab.* **2011**, *13*, 7–18. [CrossRef] [PubMed]
15. Patchett, A.A.; Harris, E.; Tristram, E.W.; Wyvratt, M.J.; Wu, M.T.; Taub, D.; Peterson, E.R.; Ikeler, T.J.; Ten Broeke, J.; Payne, L.G. A new class of angiotensin-converting enzyme inhibitors. *Nature* **1980**, *288*, 280. [CrossRef]
16. Thornberry, N.A.; Gallwitz, B. Mechanism of action of inhibitors of dipeptidyl-peptidase-4 (DPP-4). *Best Pract. Res. Clin. Endocrinol. Metab.* **2009**, *23*, 479–486. [CrossRef]
17. Fitzgerald, C.; Gallagher, E.; Tasdemir, D.; Hayes, M. Heart health peptides from macroalgae and their potential use in functional foods. *J. Agric. Food Chem.* **2011**, *59*, 6829–6836. [CrossRef]
18. Balti, R.; Bougatef, A.; Sila, A.; Guillochon, D.; Dhulster, P.; Nedjar-Arroume, N. Nine novel angiotensin I-converting enzyme (ACE) inhibitory peptides from cuttlefish (*Sepia officinalis*) muscle protein hydrolysates and antihypertensive effect of the potent active peptide in spontaneously hypertensive rats. *Food Chem.* **2015**, *170*, 519–525. [CrossRef]
19. Dong, X.-P.; Zhu, B.-W.; Zhao, H.-X.; Zhou, D.-Y.; Wu, H.-T.; Yang, J.-F.; Li, D.-M.; Murata, Y. Preparation and in vitro antioxidant activity of enzymatic hydrolysates from oyster (*Crassostrea talienwhannensis*) meat. *Int. J. Food Sci. Technol.* **2010**, *45*, 978–984. [CrossRef]
20. Lee, H.-J.; Saravana, P.S.; Cho, Y.-N.; Haq, M.; Chun, B.-S. Extraction of bioactive compounds from oyster (*Crassostrea gigas*) by pressurized hot water extraction. *J. Supercrit. Fluids* **2018**, *141*, 120–127. [CrossRef]
21. Yu, Z.; Wu, S.; Zhao, W.; Ding, L.; Shiuan, D.; Chen, F.; Li, J.; Liu, J. Identification and the molecular mechanism of a novel myosin-derived ACE inhibitory peptide. *Food Funct.* **2018**, *9*, 364–370. [CrossRef]
22. Geirsdóttir, M. Isolation, purification and investigation of peptides from fish proteins with blood pressure decreasing properties. *Food Res. Innov. Saf.* **2009**, *36*, 1–28.
23. Udenigwe, C.C.; Aluko, R.E. Food protein-derived bioactive peptides: Production, processing, and potential health benefits. *J. Food Sci.* **2012**, *77*, R11–R24. [CrossRef] [PubMed]
24. Umayaparvathi, S.; Arumugam, M.; Meenakshi, S.; Dräger, G.; Kirschning, A.; Balasubramanian, T. Purification and Characterization of Antioxidant Peptides from Oyster (*Saccostrea cucullata*) Hydrolysate and the Anticancer Activity of Hydrolysate on Human Colon Cancer Cell Lines. *Int. J. Pept. Res. Ther.* **2013**, *20*, 231–243. [CrossRef]
25. Umayaparvathi, S.; Meenakshi, S.; Vimalraj, V.; Arumugam, M.; Sivagami, G.; Balasubramanian, T. Antioxidant activity and anticancer effect of bioactive peptide from enzymatic hydrolysate of oyster (*Saccostrea cucullata*). *Biomed. Prev. Nutr.* **2014**, *4*, 343–353. [CrossRef]
26. Wang, J.; Hu, J.; Cui, J.; Bai, X.; Du, Y.; Miyaguchi, Y.; Lin, B. Purification and identification of a ACE inhibitory peptide from oyster proteins hydrolysate and the antihypertensive effect of hydrolysate in spontaneously hypertensive rats. *Food Chem.* **2008**, *111*, 302–308. [CrossRef] [PubMed]

27. Zhang, L.; Liu, Y.; Tian, X.; Tian, Z. Antimicrobial Capacity and Antioxidant Activity of Enzymatic Hydrolysates of Protein from Rushan Bay Oyster (*Crassostrea gigas*). *J. Food Process. Preserv.* **2015**, *39*, 404–412. [CrossRef]

28. Cheung, I.W.; Nakayama, S.; Hsu, M.N.; Samaranayaka, A.G.; Li-Chan, E.C. Angiotensin-I converting enzyme inhibitory activity of hydrolysates from oat (*Avena sativa*) proteins by in silico and in vitro analyses. *J. Agric. Food Chem.* **2009**, *57*, 9234–9242. [CrossRef]

29. Udenigwe, C.C. Bioinformatics approaches, prospects and challenges of food bioactive peptide research. *Trends Food Sci. Technol.* **2014**, *36*, 137–143. [CrossRef]

30. Fu, Y.; Therkildsen, M.; Aluko, R.E.; Lametsch, R. Exploration of collagen recovered from animal by-products as a precursor of bioactive peptides: Successes and challenges. *Crit. Rev. Food Sci. Nutr.* **2019**, *59*, 2011–2027. [CrossRef]

31. Coluccio, L.M. *Myosins: A Superfamily of Molecular Motors*; Springer Science & Business Media: Dordrecht, The Netherlands, 2007; Volume 7.

32. Katayama, S.; Shima, J.; Saeki, H. Solubility improvement of shellfish muscle proteins by reaction with glucose and its soluble state in low-ionic-strength medium. *J. Agric. Food Chem.* **2002**, *50*, 4327–4332. [CrossRef]

33. Lacroix, I.M.; Li-Chan, E.C. Evaluation of the potential of dietary proteins as precursors of dipeptidyl peptidase (DPP)-IV inhibitors by an in silico approach. *J. Funct. Foods* **2012**, *4*, 403–422. [CrossRef]

34. Chen, D.W.; Su, J.; Liu, X.L.; Yan, D.M.; Lin, Y.; Jiang, W.M.; Chen, X.H. Amino Acid Profiles of Bivalve Mollusks from Beibu Gulf, China. *J. Aquat. Food Prod. Technol.* **2012**, *21*, 369–379. [CrossRef]

35. Worthington, K. Worthington Enzyme Manual. Worthington Biochemical Corporation. *Lakewood NJ* **2011**.

36. Menard, R.; Khouri, H.E.; Plouffe, C.; Dupras, R.; Ripoll, D.; Vernet, T.; Tessier, D.C.; Laliberte, F.; Thomas, D.Y.; Storer, A.C. A protein engineering study of the role of aspartate 158 in the catalytic mechanism of papain. *Biochemistry* **1990**, *29*, 6706–6713. [CrossRef]

37. Ramalingam, C.; Srinath, R.; Islam, N.N. Isolation and characterization of bromelain from pineapple (*Ananas comosus*) and comparing its anti-browning activity on apple juice with commercial anti-browning agents. *Elixir. Food Sci.* **2012**, *45*, 7822–7826.

38. Alarcón, F.; Moyano, F.; Díaz, M. Use of SDS-page in the assessment of protein hydrolysis by fish digestive enzymes. *Aquac. Int.* **2001**, *9*, 255–267. [CrossRef]

39. Lee, S.Y.; Hur, S.J. Antihypertensive peptides from animal products, marine organisms, and plants. *Food Chem.* **2017**, *228*, 506–517. [CrossRef]

40. Sarmadi, B.H.; Ismail, A. Antioxidative peptides from food proteins: A review. *Peptides* **2010**, *31*, 1949–1956. [CrossRef]

41. Asha, K.K.; Remya Kumari, K.R.; Ashok Kumar, K.; Chatterjee, N.S.; Anandan, R.; Mathew, S. Sequence Determination of an Antioxidant Peptide Obtained by Enzymatic Hydrolysis of Oyster Crassostrea madrasensis (Preston). *Int. J. Pept. Res. Ther.* **2016**, *22*, 421–433. [CrossRef]

42. Aluko, R. Bioactive Peptides. *Funct. Foods Nutraceuticals* **2012**, 37–61. [CrossRef]

43. Tsai, J.-S.; Chen, J.-L.; Pan, B.S. ACE-inhibitory peptides identified from the muscle protein hydrolysate of hard clam (*Meretrix lusoria*). *Process. Biochem.* **2008**, *43*, 743–747. [CrossRef]

44. Byun, H.-G.; Kim, S.-K. Structure and activity of angiotensin I converting enzyme inhibitory peptides derived from Alaskan pollack skin. *BMB Rep.* **2002**, *35*, 239–243. [CrossRef]

45. Lee, S.-H.; Qian, Z.-J.; Kim, S.-K. A novel angiotensin I converting enzyme inhibitory peptide from tuna frame protein hydrolysate and its antihypertensive effect in spontaneously hypertensive rats. *Food Chem.* **2010**, *118*, 96–102. [CrossRef]

46. Lacroix, I.M.E.; Li-Chan, E.C.Y. Food-derived dipeptidyl-peptidase IV inhibitors as a potential approach for glycemic regulation–current knowledge and future research considerations. *Trends Food Sci. Technol.* **2016**, *54*, 1–16. [CrossRef]

47. Ketnawa, S.; Suwal, S.; Huang, J.Y.; Liceaga, A.M. Selective separation and characterisation of dual ACE and DPP-IV inhibitory peptides from rainbow trout (*Oncorhynchus mykiss*) protein hydrolysates. *Int. J. Food Sci. Technol.* **2018**, *54*, 1062–1073. [CrossRef]

48. Laemmli, U.K. Cleavage of structural proteins during the assembly of the head of bacteriophage T4. *Nature* **1970**, *227*, 680. [CrossRef] [PubMed]

49. Shevchenko, A.; Tomas, H.; Havlis, J.; Olsen, J.V.; Mann, M. In-gel digestion for mass spectrometric characterization of proteins and proteomes. *Nat. Protoc.* **2006**, *1*, 2856–2860. [CrossRef]

50. Huang, B.-B.; Lin, H.-C.; Chang, Y.-W. Analysis of proteins and potential bioactive peptides from tilapia (*Oreochromis* spp.) processing co-products using proteomic techniques coupled with BIOPEP database. *J. Funct. Foods* **2015**, *19*, 629–640. [CrossRef]

51. Jun, S.Y.; Park, P.J.; Jung, W.K.; Kim, S.K. Purification and characterization of an antioxidative peptide from enzymatic hydrolysate of yellowfin sole (*Limanda aspera*) frame protein. *Eur. Food Res. Technol.* **2004**, *219*, 20–26.

52. Charoenphun, N.; Cheirsilp, B.; Sirinupong, N.; Youravong, W. Calcium-binding peptides derived from tilapia (*Oreochromis niloticus*) protein hydrolysate. *Eur. Food Res. Technol.* **2013**, *236*, 57–63. [CrossRef]

53. Raghavan, S.; Kristinsson, H.G. ACE-inhibitory activity of tilapia protein hydrolysates. *Food Chem.* **2009**, *117*, 582–588. [CrossRef]

54. Udenigwe, C.C.; Lin, Y.-S.; Hou, W.-C.; Aluko, R.E. Kinetics of the inhibition of renin and angiotensin I-converting enzyme by flaxseed protein hydrolysate fractions. *J. Funct. Foods* **2009**, *1*, 199–207. [CrossRef]

55. Lacroix, I.M.; Li-Chan, E.C. Inhibition of dipeptidyl peptidase (DPP)-IV and alpha-glucosidase activities by pepsin-treated whey proteins. *J. Agric. Food Chem.* **2013**, *61*, 7500–7506. [CrossRef] [PubMed]

56. Zhang, Y.; Chen, R.; Chen, X.; Zeng, Z.; Ma, H.; Chen, S. Dipeptidyl Peptidase IV-Inhibitory Peptides Derived from Silver Carp (*Hypophthalmichthys molitrix* Val.) Proteins. *J. Agric. Food Chem.* **2016**, *64*, 831–839. [CrossRef]

International Journal of
Molecular Sciences

Article

Protective Effects of Novel Antioxidant Peptide Purified from Alcalase Hydrolysate of Velvet Antler Against Oxidative Stress in Chang Liver Cells In Vitro and in a Zebrafish Model In Vivo

Yuling Ding [1,†], Seok-Chun Ko [2,†], Sang-Ho Moon [3] and Seung-Hong Lee [1,*

1   Department of Pharmaceutical Engineering, Soonchunhyang University, Asan 31538, Korea;
    dingyuling@naver.com
2   National Marine Bio-Resources and Information Center, National Marine Biodiversity Institute of Korea,
    Seochun 33662, Korea; scko0527@gmail.com
3   Division of Food Bioscience, Konkuk University, Chungju 27478, Korea; moon0204@kku.ac.kr
*   Correspondence: seunghong0815@gmail.com; Tel.: +82-41-530-4980; Fax: +82-41-530-3085
†   These authors contributed equally to this study.

Received: 23 September 2019; Accepted: 18 October 2019; Published: 19 October 2019

Abstract: Velvet antler has a long history in traditional medicine. It is also an important healthy ingredient in food as it is rich in protein. However, there has been no report about antioxidant peptides extracted from velvet antler by enzymatic hydrolysis. Thus, the objective of this study was to hydrolyze velvet antler using different commercial proteases (Acalase, Neutrase, trypsin, pepsin, and α-chymotrypsin). Antioxidant activities of different hydrolysates were investigated using peroxyl radical scavenging assay by electron spin resonance spectrometry. Among all enzymatic hydrolysates, Alcalase hydrolysate exhibited the highest peroxyl radical scavenging activity. Alcalase hydrolysate was then purified using ultrafiltration, gel filtration, and reverse-phase high performance liquid chromatography. The purified peptide was identified to be Trp-Asp-Val-Lys (tetrapeptide) with molecular weight of 547.29 Da by Q-TOF ESI mass spectroscopy. This purified peptide exhibited strong scavenging activity against peroxyl radical (IC_{50} value, 0.028 mg/mL). In addition, this tetrapeptide showed significant protection ability against AAPH-induced oxidative stress by inhibiting of reactive oxygen species (ROS) generation in Chang liver cells in vitro and in a zebrafish model in vivo. This research suggests that the tetrapeptide derived from Alcalase-proteolytic hydrolysate of velvet antler are excellent antioxidants and could be effectively applied as functional food ingredients and pharmaceuticals.

Keywords: velvet antler; alcalase hydrolysate; antioxidant peptide; protection ability; oxidative stress

---

1. Introduction

Reactive oxygen species (ROS) are chemically reactive species containing oxygen. ROS are normally produced in living organisms during metabolism of oxygen. Under normal conditions in our body, ROS can be effectively eliminated by antioxidant defense systems such as endogenous antioxidant enzymes and non-enzymatic factors [1]. However, overproduction of ROS by various factors can cause oxidative stress and lead to a variety of pathological conditions, including metabolic impairments such as inflammation, aging, cancer, and cardiovascular diseases [2]. Therefore, sufficient amount of antioxidants need to be consumed to prevent or slow down oxidative stress induced by ROS. The amount of synthetic antioxidants used by humans is under strict regulation due to their potential health hazards [3]. Thus, natural antioxidants without side effects or toxicity have attracted great interest.

Food-derived peptides have shown to be potent antioxidants without serious side effects [4]. As such, to discover bioactive peptides from food proteins and to develop the peptides as alternatives to synthesis antioxidants has been considered by many researchers. In addition, the peptides from gastrointestinal digested food proteins may act as potential physiological modulators of metabolism during gastrointestinal digestion [5,6]. Several recent studies have suggested that animal proteins are good sources to produce antioxidant peptides and demonstrated that animal proteins hydrolystes and/or its antioxidant peptides may promote health by decreasing oxidative stress [7–9].

Velvet antler is a typical traditional medicine from animal origin that is recognized in the pharmacopeias of Korea, China, and Japan. In has been used as traditional medicine for over 2000 years. It also used as a functional foods or nutraceutical supplement in New Zealand, Canada, and the USA [10]. The reports support that main prominent bioactive components of velvet antler are polypeptides and proteins [11]. The traditional extraction of bioactive components from velvet antler is generally done via simmering in hot water. However, there is some controversy surrounding this approach, due largely to the extremely limited recovery of bioactive components in water extractions. In recent years, enzymatic hydrolysis using commercial proteases has been successfully applied to extraction of numerous biologically active peptides from a wide variety of food proteins and organisms. Recently, several studies have reported that enzymatic hydrolysate extracted from velvet antler using commercial proteases such as Alcalase, Protamex, pepsin, and Neutrase show a variety of biological benefits, including anti-obesity, anti-inflammatory, and antioxidant effects [12–14]. Anti-inflammatory peptides derived from velvet antler protein have also been reported [15].

However, to the best of our knowledge, there have been no reports about antioxidant peptides extracted from velvet antler by enzymatic hydrolysis. Therefore, the objective of the present study was to evaluate antioxidant activities of hydrolysates from velvet antler prepared with five commercial proteases (Acalase, Neutrase, trypsin, pepsin, and α-chymotrypsin) and identify amino acid sequences of purified peptides with the strongest antioxidant activity. Protective effects of purified peptides against 2,2′-Azobis(2-amidinopropane) dihydrochloride (AAPH)-induced oxidative stress in Chang liver cells and a zebrafish model were also investigated.

2. Results

2.1. Preparation of Enzymatic Hydrolysates from Velvet Antler and Their Peroxyl Radical Scavenging Activities

Velvet antler was successfully hydrolyzed with various commercial proteases such as trypsin, pepsin, α-chymotrypsin, Neutrase, and Alcalase to produce potent antioxidant peptides. Yields of velvet antler enzymatic hydrolysates measured by dry weight were observed to be 34.09%, 12.39%, 38.96%, 23.81%, and 29.75% for trypsin, pepsin, α-chymotrypsin, Neutrase, and Alcalase, respectively (Table 1). Antioxidant activities of these hydrolysates against peroxyl radical were examined using an ESR spectrometer. Their scavenging activities are shown in Table 1. Among these hydrolysates, Alcalase-derived hydrolysate possessed the highest peroxyl radical scavenging activity. Although the trypsin hydrolysate also showed a potent peroxyl radical scavenging activity, Alcalase can produce shorter peptide sequences and terminal amino acid sequences responsible for various bioactivities as well as useful for the production of bioactive peptide [16–18]. Therefore, Alcalase hydrolysate was selected to identify antioxidant peptide for further studies.

Table 1. Extraction yield and peroxyl radical scavenging activities of enzymatic hydrolysates from velvet antler.

Enzyme	Extraction Yield (%)	Radical Scavenging Activity (%) [1]
Trypsin	34.09	90.00 ± 1.69
Pepsin	12.39	78.58 ± 1.08
α-Chymotrypsin	38.96	89.54 ± 0.42
Neutrase	23.81	89.97 ± 1.05
Alcalase	29.75	93.78 ± 0.45

These values are expressed as mean $\pm$ S.E. from triplicate experiments. [1] Radical scavenging activity was measured at 1 mg/mL by ESR spectrometry.

2.2. Purification and Identification of Antioxidant Peptide

Initially, the Alcalase hydrolysate of velvet antler was cut off by two kinds of ultrafiltration membranes (5 and 10 kDa, MWCO). Three fractions with different molecular weights (>10 kDa, 5–10 kDa, and <5 kDa) were obtained. Peroxyl radical scavenging activities of these three separated fractions are shown in Table 2. The <5 kDa fraction possessed the highest peroxyl radical scavenging activity. IC_{50} (the half maximal inhibitory concentration) value of the <5 kDa fraction was 0.26 mg/mL, lower than that of the >10 kDa fraction (0.30 mg/mL) or the 5–10 kDa fraction (0.35 mg/mL). Accordingly, the <5 kDa fraction was further purified and separated using a Sephadex G-25 column. As shown in Figure 1A, four fractions were obtained. Their peroxyl radical scavenging activities were then determined. Fraction 3 (Fr. 3) exhibited the strongest peroxyl radical scavenging activity, with IC_{50} value of 0.12 mg/mL. Thus, Fr. 3 was further separated by RP-HPLC and five main fractions were obtained (Figure 1B). Fraction 3-3 (Fr. 3-3) showed the strongest peroxyl radical scavenging activity with an IC_{50} value of 0.028 mg/mL (Figure 1B), suggesting that Fr. 3-3 could possess potent antioxidant activity by scavenging peroxyl radicals. Thus, amino acid sequences of Fr. 3-3 was determined using Q-TOF ESI mass spectrometer. The purified peptide was identified as a tetrapeptide Trp-Asp-Val-Lys (named TAVL). The molecular weight of this tetrapeptide was 547.29 Da (Figure 1C). As shown in Table 3, the IC_{50} value of the TAVL was 51.16 µM for peroxyl radical scavenging activity, whereas positive control as ascorbic acid showed 19.26 µM of IC_{50} value, indicating the ascorbic acid is more potent scavenging activity. However, based on this results, we can confirm the superior antioxidant activity of TAVL.

Table 2. Peroxyl radical scavenging activities of different molecular weight fractions from Alcalase hydrolysate of velvet antler.

Molecular Sizes	IC_{50} Value (mg/mL)
>10 kDa	0.30 ± 0.01 [b]
5–10 kDa	0.35 ± 0.01 [b]
<5 kDa	0.26 ± 0.02 [a]

These values are expressed as mean $\pm$ S.E. from triplicate experiments. [a,b] Values with different alphabets are significantly different at $p < 0.05$ as analyzed by Duncan's multiple range test.

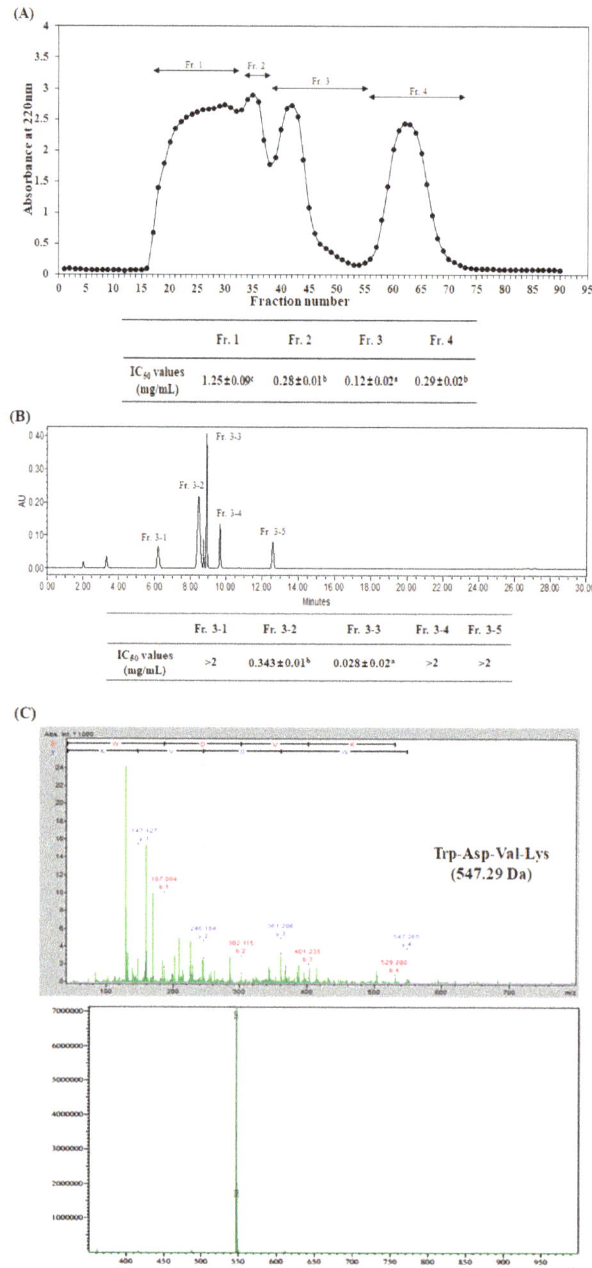

Figure 1. Purification and identification of antioxidant peptide. (**A**) Sephadex G-25 gel filtration chromatogram of <5 kDa fraction from Alcalase hydrolysate (upper panel) and its peroxyl radical scavenging activity (lower panel). (**B**) RP-HPLC chromatogram of the potent peroxyl radical scavenging activity fraction (Fr. 3) isolated from G-25 (upper panel) and its peroxyl radical scavenging activity (lower panel). (**C**) Identification of amino acid sequence and (left panel) molecular weight (right panel) of the purified peptide (TAVL) from Alcalase hydrolysate of velvet antler with a Q-TOF ESI mass spectrometer. These values are expressed as mean ± S.E. from triplicate experiments. [a–c] Values with different alphabets are significantly different at $p < 0.05$ as analyzed by Duncan's multiple range test.

Table 3. Comparison with peroxyl radical scavenging activity by the tetrapeptide (TAVL) and ascorbic acid.

	IC_{50} Value (μM)
TAVL	51.16 ± 0.2
Ascorbic acid	19.26 ± 0.1

These values are expressed as mean $\pm$ S.E. from triplicate experiments.

2.3. Intracellular Antioxidant Activities of the Purified Peptide

The cytotoxicity of the purified peptide (TAVL) was determined by 3-(4,5-Dimethylthiazol-2-yl)-2,5-diphenyltetrazolium bromide (MTT) assay at multiple TAVL concentrations (25, 50, 100, 200, 400, 800, and 1600 μg/mL) prior to evaluating its intracellular antioxidant activities. Results revealed that TAVL did not exhibit cytotoxicity at concentrations of up to 400 μg/mL, as compared with control survival (Figure 2A). Therefore, TAVL of non-toxic concentration was used to examine its protective effect against AAPH-induced cell damage in Chang liver cells. As shown in Figure 2B, AAPH treatment without TAVL significantly decreased cell viability. However, TAVL protected cells against cellular damage induced by AAPH in a dose-dependent manner. The generation of intracellular ROS could be measured by analyzing DCF fluorescence intensity levels. As Figure 2C shows, the fluorescence intensity of control group (AAPH and TAVL-untreated negative control) was recorded as 137, and the fluorescence intensity of only AAPH-treated cells was recorded as 3272. However, pretreatment of TAVL at 25, 50, 100, 200, and 400 μg/mL to cells mixed with AAPH reduced fluorescence intensities (intracellular ROS production levels) to 3202, 2899, 2806, 2719, and 2420, respectively. These results suggest that this antioxidant peptide (TAVL) could be developed into a potential bio-molecular candidate to inhibit cellular damage and intracellular ROS formation. Since this antioxidant peptide (TAVL) was found to exert antioxidant effects, its protective effect against AAPH-induced oxidative stress was further investigated using a zebrafish model in vivo.

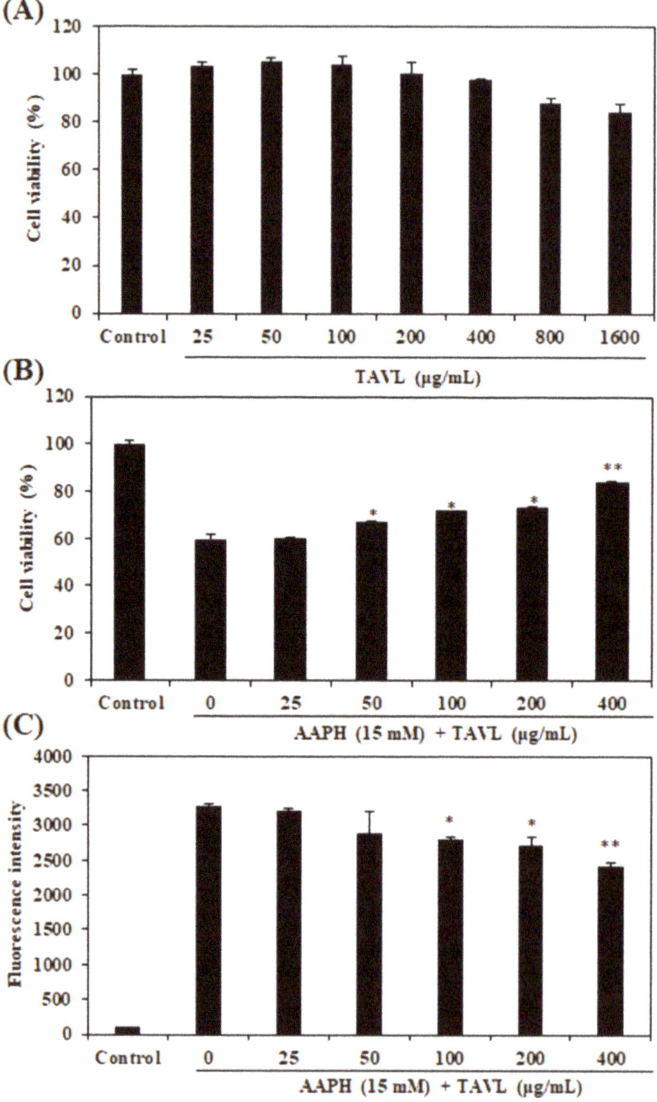

Figure 2. Protective effects of purified peptide (TAVL) against 2,2-Azobis-(2-amidinopropane) dihydrochloride (AAPH)-induced oxidative stress in Chang liver cells. Cells were treated with TAVL at indicated concentrations. (**A**) Cytotoxic effect of TAVL on viability of normal cells. After 24 of treatment with TAVL, cell viabilities were assessed by MTT assay. (**B**) Effect of TAVL on cell viability of AAPH-treated Chang liver cells. Cell viabilities were assessed by MTT assay. (**C**) Effect of TAVL on intracellular ROS generation in AAPH-treated Chang liver cells. Intracellular ROS generated was detected by 2′,7′-dichlorodihydrofluorescein diacetae (DCFH-DA) assay. Values are expressed as mean ± S.E. from triplicate experiments. Significant differences from only AAPH-treated group (positive control) were identified at * $p < 0.05$, ** $p < 0.01$ as analyzed by Duncan's multiple range test. The control group represents the negative control that does not receive treatment of AAPH and sample in an experiment.

2.4. Protective Effects of Antioxidant Peptide (TAVL) against AAPH-Induced Oxidative Stress in a Zebrafish Model In Vivo

AAPH-induced oxidative stress could eventually lead to cell death and overproduction of ROS and lipid peroxidation. In the present study, the protective effects of TAVL against AAPH-induced cell death, ROS generation, and lipid peroxidation in the zebrafish were investigated. As shown in Figure 3A, cell death of zebrafish was significantly elevated by AAPH treatment compared to non-AAPH-treated zebrafish. However, cell death induced by AAPH in zebrafish was remarkably reduced by treatment with TAVL in a dose-dependent manner. Effects of TAVL on AAPH-induced ROS generation and lipid peroxidation level are shown in Figure 3B,C, respectively. The control, which contained no AAPH or TAVL, generated a clear image. After treatment with only AAPH, a fluorescence image was generated, suggesting that generation of ROS and lipid peroxidation had taken place in zebrafish embryos in the presence of AAPH. However, when zebrafish embryos were treated with TAVL prior to AAPH treatment, dose-dependent reductions in the generation of ROS and lipid peroxidation were observed. These results demonstrated that the antioxidant peptide (TAVL) could have a protective effect against oxidative stress through its antioxidant activity.

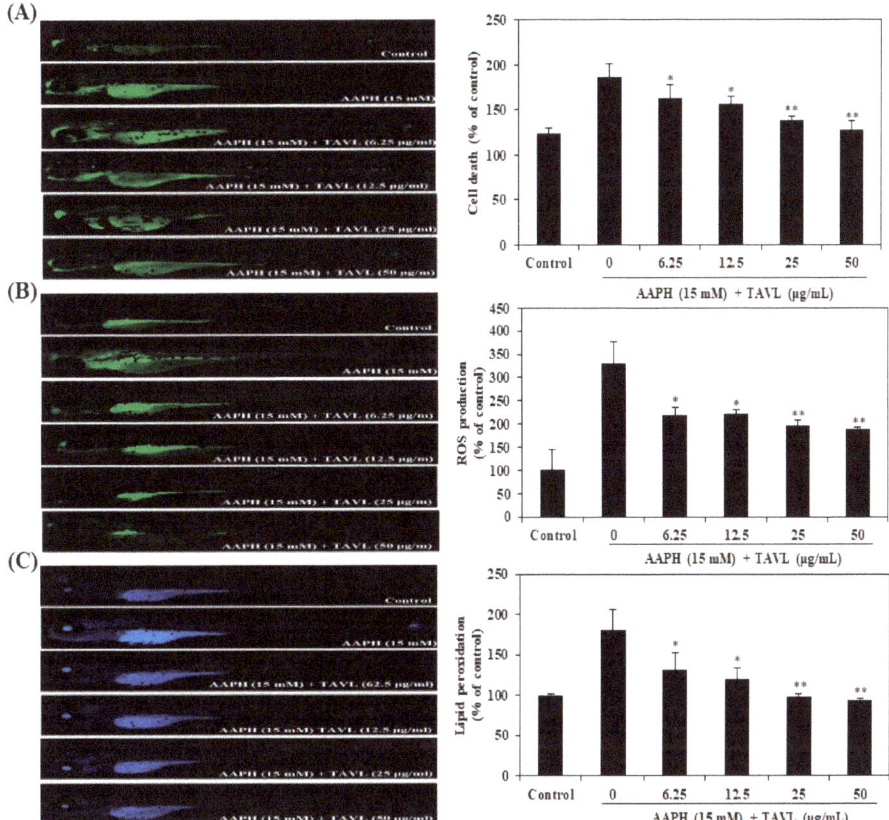

Figure 3. Protective effects of antioxidant peptide (TAVL) against AAPH-induced oxidative stress in zebrafish model. (**A**) Protective effect of TAVL on AAPH-induced cell death in zebrafish embryos. Cell death levels were measured after staining with acridine orange followed by image analysis and fluorescence microscopy. (**B**) Inhibitory effect of TAVL on AAPH-induced ROS production in zebrafish

embryos. ROS levels were measured after staining with 2′,7′-dichlorodihydrofluorescein diacetae (DCF-DA) followed by image analysis and fluorescence microscopy. (**C**) Inhibitory effect of TAVL on AAPH-induced lipid peroxidation in zebrafish. Lipid peroxidation levels were by DPPP staining. The fluorescence intensity of individual zebrafish was quantified using Image J program. Values are expressed as mean ± S.E. Significant differences from only AAPH-treated group (positive control) were identified at * $p < 0.05$, ** $p < 0.01$ as analyzed by Duncan's multiple range test. The control group represents the negative control that does not receive treatment of AAPH and sample in an experiment.

3. Discussion

Velvet antler is rich in proteins that may account for 60% (*w/w*) of dry matter [19]. However, velvet antler proteins have only received limited attention as a potential bioactive resource. Recently, the proteases have been successfully applied to extraction bioactive compounds from Velvet antler. Several studies have also reported that enzymatic hydrolysate extracted from velvet antler using protease show a variety of biological benefits, including anti-obesity, anti-inflammatory, and antioxidant effects [12–15]. However, there have been no reports about antioxidant peptides that can be extracted from velvet antler by enzymatic hydrolysis. Therefore, the aim of this study was to purify and identify antioxidant peptides from velvet antler enzymatic hydrolysates and to evaluate their antioxidant properties using peroxyl radicals scavenging assay. Protective effects of the purified peptide against AAPH-induced oxidative stress in Chang liver cells and in zebrafish model in vivo were also determined.

To obtain novel active antioxidant peptides, five commercial proteases were used under optimal conditions to hydrolyze velvet antler. Alcalase hydrolysate showed the highest peroxyl radical scavenging activity. In addition, several studies have suggested that Alcalase is useful for the production of bioactive peptide from food proteins [16–18,20,21]. Moreover, Alcalase can produce shorter peptide sequences and terminal amino acid sequences responsible for various bioactivities [16,18]. Thus, Alcalase hydrolysate was selected to identify antioxidant peptide for further studies.

The molecular weight of peptide is an important factor for its function [22]. Ultrafiltration (UF) is a simple and efficient technology for separating different molecular weights of molecules based on their molecular weights [23]. In this study, the Alcalase-proteolytic hydrolysate of velvet antler was separated to three fractions with different molecular weight (MW < 5 kDa, MW of 5–10 kDa, and MW > 10 kDa) by an UF system. Among these three MW groups, the <5 kDa fraction showed the strongest peroxyl radical scavenging activity. Previous reports have found that food proteins hydrolysates can be separated into three fractions (>10 kDa, 5–10 kDa and <5 kDa) by UF according to MW and that the <5 kDa fraction exhibits the strongest free radical scavenging activity [6,17,24]. Results of the present study also demonstrated that low MW fraction of Alcalase-proteolytic hydrolysate had higher free radical scavenging activity than higher MW fractions. Therefore, the <5 kDa fraction was selected for purification and identification of antioxidant peptide.

Sequential chromatography was used to purify antioxidant peptide from the active <5 kDa fraction, including Sephadex G-25 column gel filtration chromatography and RP-HPLC. After two-step isolation, we finally obtained the purified active peptide. Its amino acid sequence was determined with a Q-TOF ESI mass spectrometer. The purified antioxidant peptide was identified as a tetrapeptide Trp-Asp-Val-Lys (TAVL). The antioxidant peptide's properties are based on its molecular weight, amino acid sequence, and composition [25]. As reported previously, peptides showing antioxidant activities with a lower molecular weight can more easily pass the intestinal barrier and exert biological effects [25–27]. The purified tetrapeptide Trp-Asp-Val-Lys in the present study had a low molecular weight of 547.29 Da. It showed good antioxidant activities in the present experiments, in agreement with previous reports. In addition, the composition of amino acids within sequences of the peptide is another important factor for its antioxidant effects [4]. Hydrophobic amino acids, including Trp, Pro, Tyr, Lys, Leu, Val, and His, play an important role in the radical scavenging effects of peptides [26]. In addition, it has been reported and proven that antioxidant peptides containing aromatic amino acid residues (Trp and Tyr) have strong antioxidative capacities because they can make active oxygen stable through direct electron transfer [1,4]. In this study, the identified tetrapeptide has three amino acid residues (Trp, Val, and Lys) responsible for it antioxidant activity. This represents 3/4 of its compositions.

Overproduction of ROS can induce oxidative stress, which can cause numerous diseases and disorders. Also, cellular damage by ROS-induced oxidative stress often impairs biomolecules function and leads to cell death [28]. AAPH is a free radical–generating compound widely used to mimic the oxidative stress state [24,29,30]. Hence, in order to assess the intracellular antioxidant activity of the purified antioxidant peptide, in this study, AAPH was used to induce oxidative stress. The level of ROS production in cells was detected via oxidant sensitive fluorescent probe DCFH-DA to measure whether purified antioxidant peptide could prevent AAPH-induced ROS generation and the resulting oxidative stressors. Our results showed that treatment of Chang liver cells with AAPH significantly increased intracellular ROS level. However, purified antioxidant peptide inhibited such ROS generation induced by AAPH. AAPH generates free radicals through reacting with oxygen, which resulting in rapid formation of peroxyl radicals. The presently demonstrated inhibitory action of purified antioxidant peptide on ROS production can be attributed to its peroxyl radical scavenging activity. This purified antioxidant peptide was evaluated further with regard to its protective effects against AAPH-induced cellar damage. Exposure of cells to AAPH resulted in a significant decrease of cell viability. However, treatment with the purified antioxidant peptide inhibited cell death, suggesting that the purified antioxidant peptide could protect cells against AAPH-induced cytotoxicity. These results suggest that the purified antioxidant peptide could have a protective effect against ROS-induced oxidative stress, thus leading to reduced cellular injuries.

Recent reports indicated that zebrafish can be used as a rapid and simple model to assess the antioxidant activity against oxidative stress in vivo [29,30]. Therefore, in the present study, we investigated the antioxidant effect of purified antioxidant peptide in vivo using the zebrafish model. In the current study, antioxidant effects of purified antioxidant peptide against AAPH-induced oxidative stress in zebrafish model were investigated. Our results showed that treating zebrafish embryos with AAPH-treatment significantly increased cell death and ROS levels. However, the purified antioxidant peptide inhibited such AAPH-induced cell death and ROS generation. Lipid peroxidation may be a form of free radical–caused cellular damage [31]. In the present study, lipid peroxidation significantly increased by AAPH treatment in zebrafish embryos. However, the purified antioxidant peptide inhibited such lipid peroxidation formation effectively. The protective effect of the purified antioxidant peptide against lipid peroxidation formation can be attributed to its antiperoxidative effect. Taken together, these results further support that the purified antioxidant peptide could be utilized as a natural antioxidant to potentially protect cells against oxidative stress.

In conclusion, an antioxidant tetrapeptide was purified and identified from Alcalase-proteolytic hydrolysate of velvet antler. The identified antioxidant peptide (Trp-Asp-Val-Lys, 547.29 Da) exhibited great antioxidant activity based on peroxyl radical scavenging assay. In addition, this tetrapeptide significantly inhibited AAPH-induced ROS production in Chang liver cells and in a zebrafish model in vivo. These results demonstrate that ROS reduction by tetrapeptide may contribute to attenuation of intracellular oxidative stress. This tetrapeptide could have potential applications in functional foods, nutraceutical, and pharmaceutical industries.

4. Materials and Methods

4.1. Chemicals and Reagents

2,2-Azobis-(2-amidinopropane) dihydrochloride (AAPH) and a-(4-pyridyl-1-oxide)-N-t-butylnitrone (4-POBN) were purchased from Sigma Chemical Co. (St. Louis, MO, USA). Protein proteases including pepsin, trypsin, and α-chymotrypsin were purchased from Sigma-Aldrich (St. Louis, MO, USA). Neutrase and Alcalase were purchased from Novozyme Co. (Novo Nordisk, Bagsvaerd, Denmark). Penicillin-streptomycin and trypsin-EDTA were purchased from Gibco-BRL (Burlington, ON, Canada). 2′,7′-dichlorodihydrofluorescein diacetae (DCFH-DA) and 3-(4,5-Dimethylthiazol-2-yl)-2,5-diphenyltetrazolium bromide (MTT) were obtained from Sigma-Aldrich (St. Louis, MO, USA). All other chemicals and reagents used were of analytical grade and obtained from commercial sources.

4.2. Sample Preparation

Velvet antler was obtained from a farmed elk Daesungsan Deer Farm (Daegwallyeong, Korea) at 75 days after casting. The fresh velvet antler was immediately sliced and lyophilized. The lyophilized velvet antler was ground into a fine powder and stored at −20 °C until use.

4.3. Preparation of Enzymatic Hydrolysates from Velvet Antler

Enzymatic hydrolysis was performed using five commercial proteases (trypsin, pepsin, α-chymotrypsin, Neutrase, and Alcalase) at their optimal conditions (pH and temperature) as described previously [24]. Briefly, one gram of dried velvet antler powder was added into 100 mL of distilled water. Each enzyme was then added to have a substrate to enzyme ratio of 100:1. Enzymatic hydrolysis was conducted under optimal conditions for 24 h, after which the hydrolysate was boiled at 100 °C for 10 min to inactivate the enzyme. These hydrolysates were clarified by centrifugation at 3000× *g* for 20 min to remove any unhydrolyzed residue. The supernatant of each hydrolysate was filtered, adjusted to pH 7.0, and stored for subsequent use in experiments.

4.4. Peroxyl Radical Scavenging Activity

Peroxyl radicals were generated by AAPH and their scavenging activities were measured using an electron spin resonance (ESR) spectrometer (JEOL, Tokyo, Japan) in accordance with the method described by Hiramoto et al. [32]. Briefly, 20 μL of 40 mM AAPH and 20 μL of 40 mM 4-POBN were mixed with 20 μL of PBS and 20 μL of indicated concentration of tested sample. The mixture solution was incubated at 37 °C in a water bath for 30 min and then transferred into a capillary tube. Experimental conditions were as follows: power, 10 mW; amplitude, 1 × 1000; modulation width, 0.2 mT; sweep width, 10 mT; sweep time, 30 s; and time constant, 0.03 s.

4.5. Isolation of Antioxidant Peptides from the Enzymatic Hydrolysate of Velvet Antler

4.5.1. Fractionation According to the Molecular Weight

The enzymatic hydrolysate, which possess the highest peroxyl radical scavenging activity, was fractionated using the Millipore's Lab scale TFF system (Millipore Corporation, Bedford, MA, USA) equated with ultrafiltration membranes (MWCO: 5 and 10 kDa) at 4 °C. Then, three fractions (>10 kDa, 5–10 kDa, and <5 kDa) were obtained.

4.5.2. Purification of Antioxidant Peptides

The target fraction (500 mg) was loaded onto a Sephadex G-25 column (2.5 × 100 cm) pre-equilibrated with filtered distilled water. Elution was then carried out with filtered distilled water at a flow rate of 1.5 mL/min. Absorbance of each fraction at 220 nm was read and the sub-fractions were collected. The fraction with the highest peroxyl radical scavenging activity obtained was then subjected to reverse-phase high performance liquid chromatography (RP-HPLC) on an Atlantis T3 column (3 μm, 3.0 × 150 mm, Waters, NY, USA) with a linear gradient of acetonitrile (0–100% *v/v*, 30 min) at a flow rate of 1.0 mL/min. Elution peaks were detected at 220 nm.

4.5.3. Identification of Purified Antioxidant Peptide

Molecular weight and amino acid sequences of antioxidant peptides purified from velvet antler were determined using a MicroQ-TOFIII mass spectrometer (Bruker Daltonics, Hamburg, Germany) coupled with electrospray ionization (ESI) source. The purified peptide was dissolved in distilled water and infused into the ESI source. Its molecular weight was determined by singly charged (M + H) state analysis in mass spectrum.

4.6. Experiments for Antioxidant Activity Assay Using Cells

4.6.1. Cell Culture

The human hepatocyte–derived cell line termed Chang Liver were obtained from American Type Culture Collection (ATCC, Manassas, VA, USA) and it is well-known cell line used in various biological activities experiments including cellular antioxidant activity. Chang liver cells were cultured in Dulbecco's modified Eagle's medium (DMEM, Gibco-BRL, Burlington, ON, Canada) supplemented with 10% (*v/v*) heat-inactivated bovin serum (FBS, Gibco-BRL, Burlington, ON, Canada) and 1% (*v/v*) antibiotic. Cultures were maintained at 37 °C in a 5% CO_2 incubator.

4.6.2. Measuring Cytoprotective Effect by MTT Assay

Cytoprotective effect of the purified peptide was determined by a colorimetric MTT assay using Chang liver cells. Briefly, cells were seeded into a 96-well culture plates at cell density of 1×10^5 cells/mL. After incubation for 16 h, cells were treated with various concentrations (25, 50, 100, 200, 400, 800, and 1600 μg/mL) of purified peptide. One hour later, 15 mM of AAPH was added to each well. Cells were then incubated for an additional 24 h at 37 °C. After incubation, 50 μL of MTT solution (stock concentration: 5 mg/mL in DPBS) was added into each well, and cells were incubated at 37 °C for 4 h. Supernatants were aspirated and formazan crystals in each well were dissolved in DMSO. Absorbance at 540 nm was then measured.

4.6.3. Intracellular ROS Measurement

To detect levels of intracellular ROS, the DCFH-DA method was used as described previously [33]. Briefly, Chang liver cells were seeded into 96-well culture plates at cell density of 1×10^5 cells/mL. After 16 h, cells were treated with various concentrations of purified peptide and then incubated at 37 °C. One hour later, 15 mM of AAPH was added to the culture. Cells were then incubated for an additional 30 min at 37 °C. DCFH-DA solution (5 μg/mL) was then introduced to cells. DCF-DA fluorescence was detected at an excitation wavelength of 485 nm and an emission wavelength of 535 nm using a Perkin-Elmer LS-5B spectrofluorometer.

4.7. In Vivo Zebrafish Model for Antioxidant Activity Assay

The adult zebrafish were maintained following our previous study [30,34]. At 7–9 h post-fertilization (hpf), zebrafish embryos were collected and arrayed in a 12-well plate (15 embryos/well) containing 2 mL embryo medium. The embryos were incubated with or without purified peptide for 1 h and then exposed to AAPH (15 mM) for 24 hpf. Thereafter, zebrafish embryos were transferred into fresh embryo medium and allowed to develop up to 72 hpf. Cell death, intracellular ROS, and lipid peroxidation in zebrafish were estimated according to previously reported methods [33,34]. Briefly, at 72 hpf, zebrafish embryos were transferred into 24-well plates and separately stained with specific fluorescent probe dyes to determine cell death (acridine orange), intracellular ROS (2′,7′-dichlorodihydrofluorescein diacetate (DCFH-DA)), and lipid peroxidation generation (diphenyl-1-pyrenylphosphine (DPPP). Following incubation for a specified period in the dye-containing media, embryos were rinsed with fresh embryo media, anesthetized, and then observed under a fluorescence microscope equipped with a CoolSNAP-Pro color digital camera (Olympus, Tokyo, Japan). The fluorescence intensities of individual zebrafish were quantified using Image J 1.46r software (Wayne Rasband, National Institutes of Health, Bethesda, MD, USA). Cell death, intracellular ROS, and lipid peroxidation generation were calculated by comparing fluorescence intensities of treated embryos to those of controls.

4.8. Statistical Analysis

Data are presented as means ± standard error (SE). Statistical comparisons of mean values were performed by analysis of variance (ANOVA) followed by a Duncan's multiple range test using SPSS software.

Author Contributions: Conceptualization, S.-H.M. and S.-H.L.; Formal analysis, Y.D., S.-C.K. and S.-H.L.; Investigation, Y.D. and S.-C.K.; Writing – original draft, Y.D. and S.-H.L.

Acknowledgments: This research was supported by the Basic Science Research Program through the National Research Foundation of Korea (NRF) funded by the Ministry of Education (NRF-2018R1D1A1B07046262) and was supported by the Soonchunhyang University Research Fund.

Conflicts of Interest: The authors declare no conflict of interest.

References

1. Qian, Z.J.; Jung, W.K.; Kim, S.K. Free radical scavenging activity of a novel antioxidative peptide purified from hydrolysate of bullfrog skin, Rana catesbeiana Shaw. *Bioresour. Technol.* **2008**, *99*, 1690–1698. [CrossRef]

2. Halliwell, B. Free radicals, antioxidants, and human disease: Curiosity, cause, or consequence. *Lancet* **1994**, *344*, 721–724. [CrossRef]

3. Ito, N.; Hirose, M.; Fukushima, S.; Tsuda, H.; Shirai, T.; Tatematsu, M. Studies on antioxidants: Their carcinogenic and modifying effects on chemical carcinogenesis. *Food Chem. Toxicol.* **1986**, *24*, 1071–1082. [CrossRef]

4. Sarmadi, B.H.; Ismail, A. Antioxidative peptides from food proteins: A review. *Peptides* **2010**, *31*, 1949–1956. [CrossRef]

5. Byun, H.G.; Lee, J.K.; Park, H.G.; Jeon, J.K.; Kim, S.K. Antioxidant peptides isolated from the marine rotifer, *Brachionus rotundiformis*. *Process Biochem.* **2009**, *44*, 842–846. [CrossRef]

6. Ko, S.C.; Kim, D.; Jeon, Y.J. Protective effect of a novel antioxidative peptide purified from a marine *Chlorella ellipsoidea* protein against free radical-induced oxidative stress. *Food Chem. Toxicol.* **2012**, *50*, 2294–2302. [CrossRef]

7. Davalos, A.; Miguel, M.; Bartolome, B.; Lopez-Fandino, R. Antioxidant activity of peptides derived from egg white proteins by enzymatic hydrolysis. *J. Food Prot.* **2004**, *67*, 1939–1944. [CrossRef]

8. Guo, H.; Kouzuma, Y.; Yonekura, M. Structures and properties of antioxidative peptides derived from royal jelly protein. *Food Chem.* **2009**, *113*, 238–245. [CrossRef]

9. Lafarga, T.; Hayes, M. Bioactive peptides from meat muscle and by-products: Generation functionality and application as functional ingredients. *Meat Sci.* **2014**, *98*, 227–239. [CrossRef]

10. Wu, F.; Li, H.; Jin, L.; Li, X.; Ma, Y.; You, J.; Xu, Y. Deer antler base as a traditional Chinese medicine: A review of its traditional uses, chemistry and pharmacology. *J. Ethnopharmacol.* **2013**, *145*, 403–415. [CrossRef]

11. Sui, Z.; Zhang, L.; Huo, Y.; Zhang, Y. Bioactive components of velvet antlers and their pharmacological properties. *J. Pharm. Biomed. Anal.* **2014**, *87*, 229–240. [CrossRef] [PubMed]

12. Zhao, L.; Luo, Y.C.; Wang, C.T.; Ji, B.P. Antioxidant activity of protein hydrolysates from aqueous extract of velvet antler (*Cervus elaphus*) as influenced by molecular weight and enzymes. *Nat. Prod. Commun.* **2011**, *6*, 1683–1688. [CrossRef] [PubMed]

13. Lee, S.H.; Yang, H.W.; Ding, Y.; Wang, Y.; Jeon, Y.J.; Moon, S.H.; Jeon, B.T.; Sung, S.H. Anti-inflammatory effects of enzymatic hydrolysates of velvet antler in Raw 264.7 cells in vitro and zebrafish model. *EXCLI J.* **2015**, *14*, 1122–1132. [PubMed]

14. Ding, Y.; Wang, Y.; Jeon, B.T.; Moon, S.H.; Lee, S.H. Enzymatic hydrolysate from velvet antler suppresses adipogenesis in 3T3-L1 cells and attenuates obesity in high-fat diet mice. *EXCLI J.* **2017**, *16*, 328–339. [PubMed]

15. Zhao, L.; Wang, X.; Zhang, X.L.; Xie, Q.F. Purification and identification of anti-inflammatory peptides derived from simulated gastrointestinal digests of velvet antler protein (*Cervus elaphus* Linnaeus). *J. Food Drug Anal.* **2016**, *24*, 376–384. [CrossRef]

16. Lee, J.K.; Hong, S.; Jeon, J.K.; Kim, S.K.; Byun, H.G. Purification and characterization of angiotensin I converting enzyme inhibitory peptides from the rotifer, *brachionus rotundiformis*. *Bioresour. Technol.* **2009**, *100*, 5255–5259. [CrossRef]

17. Kim, H.S.; Kim, S.Y.; Fernando, I.P.S.; Sanjeewa, K.K.A.; Wang, L.; Lee, S.H.; Ko, S.C.; Kang, M.C.; Jayawardena, T.U.; Jeon, Y.J. Free radical scavenging activity of the peptide from the Alcalase hydrolysate of the edible aquacultural seahorse (*Hippocampus abdominalis*). *J. Food Biochem.* **2019**, *43*, e12833. [CrossRef]

18. Saito, Y.; Wanezaki, K.; Kawato, A.; Imayasu, S. Structure and activity of angiotensin I converting enzyme inhibitory peptides from sake and sake lees. *Biosci. Biotechnol. Biochem.* **1994**, *58*, 1767–1771. [CrossRef]

19. Sunwoo, H.H.; Nakano, T.; Hudson, R.J.; Sim, J.S. Chemical composition of antlers from wapiti (*Cervus elaphus*). *J. Agric. Food Chem.* **1995**, *43*, 2846–2849. [CrossRef]

20. Li, G.H.; Wan, J.Z.; Le, G.W.; Shi, Y.H. Novel angiotensin I-converting enzyme inhibitory peptides isolated from alcalase hydrolysate of mung bean protein. *J. Pept. Sci.* **2006**, *12*, 509–514. [CrossRef]

21. Wijesekara, I.; Qian, Z.J.; Ryu, B.M.; Ngo, D.H.; Kim, S.K. Purification and identification of antihypertensive peptides from seaweed pipefish (*Syngnathus schlegeli*) muscle protein hydrolysate. *Food Res. Int.* **2011**, *44*, 703–707. [CrossRef]

22. Byun, H.G.; Kim, S.K. Purification and chracterization of angiotensin I converting enzyme (ACE) inhibitory peptides from Alaska pollack (*Theragra chalogramma*) skin. *Process Biochem.* **2001**, *36*, 1155–1162. [CrossRef]

23. Sivakumar, R.; Hordur, G.K. ACE-inhibitory activity of tilapia protein hydrolysates. *Food Chem.* **2009**, *117*, 582–588.

24. Ko, J.Y.; Lee, J.H.; Samarakoon, K.; Kim, J.S.; Jeon, Y.J. Purification and determination of two novel antioxidant peptides from flounder fish (*Paralichthys olivaceus*) using digestive proteases. *Food Chem. Toxicol.* **2013**, *52*, 113–120. [CrossRef] [PubMed]

25. Sheih, I.C.; Wu, T.K.; Fang, T.J. Antioxidant properties of a new antioxidative peptide from algae protein waste hydrolysate in different oxidation systems. *Bioresour. Technol.* **2009**, *100*, 3419–3425. [CrossRef]

26. Rajapakse, N.; Mendis, E.; Byun, H.G.; Kim, S.K. Purification and in vitro antioxidative effects of giant squid muscle peptides on free radical-mediated oxidative systems. *J. Nutr. Biochem.* **2005**, *16*, 562–569. [CrossRef]

27. Roberts, P.R.; Burney, J.D.; Black, K.W.; Zaloga, G.P. Effect of chain length on absorption of biologically active peptides from the gastrointestinal tract. *Digestion* **1999**, *60*, 332–337. [CrossRef]

28. Finkel, T.; Holbrook, N.J. Oxidants, oxidative stress and the biology of aging. *Nature* **2000**, *408*, 239–247. [CrossRef]

29. Kim, E.A.; Lee, S.H.; Ko, C.I.; Cha, S.H.; Kang, M.C.; Kang, S.M.; Ko, S.C.; Lee, W.W.; Ko, J.Y.; Lee, J.H.; et al. Protective effect of fucoidan against AAPH-induced oxidative stress in zebrafish model. *Carbohydr. Polym.* **2014**, *102*, 185–191. [CrossRef]

30. Kang, M.C.; Cha, S.H.; Wijesinghe, W.A.J.P.; Kang, S.M.; Lee, S.H.; Kim, E.A.; Jeon, Y.J. Protective effect of marine algae phlorotannins against AAPH-induced oxidative stress in zebrafish embryo. *Food Chem.* **2013**, *138*, 950–955. [CrossRef]

31. Sevanian, A.; Hochstein, P. Mechanism and consequence of lipid peroxidation in biological systems. *Annu. Rev. Nutr.* **1985**, *5*, 365–390. [CrossRef] [PubMed]

32. Hiramoto, K.; Johkoh, H.; Sako, K.I.; Kikugawa, K. DNA breaking activity of the carbon-centered radical generated from 2, 2′-azobis (2-amidinopropane) hydrochloride (AAPH). *Free Radic. Res. Commun.* **1993**, *19*, 323–332. [CrossRef]

33. Rosenkranz, A.R.; Schmaldienst, S.; Stuhlmeier, K.M.; Chen, W.; Knapp, W.; Zlabinger, G.J. A microplate assay for the detection of oxidative products using 2′,7′-dichlorofluorescin-diacetate. *J. Immunol. Methods* **1992**, *156*, 39–45. [CrossRef]

34. Lee, S.H.; Ko, C.I.; Jee, Y.; Jeong, Y.; Kim, M.; Kim, J.S.; Jeon, Y.J. Anti-inflammatory effect of fucoidan extracted from *Ecklonia cava* in zebrafish model. *Carbohydr. Polym.* **2013**, *92*, 84–89. [CrossRef]

International Journal of
Molecular Sciences

Article

Hydrolysed Collagen from Sheepskins as a Source of Functional Peptides with Antioxidant Activity

Arely León-López [1], Lucía Fuentes-Jiménez [2], Alma Delia Hernández-Fuentes [1], Rafael G. Campos-Montiel [1] and Gabriel Aguirre-Álvarez [1,*]

[1]   Universidad Autónoma del Estado de Hidalgo, Instituto de Ciencias Agropecuarias, Av. Universidad Km. 1 Rancho Universitario, Tulancingo, Hidalgo 43600, Mexico

[2]   Instituto Tecnológico Superior del Oriente del Estado de Hidalgo, Carretera Apan-Tepeapulco Km 3.5, Las Peñitas, Ápan, Hidalgo 43900, Mexico

*   Correspondence: aguirre@uaeh.edu.mx; Tel.: +52-775-145-9265

Received: 5 July 2019; Accepted: 29 July 2019; Published: 13 August 2019

Abstract: The extraction and enzymatic hydrolysis of collagen from sheepskins at different times of hydrolysis (0, 10, 15, 20, 30 min, 1, 2, 3 and 4 h) were investigated in terms of amino acid content (hydroxyproline), isoelectric point, molecular weight (Mw) by sodium dodecyl sulphate polyacrylamide gel electrophoresis (SDS-PAGE) method, viscosity, Fourier-transform infrared (FTIR) spectroscopy, antioxidant capacity by 2,2′-azino-bis(3-ethylbenzothiazoline-6-sulphonic acid) (ABTS) and 2,2-diphenyl-1-picrylhydrazyl (DPPH) assays, thermal properties (Differential Scanning Calorimetry) and morphology by scanning electron microscopy (SEM) technique. The kinetics of hydrolysis showed an increase in the protein and hydroxyproline concentration as the hydrolysis time increased to 4 h. FTIR spectra allowed us to identify the functional groups of hydrolysed collagen (HC) in the amide I region for collagen. The isoelectric point shifted to lower values compared to the native collagen precursor. The change in molecular weight and viscosity from time 0 min to 4 h promoted important antioxidant activity in the resulting HC. The lower the Mw, the greater the ability to donate an electron or hydrogen to stabilize radicals. From the SEM images it was evident that HC after 2 h had a porous and spongy structure. These results suggest that HC could be a good alternative to replace HC from typical sources like pigs, cows and fish.

Keywords: collagen; hydrolysis; enzyme; molecular weight; sheepskin

---

1. Introduction

Collagen is the most abundant protein in bones and connective tissue in vertebrates, and there are at least 29 types. They are different in terms of their amino acid sequence and composition, the function in the organism, and the structure [1,2]. The structure of collagen is a triple helix formed for 3 α chains Gly-X-Y, where X is proline, Y is mainly hydroxyproline, and the triple helix is stabilized for hydrogen bonds with continuous repetition of the Gly-X-Y depending on the collagen type [3]. Collagen can be extracted from the skins, bones, tendons and cartilages of pigs [4], cows [5], marine organisms [6–8] and rabbits [9]. Hydrolysed collagen (HC) refers to a group of peptides that results from the proteolysis of native collagen type 1; its molecular weight (Mw) varies from 0.3 to 8 KDa [10]. It does not jellify in solution at room temperature and is soluble in cold water, so it can mix easily with other products [11–13]. Additionally, hydrolysed collagen has a neutral smell, is colourless, and can be used in emulsions as a stabilizer. It is widely used in the pharmaceutical industry for the treatment of diseases like osteoarthritis and osteoporosis. Also, in the cosmetics and food industries it is applied for the preparation of fruity beverages and nutritional supplements [14–17].

The antioxidant activity is the capacity of a substance to inhibit oxidative degradation by reacting with free radicals. There are natural antioxidants such as HC that exhibit mechanisms to exert antioxidant activity hydrogen transfer or electron donation [18]. The antioxidant activity of hydrolysed collagen is generally associated with the molecular weight. Peptides with 2 to 10 amino acid residues have a molecular weight of around 10 KDa. They show high radical scavenging because of their accessibility to active radicals. The amino acid content as well as the Mw of HC are properties closely related to the antioxidant activity [19,20].

Several studies on antioxidant activity have been conducted with HC from different sources such as pigs [4,21], cows [22], fish [23–25], and invertebrates like jellyfishes or sponges [26,27]. However, less is known about the properties of hydrolysed collagen extracted from ovine sources and its possible applications. The objective of this research is the extraction and hydrolysis of collagen from sheepskins to establish the antioxidant activity as well as the physicochemical properties of the obtained peptides as a function of the hydrolysis time. These results could be of interest in developing an alternative source to fish, cows and pigs.

2. Results and Discussion

2.1. Protein Content

From Figure 1, it can be seen that the protein concentration of HC was affected by the hydrolysis time. Before hydrolysis (at 0 min), the protein concentration was 1.61 mg/mL. However, in the first 20 min the lowest concentration was reported, up to approximately 1.1 mg/mL. The concentration remained constant around 1.21 mg/mL and there were no statistical differences ($p > 0.05$) from 30 min treatment to the end of the experiment (4 h). This behaviour could be due to the decrease in available substrate and/or enzyme autodigestion [28]. All the treatments in this experiment reported higher protein concentrations compared to those reported by Paul and co-workers [29]. They obtained hydrolysed collagen from cowhide using an enzymatic treatment. The protein concentration was reported with 0.11 mg/mL. Other works [30] also obtained a low protein concentration (around 0.76 mg/mL) from chicken connective tissue with enzymatic treatment at pH 7.5. These results suggest that hydrolysis of collagen from sheepskin under the conditions described above was efficient compared to chicken and bovine sources.

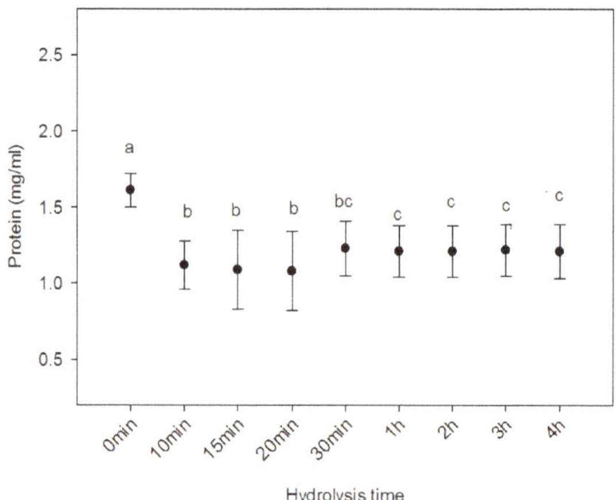

Figure 1. Protein concentration for different hydrolysis times in ovine collagen. Different letters represent the average of three replicates and indicate significant difference at $p \le 0.05$.

2.2. Hydroxyproline Content

Collagen is different from other proteins due to its high concentration of hydroxyproline. This amino acid provides thermal stability to collagen molecules because of the hydrogen bond formation and the presence of an hydroxyl group (OH), limiting the rotation of the peptide chain [15,31]. The hydrolysis of collagen reported in Figure 2 indicates that the longer the hydrolysis time, the more hydroxyproline is obtained. After 4 h of enzymatic and thermal treatment, HC reported 24.47 mg/L. This trend agrees well with previous works carried out on fish bone gelatine [32] and pig collagen [33]. They found a significant increment in hydroxyproline content as the hydrolysis time increased. Also, their maximum yield of this amino acid was found at 4 h of thermal treatment. This increment in hydroxyproline content could be attributed to the hydrolysis of the polypeptide chain. These thermal and enzymatic treatments increased the detectable amount of hydroxyproline. This could be the case with some marine sources like Atlantic salmon skin [34] and bigeye snapper skin [35], for which values of 88.24 mg/mL and 87.75–90.86 mg/mL, respectively, were reported. However, Gómez-Lizárraga and co-workers [36] obtained 1.18 mg/L from bovine tendons.

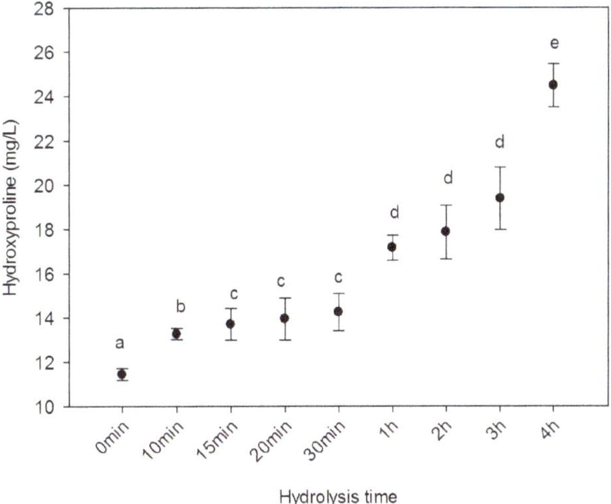

Figure 2. Hydroxyproline concentration in hydrolysed collagen. Different letters indicate significant difference at $p \leq 0.05$.

2.3. Amino Acid Content in Hydrolysed Collagen as a Function of Hydrolysis Time

The amino acid composition of hydrolysed collagen (HC) from sheepskin was very similar to other vertebrate collagen sources such as pigs [33], chicken [37], calves [38], cows [39] and fish [40]. Seventeen amino acids were identified and quantified as structural components of ovine collagen. These amino acids were monitored during the enzymatic hydrolysis process of collagen. Results after 1 h were not included in Table 1 because no significant differences ($p > 0.05$) were observed. After 1 m of hydrolysis, serine was detected as the major component (19.33 mg/g of protein). The enzymatic treatment of collagen showed significant increments in aspartic acid and glutamic acid due to enzymatic cleavage of the polypeptide chains of collagen fibres [41]. These amino acids increased their concentrations considerably as a result of the deamidation process of asparagine and glutamine, respectively [42]. Some amino acids such as lysine, proline, cysteine, tyrosine, valine, methionine, isoleucine, leucine and phenylalanine were sensitive to hydrolysis and their concentrations decreased considerably. The same behaviour was observed with HC from fish skin [43].

Table 1. Amino acid content (mg of amino acid/g of protein) of hydrolysed collagen as a function of hydrolysis time.

Amino Acid	0 min	10 min	20 min	30 min	1 h
Aspartic acid	4.04 ± 0.03 [a]	10.86 ± 0.05 [d]	5.06 ± 0.04 [f]	28.03 ± 0.036 [g]	32.85 ± 0.16 [a]
Glutamic acid	7.99 ± 0.06 [a]	7.71 ± 0.04 [a]	16.19 ± 0.05 [e]	38.21 ± 0.048 [f]	16.58 ± 0.03 [a]
Serine	19.33 ± 0.16 [a]	25.99 ± 0.03 [c]	17.45 ± 0.03 [a]	13.72 ± 0.03 [a]	17.87 ± 0.05 [a]
Glycine	4.86 ± 0.01 [a]	4.31 ± 0.04 [a]	16.71 ± 0.04 [a]	1.92 ± 0.00 [a]	3.10 ± 0.00 [a]
Lysine	6.60 ± 0.04 [a]	5.72 ± 0.02 [a]	3.32 ± 0.06 [a]	2.05 ± 0.00 [a]	1.92 ± 0.00 [a]
Histidine	3.38 ± 0.03 [a]	3.25 ± 0.05 [a]	2.41 ± 0.01 [a]	3.95 ± 0.02 [a]	1.59 ± 0.00 [a]
Threonine	2.72 ± 0.06 [a]	4.50 ± 0.02 [a]	2.90 ± 0.01 [a]	3.18 ± 0.01 [a]	3.12 ± 0.00 [a]
Arginine	8.91 ± 0.01 [a]	3.46 ± 0.05 [a]	18.62 ± 0.03 [b]	2.68 ± 0.02 [a]	11.02 ± 0.02 [a]
Alanine	1.39 ± 0.04 [a]	0.40 ± 0.04 [a]	1.26 ± 0.00 [a]	0.14 ± 0.00 [a]	1.39 ± 0.00 [a]
Proline	2.68 ± 0.04 [a]	1.23 ± 0.03 [a]	0.86 ± 0.00 [a]	0.13 ± 0.00 [a]	0.49 ± 0.00 [a]
Cysteine	1.53 ± 0.02 [a]	1.84 ± 0.03 [a]	1.00 ± 0.00 [a]	0.13 ± 0.00 [a]	0.36 ± 0.00 [a]
Tyrosine	12.36 ± 0.18 [a]	3.02 ± 0.02 [d]	7.24 ± 0.05 [f]	0.78 ± 0.00 [g]	2.9 ± 0.01 [a]
Valine	3.17 ± 0.05 [a]	2.73 ± 0.01 [a]	1.67 ± 0.00 [a]	1.53 ± 0.00 [a]	1.26 ± 0.01 [a]
Methionine	7.80 ± 0.04 [a]	7.48 ± 0.04 [a]	0.75 ± 0.00 [c]	1.36 ± 0.00 [d]	1.65 ± 0.00 [a]
Isoleucine	6.13 ± 0.03 [a]	6.49 ± 0.05 [a]	0.52 ± 0.00 [b]	0.27 ± 0.00 [c]	1.21 ± 0.00 [a]
Leucine	3.41 ± 0.01 [a]	6.35 ± 0.03 [b]	1.00 ± 0.00 [d]	0.24 ± 0.00 [e]	1.30 ± 0.00 [a]
Phenylalanine	2.66 ± 0.07 [a]	3.31 ± 0.04 [a]	1.66 ± 0.00 [a]	0.87 ± 0.00 [a]	0.69 ± 0.00 [a]

Results are mean values of three replicates' SD. Values followed by different letters are significantly different according to Tukey's test ($p \leq 0.05$).

2.4. Isoelectric Point

Isoelectric point (pI) is the pH of the collagen molecule at 0 charge. Looking at Figure 3, the pI shifted from 4.61 to 3.68 at the end of the hydrolysis (4 h). Native collagen (0 min) reported a pI value of about 4.7. Similar values of pI for acid (4.9) and pepsin-soluble (5.7) collagen were reported in the literature on the extraction of collagen from ovine bones [1]. Hydrolysed collagen (HC) is an amphoteric macromolecule composed of both acidic (COOH) and basic (NH_3) functional groups and the pI decrement could be due to the deamination process [10]. When HC was treated at high temperature, the asparagine groups transformed to aspartic acid and the glutamine groups into glutamic acid [42]. This leads to a loss of amino groups and a large relative increase in the carboxyl groups, or a higher content of acidic amino acids, which become dominant, shifting the pI to lower values [44,45]. Collagen is an amphoteric macromolecule that possesses different pI according to the hydrolysis time. The higher the pI, the higher the viscosity observed due to stronger electrostatic repulsions between collagen chains [46].

2.5. Molecular Weight and Viscosity

Hydrolysis of collagen is characterized by a reduction in its molecular weight (Mw). Changes in collagen Mw were monitored by SDS-PAGE methodology. Figure 4 shows that native collagen (0 min) reflected the highest Mw with 260.33 KDa. When the hydrolysis started, the Mw dropped to lower values. The first significant changes ($p \leq 0.05$) were observed at 15 min and 20 min with 160.67 KDa and 138.89 KDa, respectively. However, there was a massive decrement in Mw after treatment for 2 h (15.20 KDa). After this time, no significant changes in Mw were registered ($p \geq 0.05$) up to 4 h with 5.62 KDa. These results are in good agreement with those reported in the literature with values between 3 and 6 KDa [11,13]. Chi and co-workers [47] used trypsin for digestion to obtain fish hydrolysed collagen with Mw around 14 KDa. Also, hydrolysed collagen from turkey byproducts was obtained with Mw of 34 KDa by using different enzymes [48]. Previous works carried out on Alaska Pollack skin [43] and sea cucumber [49] found high antioxidative properties in HC with Mw around 6-8 KDa and 5 KDa, respectively. There is a close relationship between the Mw and the viscosity of the collagen [12]. At 0 min, the viscosity reported 6800 Cp. This higher viscosity could be attributed to the presence of high molecular weight and chains [50]. The viscosity decreased upon heating, in accordance with the hydrolysis time. It shifted to 0.5 Cp when the hydrolysis had been going on for 1 h. After this time, no significant changes ($p \geq 0.05$) were observed. The triple helix structure of native collagen was changed to a random coil form due to the dissociation of the hydrogen bonds [51].

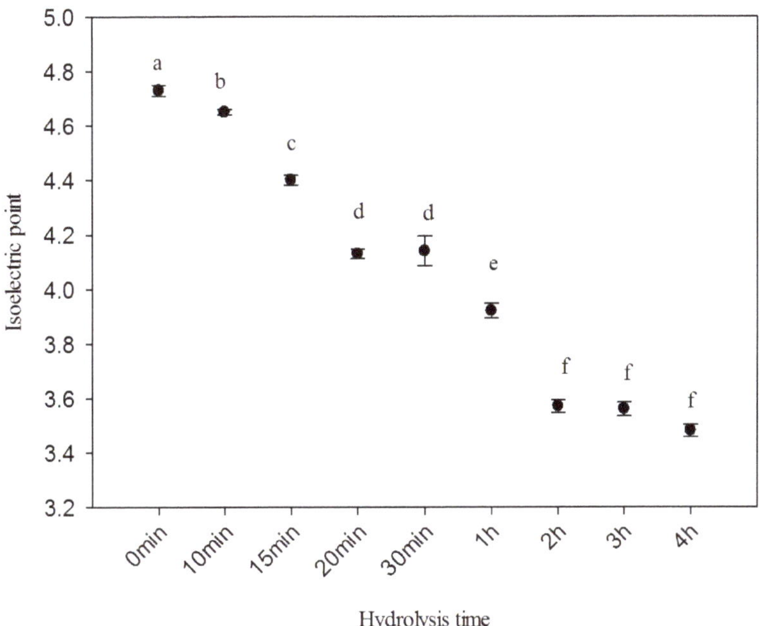

Figure 3. Isoelectric point of ovine collagen hydrolysates as a function of hydrolysis time. Different letters indicate significant difference at $p \leq 0.05$.

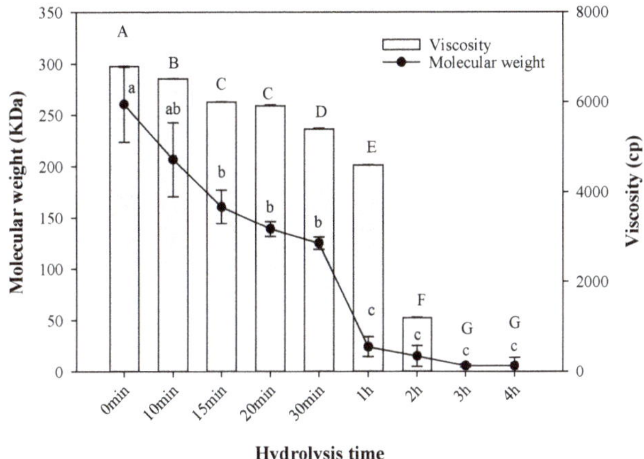

Figure 4. Co-relation between molecular weight and viscosity during the hydrolysis of ovine collagen. The different letters indicate significant difference at $p \leq 0.05$.

2.6. Fourier Transform-Infrared Spectroscopy (FTIR)

The FTIR spectra of both native collagen (0 min) and hydrolysed collagen (HC) were produced in the range of 600-4000 cm^{-1}. All FTIR spectra of HC samples overlapped each other. However, for reasons of clarity, Figure 5 only shows the range of 1000-3500 cm^{-1} for the samples at 0 min and 4 h. There were no changes in peak location for the amide bands between the control and the treated sample. However, the magnitude of amplitude in HC decreased significantly. Amide I at wavelength 1641 cm^{-1} was interpreted as the stretching vibrations of the carbonyl groups (C=O) along the polypeptide backbone. This band is characteristic of α-helix chains and is widely used to analyse the secondary structure of collagen [52]. Amide II was detected at 1548 cm^{-1} for the stretching vibrations of the CN group. Amide III was mainly associated with intermolecular interactions at 1248 cm^{-1}, representing the stretching vibrations of the C-N group and the deformation of the NH group from amide bonds [1]. The amide B (2946 cm^{-1}) and Amide A (3295 cm^{-1}) bands were related to the asymmetric stretching of the CH$_2$ groups and vibrations of tension of the NH group, respectively. The longer the hydrolysis time, the higher the vibrations of OH groups (1037 cm^{-1}) reported in the spectra. These results agree very well with the literature [7], suggesting that the HC (1 h) maintained the same characteristics of native collagen (0 min) as the peak locations of all amide bands scarcely caused changes.

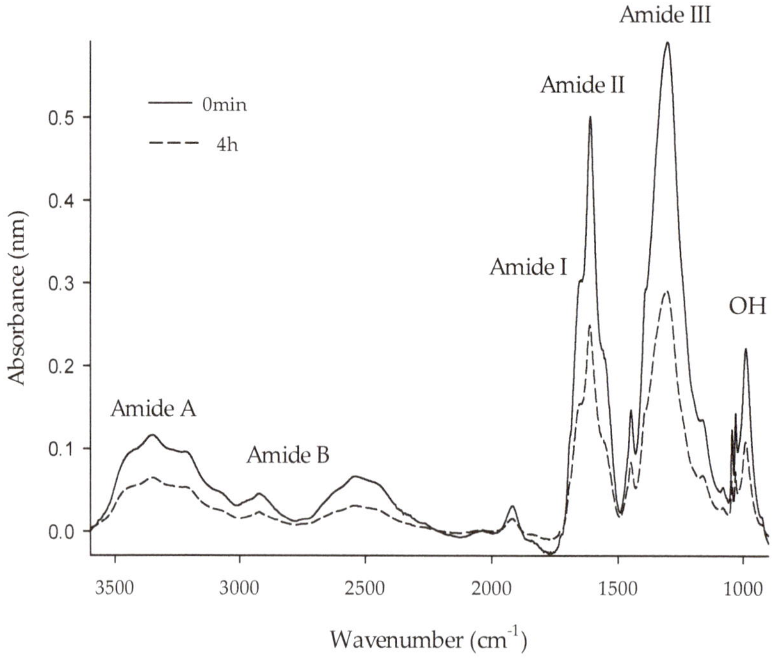

Figure 5. Fourier transform infrared spectra of collagen hydrolysates from 0 min, and 1 h of hydrolysis time.

2.7. Antioxidant Activity

2,2′-azino-bis(3-ethylbenzothiazoline-6-sulphonic acid) (ABTS) and 2,2-diphenyl-1-picrylhydrazyl (DPPH) assays are methods frequently used in the evaluation of radical scavengers to assess the antioxidant capacity of compounds. The ABTS radical can be applied in a wide range of pH, and is soluble in aqueous and organic media. It allows for the evaluation of both hydrophilic and lipophilic antioxidants [53,54]. DPPH radical is one of the most stable free radicals. It is a simple and quick method that can be used to test the ability of compounds to act as free radical scavengers or hydrogen donors [55]. The antioxidant activity of hydrolysed collagen (HC) was evaluated by the ABTS and DPPH methods as shown in Figure 6. There were differences ($p < 0.05$) between the ABTS and DPPH radical scavenging activities. The highest ABTS and DPPH radical scavenging activity was found at 4 h of hydrolysis with 67.6% and 52.75%, respectively. The ABTS technique reported higher values compared to DPPH. ABTS radical scavenging is commonly used to evaluate the ability of antioxidants to donate an electron or hydrogen atom to stabilize radicals [56]. The antioxidant activity of protein hydrolysates seemed to be affected by the amino acid composition as well as the degree of hydrolysis [33]. The longer the time, the higher the antioxidative activity was observed. It is well known that several amino acids like tyrosine, histidine [57] and lysine possess antioxidant properties [43]. Also, some hydrophobic amino acids like isoleucine and methionine could donate electrons or hydrogen, converting the radical to a more stable species and contributing to higher radical scavenging [58]. The amino acid content results of this research showed that strong hydrolysis (4 h) of collagen from sheepskin increased the concentration of these amino acids. At this time (4 h), the radical scavenging activity increased significantly because there was an increment of glutamic acid from 7.99 to 16.58 mg/g of protein. On the other hand, the enzymatic treatment of native collagen decreased its Mw by around 6 KDa. Previous works carried out with different sources such as fish [6,8] and squid [59] showed that Mw was one of the most important parameters that determined the biological activity of collagen [60]. The lower the Mw polypeptides, the higher the antioxidant activity was found to be. These results suggested that hydrolysis of collagen generated a wide variety of smaller peptides and free amino acids depending on the hydrolysis time [61]. Therefore, the composition of amino acid content, degree of hydrolysis and size of collagen chains and source of raw material could define the antioxidant capacity of the HC.

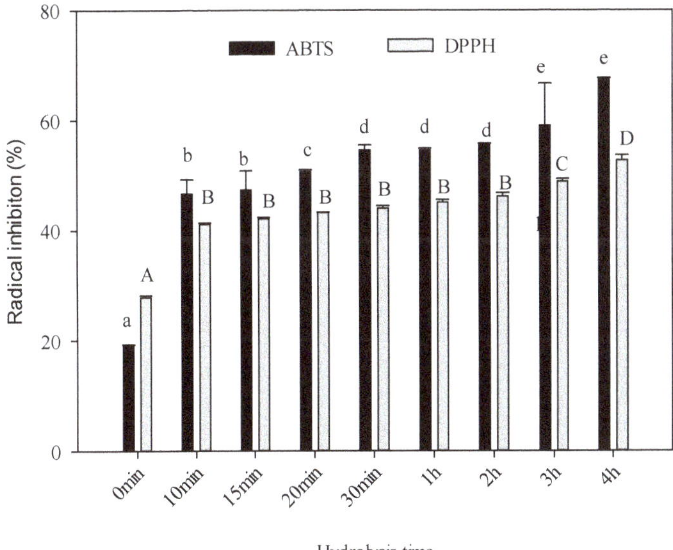

Figure 6. Antioxidant capacity of ovine collagen during hydrolysis. The activity was evaluated with TBS and DPPH methods. Values are expressed as the mean ± SD (*n* = 3). Different letters indicate significant difference at *p* ≤ 0.05.

2.8. Thermal Properties

The thermal properties of hydrolysed collagen was evaluated in dried powder with 0% of water content (d.b.) from samples obtained after different hydrolysis times. From Table 2, it can be seen that the melting temperature (Tm) reduced gradually from 153.3 °C (0 min) to 136.9 °C as the hydrolysis time increased up to 4 h. This reduction in Tm suggested that the thermal stability of the triple helical structure of collagen was affected after 1 h of thermal and enzymatic treatment (137.2 °C). It is well known that intermolecular helix formation is dependent on the molecular weight (Mw) of alpha chains [62]. The higher the Mw, the higher the Tm values that will be observed. The Tm results suggested that intermolecular helix formation reduced significantly as hydrolysis took place. This reduction originated at the bimolecular nucleation stage, which involved an intramolecular β-turn facilitated by glycine and/or proline residues [63]. This means that propagation was far less effective compared with native collagen at 0 min of treatment. The DSC thermograms of this native collagen showed a narrow melting range, suggesting a more homogenous population of longer helical segments. However, as the hydrolysis process started, the samples showed a broad melting range, indicating a wide molecular weight distribution of helix lengths by shifting the Tm towards lower values.

Table 2. Thermal properties of dried hydrolysed collagen powder at different times of hydrolysis. Average value of three replicates. Values followed by different letters are significantly different according to Tukey's test ($p \leq 0.05$).

Hydrolysis Time	Tm (°C)	ΔH (J/g)
0 min	153.38 ± 0.40 [a]	20.06 ± 0.10 [a]
15 min	147.97 ± 0.80 [a]	17.21 ± 0.47[ab]
20 min	147.72 ± 0.45 [a]	17.48 ± 0.18 [ab]
30 min	147.59 ± 0.78 [a]	16.91 ± 0.32 [ab]
1 h	137.91 ± 0.35 [b]	15.61 ± 0.21 [ab]
2 h	137.77 ± 0.13 [b]	11.46 ± 0.49 [ab]
3 h	137.24 ± 0.37 [b]	9.09 ± 0.20 [ab]
4 h	136.91 ± 0.55 [b]	8.93 ± 0.11 [b]

Considering the enthalpy as the energy required to disorganize the helical structure, it was possible to assume that native collagen possessed a more ordered structure (20.06 J/g). However, hydrolysis treatment produced low molecular weight residues with a low possibility of intramolecular refolding. In fact, previous work [63] has suggested a limit of 40-80 amino acid residues as the critical size of nuclei for renaturation. Also, we have seen that samples with Mw < 15 KDa cannot recover their helical conformation even at high concentrations [64]. Enthalpy of samples with 4 h of hydrolysis showed significant differences ($p < 0.05$) with the lowest degree of reorganization (8.93 J/g). However, it still showed some energy requirements to disorganize its structure. This structural conformation could be due to the intermolecular interaction of two or three strands with low Mw [64].

2.9. SEM Images

The morphological appearance of native collagen and their resulting hydrolysates are shown in Figure 7. It can be seen that HC showed changes in morphology across the different hydrolysis times. During the first 20 min of hydrolysis, the collagen did not appear to have pores in its structure (Figure 7a-c). However, after 30 min of treatment (Figure 7d-f), initial degradation of collagen was seen in the form of small pores in the protein structure, the result of enzymatic action leading to a partial disassembly of fibres into fibrils, and therefore, the generation of low molecular weight polypeptides.

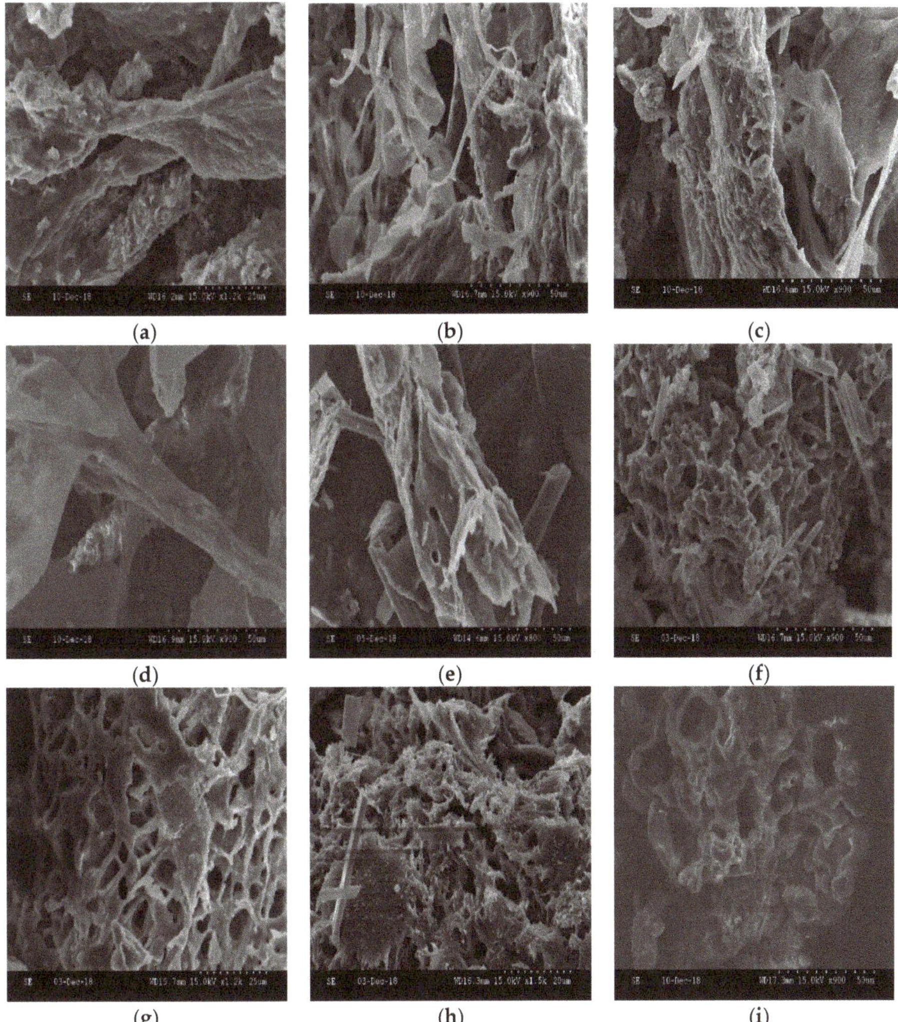

Figure 7. HC morphology changes during the hydrolysis alkaline treatment: (**a**) 0 min, (**b**) 10 min, (**c**) 15 min, (**d**) 20 min, (**e**) 30 min, (**f**) 1 h, (**g**) 2 h, (**h**) 3 h, (**i**) 4 h.

After 2 h of enzymatic hydrolysis (Figure 7h,i), the disaggregation of collagen fibrils within the collagen fibres was evidence of the autolysis of the enzymatic treatment [7]. The collagen structure of these treatments (after 3 h and 4 h) appeared to be extremely degraded, with a spongy and porous form. Previous works carried out with marine sources [65] discovered that the Mw of peptides influences the properties of HC. The lower the Mw, the more pores there are, and the more open the structure observed in the images.

2.10. Relationship between Mw, Viscosity, Antioxidant Activity and Thermal Properties

The properties of HC mainly appeared to obey the Mw of polypeptide chains obtained after hydrolysis. However, a strong relationship was observed with the other parameters evaluated; viscosity was closely related to Mw because these results suggested that the low Mw of polypeptide chains produced a low hydrodynamic volume of collagen molecules in solution [42]. This could be why the viscosity dropped to close to 0 Cp after 1 h of hydrolysis. The antioxidant activity also appeared to be dependent on the Mw parameter. The results from this research showed that the lower the Mw, the higher the antioxidant activity of HC. This behaviour was confirmed by the higher concentrations of hydroxyproline after 1 h of hydrolysis. Additionally, higher concentrations of aspartic acid and glutamic acid were obtained over the same period of time. The measurement of thermal properties in HC indicated that low-Mw samples (2 h, 3 h and 4 h) resulted in the lowest enthalpy. This means that less energy was required to disorganize the structure of HC because its low Mw avoided the organization into a triple helical form [66].

3. Materials and Methods

Sheepskins with 40–50% water content were used in this experiment. They were obtained as byproducts from a local market in Tulancingo, Hidalgo, Mexico. Reactive grade acetic acid (99% purity), sodium chloride reactive grade, porcine digestive protease (pepsin), dialysis membrane tubing with 6–8 kDa molecular weight cut-off, 4-dimethylaminobenzaldehyde at 5%, 65%, perchloric acid, 0.006 M chloramine T, 0.8 M and citrate buffer were purchased from Sigma-Aldrich Corp. (MA, USA).

3.1. Conditioning of The Skin before Collagen Extraction

The sheepskins were soaked to recover up to 70% water, followed by a fleshing process to remove the connective tissue and fat. Then, the skins were shaved to remove most of the hair.

3.2. Collagen Extraction from Ovine Skin

The methodology of Chuaychan et al. [67] was used, with some modifications. Pre-treated sheepskin was cut into small squares of approximately 1 cm^2 and suspended in 0.5 M acetic acid solution at a ratio of 1:10 (*w/v*). The sample was placed in a shaker machine (Bellco Biotechnology. NJ, USA) at 140 rpm for 3 h at room temperature (20 ± 2 °C), followed by the addition of pepsin at a concentration of 1 g/L with gentle stirring for 48 h. The sample was filtered and precipitated with a solution of 2.6 M sodium chloride. The precipitated material was centrifuged for 15 min at a relative centrifugal force of 3380× *g* in a centrifuge model Z36HK (HERMLE Labortechnik GmbH, Wehingen, Germany).

3.3. Hydrolysis of Collagen

Precipitated collagen was re-suspended in a solution of 1 M NaCO$_3$ at a ratio of 1:4 (*w/v*). The pH was adjusted to 8 ± 0.2. The hydrolysis of collagen was carried out with the enzyme trypsin at a concentration of 1:50 (*w/v*) in a water bath at 60 °C for different times, as follows: 10 min, 15 min, 20 min, 30 min, 1 h, 2 h, 3 h and 4 h. The control sample was called 0 min. All the samples were inactivated at 90 °C for 10 min and stored at 4 °C.

3.4. Protein Determination

Following Bradford determination [68], 5 mL of Bradford reagent and 100 µL of the sample were added and mixed in a vortex for 2 min. After 5 min storage in darkness, the sample was read in a spectrophotometer at 595 nm. The serum albumin at different concentrations was used to create a calibration curve.

3.5. Isoelectric Point

The isoelectric point was measured by using a Zetasizer nano ZS90 coupled to auto-titrator MPT-2. Laser Doppler and a DTS1070 cell (Worcestershire, UK) were used to determinate the electrophoretic mobility. HC was diluted in distilled water at 1:10. Different values of pH from 2 to 7 were obtained by using 0.5 M HCl and 0.75 M NaOH buffers, respectively.

3.6. Hydroxyproline Quantification

According to the AOAC methodology [69], 4 g of the sample and 30 mL of 3.5M sulphuric acid were placed in an oven at 105 °C for 12 h. The volume was adjusted to 500 mL with distilled water and filtered. Two millilitres of filtered sample were mixed with 1 mL of oxidant solution (0.006 M chloramine T in 0.8 M citrate buffer, pH 6.0) in a reaction tube. The volume was adjusted to 100 mL and stirred for 30 min at room temperature. Two millilitres of colour reagent (10 g of dimethylamine benzaldehyde in 35 mL of 65% perchloric acid) were added with stirring at 60 °C for 15 min. The absorbance of samples was measured against the blank at 558 nm in a UV-visible Jenway Genova, (Bibby Scientific, Staffordshire UK) spectrophotometer.

The collagen content was calculated with the next equation:

$$\% \, Hydroxyproline = \frac{(Y)(2.5)}{(W)(V)},\tag{1}$$

where:

Y = Hydroxyproline concentration from the standard curve
W = sample weight
V = volume in mL to adjust the 100 mL

3.7. Viscosity Analysis

Viscosity measurement was carried out using a viscometer Brookfield RTV (MA, USA) (spindle number: 5; speed of 100 rpm). Viscosity was expressed in centipoise (cP). All the samples were previously conditioned at 7 °C [70].

3.8. Molecular Weight

The sodium dodecyl sulphate polyacrylamide gel electrophoresis (SDS-PAGE) determination was performed according to Laemmli methodology [71]. One millilitre of dialyzed collagen was dissolved in 0.5M Tris–HCl buffer pH 6.8 (1% SDS, 10% glycerol and 0.01% bromophenol blue) and boiled for 5 min. Then, 10 μL of the denatured sample and 5 μL of a marker with a molecular weight from 10 kDa to 220 kDa (BenchMark Protein Ladder, Thermo Scientific, Pierce™, MA, USA) were loaded into wells at the top of the polyacrylamide gel. This gel contained a 4% stacking gel on top of the 12.5% resolving gel. A voltage of 50 V was applied for 30 min, and once the mobility of the proteins reached the resolving layer, the voltage was increased to 100 V for 4 h in order to see the separation of proteins according to size. As the electric current ran through the buffer, the negatively charged proteins migrated towards the anode and lower molecular weight proteins reached the bottom of the gel. After the migration of proteins, the gel was stained with Silver Stain Kit (Thermo Scientific, Pierce™ MA, USA).

3.9. Fourier Transform-Infrared (FTIR) Spectroscopy

The FTIR technique offers a green alternative because it allows us to quantify substances without organic solvents. The samples do not require any pretreatment, thus reducing the environmental damage caused by toxic waste. Also, it is a fast technique based on the natural vibrational frequencies of the chemical bonds present in molecules. FTIR is a non-destructive technique using a minimum amount of sample [72]. The absorption spectra by the FTIR technique were obtained with the Frontier FT-MIR (Perkin Elmer. MA, USA) equipment. The wavelength ranged from 380 to 4000 cm^{-1} at room temperature. The samples were brought into intimate contact with the diamond crystal by applying a loading pressure. For each sample, the spectrum represented an average of four scans with 4 cm^{-1} resolution. A spectrum of the empty cell was used as the background. Automatic signals were collected in 3620 scans at 1 cm^{-1} resolution. All the data were processed with Spectrum^{TM} 10 (Perkin Elmer. MA, USA) software.

3.10. Differential Scanning Calorimetry (DSC)

Thermal properties of HC were obtained with DSC series Q 2000 with intracooler RCS90 (DE, USA). It was calibrated with indium (Tm, onset $\frac{1}{4}$ 156.6 8C, ΔH $\frac{1}{4}$ 28.45 J/g). An average of 1.5 ± 0.1 mg of sample with known water content (0% db) was packed and hermetically sealed in a 50-mL stainless steel pan. An empty, hermetically sealed pan was used for a reference. Both heating and cooling scan rates were performed at 10 °C/min. Two heating scans were performed from 25 °C to 120 °C. Melting point temperature (Tm) and enthalpy (ΔH) were determined with TA 2000 analysis software (TA Instruments, DE, USA) based on the endothermic changes registered in the thermogram.

3.11. Amino Acid Determination

Amino acid content determination was performed according to Cohen [73] with some modifications. Three milligrams of freeze-dried HC were suspended in 6 M HCl and 1% *v/v* phenol at 150 °C for 1 h. Hydrolysed samples were dissolved in 2 mL of 0.5 M citrate buffer. Amino acid content was determined by high-performance liquid chromatography (HPLC) in a Hewlett Packard model GmbH (Winchester, UK) connected to a fluorescence detector (Ex. 250 Em. 395). The derivation reaction was carried out with 20 µL of the sample diluted in 60 µL buffer (borate buffer, Waters, Thermo Scientific, Pierce™, MA, USA) and 1 min stirring. Twenty microlitres of reagent AQC (Waters) were added with stirring for another 1 min, followed by the heating of the sample at 50 °C for 10 min. The amino acid separation was carried out in a Bluespher®column (100 × 2 mm ID) in reverse phase C18 octa-decyl dimethylsilane (Berlin, Germany). Conditions of work: mobile phase A: 50 mM sodium acetate, pH 5.75 and mobile phase B: 50 mM sodium acetate, pH 6/CAN 30:70 *v/v*, and 1 mL/min flow.

3.12. Antioxidant Activity

A solution of 2,2′-azino-bis(3-ethylbenzothiazoline-6-sulphonic acid) (ABTS) radical was prepared according to the literature [74] by mixing 7 mM ABTS and 2.45 mM potassium persulfate. After 16 h of stirring at room temperature in the dark, the ABTS solution was diluted with ethanol to stabilize it to 0.70 ± 0.02 at 734 nm. One millilitre of stabilized ABTS solution was mixed with 0.2 mL of the sample and the absorbance raised to 734 nm in a UV-visible Jenway Genova (Bibby Scientific, Staffordshire UK) spectrophotometer.

For assessing the antioxidant activity by 2,2-diphenyl-1-picrylhydrazyl (DPPH) radical inhibition [75], 0.5 mL of the sample was mixed with 2.5 mL of 6.1×10^{-5} M methanolic radical DPPH solution and maintained in darkness for 30 min. The absorbance was measured at 515 nm in a UV-visible Jenway Genova spectrophotometer. The antioxidant activity for ABTS and DPPH radical inhibition was calculated via the following equation:

$$\% \text{ Inhibition} = \frac{\text{Initial absorbance} - \text{Final absorbance}}{\text{Initial absorbance}} \times 100 \tag{2}$$

3.13. Scanning Electron Microscopy (SEM)

Morphology analysis was observed by a scanning electron microscope (Model S-2600N, HITACHI, Tokio, Japan). Freeze-dried HC samples were mounted on a strip of self-adhesive carbon paper and sputter-coated with gold to be observed in the scanning electron microscope at an acceleration voltage of 15 kV.

3.14. Statistical Analysis

A randomized design experiment and an analysis of variance (ANOVA) were applied to the experimental data, which included a Tukey test ($p \leq 0.05$). Data were analysed with SPSS 16.0 software (SPSS Inc., Chicago, IL, USA). Three replicates per treatment were considered in this experiment.

4. Conclusions

The study demonstrated that sheepskins are a good source of hydrolysed collagen. The best results were seen after 2 h of hydrolysis treatment. From this point, the kinetics of native collagen hydrolysis produced polypeptides with a low molecular weight and viscosity. This reduction in the polypeptide chain size affected the thermal properties of HC as the 4 h treatment produced a lower enthalpy value. Also, the antioxidant properties of HC were enhanced as the hydrolysis time increased. The functional properties of HC could be controlled by the hydrolysis time and sheepskins appeared to be a good alternative to typical sources like pigs, cows and fish.

Author Contributions: Conceptualization, R.G.C.-M and G.A.-Á.; Data curation, A.D.H.-F and R.G.C.-M.; Formal analysis, A.D.H.-F, R.G.C.-M and G.A-A; Investigation, A.L.-L and L.F.-J; Methodology, A.L.-L; Supervision, G.A.-A; Writing—original draft, A.L.-L; Writing—review & editing, G.A.-A.

Funding: This research was funded by CONACyT, grant number 621400.

Acknowledgments: The first author gratefully acknowledges Dimitrios Zeugolis for his technical support during a research stay at the University of Galway, Ireland.

Conflicts of Interest: The authors declare no conflict of interest.

Abbreviations

HC	Hydrolysed collagen
SDS-PAGE	Sodium dodecyl sulphate polyacrylamide gel Electrophoresis
Mw	Molecular weight
FTIR	Fourier transform-infrared spectroscopy
DSC	Differential scanning calorimetry
DPPH	2,2-diphenyl-1-picrylhydrazyl
ABTS	2,2′-azino-bis, 3-ethylbenzothiazoline-6-sulphonic acid
ANOVA	Analysis of variance
SEM	Scanning electron microscopy

References

1. Gao, L.-L.; Wang, Z.-Y.; Li, Z.; Zhang, C.-X.; Zhang, D.-Q. The characterization of acid and pepsin soluble collagen from ovine bones (Ujumuqin sheep). *J. Integr. Agric.* **2018**, *17*, 704–711. [CrossRef]
2. Liu, D.; Liang, L.; Regenstein, J.M.; Zhou, P. Extraction and characterisation of pepsin-solubilised collagen from fins, scales, skins, bones and swim bladders of bighead carp (Hypophthalmichthys nobilis). *Food Chem.* **2012**, *133*, 1441–1448. [CrossRef]
3. Rodriguez, M.I.A.; Barroso, L.G.R. Sanchez, M.L. Collagen: A review on its sources and potential cosmetic applications. *J. Cosmet. Dermatol.* **2018**, *17*, 20–26. [CrossRef] [PubMed]
4. Choi, D.; Min, S.G.; Jo, Y.J. Functionality of porcine skin hydrolysates produced by hydrothermal processing for liposomal delivery system. *J. Food Biochem.* **2018**, *42*, e12464. [CrossRef]
5. O'Sullivan, S.M.; Lafarga, T.; Hayes, M.; O'Brien, N.M. Bioactivity of bovine lung hydrolysates prepared using papain, pepsin, and Alcalase. *J. Food Biochem.* **2017**, *41*, e12406. [CrossRef]
6. Benjakul, S.; Karnjanapratum, S.; Visessanguan, W. Production and Characterization of Odorless Antioxidative Hydrolyzed Collagen from Seabass (Lates calcarifer) Skin without Descaling. *Waste Biomass Valori.* **2018**, *9*, 549–559. [CrossRef]
7. Liu, Z.Q.; Tuo, F.Y.; Song, L.; Liu, Y.X.; Dong, X.P.; Li, D.M.; Zhou, D.Y.; Shahidi, F. Action of trypsin on structural changes of collagen fibres from sea cucumber (Stichopus japonicus). *Food Chem.* **2018**, *256*, 113–118. [CrossRef]

8. Wu, R.; Wu, C.; Liu, D.; Yang, X.; Huang, J.; Zhang, J.; Liao, B.; He, H. Antioxidant and anti-freezing peptides from salmon collagen hydrolysate prepared by bacterial extracellular protease. *Food Chem.* **2018**, *248*, 346–352. [CrossRef]

9. Martínez-Ortiz, M.A.; Hernández-Fuentes, A.D.; Pimentel-González, D.J.; Campos-Montiel, R.G.; Vargas-Torres, A.; Aguirre-Álvarez, G. Extraction and characterization of collagen from rabbit skin: Partial characterization. *Cyta J. Food* **2015**, *13*, 253–258. [CrossRef]

10. Sibilla, S.; Godfrey, M.; Brewer, S.; Budh-Raja, A.; Genovese, L. An Overview of the Beneficial Effects of Hydrolysed Collagen as a Nutraceutical on Skin Properties: Scientific Background and Clinical Studies. *Open Nutraceuticals J.* **2015**, *8*, 29–42. [CrossRef]

11. Bilek, S.E.; Bayram, S.K. Fruit juice drink production containing hydrolyzed collagen. *J. Funct. Foods* **2015**, *14*, 562–569. [CrossRef]

12. Denis, A.; Brambati, N.; Dessauvages, B.; Guedj, S.; Ridoux, C.; Meffre, N.; Autier, C. Molecular weight determination of hydrolyzed collagens. *Food Hydrocoll.* **2008**, *22*, 989–994. [CrossRef]

13. Wang, Y.; Zhang, C.-l.; Zhang, Q.; Li, P. Composite electrospun nanomembranes of fish scale collagen peptides/chito-oligosaccharides: antibacterial properties and potential for wound dressing. *Int J Nanomed.* **2011**, *6*, 667–676.

14. Moskowitz, R.W. Role of collagen hydrolysate in bone and joint disease. *Semin. Arthritis Rheum.* **2000**, *30*, 87–99. [CrossRef] [PubMed]

15. Schrieber, R.; Gareis, H. *Gelatine Handbook: Theory and Industrial Practice*; John Wiley & Sons: Hoboken, NJ, USA, 2007.

16. Zague, V. A new view concerning the effects of collagen hydrolysate intake on skin properties. *Arch. Dermatol. Res.* **2008**, *300*, 479–483. [CrossRef] [PubMed]

17. Zhang, Z.; Li, G.; Shi, B. Physicochemical properties of collagen, gelatin and collagen hydrolysate derived from bovine limed split wastes. *J. Soc. Leather Technol. Chem.* **2006**, *90*, 23.

18. Lorenzo, J.M.; Munekata, P.E.S.; Gomez, B.; Barba, F.J.; Mora, L.; Perez-Santaescolastica, C.; Toldra, F. Bioactive peptides as natural antioxidants in food products—A review. *Trends Food Sci. Technol.* **2018**, *79*, 136–147. [CrossRef]

19. Jin, H.X.; Xu, H.P.; Li, Y.; Zhang, Q.W.; Xie, H. Preparation and Evaluation of Peptides with Potential Antioxidant Activity by Microwave Assisted Enzymatic Hydrolysis of Collagen from Sea Cucumber Acaudina Molpadioides Obtained from Zhejiang Province in China. *Mar. Drugs* **2019**, *17*, 169. [CrossRef]

20. Zhao, W.H.; Luo, Q.B.; Pan, X.; Chi, C.F.; Sun, K.L.; Wang, B. Preparation, identification, and activity evaluation of ten antioxidant peptides from protein hydrolysate of swim bladders of miiuy croaker (Miichthys miiuy). *J. Funct. Foods* **2018**, *47*, 503–511. [CrossRef]

21. Hong, G.P.; Min, S.G.; Jo, Y.J. Anti-Oxidative and Anti-Aging Activities of Porcine By-Product Collagen Hydrolysates Produced by Commercial Proteases: Effect of Hydrolysis and Ultrafiltration. *Molecules* **2019**, *24*, 1104. [CrossRef]

22. Vidal, A.R.; Ferreira, T.E.; Mello, R.D.; Schmidt, M.M.; Kubota, E.H.; Demiate, I.M.; Zielinski, A.A.F.; Dornelles, R.C.P. Effects of enzymatic hydrolysis (Flavourzyme (R)) assisted by ultrasound in the structural and functional properties of hydrolyzates from different bovine collagens. *Food Sci. Technol.* **2018**, *38*, 103–108. [CrossRef]

23. Liu, C.Y.; Ma, X.M.; Che, S.; Wang, C.W.; Li, B.F. The Effect of Hydrolysis with Neutrase on Molecular Weight, Functional Properties, and Antioxidant Activities of Alaska Pollock Protein Isolate. *J. Ocean Univ. China* **2018**, *17*, 1423–1431. [CrossRef]

24. Blanco, M.; Vazquez, J.A.; Perez-Martin, R.I.; Sotelo, C.G. Hydrolysates of Fish Skin Collagen: An Opportunity for Valorizing Fish Industry Byproducts. *Mar. Drugs* **2017**, *15*, 131. [CrossRef]

25. Pal, G.K.; Suresh, P.V. Comparative assessment of physico-chemical characteristics and fibril formation capacity of thermostable carp scales collagen. *Mater. Sci. Eng. C-Mater. Biol. Appl.* **2017**, *70*, 32–40. [CrossRef]

26. Lee, H.-J.; Chae, S.-J.; Saravana, P.S.; Chun, B.-S. Physical and functional properties of tunicate (Styela clava) hydrolysate obtained from pressurized hydrothermal process. *Fish. Aquat. Sci.* **2017**, *20*, 14. [CrossRef]

27. Felician, F.F.; Xia, C.L.; Qi, W.Y.; Xu, H.M. Collagen from Marine Biological Sources and Medical Applications. *Chem. Biodivers.* **2018**, *15*, e1700557. [CrossRef]

28. Sae-leaw, T.; Benjakul, S. Antioxidant activities of hydrolysed collagen from salmon scale ossein prepared with the aid of ultrasound. *Int. J. Food Sci. Technol.* **2018**, *53*, 2786–2795. [CrossRef]

29. Paul, R.; Adzet, J.M.; Brouta-Agnésa, M.; Balsells, S.; Esteve, H. Hydrolyzed collagen: A novel additive in cotton and leather dyeing. *Dye. Pigment.* **2012**, *94*, 475–480. [CrossRef]
30. Offengenden, M.; Chakrabarti, S.; Wu, J. Chicken collagen hydrolysates differentially mediate anti-inflammatory activity and type I collagen synthesis on human dermal fibroblasts. *Food Sci. Hum. Wellness* **2018**, *7*, 138–147. [CrossRef]
31. Bateman, J.F.; Lamande, S.R.; Ramshaw, J.A. Collagen superfamily. *Extracell. Matrix* **1996**, *2*, 22–67.
32. Atma, Y.; Hanifah, N.L.; Endang, P.; Hermawan, S.; Moh, T.; Dita, F.; Apon, Z.M. The hydroxyproline content of fish bone gelatin from Indonesian pangasius catfish by enzymatic hydrolysis for producing the bioactive peptide. *Biofarmasi. J. Nat. Prod. Biochem.* **2018**, *6*, 64–68.
33. Ao, J.; Li, B. Amino acid composition and antioxidant activities of hydrolysates and peptide fractions from porcine collagen. *Food Sci. Technol. Int.* **2012**, *18*, 425–434. [CrossRef] [PubMed]
34. Li-Chan, E.C.Y.; Hunag, S.L.; Jao, C.L.; Ho, K.P.; Hsu, K.C. Peptides Derived from Atlantic Salmon Skin Gelatin as Dipeptidyl-peptidase IV Inhibitors. *J. Agric. Food Chem.* **2012**, *60*, 973–978. [CrossRef]
35. Benjakul, S.; Oungbho, K.; Visessanguan, W.; Thiansilakul, Y.; Roytrakul, S. Characteristics of gelatin from the skins of bigeye snapper, Priacanthus tayenus and Priacanthus macracanthus. *Food Chem.* **2009**, *116*, 445–451. [CrossRef]
36. Gómez-Lizárraga, K.; Piña-Barba, C.; Rodríguez-Fuentes, N.; Romero, M. Obtención y caracterización de colágena tipo I a partir de tendón bovino. *Superf. Y Vacío* **2011**, *24*, 137–140.
37. Soladoye, O.P.; Saldo, J.; Peiro, L.; Rovira, A.; Mor-Mur, M. Antioxidant and Angiotensin 1 Converting Enzyme Inhibitory Functions from Chicken Collagen Hydrolysates. *J. Nutr. Food Sci.* **2015**, *5*, 1. [CrossRef]
38. Friess, W. Collagen–biomaterial for drug delivery. *Eur. J. Pharm. Biopharm.* **1998**, *45*, 113–136. [CrossRef]
39. Paschalis, E.; Verdelis, K.; Doty, S.; Boskey, A.; Mendelsohn, R.; Yamauchi, M. Spectroscopic characterization of collagen cross-links in bone. *J. Bone Miner. Res.* **2001**, *16*, 1821–1828. [CrossRef]
40. Kittiphattanabawon, P.; Benjakul, S.; Visessanguan, W.; Nagai, T.; Tanaka, M. Characterisation of acid-soluble collagen from skin and bone of bigeye snapper (Priacanthus tayenus). *Food Chem.* **2005**, *89*, 363–372. [CrossRef]
41. Wu, H.C.; Chen, H.M.; Shiau, C.Y. Free amino acids and peptides as related to antioxidant properties in protein hydrolysates of mackerel (Scomber austriasicus). *Food Res. Int.* **2003**, *36*, 949–957. [CrossRef]
42. Aguirre-Alvarez, G.; Foster, T.; Hill, S.E. Impact of the origin of gelatins on their intrinsic properties. *Cyta-J. Food* **2012**, *10*, 306–312. [CrossRef]
43. Kim, S.K.; Kim, Y.T.; Byun, H.G.; Nam, K.S.; Joo, D.S.; Shahidi, F. Isolation and characterization of antioxidative peptides from gelatin hydrolysate of Alaska pollack skin. *J. Agric. Food Chem.* **2001**, *49*, 1984–1989. [CrossRef] [PubMed]
44. Caessens, P.W.; Daamen, W.F.; Gruppen, H.; Visser, S.; Voragen, A.G. β-Lactoglobulin hydrolysis. 2. Peptide identification, SH/SS exchange, and functional properties of hydrolysate fractions formed by the action of plasmin. *J. Agric. Food Chem.* **1999**, *47*, 2980–2990. [CrossRef] [PubMed]
45. Chalamaiah, M.; Hemalatha, R.; Jyothirmayi, T. Fish protein hydrolysates: proximate composition, amino acid composition, antioxidant activities and applications: a review. *Food Chem.* **2012**, *135*, 3020–3038. [CrossRef] [PubMed]
46. Zhang, M.; Liu, W.T.; Li, G.Y. Isolation and characterisation of collagens from the skin of largefin longbarbel catfish (Mystus macropterus). *Food Chem.* **2009**, *115*, 826–831. [CrossRef]
47. Chi, C.; Hu, F.; Li, Z.; Wang, B.; Luo, H. Influence of Different Hydrolysis Processes by Trypsin on the Physicochemical, Antioxidant, and Functional Properties of Collagen Hydrolysates from Sphyrna lewini, Dasyatis akjei, and Raja porosa. *J. Aquat. Food Prod. Technol.* **2016**, *25*, 616–632. [CrossRef]
48. Khiari, Z.; Ndagijimana, M.; Betti, M. Low molecular weight bioactive peptides derived from the enzymatic hydrolysis of collagen after isoelectric solubilization/precipitation process of turkey by-products. *Poult. Sci.* **2014**, *93*, 2347–2362. [CrossRef]
49. Yan, L.J.; Jin, T.C.; Chen, Y.L.; Zhan, C.L.; Zhang, L.J.; Weng, L.; Liu, G.M.; Cao, M.J. Characterization of a recombinant matrix metalloproteinase-2 from sea cucumber (Stichopus japonicas) and its application to prepare bioactive collagen hydrolysate. *Process Biochem.* **2018**, *72*, 63–70. [CrossRef]
50. Ogawa, M.; Portier, R.J.; Moody, M.W.; Bell, J.; Schexnayder, M.A.; Losso, J.N. Biochemical properties of bone and scale collagens isolated from the subtropical fish black drum (Pogonia cromis) and sheepshead seabream (Archosargus probatocephalus). *Food Chem.* **2004**, *88*, 495–501. [CrossRef]

51. Pan, B.S.; Chen, H.E.; Sung, W.C. Molecular and thermal characteristics of acid-soluble collagen from orbicular batfish: effects of deep-sea water culturing. *Int. J. Food Prop.* **2018**, *21*, 1080–1090. [CrossRef]

52. Barth, A.; Zscherp, C. What vibrations tell us about proteins. *Q. Rev. Biophys.* **2002**, *35*, 369–430. [CrossRef]

53. Shalaby, E.A.; Shanab, S.M.M. Comparison of DPPH and ABTS assays for determining antioxidant potential of water and methanol extracts of Spirulina platensis. *Indian J. Geo-Mar. Sci.* **2013**, *42*, 556–564.

54. Amorati, R.; Valgimigli, L. Advantages and limitations of common testing methods for antioxidants. *Free Radic. Res.* **2015**, *49*, 633–649. [CrossRef]

55. Foh, M.B.K.; Amadou, I.; Foh, B.M.; Kamara, M.T.; Xia, W. Functionality and antioxidant properties of Tilapia (*Oreochromis niloticus*) as influenced by the degree of hydrolysis. *Int. J. Mol. Sci.* **2010**, *11*, 1851–1869. [CrossRef]

56. Leong, L.P.; Shui, G. An investigation of antioxidant capacity of fruits in Singapore markets. *Food Chem.* **2002**, *76*, 69–75. [CrossRef]

57. Chen, H.M.; Muramoto, K.; Yamauchi, F.; Fujimoto, K.; Nokihara, K. Antioxidative properties of histidine-containing peptides designed from peptide fragments found in the digests of a soybean protein. *J. Agric. Food Chem.* **1998**, *46*, 49–53. [CrossRef]

58. Pownall, T.L.; Udenigwe, C.C.; Aluko, R.E. Amino Acid Composition and Antioxidant Properties of Pea Seed (*Pisum sativum* L.) Enzymatic Protein Hydrolysate Fractions. *J. Agric. Food Chem.* **2010**, *58*, 4712–4718. [CrossRef]

59. Nam, K.A.; You, S.G.; Kim, S.M. Molecular and physical characteristics of squid (Todarodes pacificus) skin collagens and biological properties of their enzymatic hydrolysates. *J. Food Sci.* **2008**, *73*, C249–C255. [CrossRef]

60. Gómez-Guillén, M.C.; Giménez, B.; López-Caballero, M.E.; Montero, M.P. Functional and bioactive properties of collagen and gelatin from alternative sources: A review. *Food Hydrocoll.* **2011**, *25*, 1813–1827. [CrossRef]

61. Khantaphant, S.; Benjakul, S. Comparative study on the proteases from fish pyloric caeca and the use for production of gelatin hydrolysate with antioxidative activity. *Comp. Biochem. Physiol. B-Biochem. Mol. Biol.* **2008**, *151*, 410–419. [CrossRef]

62. Busnel, J.P.; Morris, E.R.; Rossmurphy, S.B. Interpretation of the renaturation kinetics of gelatin solutions. *Int. J. Biol. Macromol.* **1989**, *11*, 119–125. [CrossRef]

63. Harrington, W.F.; Rao, N.V. Collagen structure in solution. I. Kinetics of helix regeneration in single-chain gelatins. *Biochemistry* **1970**, *9*, 3714–3724. [CrossRef]

64. Djabourov, M.; Nishinari, K.; Ross-Murphy, S.B.; Djabourov, M.; Nishinari, K.; RossMurphy, S.B. *Helical Structures from Neutral Biopolymers*; Cambridge Univ Press: Cambridge, UK, 2013; pp. 182–221.

65. Chi, C.F.; Cao, Z.H.; Wang, B.; Hu, F.Y.; Li, Z.R.; Zhang, B. Antioxidant and functional properties of collagen hydrolysates from Spanish mackerel skin as influenced by average molecular weight. *Molecules* **2014**, *19*, 11211–11230. [CrossRef]

66. Aguirre-Alvarez, G.; Pimentel-Gonzalez, D.J.; Campos-Montiel, R.G.; Foster, T.; Hill, S.E. The effect of drying temperature on mechanical properties of pig skin gelatin films. *Cyta-J. Food* **2011**, *9*, 243–249. [CrossRef]

67. Chuaychan, S.; Benjakul, S.; Kishimura, H. Characteristics of acid- and pepsin-soluble collagens from scale of seabass (Lates calcarifer). *Lwt Food Sci. Technol.* **2015**, *63*, 71–76. [CrossRef]

68. Bradford, M.M. A rapid and sensitive method for the quantitation of microgram quantities of protein utilizing the principle of protein-dye binding. *Anal. Biochem.* **1976**, *72*, 248–254. [CrossRef]

69. AOAC Authors. Official methods of analysis Amino Acids. Analysis Acid hydrolysis Hydroxyproline—Item 77. In *Food Analysis Methods*, 17th ed.; Association of Analytical Communities: Gaithersburg, MD, USA, 2006; Reference data: Method 990.26 (39.1.27); NFNAP; NITR; HYP.

70. Yu, D.; Chi, C.-F.; Wang, B.; Ding, G.-F.; Li, Z.-R. Characterization of acid-and pepsin-soluble collagens from spines and skulls of skipjack tuna (Katsuwonus pelamis). *Chin. J. Nat. Med.* **2014**, *12*, 712–720. [CrossRef]

71. Laemmli, U.K. Cleavage of Structural Proteins during the Assembly of the Head of Bacteriophage T4. *Nature* **1970**, *227*, 680–685. [CrossRef]

72. Fanelli, S.; Zimmermann, A.; Totoli, E.G.; Salgado, H.R.N. FTIR Spectrophotometry as a Green Tool for Quantitative Analysis of Drugs: Practical Application to Amoxicillin. *J. Chem.* **2018**. [CrossRef]

73. Cohen, S.A. Amino acids analysis using precolumn derivatization with 6-Aminoquinolyl-N-Hidroxysuccinimidyl Carbamate. In *Amino Acid Analysis Protocols. Methods in Molecular Biology*; Cooper, C., Packer, N., Williams, K., Eds.; Humana Press: Totowa, NJ, USA, 2001; Volume 159, pp. 39–47.

74. Re, R.; Pellegrini, N.; Proteggente, A.; Pannala, A.; Yang, M.; Rice-Evans, C. Antioxidant activity applying and improve ABTS radical cation decolorization assay. *Free Radic. Biol. Med.* **1999**, *26*, 1231–1237. [CrossRef]

75. Brand-Williams, W.; Cuvelier, M.E.; Berse, C. Use of a free radical method to evaluate antioxidant activity. *Lwt Food Sci. Technol.* **1995**, *28*, 25–30. [CrossRef]

International Journal of
Molecular Sciences

Article

Development of a Soy Protein Hydrolysate with an Antihypertensive Effect

Eric Banan-Mwine Daliri [1], **Fred Kwame Ofosu** [1], **Ramachandran Chelliah** [1], **Mi Houn Park** [2],
Jong-Hak Kim [2] **and Deog-Hwan Oh** [1],[*]

[1]   Department of Food Science and Biotechnology, Kangwon National University, Chuncheon 200-701, Korea;
     ericdaliri@yahoo.com (E.B.-M.D.); fkofosu17@gmail.com (F.K.O.); ramachandran865@gmail.com (R.C.)

[2]   Erom Company Limited, R&D Center, 111, Toegye Nonggong-ro, Chuncheon-si, Gangwon-do 24427, Korea;
     mhpark@yahoo.com (M.H.P.); jhkim@yahoo.com (J.-H.K.)

[*]   Correspondence: deoghwa@kangwon.ac.kr; Tel.: +82-33-250-6457 (cp): +82-10-5118-6457

Received: 15 February 2019; Accepted: 21 March 2019; Published: 25 March 2019

Abstract: In this study, we combined enzymatic hydrolysis and lactic acid fermentation to
generate an antihypertensive product. Soybean protein isolates were first hydrolyzed by Prozyme
and subsequently fermented with *Lactobacillus rhamnosus* EBD1. After fermentation, the in vitro
angiotensin-converting enzyme (ACE) inhibitory activity of the product (P-SPI) increased from 60.8
$\pm$ 2.0% to 88.24 $\pm$ 3.2%, while captopril (a positive control) had an inhibitory activity of 94.20 $\pm$ 5.4%.
Mass spectrometry revealed the presence of three potent and abundant ACE inhibitory peptides,
PPNNNPASPSFSSSS, GPKALPII, and IIRCTGC in P-SPI. Hydrolyzing P-SPI with gastrointestinal
proteases did not significantly affect its ACE inhibitory ability. Also, oral administration of P-SPI
(200 mg/kg body weight) to spontaneous hypertensive rats (SHRs) for 6 weeks significantly lowered
systolic blood pressure (-19 ± 4 mm Hg, $p < 0.05$) and controlled body weight gain relative to control
SHRs that were fed with physiological saline. Overall, P-SPI could be used as an antihypertensive
functional food.

Keywords: antihypertensive peptides; functional food; food-derived; fermentation

1. Introduction

High blood pressure (hypertension) is a chronic degenerative disease and the leading risk factor
for chronic kidney disease and cardiovascular diseases [1]. Uncontrolled hypertension can result
in chronic damage to the vascular system, myocardial strokes, and even death [2]. For this reason,
several pharmacological and nonpharmacological strategies aimed at reducing the incidence of this
disease have been implemented. Over the years, several functional foods have been developed from
different food proteins to be used as nonpharmacological treatments of high blood pressure [3]. More
so, many food-derived bioactive peptides have demonstrated antihypertensive effects. Such peptides
commonly inhibit angiotensin-converting enzyme (ACE) activity and/ or reduce renin activity [4,5].
Soybean proteins are among the most common plant substrates used for food-derived antihypertensive
peptide development [6,7]. Soy proteins constitute about 35–40% of the total dry weight of the bean
and the major storage proteins are glycinin (11S globulin) and β-conglycinin (7S globulin). These
storage proteins account for about 65–85% of the total soy proteins [8]. Recent studies have shown
that consumption of fermented soybean meal can reduce the risk of cardiovascular disease mortality,
cardiovascular disease, stroke, and coronary heart disease risk [9,10]. However, to release bioactive
peptides from parent proteins, enzyme treatment and fermentation are the two most common methods
used. While enzyme hydrolysis saves time and enhances scalability and predictability of peptides,
fermentation (though a relatively slow process) is a cheaper strategy to generate various bioactive
peptides with diverse activities. Combining the two methods could enhance the number and kinds

of peptides released from the parent protein. A number of lactic acid bacteria such as *Pediococcus pentosaceus* and *Lactobacillus casei* have been used to ferment soy proteins to release antihypertensive peptides of high potency [2,11,12]. In an earlier study, we found that *Lactobacillus rhamnosus* EBD1 isolated from Korean fermented soybean (doenjang) had strong proteolytic activity and could be helpful in generating bioactive peptides.

Therefore, to develop a soybean product with a strong antihypertensive effect, we first hydrolyzed soy protein isolates (SPI) with Prozyme and subsequently fermented the hydrolysate with *Lactobacillus rhamnosus* EBD1 to obtain a product we named P-SPI. This combined strategy enhanced the degree of hydrolysis of the proteins. The long term effect of P-SPI consumption on the systolic blood pressure of spontaneous hypertensive rats (SHR) was studied.

2. Results

2.1. The Extent of Hydrolysis

L. *rhamnosus* EBD1 was able to grow on SPI as a sole nitrogen source and both Prozyme and *L. rhamnosus* digested SPI to various extents. Analysis of hydrolysates by RP-HPLC showed that approximately 15% of the substrate was hydrolyzed by Prozyme after treatment for 1 h. However, subsequent fermentation of the hydrolysate with *L. rhamnosus* EBD1 for 48 h resulted in hydrolysis of approximately 55% of the initial SPI concentration (Figure 1).

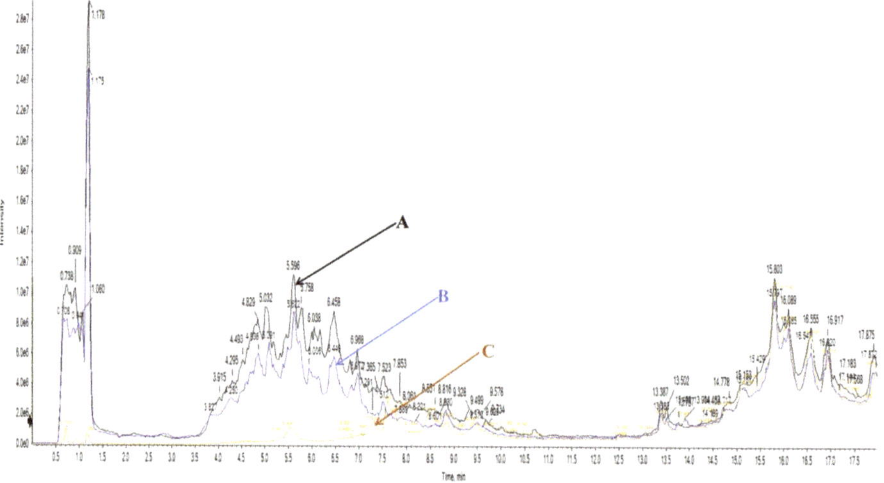

Figure 1. RP-HPLC chromatograms of various stages of sample preparation: (**A**) represents the chromatogram of raw soy protein isolates (SPI), (**B**) represents the Prozyme treated SPI and (**C**) represents the product (P-SPI).

2.2. ACE-Inhibitory Ability of Hydrolysates and Fermentates

Both the enzyme hydrolyzed and fermented hydrolysates inhibited ACE to various extents (Table 1). As shown, fermentation of the enzyme hydrolyzed samples improved the ACE inhibitory activity from 60.8% to 88.24%.

Table 1. Angiotensin-converting enzyme (ACE) inhibitory activity of processed SPI

Sample	Inhibitory Activity (%)	IC$_{50}$ (mg/mL)
Raw SPI	10.21 ± 4.0	n.d
Prozyme treated SPI	60.8 ± 2.0	0.980
P-SPI	88.24 ± 3.2	0.592
Captopril	94.20 ± 5.4	0.005

Data shows mean ± SD (n = 3). Values represent the means of three replicates ± S.D. n.d: Not determined.

2.3. ACE Inhibitory Peptides from P-SPI

All the peptides identified in the peptide profile are displayed in Supplementary Table S1. Since 3008 peptides were identified and could not be individually synthesized/eluted and tested in vitro for ACE inhibitory activity, the peptide sequences were screened using an in silico platform (http://crdd.osdd.net/raghava/ahtpin/index.php) developed by Kumar et al. [13] to predict potential ACE inhibitory peptides. Although many potential ACE inhibitory peptides were identified, peptides IAKKLVLP, PDIGGFGC, PPNNNPASPSFSSSS, GPKALPII and IIRCTGC were most abundant. The peptides were synthesized and their ACE inhibitory activities were confirmed in vitro. Among the peptides tested, IIRCTGC showed the strongest inhibitory activity of 83 ± 0.9%, followed by PPNNNPASPSFSSS and GPKALPII, which were not significantly different in their inhibitory abilities ($p > 0.05$) (Figure 2). Meanwhile, peptide PPNNNPASPSFSSSS showed an inhibitory activity of 18 ± 7%, while IAKKLVLP displayed the least inhibitory activity of 10 ± 3%. Captopril (the positive control) showed the strongest inhibitory activity of 94 ± 4%.

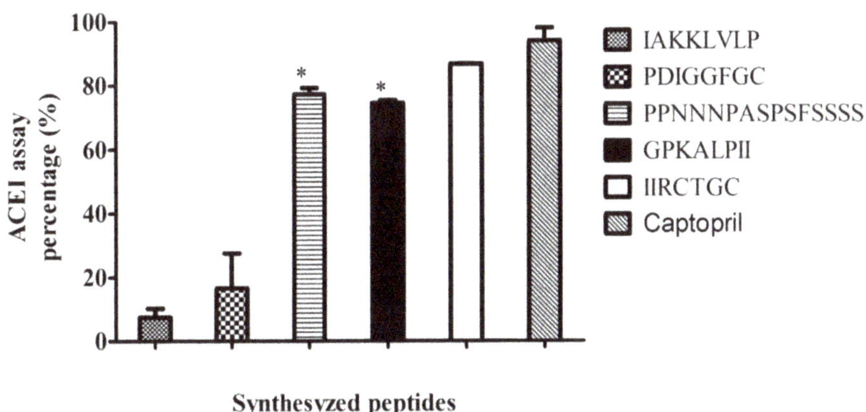

Figure 2. The ACE inhibitory ability of selected peptides from P-SPI. Bars represent the means of three replicates (n = 3) ± SD, *$p < 0.05$.

2.4. The Effects of Gastrointestinal Enzymes on ACE Inhibitory Activity of P-SPI

When P-SPI was subjected to pepsin digestion, its ACE inhibitory activity was not significantly altered. Also, subsequent treatment of the peptides with pancreatin did not affect the ACE inhibitory ability ($p > 0.05$) as shown in Figure 3.

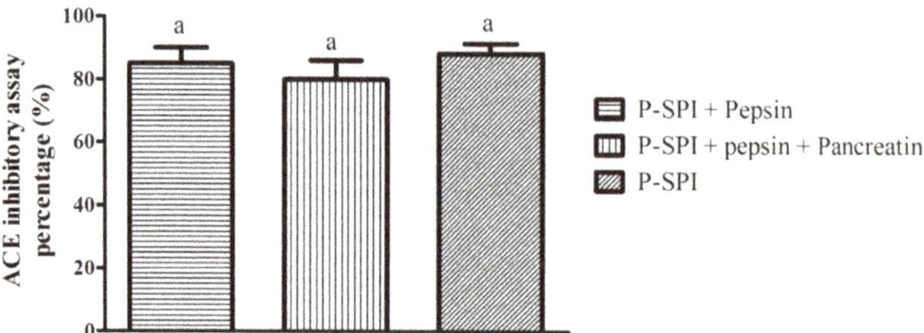

Figure 3. The effects of gastrointestinal enzymes on ACE inhibitory activity. Bars represent means of three replicates ($n = 3$) $\pm$ SD, [a] $p < 0.05$.

2.5. Blood Pressure Reducing Effects of P-SPI

The mean SBP measured for all the SHRs from the four experimental groups prior to treatment (zero time) was 179 ± 5.6 mm Hg ($n = 20$). Oral administration of P-SPI at a dose of 100 mg/kg resulted in a significant reduction in SBP when compared to the lack of SBP reduction by physiological saline. The decrease in SBP was maximal at the 4th week of oral administration (-19 ± 4 mm Hg). By contrast, oral administration of 10 mg/kg P-SPI induced a slight decrease in SBP which was maximum at the 4th-week post-administration (-11.2 ± 2 mm Hg). However, all the P-SPI doses reduced SBP significantly when compared to the negative control group and maintained lower blood pressures throughout the feeding period. 50 mg/kg captopril administration resulted in an SBP reduction of 22 mm Hg (Figure 4).

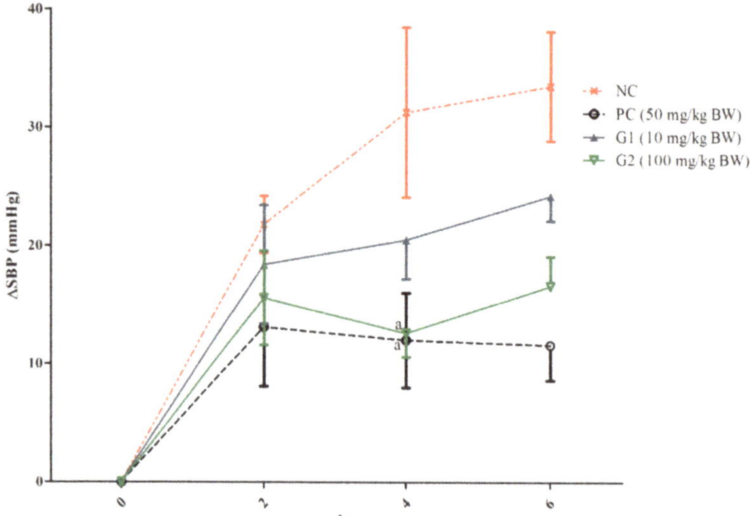

Figure 4. Systolic blood pressure changes from the baseline are expressed in absolute values (mmHg) and data are mean $\pm$ SEM from 5 determinations. Data points with the same alphabets are not significantly different ([a] $p > 0.05$) using one-way ANOVA followed by Duncan tests. NC: Negative control, PC: Positive control, G1: Group 1, G2: Group 2.

2.6. Effects of P-SPI Consumption on Body Weight Gain

P-SPI (10 mg/kg and 100 mg/kg) and captopril administration significantly reduced weight gain from the 2nd week to the 4th week of feeding ($p < 0.05$) relative to SHRs that were given physiological saline. The weights of the rats in these groups were however maintained after the 4th week up to the 6th week of feeding (Figure 5).

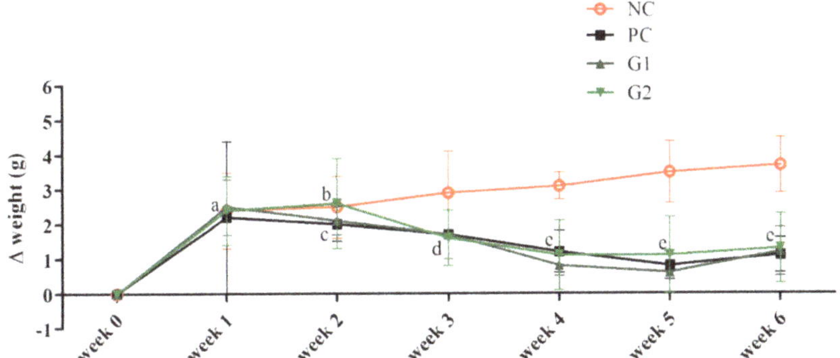

Figure 5. Time course of weight changes (Δ weight) after oral administration of physiological saline, 50 mg/kg, and 100 mg/kg body weight of P-SPI. Each data point represents mean ± SEM from 5 determinations, and data points with the different alphabets are significantly different ($p < 0.05$) using one-way ANOVA followed by Duncan tests. NC: Negative control, PC: Positive control, G1: Group 1, G2: Group 2.

2.7. Effects of P-SPI Consumption on Feed Intake

Generally, neither the administration of P-SPI nor captopril affected the quantity of feed intake of SHRs (Figure 6). The quantity of feed consumed by the rats in the test and control groups was not significantly different throughout the course of study ($p > 0.05$).

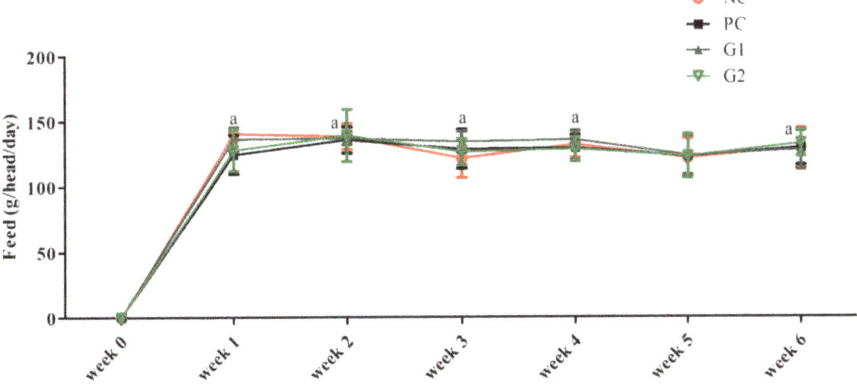

Figure 6. Each data point represents mean ± SEM from 5 determinations, and data points with the different alphabets are significantly different ($p < 0.05$) using one-way ANOVA followed by Duncan tests. NC: Negative control, PC: Positive control, G1: Group 1, G2: Group 2.

3. Discussion

To generate bioactive peptides from soy proteins, proteolytic enzymes or microorganisms would be required to release the peptides from the parent proteins. The use of fermented soybean or enzyme hydrolyzed soy proteins for reducing high blood pressure is well recognized. However, very few studies (if any) have exploited the combined effects of proteolytic enzymes and fermentation for developing antihypertensive foods.

3.1. Hydrolysis and Fermentation of Soy Proteins

Hydrolysis with proteases and subsequent fermentation reduces the time needed for efficient substrate hydrolysis when only fermentation is employed. Also, this strategy allows the generation of new antihypertensive peptide sequences that would not have been generated if only a single method was applied. The combined processing method in this work enhanced hydrolysis of the soy protein compared to the enzyme treatment alone. *Lactobacillus rhamnosus* is known for its well-developed protein degradation machineries with which it hydrolyzes proteins. Peptides generated by the cell-envelope proteinase hydrolysis are transported into the bacterial cell for further hydrolysis by peptidases so as to meet its nitrogen requirements [14,15]. The cell number increases from 2×10^8 cells to 10^9 cells after 48 h incubation.

3.2. Effects of P-ISP on ACE Activity

The product obtained (P-SPI) displayed strong ACE inhibitory ability of 88.24 $\pm$ 3.2% (IC_{50} = 0.592 mg/mL) compared to raw SPI. Using LC-ESI-TOF-MS/MS, we identified 3008 peptides in P-SPI samples which were generated by the processing method. In silico screening of peptides using the AHTpin software available at http://crdd.osdd.net/raghava/ahtpin/index.php revealed many potential ACE inhibitory peptides, among which IAKKLVLP, PDIGGFGC, PPNNNPASPSFSSSS, GPKALPII, and IIRCTGC were most abundant. Since these 5 peptides satisfied some of the common structural features described for many food-derived ACE inhibitory peptides [16–18], they were synthesized and their inhibitory activities were tested in vitro. As seen in Figure 2, only PPNNNPASPSFSSSS, GPKALPII, and IIRCTGC were strong ACE inhibitors, while IAKKLVLP and PDIGGFGC were weak inhibitors. Earlier studies about structural-activity relationships between peptides and ACE inhibition indicated that peptides whose C-terminal tripeptides are hydrophobic show a stronger binding ability to ACE [16], and this could account for why GPKALPII showed good ACE inhibition. Also, peptides with branched-chain aliphatic amino acids or hydrophobic amino acid at the N-terminal have been shown to be good competitive inhibitors of ACE. These criteria make IIRCTGC and PPNNNPASPSFSSSS good ACE inhibitory candidates [17,18].

3.3. Effect of Gastrointestinal Enzymes on P-ISP Activity

In the gut, ingested peptides encounter gastrointestinal enzymes and may be hydrolyzed. This may either result in loss of activity or generation of other potent peptides. Treatment of P-SPI with gastrointestinal enzymes in vitro, however, did not significantly affect ACE inhibitory activity ($p > 0.05$), indicating that the ACE inhibitory peptides were either resistant to gastrointestinal enzyme digestion or retained their activity even after digestion.

3.4. The Effect of P-ISP Consumption on Systolic Blood Pressure

Recent studies have indicated that systolic blood pressure is a better factor for predicting cardiovascular disease than diastolic blood pressure [19,20]; hence, reducing SBP reduces the risk of CVD. Results from this study showed a clear reduction in SBP when SHR were fed with P-SPI (10 mg/kg BW and 100 mg/kg BW) for six weeks. Nevertheless, the effect of P-SPI was less pronounced than the effect of captopril (a standard antihypertensive drug). However, this study was aimed at developing a functional food that could prevent or reduce high blood pressure rather than curing the condition. Compared to synthetic antihypertensive drugs, food-derived antihypertensive peptides have been reported to have no side effects, have higher tissue affinities, and maybe more slowly cleared from tissues [21].

3.5. The Effect of P-ISP Consumption on Feed Satiety and Body Weight Gain

Many studies have found a strong association between obesity and hypertension in humans [22–24]. This is because an increase in body weight seems to be followed by an increase in blood pressure. However, whether obesity precedes hypertension or hypertension leads to obesity still remains unclear. Yet, due to the association between these two conditions, we studied the effect of P-SPI consumption on SHR body weight. Relative to the untreated group, P-SPI reduced SHR body weight gain significantly from the second week to the fourth week of feeding. SHR body weight was, however, maintained ($p > 0.05$), though they were continuously fed with P-SPI from the 4th–6th weeks. A similar observation was made among SHRs administered with captopril.

Some studies have shown that certain bioactive peptides decrease appetite and lead to reduced food intake, resulting in reduced weight gain [25–27]. For this reason, we checked whether P-SPI consumption affected feed intake relative to control groups. It was observed that P-SPI consumption did not have any significant effect on the quantity of feed consumed by the rats (Figure 6). It is therefore possible that the reduction in weight gain might have been caused by other reasons apart from increased satiety.

The relationship between long term soy protein hydrolysate consumption and blood pressure has been discussed in many previous reports. For instance, Yang et al. [28] reported that pepsin hydrolyzed soy peptides reduced SBP (up to −35 mmHg) after SHR were fed for twelve weeks. Rhyu [29] also observed an SBP reduction of about −13 mmHg when SHR were fed with fermented soybean paste for seven weeks. These studies, however, used low molecular weight peptides mixed with some other foods. We believe our study is a better representation of how food could be processed and directly consumed as a functional food for reducing high blood pressure. Our results are similar to Wu et al. [30], who recorded an SBP reduction of about −20 mm Hg when SHRs were fed with 100 mg/kg BW of soy proteins hydrolyzed with Alcalase. It is, however, obvious that the different enzymes used for SPI hydrolysis result in different peptides with different potencies for ACE inhibition. For this reason, different treatments would result in different abilities to lowering blood pressure. Meanwhile, any small reduction in high blood pressure could beneficially reduce the risk of cardiovascular diseases [31].

In conclusion, our data demonstrates that Prozyme hydrolysis followed by *Lactobacillus rhamnosus* EBD1 fermentation enhanced bioactive peptide generation and improved ACE inhibition. Consumption of P-SPI could therefore be helpful in reducing high blood pressure in humans. Meanwhile, studies about the mechanism(s) by which P-SPI reduces blood pressure are warranted.

4. Materials and Methods

4.1. Chemicals and Cultures

Soybean protein isolates (Pro-Fam®) were obtained from Archer Daniels Midland Company (ADM, Decatur, Illinois, USA). Hip-His-Leu, ACE (from rabbit lung), Pepsin (from porcine gastric mucosa), and Pancreatin (from porcine pancreas) were obtained from Sigma-Aldrich (Yongin, Korea). Prozyme 2000P was obtained from Bison Corporation, Gyunggi-Do, Korea. *Lactobacillus rhamnosus* EBD1 was obtained from the Department of Food Science and Biotechnology (Chuncheon-si, Gangwon-do, Korea) and used for this study because it showed strong proteolytic ability in our earlier study [32]. The bacteria stock culture was maintained at −80 °C in de Man, Rogosa, and Sharpe (MRS) broth (Difco, Hongcheon, Korea), containing 20% glycerol (v/v). The culture was streaked on MRS agar and cultured at 37 °C for 24 h. A single colony was then transferred into MRS broth at 37 °C and harvested at the exponential phase of growth.

4.2. Preparation of Protein Hydrolysates and Fermentation

SPI was hydrolyzed with Prozyme according to the enzyme manufacturer's instructions. Briefly, 20% (w/v) of SPI in distilled water was prepared, and the pH was adjusted to 7. The protein was digested by 3% Prozyme at 55 °C for 1 h. The sample was then autoclaved at 121 °C to stop the enzyme activity and to sterilize the sample. *Lactobacillus rhamnosus* EBD1 (2×10^8 cfu/mL) in the starter culture was inoculated into a 500 mL Erlenmeyer flask containing the hydrolyzed SPI. Cultivation was carried out at 37 °C with 150 rpm of agitation. After 48 h incubation, the fermented sample (P-SPI) was freeze-dried with a TFD5505 table top freeze dryer (ilshinBioBase Co. Ltd., Dongducheon-si, Korea) and stored at -20 °C for further analysis.

4.3. Determination of Extent of Proteolysis

The degree of proteolysis of the SPI samples (raw SPI and P-SPI) were analyzed as reported earlier [33] with slight modifications. Briefly, SPI hydrolysates were analyzed by reversed-phase high-performance liquid chromatography (RP-HPLC) using a Waters system (Waters Corporation, Milford, MA, USA) equipped with a 1525 Binary HPLC pump, a 2996 Photodiode Array Detector, and a 717 plus Autosampler. An aliquot (90 µL, 10 mg/mL) of the sample was applied to a Symmetry$^{®}$ C_{18} 5 µm, 4.6 × 150 mm column (Waters, Milford, MA, USA). The column was developed at a flow rate of 1 mL/min at 40 °C. Elution was performed with a linear gradient of solvent B (acetonitrile with 1% TFA) in solvent A (water with 1% TFA) from 0–80% in 60 min. Detection of peptides and proteins was carried out at 214 nm. The extent of proteolysis was calculated by expressing the chromatographic peak areas of either enzyme treated alone or enzyme treated and fermented ISP hydrolysates as a percentage of that of raw SPI.

The degree of hydrolysis after enzyme treatment was calculated as:

$$\text{Degree of hydrolysis} \; = \; \frac{100\% \; \times \; (\text{Peak area of A} - \text{Peak area of B})}{(\text{Peak area of A})} \tag{1}$$

The degree of hydrolysis after enzyme treatment and fermentation was calculated as:

$$\text{Degree of hydrolysis} \; = \; \frac{100\% \; \times \; (\text{Peak area of A} - \text{Peak area of C})}{(\text{Peak area of raw A})} \tag{2}$$

where A represents the chromatogram of raw SPI, B represents the Prozyme treated SPI, and C represents P-SPI.

4.4. In-Vitro Assay for ACE Inhibitory Activity

ACE inhibitory activity was determined by the procedure described by Cushman & Cheung, [34]. Briefly, 20 µL of ACE inhibitor solution with 50 µL of 5mM HHL in 100mM sodium borate buffer (pH 8.3) containing 0.3M NaCl was incubated at 37 °C for 5 min. To initiate the reaction, 10 µL of 0.1 U/mL ACE solution was added, and the mixture was incubated at 37 °C for 30 min. The reaction was terminated by adding 100 µL of 1M HCl, and the reaction mixture was mixed with 1 mL ethyl acetate. The mixture was vortexed for 60 s and centrifuged at $2000 \times g$ for 5 min. The ethyl acetate layer (0.8 mL) was transferred to a 1.5 mL Eppendorf tube and evaporated in a water bath. The hippuric acid (HA) in the tube was dissolved with distilled water (0.8 mL). The amount of HA formed was measured at 228 nm using a biospectrometer (Eppendorf Biospectrometer$^{®}$ fluorescence, Eppendorf Korea Ltd. Korea). The amount of HA liberated from Hip-His-Leu under this reaction conditions without an inhibitor was used as a control. The extent of inhibition was calculated as

$$\text{ACE inhibition} = 100\,\% \times [(\text{B} - \text{A})/\text{B}]$$

where A is the optical density in the presence of ACE and ACE inhibitory component and B is the optical density without ACE inhibitory component.

For the determination of IC_{50}, series of dilutions containing 5000 µg/mL, 500 µg/mL, 50 µg/mL, 5 µg/mL, 0.5 µg/mL, and 0.05 µg/mL of P-SPI samples were prepared. The amount of peptides required to suppress 50% ACE activity was calculated from the regression curves observed for each fraction.

4.5. Identification of Peptides by Mass Spectrometry

Liquid chromatography-electrospray ionization-quantitative time-of-flight tandem mass spectrometry experiments (LC-ESI-TOF-MS/MS) were carried out at the National Instrumentation Center for Environmental Management of Seoul National University in Korea, according to an earlier method [35]. Analysis was done using high-performance liquid chromatography (UltiMate 3000 Series system, DIONEX Technologies, Sunnyvale, CA, USA), an integrated system comprising an auto-switching nano pump, an autosampler (TempoTM nano LC system; MDS SCIEX, Seoul, Korea), and a hybrid quadrupole-time-of-flight (TOF) mass spectrometer (QStar Elite; Applied Biosystems, USA) fitted with a fused silica emitter tip (New Objective, Woburn, MA, USA). To ionize the samples, nano-electrospray ionization was used. 1.5 g of the P-SPI was dissolved in 50 mL of double distilled water. Fractions (1.5 µL) of the sample were injected into the LC-nano ESI-MS/MS system. The sample was trapped on a ZORBAX 300SB-C18 trap column (300-µm i.d × 5 mm, 5-µm particle size, 100 pore size, Agilent Technologies, Santa Clara, California, USA, part number 5065-9913) and washed for 6 min with gradient with 98% solvent A [water/acetonitrile (98:2, v/v), 0.1% formic acid] and 2% solvent B [Water/acetonitrile (2:98, v/v), 0.1% formic acid] at a flow rate of 5 µL/min. The peptides were separated on a ZORBAX 300SB-C18 capillary column (75-µm i. d × 150 mm, 3.5 µm particle size, 100 pore size, part number 5065-9911) at a flow rate of 300 nL/min with a gradient at 2%–35% solvent B over 30 min, then from 35%–90% over 10 min, followed by 90% solvent B for 5 min, and finally 5% solvent B for 15 min. Electrospray was performed at an ion spray voltage of 2000 eV through a coated silica tip (FS360-20-10- N20-C12, PicoTip emitter, New Objective). The peptides were analyzed automatically using Analyst QS 2.0 software (Applied Biosystems, Seoul, Korea). The range of m/z values was 200–2000. Peptides of interest were ordered at >95% purity from GL Biochem (Shanghai, China) Limited (Shanghai, China).

4.6. Effects of Gastrointestinal Enzymes on P-SPI (In Vitro)

A two-stage simulated gastrointestinal digestion was carried out on P-SPI similar to an earlier report [2]. Pepsin (0.2 mg) was added to 10 mL of 1 mg/mL P-SPI solutions and adjusted to pH 2.0 using 1 M HCl. The samples were incubated at 37 °C. After 120 min, the pH was raised to 7.5 by adding 1 M NaOH. Pancreatin (0.2 mg) was added and the samples were further incubated at 37 °C for 180 min. The reaction was stopped by heating at 80 °C for 10 min in a water bath, followed by cooling at room temperature. The samples were analyzed for their ACE inhibitory abilities.

4.7. Long-Term Effect of P-SPI Consumption on SHR Blood Pressure

All animal experimental procedures were in accordance with the ethical procedures and scientific care by Kangwon National University-Institutional Animal Care and Use Committee (approval no. KW-151127-1, 13 August 2018). Twenty male SHRs weighing 250–300 g were used (Charles River Laboratories, Barcelona, Spain). The rats were divided randomly into four groups. Rats were housed in temperature-controlled rooms (23 °C) with 12 h light/dark cycles and consumed tap water and standard diets ad libitum. Experimental procedures were conducted in accordance with the Kangwon National University animal ethics committee guidelines. Indirect measurement of systolic blood pressure (SBP) in awake restrained rats was carried out by the non-invasive tail-cuff method using computer-assisted non-invasive blood pressure equipment (NIBP 76-0173 unit with LE5160R cuff & transducer, Sang Chung Commercial Co., Ltd., Kangnam-Ku, Korea). The rats were kept at 37 °C for 15 min to make the pulsations of the tail artery detectable. By gastric intubation, each group of rats was either administered with 10 mg of P-SPI per kg body weight (BW), 100 mg of P-SPI per kg BW, captopril (50 mg/kg BW), or 750 μL physiological saline once daily. SBP was measured before peptide intake (week 0), at the 2nd, 4th and 6th weeks after intake. Each value of SBP was obtained by averaging five successful measurements without disturbance of the signal. Changes in SBP were calculated as the absolute difference (in mmHg) with respect to the basal values of measurements obtained just before starting the treatments.

4.8. Effect of P-SPI Consumption on Feed Consumption and Weight Gain

To assess the amount of feed consumed by the rats, the mass of feed supplied to the rats each morning was recorded, and the remaining feed in the feeding trough was weighed the next morning. The difference in mass was recorded as the amount of feed consumed by the rats.

For weight gain assessment, the weight of each rat in each group was recorded once every week (from week 0–6) and the changes in body weight were noted as weight gain or loss.

4.9. Statistical Analysis

Baseline systolic blood pressure was defined as the mean of the values measured in the first run-in period. Blood pressures, weight, and the amount of feed consumed were presented as the mean value ± standard deviations (SD) for all SHR in each group. The outcomes for each week between groups were analyzed with one-way ANOVA followed by Duncan tests. Differences were considered significant when $p < 0.05$. All statistical analysis was done using GraphPad Prism version 5.01 (GraphPad Software, Inc, La Jolla, CA, USA).

Supplementary Materials: Supplementary materials can be found at http://www.mdpi.com/1422-0067/20/6/1496/s1. Table S1: LC-ESI-TOF-MS/MS analysis of P-SPI-derived peptides.

Author Contributions: Conceptualization: D.-H.O. and M.H.P.; methodology: E.B.-M.D., F.K.O., R.C. and J.-H.K.; writing, reviewing and editing: E.B.-M.D. and D.-H.O.; all authors gave their feedback, edited and approved the final manuscript.

Funding: This work was funded by the Korean Ministry of Small and Medium scale Enterprises and Startups under the "Regional Specialized Industry Development Program (R&D, R0006438)" supervised by the Korea Institute for Advancement of Technology (KIAT). Grant number C0502529.

Acknowledgments: We thank the central laboratory of Kangwon National University for their assistance in LC-MS analysis.

Conflicts of Interest: The authors declare that they have no conflict of interest.

References

1. Zhou, B.; Bentham, J.; Di Cesare, M.; Bixby, H.; Danaei, G.; Cowan, M.J.; Paciorek, C.J.; Singh, G.; Hajifathalian, K.; Bennett, J.E. Worldwide trends in blood pressure from 1975 to 2015: A pooled analysis of 1479 population-based measurement studies with 19·1 million participants. *Lancet* **2017**, *389*, 37–55. [CrossRef]

2. Daliri, E.B.-M.; Lee, B.H.; Park, M.H.; Kim, J.-H.; Oh, D.-H. Novel angiotensin I-converting enzyme inhibitory peptides from soybean protein isolates fermented by *Pediococcus pentosaceus* SDL1409. *LWT* **2018**, *93*, 88–93. [CrossRef]

3. Daliri, E.B.-M.; Lee, B.H.; Oh, D.H. Current trends and perspectives of bioactive peptides. *Crit. Rev. Food Sci. Nutr.* **2018**, *58*, 2273–2284. [CrossRef] [PubMed]

4. Ciau-Solís, N.A.; Acevedo-Fernández, J.J.; Betancur-Ancona, D. In vitro renin–angiotensin system inhibition and in vivo antihypertensive activity of peptide fractions from lima bean (*Phaseolus lunatus* l.). *J. Sci. Food Agric.* **2018**, *98*, 781–786. [CrossRef]

5. Bleakley, S.; Hayes, M.; O'Shea, N.; Gallagher, E.; Lafarga, T. Predicted release and analysis of novel ACE-I, renin, and DPP-IV inhibitory peptides from common oat (*Avena sativa*) protein hydrolysates using in silico analysis. *Foods* **2017**, *6*, 108. [CrossRef]

6. Chatterjee, C.; Gleddie, S.; Xiao, C.-W. Soybean bioactive peptides and their functional properties. *Nutrients* **2018**, *10*, 1211. [CrossRef]

7. Lin, Q.; Xu, Q.; Bai, J.; Wu, W.; Hong, H.; Wu, J. Transport of soybean protein-derived antihypertensive peptide LSW across Caco-2 monolayers. *J. Funct. Foods* **2017**, *39*, 96–102. [CrossRef]

8. Sanjukta, S.; Rai, A.K. Production of bioactive peptides during soybean fermentation and their potential health benefits. *Trends Food Sci. Technol.* **2016**, *50*, 1–10. [CrossRef]

9. Nagata, C.; Wada, K.; Tamura, T.; Konishi, K.; Goto, Y.; Koda, S.; Kawachi, T.; Tsuji, M.; Nakamura, K. Dietary soy and natto intake and cardiovascular disease mortality in Japanese adults: The Takayama study. *Am. J. Clin. Nutr.* **2016**, *105*, 426–431. [CrossRef]

10. Yan, Z.; Zhang, X.; Li, C.; Jiao, S.; Dong, W. Association between consumption of soy and risk of cardiovascular disease: A meta-analysis of observational studies. *Eur. J. Prev. Cardiol.* **2017**, *24*, 735–747. [CrossRef] [PubMed]

11. Vallabha, V.S.; Tiku, P.K. Antihypertensive peptides derived from soy protein by fermentation. *Int. J. Pept. Res. Ther.* **2014**, *20*, 161–168. [CrossRef]

12. Thakkar, P.; Patel, A.; Modi, H.; Prajapati, J. Evaluation of antioxidative, proteolytic, and ACE inhibitory activities of potential probiotic lactic acid bacteria isolated from traditional fermented food products. *Acta Aliment.* **2018**, *47*, 113–121. [CrossRef]

13. Kumar, R.; Chaudhary, K.; Chauhan, J.S.; Nagpal, G.; Kumar, R.; Sharma, M.; Raghava, G.P. An in silico platform for predicting, screening and designing of antihypertensive peptides. *Sci. Rep.* **2015**, *5*, 12512. [CrossRef]

14. Daliri, E.B.-M.; Lee, B.H.; Oh, D.H. Current perspectives on antihypertensive probiotics. *Probiot. Antimicrob. Proteins* **2017**, *9*, 91–101. [CrossRef] [PubMed]

15. Ceapa, C.; Davids, M.; Ritari, J.; Lambert, J.; Wels, M.; Douillard, F.P.; Smokvina, T.; de Vos, W.M.; Knol, J.; Kleerebezem, M. The variable regions of *Lactobacillus rhamnosus* genomes reveal the dynamic evolution of metabolic and host-adaptation repertoires. *Genome Biol. Evol.* **2016**, *8*, 1889–1905. [CrossRef]

16. Wijesekara, I.; Kim, S.-K. Angiotensin-i-converting enzyme (ACE) inhibitors from marine resources: Prospects in the pharmaceutical industry. *Mar. Drugs* **2010**, *8*, 1080–1093. [CrossRef]

17. Wu, J.; Aluko, R.E.; Nakai, S. Structural requirements of angiotensin I-converting enzyme inhibitory peptides: Quantitative structure– activity relationship study of di-and tripeptides. *J. Agric. Food Chem.* **2006**, *54*, 732–738. [CrossRef] [PubMed]

18. Wu, J.; Aluko, R.E.; Nakai, S. Structural requirements of angiotensin I-converting enzyme inhibitory peptides: Quantitative structure-activity relationship modeling of peptides containing 4-10 amino acid residues. *QSAR Comb. Sci.* **2006**, *25*, 873–880. [CrossRef]

19. Mourad, J.-J. The evolution of systolic blood pressure as a strong predictor of cardiovascular risk and the effectiveness of fixed-dose arb/ccb combinations in lowering levels of this preferential target. *Vasc. Health Risk Manag.* **2008**, *4*, 1315. [CrossRef] [PubMed]

20. Rapsomaniki, E.; Timmis, A.; George, J.; Pujades-Rodriguez, M.; Shah, A.D.; Denaxas, S.; White, I.R.; Caulfield, M.J.; Deanfield, J.E.; Smeeth, L. Blood pressure and incidence of twelve cardiovascular diseases: Lifetime risks, healthy life-years lost, and age-specific associations in 1·25 million people. *Lancet* **2014**, *383*, 1899–1911. [CrossRef]

21. Koyama, M.; Hattori, S.; Amano, Y.; Watanabe, M.; Nakamura, K. Blood pressure-lowering peptides from neo-fermented buckwheat sprouts: A new approach to estimating ACE-inhibitory activity. *PLoS ONE* **2014**, *9*, e105802. [CrossRef]

22. Wang, S.-K.; Ma, W.; Wang, S.; Yi, X.-R.; Jia, H.-Y.; Xue, F. Obesity and its relationship with hypertension among adults 50 years and older in Jinan, China. *PLoS ONE* **2014**, *9*, e114424. [CrossRef]

23. Leggio, M.; Lombardi, M.; Caldarone, E.; Severi, P.; D'Emidio, S.; Armeni, M.; Bravi, V.; Bendini, M.G.; Mazza, A. The relationship between obesity and hypertension: An updated comprehensive overview on vicious twins. *Hypertens. Res.* **2017**, *40*, 947. [CrossRef] [PubMed]

24. Nurdiantami, Y.; Watanabe, K.; Tanaka, E.; Pradono, J.; Anme, T. Association of general and central obesity with hypertension. *Clin. Nutr.* **2018**, *37*, 1259–1263. [CrossRef]

25. Oh, S.; Shimizu, H.; Satoh, T.; Okada, S.; Adachi, S.; Inoue, K.; Eguchi, H.; Yamamoto, M.; Imaki, T.; Hashimoto, K. Identification of nesfatin-1 as a satiety molecule in the hypothalamus. *Nature* **2006**, *443*, 709. [CrossRef] [PubMed]

26. Shimizu, H.; Oh-i, S.; Hashimoto, K.; Nakata, M.; Yamamoto, S.; Yoshida, N.; Eguchi, H.; Kato, I.; Inoue, K.; Satoh, T. Peripheral administration of nesfatin-1 reduces food intake in mice: The leptin-independent mechanism. *Endocrinology* **2009**, *150*, 662–671. [CrossRef]

27. Zapata, R.C.; Singh, A.; Chelikani, P.K. Peptide YY mediates the satiety effects of diets enriched with whey protein fractions in male rats. *FASEB J.* **2017**, *32*, 850–861. [CrossRef]

28. Yang, H.-Y.; Yang, S.-C.; Chen, J.-R.; Tzeng, Y.-H.; Han, B.-C. Soyabean protein hydrolysate prevents the development of hypertension in spontaneously hypertensive rats. *Br. J. Nutr.* **2004**, *92*, 507–512. [CrossRef]

29. Rhyu, M.R.; Kim, E.Y.; Han, J.S. Antihypertensive effect of the soybean paste fermented with the fungus Monascus. *Int. J. Food Sci. Technol.* **2002**, *37*, 585–588. [CrossRef]

30. Wu, J.; Ding, X. Hypotensive and physiological effect of angiotensin converting enzyme inhibitory peptides derived from soy protein on spontaneously hypertensive rats. *J. Agric. Food Chem.* **2001**, *49*, 501–506. [CrossRef] [PubMed]

31. Barnes, V.A.; Orme-Johnson, D.W. Prevention and treatment of cardiovascular disease in adolescents and adults through the transcendental meditation®program: A research review update. *Curr. Hypertens. Rev.* **2012**, *8*, 227–242. [CrossRef] [PubMed]

32. Daliri E. Kangwon National University, Chuncheon-si, Korea. 2015.

33. García-Tejedor, A.; Padilla, B.; Salom, J.B.; Belloch, C.; Manzanares, P. Dairy yeasts produce milk protein-derived antihypertensive hydrolysates. *Food Res. Int.* **2013**, *53*, 203–208. [CrossRef]

34. Cushman, D.W.; Cheung, H.S. Spectrophotometric assay and properties of the angiotensin-converting enzyme of rabbit lung. *Biochem. Pharmacol* **1971**, *7*, 1637–1648. [CrossRef]

35. Chang, O.K.; Roux, É.; Awussi, A.A.; Miclo, L.; Jardin, J.; Jameh, N.; Dary, A.; Humbert, G.; Perrin, C. Use of a free form of the *Streptococcus thermophilus* cell envelope protease *prts* as a tool to produce bioactive peptides. *Int. Dairy J.* **2014**, *38*, 104–115. [CrossRef]

International Journal of
Molecular Sciences

Review

Insect Cecropins, Antimicrobial Peptides with Potential Therapeutic Applications

Daniel Brady [1], **Alessandro Grapputo** [1], **Ottavia Romoli** [1,2] **and Federica Sandrelli** [1,*]

[1] Department of Biology, University of Padova, via U. Bassi 58/B, 35131 Padova, Italy;
    daniel.brady@studenti.unipd.it (D.B.); alessandro.grapputo@unipd.it (A.G.);
    oromoli@pasteur-cayenne.fr (O.R.)

[2] Institut Pasteur de la Guyane, 23 Avenue Pasteur, 97306 Cayenne, French Guiana, France

* Correspondence: federica.sandrelli@unipd.it; Tel.: +39-049-8276216

Received: 31 October 2019; Accepted: 20 November 2019; Published: 22 November 2019

Abstract: The alarming escalation of infectious diseases resistant to conventional antibiotics requires urgent global actions, including the development of new therapeutics. Antimicrobial peptides (AMPs) represent potential alternatives in the treatment of multi-drug resistant (MDR) infections. Here, we focus on Cecropins (Cecs), a group of naturally occurring AMPs in insects, and on synthetic Cec-analogs. We describe their action mechanisms and antimicrobial activity against MDR bacteria and other pathogens. We report several data suggesting that Cec and Cec-analog peptides are promising antibacterial therapeutic candidates, including their low toxicity against mammalian cells, and anti-inflammatory activity. We highlight limitations linked to the use of peptides as therapeutics and discuss methods overcoming these constraints, particularly regarding the introduction of nanotechnologies. New formulations based on natural Cecs would allow the development of drugs active against Gram-negative bacteria, and those based on Cec-analogs would give rise to therapeutics effective against both Gram-positive and Gram-negative pathogens. Cecs and Cec-analogs might be also employed to coat biomaterials for medical devices as an approach to prevent biomaterial-associated infections. The cost of large-scale production is discussed in comparison with the economic and social burden resulting from the progressive diffusion of MDR infectious diseases.

Keywords: antimicrobial peptides; insects; Cecropins; Cec-analogs; MDR infectious diseases

---

1. Introduction

The spread of infectious diseases resistant to conventional treatments has become an alarming phenomenon worldwide, prompting the United Nations and international agencies to call for immediate and coordinated actions to avoid a possible global drug-resistance crisis [1]. Drug-resistance phenomena involve not only antibacterial compounds, but also antiviral, antifungal, and antiprotozoal therapeutics in all countries, independent of their economic level. Currently, estimates indicate that drug-resistance cases result in 700,000 deaths per year worldwide, and without direct action, annual death tolls could reach 10 million by 2050 [1]. Research and development of new therapeutics have been included at the forefront of the proposed actions to tackle the global antimicrobial resistance phenomenon [1]. Several lines of evidence indicate that the utilization of antimicrobial peptides (AMPs) represents a compelling option [2,3].

AMPs are naturally occurring peptides produced as a first line of defense against pathogenic infections by virtually all living species, from bacteria to mammals [2]. AMPs play an essential role in those organisms that lack an adaptive immune system and base their defense only on the innate immune response, such as invertebrates. Of these, Insecta is the largest animal class on Earth, containing 50% of all known animal species, and represents a wide source of AMPs. To date, 305 out of the 3087 AMPs listed in the Antimicrobial Peptide Database (APD; Available online:

http://aps.unmc.edu/AP [4]) are derived from insects. Notwithstanding, these numbers are likely to increase extensively given the current growth of accessible genomic, transcriptomic, and proteomic insect datasets, which will accelerate the identification of new putative AMPs available for subsequent analyses and characterization.

First identified about 40 years ago, a wide variety of insect AMPs has since been characterized. These molecules have been intensively studied, not only for their physiological role in insect immunity, but also as potential alternatives to conventional antibiotics in the treatment of infectious diseases [5–7]. Moreover, some insect AMPs have been shown to possess immunomodulatory functions as well as anticancer activity [5,6]. These biological properties, combined with modern advances in biotechnology, have resulted in a renewed interest in insect AMPs and their potential to combat modern biomedical challenges.

Insect AMPs can be classified on the basis of their sequence and structure into three groups: (i) α-helical peptides, lacking in cysteine residues (e.g., Cecropins (Cecs) and Moricins); (ii) β-sheet cysteine-rich peptides (e.g., Defensins and Drosomycins); and (iii) linear-extended peptides, often characterized by high proportions of peculiar amino acids (aa) such as proline, arginine, tryptophan, glycine, and histidine. Both proline-rich peptides (e.g., Apidaecins, Drosocins, and Lebocins) and glycine-rich AMPs (e.g., Attacins and Gloverins) belong to this group. As the different classes of insect AMPs have been recently reviewed in [5–7], here we focus on Cecs, one of the largest groups of insect AMPs. We report a comprehensive overview of the Cec family in insects, and provide up-to-date models explaining their mode of action. We then highlight the antimicrobial, anti-inflammatory, and antitumor activities of natural Cecs and Cec-like peptides, as well as of synthetic Cec-analogs, which carry different types of sequence modifications. The potential benefits and limitations in the development of Cec-based antibacterial therapeutics are also presented.

2. The Family of Cecropins in Insects

Cecs and other Cec-like peptides, including Sarcotoxins, Stomoxins, Papiliocin, Enbocins, and Spodopsins, form the most abundant family of linear α-helical AMPs in insects (Table 1). Cec AMPs were first isolated from the hemolymph (insect blood) of the lepidopteran *Hyalophora cecropia* and were characterized for their antimicrobial activity against several Gram-positive and negative bacteria [8–10]. Subsequently, these peptides have been identified in two other orders of Hexapoda, Coleoptera and Diptera, as well as in other species of Lepidoptera [7,11]. In evaluating several genomes, the identification of Cec and Cec-like peptide sequences was not successful in other insect orders ([11]; this review), including Hymenoptera, which is considered the sister clade of the other holometabolous insects [12]. However, Cecs have been identified in other animals, such as Styelin in tunicates [13], and Cec P1, first isolated from pigs [14], but then found to belong to the Nematode *Ascaris suum* [15]. Cec-like peptides have been also identified in the bacterium *Helicobacter pylori* [16]. Since these peptides derive from the N-terminal part of ribosomal protein L1 (RpL1) and are similar to Cecs from *H. cecropia*, Pütsep and colleagues suggested that Cecs may have evolved from an early prokaryote RpL1 gene [16]. Indeed, the homology of Cecs has been debated and some authors consider them a single family, with Dermasptin (amphibians), Ceratotoxin (insects), and Pleurocidin (fish) forming a Cec superfamily [17]. Indeed, the members of the Cec family show sequence similarity that enabled the identification of a first sequence signature of Cecs from some species of Brachycera (Diptera) and Lepidoptera (i.e., [KR]-[KRE]-[LI]-[ED]-[RKGH]-[IVMA]-[GV]-[QRK]-[NHQR]-[IVT]-[RK]-[DN]-[GAS]-[LIVSAT][LIVE]-[RKQS]-[ATGV]-[GALIV]-[PAG]) [17]. This has been updated to include Culicomorpha (mosquitoes) and nematodes (i.e., [KRDEN]-[KRED]-[LIVMR]-[ED]-[RKGHN]-X(0,1)[IVMALT]-[GVIK]-[QRKHA]-[NHQRK]-[IVTA][RKFAS]-[DNQKE]-[GASV]-[LIVSATG][LIVEAQKG]-[RKQSGIL]-[ATGVSFIY]-[GALIVQN]) [18]. However, the lower similarity between Cecs from insects and other organisms and the lack of Cec-like peptides outside the clade of Coleoptera, Diptera, and Lepidoptera prompted other authors to suggest that insect *Cec* genes may have evolved just once in the common ancestor of these holometabolous orders, implying that insect and non-insect Cecs are not homologous [11].

Table 1. In vitro antimicrobial activity and toxicity against mammalian cells of natural Cecropins (Cecs) and Cec-like peptides.

Insect	Species	Active Peptide (aa)	Antimicrobial Activity			Peptide conc. (µM)	
Order			Virus	Bacteria	Fungi	Cytotox.	Hem Act.
Coleoptera	*Oxysternon conspicillatum*	Oxysterlin 1 (39) [19]	-	G+, G−	weak	>28	>14
		Oxysterlin 2 (55) [19]	-	G−	NA	>19.75	>19.75
		Oxysterlin 3 (39) [19]	-	G−	NA	>28	>28
	Acalolepta luxuriosa	Cec (35) [20]	-	*M. luteus, E. coli*	-	-	-
	Paederus dermatitis	Sarcotoxin Pd (34) [21]	-	G+, G−	weak	-	16
Diptera	*Simulium bannaense*	SibaCec (35) [22]	-	G+, G−	-	58	58
	Anopheles gambiae	AngCec A (35) [23]	-	G+, G−	A	-	-
	Aedes aegypti	AeaeCec 1 (34) [24–26]	-	G+, G−	A	50 [26]	50 [26]
		AeaeCec 2–4 (34) [26]	-	-	-	50	50
		AeaeCec 5 (34) [26]	-	-	-	12.5	12.5
	Aedes albopictus	Cec A1 (35) [27,28]	-	*E. coli, Francisella*	-	-	-
		Cec B (35) [28]	-	*Francisella*	-	-	-
	Culex pipens	Cec A (34) [28]	-	*Francisella*	-	-	-
		Cec B2 (34) [28]	-	*Francisella*	-	-	-
	Tabanus yao	Cec TY1 (41) [29]	-	*B. subtilis S. aureus E. coli*	A	-	-
	Hermetia illucens	CLP1 (45) [30]	-	G−	-	-	-
	Drosophila melanogaster	Cec A (34) [23,24,31,32]	-	G+, G−	A	-	-
		Cec B (34) [31,32]	-	G−	A	-	-
	Musca domestica	Mdc (40) [33–35]	-	G+, G−	-	-	-
	Glossina morsitans	Cec (39) [36]	-	*M. luteus, E. coli*	-	-	-
	Stomoxys calcitrans	Stomoxyn (42) [37]	-	G+, G−	A	-	>10
	Sarcophaga peregrina	Sarcotoxins I A, B, C (39) [38–40]	-	G+, G−	-	-	-
	Lucilia sericata	Lser Cecs 1–6 (40) [41]	-	G−	NA	-	-
		LSerStomox1 (43) [41]	-	G−	NA	-	-
		LSerStomox 2 (42) [41]	-	G−	NA	-	-

Table 1. *Cont.*

Insect Order	Species	Active Peptide (aa)	Antimicrobial Activity			Peptide conc. (µM)	
			Virus	Bacteria	Fungi	Cytotox.	Hem Act.
Lepidoptera	*Hyalophora cecropia*	Cec A (37) [8–10,42–45]	HIV	G+, G−	A	[44,45]	100 [45]
		Cec B (35) [9,42,44,46]	-	G+, G−	A	30 [44]	500 [46]
		Cec D (36) [9,47]	PRRSV	G+, G−	-	-	-
	Antheraea pernyi	Cec B (35) [48,49]	-	G+, G−		25 [49]	200 [49]
		Cec D (36) [48]	-	G+, G−	-	-	-
		ApCec (38) [50]	-	B. subtilis, E. coli	-	-	62.5
	Bombyx mori	Cec A (35) [51,52]	-	G+, G−	A	-	-
		Cec B (35) [51,53]	-	G+, G−	NA	200 [53]	200 [53]
		Cec D (36) [51]	-	G+, G−	-	-	-
		Cec E (?) [51]	-	B. thuringiensis, G−	-	-	-
	Galleria mellonella	Cec D (39) [54,55]	-	L. monocytogenes	-	-	>115 [55]
	Papilio xuthus	Papiliocin (38) [56–58]	-	G+, G−	A	12.5 [58]	100 [58]
	Spodoptera litura	Spodopsin Ia (35) [59]	-	G+, G−	NA	-	-
		Spodopsin Ib (35) [59]	-	G+, G−	NA	-	-
		Cec A (35) [60]	-	G+, G−	-	-	-
		Cec B (35) [60]	-	G+, G−	-	-	-
	Helicoverpa armigera	Cec D (42) [61]	-	G+, G−	-	-	-
	Heliothis virescens	Cec B (35) [62]	-	E. coli	-	-	-
	Agrius convolvuli	AcCec D 1-3 (38) [63,64]	-	G+, G−	-	-	-
	Artogeia rapa (*Pieris rapae*)	Hinnavin I (40) [65]	-	G+, G−	A	-	-
		Hinnavin II (38) [66]	-	G+, G−	A	-	-
	Danaus plexippus	DAN1 (37) [67]	-	G+ (weak), G−	-	-	49.56
		DAN2 (37) [67]	-	G+ (weak), G−	weak	-	48.97

Peptide conc. (µM): Peptide concentration showing no or weak toxicity in mammalian cells. Cytotox.: Cytotoxicity; Hem act.: Hemolytic activity against mammalian red blood cells; G+ Gram-positive bacteria; G−: Gram-negative bacteria; A: Active against tested species; NA: Not active against the tested species; -: Not determined; (?): Undetermined aa length.

Although most Cec diversity is found in insect taxa with whole genome sequences, phylogenetic analysis suggests that there is a significant undiscovered diversity in other holometabolous insects. Within different species, *Cec* genes are generally present in a variable number of copies organized in clusters or dispersed in the genome and can include both functional and non-functional elements (pseudogenes). For example, among Diptera, *Drosophila melanogaster* shows four functional genes (*Cec A1*, *A2*, *B*, and *C*) and two pseudogenes (Cec $\psi1$ and Cec $\psi2$), clustered in a ~7-kb region [68,69]; to date, *Musca domestica* displays the largest gene family, characterized by 12 *Cec* members [70]. Among Lepidoptera, the *H. cecropia Cec* locus spans ~20 kb and contains three *Cec* genes (A, B, and D) [9,71], coding for three Cec A, B, and D functional peptides. Moreover, *H. cecropia* shows the additional Cec forms C, E, and F, that have been isolated in low amounts, and classified as allelic variants or degradation products of the three main A, B, and D forms [9]. In the domesticated silkworm *Bombyx mori*, the *Cec* gene family is composed of at least 14 elements (two *Cec A* (*A1* and *A2*), six *Cec B* (*B1–B6*), one *Cec C*, two *Cec D* (*D* and *D2*), one *Cec E*, and two *enbocins* (*enb* 1 and 2)), organized in two clusters, mapping on two different chromosomes [72]. In Coleoptera, functional *Cec* genes have been identified in species like *Acalolepta luxuriosa* (Cec; [20]), *Oxysternon conspicillatum* (Oxysterlins; [19]), and *Paederus dermatitis* (Sarcotoxin Pd; [21]), whereas only non-functional *Cec* pseudogenes have been reported in the coleopteran model *Tribolium castaneum* [73,74].

Phylogenetic analyses and single genome sequencing revealed that insect Cec and Cec-like peptides originated via gene duplication and evolved via a birth and death model of gene evolution [72,75]. The occurrence of gene duplication events is confirmed by the presence of transposable elements in both 5′ and 3′ flanking regions, and repeated gene duplication within species. Furthermore, tandem gene arrangement within the genome, non-functionalization, and loss of some *Cec* gene copies, and the presence of highly divergent and highly similar gene copies within species all support the gene duplication hypothesis [75,76]. Compared to other AMPs, Cecs show no sites under positive selection [77,78], but frequent duplication events may be adaptive, enabling new gene copies to mutate and acquire novel antimicrobial properties [79].

Phylogenetic analysis (Figure 1) [11,72,76] shows that Cecs from Lepidoptera form a monophyletic group (derived from a single ancestral gene) and evolved independently in this order of insects [22]. In contrast, the phylogenetic relationships of Cecs from Diptera and Coleoptera are more complex. Complementing previous phylogenetic analyses [76,80], we included new data from mosquitos and several Coleoptera species. Cecs from Diptera and Coleoptera are both paraphyletic, suggesting that Cecs originated before these lineages diverged. Within Diptera, Cecs from Brachycera (which include *Drosophila*) form a monophyletic group, which is closely related to that of Lepidoptera and is distinct from that of Culicomorpha (mosquitos) (Figure 1).

Figure 1. Phylogenetic tree of insect Cecropins (Cecs) and Cec-like peptides. Maximum likelihood mid-point rooted phylogenetic tree showing the relationships of insect Cecs and Cec-like peptides. The tree was obtained with FastTree 2.1.5 software with the WAG + Γ model [81]. Lepidoptera peptides are shown in red, Trichoptera in orange, Diptera Brachycera in dark blue, Diptera Culicomorpha in light blue, and Coleoptera in green. Full-length Cecs and Cec-like peptides were downloaded from the OrthoDB database (Available online: https://www.orthodb.org/), which contains 230 sequences in 61 species of Lepidoptera and Diptera. Other sequences, including those from *Simulium*, Trichoptera, and Coleoptera, were downloaded from NCBI and UniprotKB. Sequences were aligned with ClustalW using default parameters in Geneious 8.1.9 (BioMatters). Identical sequences within species were removed leaving a total of 254 Cecs and Cec-like peptides. Either UniprotKB or NCBI accession number are reported for each sequence in the tree. Circle at the nodes indicate node support obtained with Shimodaira–Hasegawa-like local support. Shade (as shown in the legend) and circle size are proportional to the node support value (0–1). The scale bar corresponds to estimated amino acid substitutions per site.

3. Cec Gene Expression and Mechanism of Action Against Microorganisms

In the absence of any infections, *Cec* genes can be constitutively expressed at low levels in different body compartments, as demonstrated in the *Drosophila* reproductive tract [82] or in the silkworm *B. mori* midgut or fat body (a structure equivalent to the mammalian liver) [83]. Following an immune challenge, *Cecs* become highly transcribed in several tissues, such as gut epithelia or epidermis during local infections, and the fat body and hemocytes, during systemic infections (e.g., [51,82,83]). Like other

AMPs, Cecs are translated as immature pre-peptides, undergo proteolytic cleavage of the N-terminal signal peptide, and are secreted in a mature and active form [5,7]. Before maturation, Cec sizes range between 58 and 79 aa, while active forms contain between 34 and 55 residues (Table 1). Experimental and computational analyses indicated that Cec and Cec-like peptides are structurally related and are characterized by an N-terminal basic, amphipathic domain linked to a more hydrophobic C-terminal segment, through a flexible proline- and glycine-rich hinge region (Figure 2A; [5,7,84]).

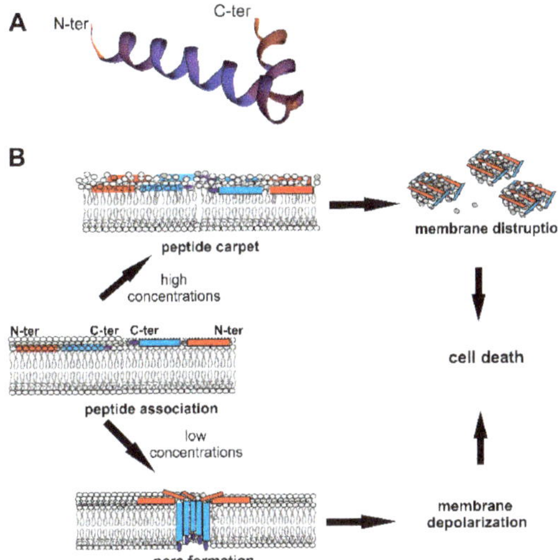

Figure 2. Cecropin (Cec) structure and mechanisms of action against bacteria. (**A**) Structure of the mature 35 aa *B. mori* Q53 Cec B natural variant [53] obtained using SWISS-MODEL (Available online: https://swissmodel.expasy.org/), showing N- and C-terminal α-helices linked through a flexible hinge region. (**B**) Model of action against bacteria. Cecs associate with the bacterial membrane, with the long axes of the α−helical domains parallel to the lipid bilayer surface. Polar residues interact with the lipid phosphates; non-polar residues bury in the hydrophobic core of the membrane. At high concentrations (upper part), Cecs form a carpet-like structure with detergent-like properties, disrupting membranes. At lower concentrations (lower part), Cecs form pores, which affect the cellular electrolyte balance, causing bacterial death [85]. The pore is formed of different Cec molecules organized as oligomers, with C-terminal hydrophobic domains submerged into the phospholipidic hydrophobic chains [86]. The red rectangle represents the N-terminal helix, the blue one the C-terminal helix; the dark blue ellipse indicates the C-terminal amidated residue.

Insect Cecs and Cec-like peptides are generally active against Gram-negative bacteria and to a lesser extent, Gram-positive bacteria (Table 1). Some have been demonstrated to also exhibit antifungal activity (Table 1). Moreover, Cec and Cec-like peptides were shown to have a low toxicity against normal mammalian cells and a weak or absent hemolytic effect against mammalian erythrocytes (Table 1). As for other cationic AMPs, the ability of these peptides to target microorganisms without interacting with host eukaryotic cells relies on the difference in composition of the respective cell membranes. Bacterial membranes are predominantly composed of negatively charged compounds (e.g., phosphatidylglycerol, cardiolipin, and phosphatidylserine), while eukaryotic membranes are positively charged by the presence of zwitterionic phospholipids and cholesterol [87]. Furthermore, Gram-negative bacteria possess an external membrane rich in negatively charged Lipopolysaccharides (LPS, also known as endotoxin), whereas in Gram-positive bacteria, the peptidoglycan is anchored to the cytoplasmic membrane by negatively charged teichoic acids. It is also generally thought that the discrimination between fungi and other eukaryotic host membranes is due to the different sterol compositions of their respective membranes [87].

Using chemically synthetized natural Cec variants and modified analogs, several studies have been performed to explain the Cec action mechanism against pathogens, as well as to identify the functions of specific residues within the peptide. Most mature Cec peptides contain a tryptophan residue in the first or second positions, which is considered important in conferring full antimicrobial activity to the peptide [5,7,84,88]. A study performed on Papiliocin, from the lepidopteran *Papilio xuthus*, suggested that the presence of tryptophan[2] and phenylalanine[5] aromatic residues in the N-terminal region are essential for the full-length peptide to interact with LPS in the outer membrane, and permeabilize the inner membrane of Gram-negative bacteria [58]. However, some dipteran Cecs, such as those from the black fly *Simulium bannaense* and the mosquito *Aedes aegypti* have been shown to be highly effective against different bacteria, although lacking an N-terminal tryptophan residue [22,25].

In several cases, Cec peptides undergo amidation of the C-terminal residue, a post-translational modification, which increases both antimicrobial activity and the action spectrum of the peptide [6,7]. It has been demonstrated that the antimicrobial activity of Cec AMPs relies on the structure they assume in the presence of bacterial cells. Circular dichroism analyses showed that in aqueous solution, Cecs have a random coiled structure but adopt α-helical conformations upon interaction with microbial membranes, where they exert a lytic effect [53,58,84,86]. Although some aspects remain unclear, it is currently accepted that Cec peptides do not interact with specific receptors but initially associate with the bacterial membrane along the axes of the α-helical domains parallel to the lipid bilayer surface. At this level, the polar residues of the peptide interact with the lipid phosphates, while the non-polar side chains burrow in the hydrophobic core of the membrane [84] (Figure 2B). In a first model of action, the continuous accumulation of peptides at the bacterial lipid bilayer leads to the formation of a peptide "carpet" on the membrane surface. This "carpet" structure possesses intrinsic detergent-like lytic properties, which disintegrate the membranes [84]. Cec P1 [14,15] and *H. cecropia* Cecs, when administrated at high concentrations (Cec P1 > 25 μM; *H. cecropia* Cecs > 5 μM), appear to act through this carpet-like mechanism (Figure 2B) [84,85]. However, at lower concentrations (2–5 μM), *H. cecropia* Cecs are able to associate with membranes and form channels or pores, which affect cellular electrolyte balance and in turn cause the death of the microorganism (Figure 2B) [84–86]. Initially, it was postulated that the N-terminal amphipathic regions of the peptides were involved in the formation of the pore (called "type II channel"), with the positively charged residues forming the inner channel [89,90]. Subsequent authors have hypothesized that the C-terminal hydrophobic domains of the peptides insert into the membrane giving rise to a more stable pore (type I channel), in which the polar aa of the C-terminal helices are oriented toward the center of the pore [85,86,90]. Efimova and colleagues analyzed the effect of *H. cecropia* Cecs A and B in model lipid membranes, with or without small molecules capable of modifying the membrane physical-chemical properties [85,86]. Using these data, they developed a model in which Cec peptides first interact as monomers with the hydrophilic heads of the lipid bilayer surface, acting parallel to the membrane plane. Next, the peptides submerge their C-terminal hydrophobic domains into the phospholipidic hydrophobic chain. Individual Cec molecules then organize into oligomers forming ion-permeable pores in the cell membrane (Figure 2B). Other monomers can then insert into the pores, increasing the ion channels' conductance. The authors also postulated that all the steps of this process are reversible and in equilibrium [86]. This pore model therefore resembles the "barrel-stave" model, in which the different C-terminal regions of the *H. cecropia* Cec peptides are organized to form a barrel penetrating the bacterial membrane. However, in cases where the peptide is shorter than ~ 22 aa (e.g., synthetically Cec-derived analogs, see below), the structure of the pore might be more similar to the so-called "toroidal-pore" model, in which the pore is composed by both peptides and lipids [84].

As mentioned above, natural Cec and Cec-like peptides show a higher activity against Gram-negative compared to Gram-positive bacteria. This feature has been related to the difference in the intrinsic properties of bacterial membranes (i.e., lipid composition, charge density, and electrochemical potential across the membrane), as demonstrated when evaluating *H. cecropia* Cec B against protoplasts obtained from Gram-negative *Escherichia coli* and Gram-positive *Staphylococcus aureus* or *S. epidermidis* [46]. Moreover, a recent study on natural Papiliocin and its modified derivatives associated the Cec's preferential activity against Gram-negative bacteria specifically with the presence of the C-terminal helix. In fact, compared to the full-length natural form, a truncated Papiliocin carrying only the N-terminal portion was less effective against Gram-negative, and more active against Gram-positive bacteria [58].

Finally, in a study evaluating the interaction between different *B. mori* natural Cec B variants and live Gram-negative *Pseudomonas aeruginosa*, it was suggested that Cecs might first affect the outer bacterial membrane, enabling the translocation of the peptide to the inner membrane, resulting in the disorganization of both lipid bilayers [53].

4. In Vitro Antimicrobial Activity of Natural Cecs and Synthetic Cec-Analogs

Numerous basic research studies have shown that natural Cecs or synthetic Cec-analogs can have antibacterial, antifungal, antiviral, and antiprotozoal properties (Tables 1 and 2 and reference herein). Although there is a lack of uniformity among these studies, the peptides have generally exhibited a high in vitro activity against Gram-negative bacteria. These also included multidrug resistance (MDR) strains listed by the World Health Organization (WHO) in the three "critical, high and medium" priority groups, requiring the development of new antibiotics [91].

Table 2. In vitro antimicrobial activity and toxicity against mammalian cells of Cec-analogs.

Peptide (aa)	Source	Modification	Antimicrobial Activity				Peptide Conc. (µM)	
			Virus	Bacteria	Fungi	Protozoa	Cytotox.	Hem Act.
SB-37 (38) [92]	*H. cecropia* Cec B	aa add./sub.	-	-	-	*P. falciparum, T. cruzi*	-	-
Shiva-1 (38) [46,92]	*H. cecropia* Cec B	aa add./sub.	-	G+, G−	NA	*P. falciparum, T. cruzi*	-	-
D-Cec B (35) [93]	*A. pernyi* Cec B	D-enantiomer	-	-	A	-	-	-
CecDH (32) [49]	*A. pernyi* Cec B	aa del.	-	G+, G−	-	-	25	100
ΔM1 (39) [55]	*G melonella* Cec D	N-term aa sub.	-	*Sa* (weak), *Ec, Pa*	-	-	-	115
ΔM2 (39) [55]	*G melonella* Cec D	N-term aa sub.	-	*Sa, Ec, Pa*	-	-	-	~60
Mdc–hly (?) [34]	*M. domestica* Mdc; human Lysozyme	Hybrid	-	G+, G−	-	-	-	-
CAMs (≤26) [94–98]	*H. cecropia* Cec A; *A. mellifera* Mellitin	Hybrids	-	G+, G−	A	*Plasmodium*	9 [96]	[98]
Ac-CAMs (15) [99]	*H. cecropia* Cec A; *A. mellifera* Mellitin	N-term fatty acid acylation	-	*Sa, Ec, Ab*	-	*L. pifanoi*	-	-
CAM-W (26) [98]	*H. cecropia* Cec A; *A. mellifera* Mellitin	aa sub.	-	G+, G−	A	-	-	3.12
CA-MAs (≤20) [100–105]	*H-cecropia* Cec A; *X. laevis* Magainin 2	Hybrids with aa sub.	virus–cell fusion inhibition	G+, G−	A	-	[105]	[107]
CA-LL37 (22) [106]	*H-cecropia* Cec A; human LL37	Hybrid	-	G+, G−	-	-	-	[106]

Table 2. *Cont.*

Peptide (aa)	Source	Modification	Antimicrobial Activity				Peptide Conc. (μM)	
			Virus	Bacteria	Fungi	Protozoa	Cytotox.	Hem Act.
CecXJ-37C (37) [107]	*B. mori* Cec B	C-term aa add.	-	G+, G−	-	-	20	19
CecXJ-37N (37) [107]	*B. mori* Cec B	C-term aa add.	-	G+, G−	-	-	20	33

Peptide conc. (μM): Peptide concentration showing no or weak toxicity in mammalian cells; Cytotox.: Cytotoxicity; Hem act.: Hemolytic activity; sub.: Substitution; add.: Addition; del: Deletion; G+ Gram-positive bacteria; G−: Gram-negative bacteria; *Sa*: *S. aureus*; *Ec*: *E. coli*; *Pa*: *P. aeruginosa*; *Ab*: *A. baumannii*; A: Active against tested species; NA: Not active against the tested species; -: Not determined; (?): Not reported.

Cec peptides were effective against laboratory strains of *P. aeruginosa* and different *Enterobacteriacae* spp. (including *K. pneumoniae* and *E. coli*), bacterial species belonging to the WHO first critical group. *M. domestica* Mdc, black fly SibaCec, and dung beetle Oxysterlins were active against MDR and clinically isolated *E. coli* strains [19,33,108], while *H. cecropia* Cec A and *P. xuthus* Papiliocin efficiently killed MDR *P. aeruginosa* isolates [45,58]. Mdc and SibaCec were also active against reference strains belonging to *Acinetobacter baumanii*, also critical on the WHO list [22,35]. Lepidopteran *H. cecropia* Cec A, *P. xuthus* Papiliocin, Cec D from the mosquito *A. aegypti*, and different synthetic CAM hybrids (formed from the fusion of the N-terminal regions of *H. cecropia* Cec A and *Apis mellifera* Mellitin) were effective against MDR *A. baumanii* strains [25,45,58,95].

Several natural Cecs and Cec-analogs have also shown activity against the food-borne Gram-negative pathogen *Salmonella typhimurium*, included in the high priority group of the WHO list (e.g., [19,23,25,58,80,98]). In addition, some dipteran Cec AMPs, such as those from the mosquitos *Aedes albopictus* and *Culex pipens*, were active against *Francisella novicida*, a facultative Gram-negative bacterium used as reference species to model *F. tularensis*, a zoonotic pathogen causing tularemia in humans and animals [28].

It is important to note that, although natural Cecs and Cec-like peptides generally demonstrated an antimicrobial activity against Gram-positive bacteria such as *Bacillus* spp and *Micrococcus luteus*, the vast majority were not or weakly active against *S. aureus*, which belongs to the high priority group on the WHO list (an exception appears to be the horse fly Cec TY1, which is reported to be more active against *S. aureus* than *E. coli*; [29]). Interestingly, synthetic Cec-analogs were active against *S. aureus*. In particular, an anti-*S. aureus* activity characterized CAM peptides [94,96,98,99], and other chimeric hybrids, such as CA-MA or CA-LL37, obtained from the fusion of *H. cecropia* Cec A N-terminal fragments with portions of *Xenopus laevis* Magainin [102] or human LL-37 AMP [106], respectively (Table 2). Similarly, ΔM2 (a synthetic variant of *Galleria melonella* Cec D with modified residues in the N-terminal region; [54]) and Cec XJ forms (2-aa longer variants of *B. mori* Cec B; [107]) were also effective against *S. aureus* (Table 2).

Moreover, Cec D from the lepidopteran *G. mellonella* showed antibacterial activity against *Listeria monocytogenes*, a Gram-positive bacterium causing listeriosis, a food-borne infection, which can cause meningitis, meningoencephalitis, and fatal sepsis [54,109].

Several natural Cecs and analog derivatives have also been tested against a variety of fungi (Tables 1 and 2). Although the peptides were not all effective against these microorganisms, *H. cecropia* Cecs A and B [44], *P. xuthus* Papiliocin [58], *Artogeia rapae* Hinnavins [65,66], Cec A from the mosquito *Anopheles gambiae* [23], and a Cec-analog derived from the D-enantiomerization of *Antheraea pernyi* Cec B [93], were active against *Candida albicans*, an opportunistic pathogen responsible for candidiasis in human hosts [110]. Synthetic analogs also showed in vitro antiprotozoal activities, as demonstrated for SB-37 and Shiva, which were effective against *Trypanosoma cruzi* and *Plasmodium falciparum* [92], and a chimeric CAM hybrid active against *P. falciparum* [94] (Table 2). Finally, several Cec and Cec-analog peptides have also been tested for their potential antiviral activity (Tables 1 and 2). *H. cecropia* Cec A was able to suppress replication of human immunodeficiency virus 1 (HIV) by inhibiting viral gene expression [43], while Cec D was active against the porcine reproductive and respiratory syndrome virus (PRRSV) [47]. Additionally, engineered CA-MA hybrids were shown to inhibit virus–cell fusion activity [104].

5. Anti-Inflammatory Properties of Natural Cecs and Synthetic Cec-Analogs

Some Cec AMPs have been explored for their potential anti-inflammatory activity. Inflammation is an organism-protective response against different factors, including pathogens, which contributes to the removal of harmful foreign agents and to the initiation of reparative processes. An uncontrolled inflammatory response can however be dangerous, eliciting different acute or chronic diseases (reviewed in [111]). During Gram-negative infections, the release of LPS can overstimulate the innate immune system resulting in septic shock [112]. Several Cecs and Cec-analogs are able to bind LPS and have shown both in vitro and in vivo anti-inflammatory properties. Specifically, peptides derived from Lepidoptera (*H. cecropia* CecA [45], Papiliocin and derivatives from *Papilio xuthus* [57,58,113], Cec B, and a synthetic analog from *A. pernyi* [49]) were able to inhibit the production of nitric oxide and the transcription of several pro-inflammatory genes in LPS-treated murine cells, in vitro. Similar properties characterized natural Cecs from Diptera, such as Cec TY from the horsefly *Tabanus yao* [108], SibaCec from the black fly *S. bannaense* [22], and AeaeCec 1 from the mosquito *A. aegypti* [26]. In addition, an in vivo study showed that an intraperitoneal administration of *H. cecropia* Cecs A and B or a Papiliocin analog were able to reduce bacterial concentrations, plasma endotoxin levels, and mortality in *E. coli*-infected rodent models [113,114]. Finally, *M. domestica* Mdc was shown to alleviate colonic mucosal barrier impairments induced in mice by a *Salmonella typhimurium* infection, with a reduction in the colonic inflammation and oxidative stress response [115]. These studies demonstrate the dual antimicrobial and anti-inflammatory functions of Cec AMPs, underpinning their potential utilization in biomedical applications.

6. Antitumor Activity of Natural Cecs and Synthetic Cec-Analogs

Although the antitumor activities of Cecs and Cec-analogs have been less widely studied than their antimicrobial activities, these peptides indeed possess antitumor properties. These characteristics, for example, refer to *H. cecropia* Cecs A and B, *M. domestica* Mdc, *B. mori* Cec XJ derivatives, and the chimeric CAM and CA-MA hybrids, which were active against different types of human and rodent cancer cell lines in vitro [80,100,116–121]. Cec XJ and Mdc were also shown to inhibit proliferation and promote apoptosis of transformed cells in vitro [80,120]. Interestingly, when tested at the same concentrations, none of the analyzed AMPs showed any cytotoxic effects against normal cell lines. This selective antitumor activity might in part depend on the variable membrane compositions and fluidity of transformed compared to non-transformed cells [122]. Finally, Cec antitumor activity was also demonstrated in in vivo mammalian models, as shown for the *H. cecropia* Cec B and *B. mori*-derived Cec XJ, both improving the survival of mice bearing malignant ascites [117,123], indicating the potential of these AMPs as anticancer therapeutics.

7. Health Benefits of Natural Cecs and Synthetic Cec-analogs: Future Potential and Limitations

Several studies have suggested that some natural Cecs and synthetic-derived Cec peptides represent promising molecules for the development of new antibacterial drugs. Resistance to conventional antibiotics is a global phenomenon, involving not only the health system, but also livestock production [124]. The potential of insect AMPs as antimicrobial dietary supplements has been recently reviewed [125]. In addition, different studies reported the use of transgenesis to produce Cec-overexpressing plants and animals exhibiting greater resistance to pathogenic infections compared to non-transformed controls (e.g., [126,127]). Although effective, the use of transgenic strategies is limited by the regulatory laws of different countries and is not discussed in detail in this review. In the following paragraphs, we consider the potential of peptides belonging to the Cec family as therapeutics for clinical applications.

7.1. Potential of Natural Cecs and Cec-analogs as Antibacterial Drugs

Unlike other AMPs, Cec and Cec-analog peptides have generally shown low in vitro toxicity, evaluated as cytotoxicity against normal mammalian cell lines and/or hemolytic activity against human or rodent erythrocytes (Tables 1 and 2). Although there is variation among the analyzed Cecs, the peptide concentrations showing initial toxicity against mammalian cells were one or two orders of magnitude higher than the minimum inhibitory concentration (MIC) values against the analyzed bacteria. Interestingly, a low toxicity was also typical of different Cec chimeric hybrids, including some CAM, CA-MA, and CA-LL37 peptides [100,102,106], which generally showed a wide action spectrum against both Gram-positive and-negative bacteria (Table 2).

Several natural Cecs and synthetic derivatives have shown a high stability to heat treatments and/or pH variations (e.g., [53,98,107]). In addition, they usually maintained their antimicrobial activity in complex biological fluids, mimicked in vitro by using high concentrations of serum, as well as in the presence of elevated levels of divalent cations such as Ca^{2+} and Mg^{2+}, which show 1–2 and 0.5 mM concentrations in human saliva, respectively, and might reduce or inhibit AMP effectiveness (e.g., [37,53,107,128]). Similarly, natural Cecs and Cec-analogs were also active when analyzed in the presence of high concentrations of Na^+, typical of airway surface fluids from patients affected by cystic fibrosis, who often suffer lung infections from bacteria such as *P. aeruginosa*, *A. baumanni*, or *S. aureus* (e.g., [49,53,98,102,129]).

The vast majority of data on Cec antimicrobial activity is derived from in vitro analyses. However, some studies have shown the potential of these peptides in vivo. For example, single intraperitoneal administrations of *H. cecropia* Cecs A and B, and *Danaus plexipibus* DAN2 decreased mortality in acutely *E. coli*-infected rodent models [114,130]. In addition, mice subjected to DAN2 doses two-fold higher than their most effective antibacterial concentration did not display any behavioral or morphological abnormalities, demonstrating in vivo that these peptides lack toxic effects after acute treatments [130].

With the prospect of employing natural Cecs and Cec-analogs in the treatment of infectious diseases, one of the potential problems is the capability of pathogens to develop resistance to these AMPs. Antimicrobial resistance is a complex phenomenon involving the development of intrinsic and/or acquired factors able to inactivate a compound or modify a target, nullifying the action of the specific drug. Currently, most considerations about, and data on AMP resistance in the literature, refer to bacteria. Although it is generally accepted that bacteria do not develop resistance to AMPs as easily as to conventional antibiotics, cases of bacterial resistance have been reported for non-Cec AMPs [125,131]. However, a recent study on *E. coli* compared the bacterial mutation rate induced by treatments with antibiotics with those with cationic AMPs, including *H. cecropia* Cec A [132]. Unlike antibiotics, none of the analyzed AMPs increased *E. coli* mutation rates. The authors linked this phenomenon to the inability of these AMPs to activate bacterial stress pathways that promote DNA mutagenesis [132]. Since the family of Cecs act against bacteria with a similar bactericidal mechanism at a molecular/cellular level, these data suggest that these AMPs are unlikely to stimulate the development of new intrinsic resistance factors linked to a mutation rate increment, at least in the *E. coli* model.

Long-term exposure to low levels of an antimicrobial compound is an important driver of antimicrobial resistance. Promising data have shown that following long-term treatments with the hybrid CAM peptide at sub-lethal concentrations did not significantly alter the peptide MIC. Following treatment, CAM remained effective against both laboratory reference and MDR *P. aeruginosa* strains, whereas similar serial exposures to sublethal doses of gentamicin or LL-37 increased their effective MICs on the same bacterial strains [97]. These studies provide important data suggesting that treatments with Cec and Cec-analog peptides do not easily induce antimicrobial resistance. However, dedicated studies analyzing all aspects of bacterial resistance, including the possible acquisition of exogenous factors through horizontal gene transfer, should be performed for each promising Cec or Cec-analog antibacterial candidate.

An innovative approach that is gaining interest is the use of AMPs as adjuvants in combination with conventional antibiotics [133]. Simultaneous treatments of AMPs and antibiotics can determine synergistic antimicrobial effects that are able to increase therapy efficacy and lower administration doses, in turn decreasing potential toxicity side effects. This aspect should be evaluated for each Cec or Cec-analog candidate, since the indications derived from in vitro studies performed on Cec-analog peptides, such as CAM or Cec-LL37 hybrids, showed variable synergistic activity grades, depending on the types of antibiotics and bacterial species (e.g., [96,97,106,134]).

7.2. Natural Cecs and Cec-Analogs as Anti-Biofilm Compounds

Biofilms are bacterial communities embedded in an extracellular matrix of polysaccharides, proteins, lipids, and DNA [135]. The bacteria forming biofilms display numerous interesting emergent social behaviors but are less susceptible to the effectors of the human defense system and exhibit a higher tolerance to conventional antibiotics, conferred in part from the extracellular matrix [135,136]. Several bacteria responsible for infections in hospitalized and/or immunodepressed patients can form biofilms, including Gram-negative *P. aeruginosa*, *A. baumanni*, *K. pneumoniae*, and Gram-positive *S. aureus*, and *S. epidermidis* [135]. Estimates indicate that biofilm infections are associated with at least two-thirds of all clinical infections [136]. In humans, many surfaces can be infected by biofilms, such as skin, teeth, ears, bones, and the respiratory and urinary tracts. Biofilms can also grow on medical devices, such as artificial implants, valves, and catheters, frequently used in modern medicine as feasible solutions to rescue compromised organs. Medical devices are composed of different types of biomaterials, and a great effort has been made to develop safe biomaterials. However, biomaterial microbial colonization remains one of the major problems related to the use of such devices. Contaminated devices can cause biomaterial-associated infections that are difficult to treat with conventional antibiotic therapies, triggering severe consequences for patient health [137].

Innovative anti-biofilm treatments are therefore needed [136,137] and Cecs and Cec-analogs might represent a promising solution. Two in vitro studies have demonstrated anti-biofilm effects of CAM hybrids, alone and in combination with conventional antibiotics, to treat both *P. aeruginosa* and *S. aureus* biofilms [134,138]. In addition, the use of AMPs to coat biomaterials during device manufacturing is considered a promising strategy to prevent biomaterial-associated infections (reviewed in [137]). Different studies have explored the possibility of using Cec and Cec-analog peptides in the functionalization of several types of materials used in biomedicine, such as hydrogels [139], polyurethane surfaces [140], as well as silk fibroin films or fibers [141,142]. These peptide-enriched materials were able to inhibit the growth of *E. coli* [139,141,142] and *S. epidermidis* [140], supporting the potential of Cec and Cec-analog peptides in these applications.

7.3. Biomedical Applications of Natural Cecs and Cec-Analogs: Limitations and Potential Solutions

All potential treatments that aim to inhibit pathogenic infections as well as combat antimicrobial resistance suffer limitations in their overall efficacy, including AMPs. Peptides are subject to degradation by naturally occurring proteases, such as trypsin, which is abundant in the digestive tract, and trypsin-Cec degradation has been demonstrated in *B. mori* (Cec B and Cec XJ variants, specifically) [53,107]. Furthermore, Cec peptides can also be targets for human elastase, which is produced by neutrophils, defense cells recruited during infections. In addition, Cec AMPs might be inactivated by proteases secreted by pathogens, such as *Pseudomonas* elastase and *S. aureus* V8 protease [98]. However, AMP sensitivity to proteolytic degradation can be limited in a number of ways. The substitution of specific residues is one such method to inhibit proteolytic degradation; this was recently demonstrated in CAM peptides, where a four-tryptophan-substitution variant (CAM-W) lost susceptibility to degradation by each of the enzymes mentioned above (Table 2) [98]. Peptide stability against proteolysis can also be achieved by the substitution of the natural L-residues with their respective D-enantiomers. This method was used to generate a whole D-enantiomer of the *A. pernyii* L-Cec B [93]. The obtained D-Cec B peptide maintained potent biocidal activity, while resisting the proteolytic activity that degraded the L-form (Table 2) [93].

In addition, enzymatic degradation might be limited by employing novel strategies based on the use of nanotechnologies. Indeed, the use of nanoparticles (NPs) to develop new formulations for AMP delivery is considered an improvement able to enhance peptide stability, while increasing peptide bioavailability and efficiency at the desired target site, as well as reducing the risk of possible toxic side effects [3,143]. In a recent study, Rai and colleagues demonstrated that the conjugation of CAM peptides to gold NPs enhanced in vitro CAM antimicrobial activity and stability as well as in vivo efficacy in a sepsis mouse model [144]. These encouraging results are opening new prospects for the use of Cec and Cec-analog peptides (and AMPs in general) as therapeutics to treat infectious diseases. In particular, the possibility of using biodegradable and biocompatible organic materials to encapsulate the peptide should be explored to give rise to new formulations for non- or less-invasive delivery routes (e.g., nasal, buccal/sublingual, or transdermal routes).

A second drawback that has slowed down the development of AMPs as new antimicrobial drugs is associated with the costs of large-scale production, which are generally much higher than those of small antibiotic molecules. Peptide compounds can be produced using a variety of techniques, including chemical synthesis, cell-free expression systems, recombinant DNA technologies for the production in heterologous cell systems, and transgenic organisms. Since natural Cecs and Cec-analogs generally show a low molecular weight (<4 KDa), chemical synthesis appears to be the best option for their production [145]. In addition, this technology allows the substitution of natural amino acids with atypical residues such as D-enantiomers, or the introduction of aa modifications (as in C-terminal amidation), often required in natural Cec and Cec-analog peptides. Chemical synthesis is undoubtedly an expensive approach [143]; however, due to the continuous development of efficient synthesis methodologies, progressive cost reductions for reagents, and competition among companies [6,145], considerable cost reductions are expected in the future. Consideration should also be given to the cost related to the development of possible AMP-based therapies compared to the social and economic burden caused by the current progressive and alarming spread of MDR infectious diseases [1]. Highlighting the USA as an example, 23,000 Americans are estimated to die annually with antibiotic resistant infections, while in 2018, direct national costs of treating antibiotic resistant infections have been projected to exceed $2 billion annually [146]. To these costs, other indirect economic and social costs should be added.

8. Conclusions

Insect Cecs and Cec-analog peptides are a class of AMPs that appear to be promising candidates as antibacterial therapeutics. These AMPs, tested alone or in combination with conventional antibiotics, show powerful antimicrobial activity against several important human pathogens, including MDR bacterial strains. They also exhibit low toxicity against mammalian cells and anti-inflammatory activity. Preliminary indications suggest that the development of new resistance phenomena against these peptides appears unlikely. However, few preclinical and no clinical analyses have been performed to date. In particular, long-term and/or longitudinal studies exploring potential side effects such as allergenicity or immunogenicity should be completed [6].

The intrinsic nature of Cec peptides, which makes them sensitive to protease degradation, together with the cost of large-scale production has slowed down or even impeded the development of Cec-based antimicrobial drugs. However, the advance of new strategies such as nanotechnologies will considerably reduce these limitations. The use of natural Cecs might allow the production of formulations active against Gram-negative bacteria, while the employment of Cec-analogs might give rise to therapeutics with a wide spectrum, effective against both Gram-negative and Gram-positive pathogens. In addition, given their anti-biofilm activity, Cecs and Cec-analogs might be used to coat biomaterials for medical devices as a strategy to prevent biomaterial-associated infections. Although further research and development studies are required, several lines of evidence suggest that both insect Cecs and Cec-analogs represent a suitable tool to counteract the alarming global spread of MDR pathogens.

Author Contributions: Conceptualization of the manuscript, D.B., A.G., O.R., and F.S. Data analysis, A.G. and D.B. All authors wrote, read, and approved the manuscript.

Acknowledgments: We thank two anonymous reviewers for their useful comments on a previous version of the manuscript and Maxine Iversen for proofreading the final version of the manuscript. Federica Sandrelli acknowledges Cinchron, a European Union's Horizon 2020 research and innovation programme under the Marie Sklodowska-Curie (grant agreement No 765937), CARIPARO (Progetti di Eccellenza 2011/12) and Università degli Studi di Padova (CPDA154301).

Conflicts of Interest: The authors declare no conflict of interest.

References

1. *Report to the Secretary General of the Nations: No Time to Wait–Securing the Future from Drug-Resistant Infections;* Interagency Coordination Group on Antimicrobial Resistance (IACG): New York, NY, USA, 2019.
2. Zhang, L.-J.; Gallo, R.L. Antimicrobial peptides. *Curr. Biol.* **2016**, *26*, R14-9. [CrossRef] [PubMed]
3. Kumar, P.; Kizhakkedathu, J.N.; Straus, S.K. Antimicrobial peptides: Diversity, mechanism of action and strategies to improve the activity and biocompatibility in vivo. *Biomolecules* **2018**, *8*, 4. [CrossRef] [PubMed]
4. Wang, G.; Li, X.; Wang, Z. APD3: The antimicrobial peptide database as a tool for research and education. *Nucleic Acids Res.* **2015**, *44*, D1087–D1093. [CrossRef] [PubMed]
5. Wu, Q.; Patočka, J.; Kuča, K. Insect antimicrobial peptides, a mini review. *Toxins* **2018**, *10*, 461. [CrossRef]
6. Tonk, M.; Vilcinskas, A. The medical potential of antimicrobial peptides from insects. *Curr. Top. Med. Chem.* **2017**, *17*, 554–575. [CrossRef]
7. Yi, H.-Y.; Chowdhury, M.; Huang, Y.-D.; Yu, X.-Q. Insect antimicrobial peptides and their applications. *Appl. Microbiol. Biotechnol.* **2014**, *98*, 5807–5822. [CrossRef]
8. Hultmark, D.; Steiner, H.; Rasmuson, T.; Boman, H.G. Insect immunity. Purification and properties of three inducible bactericidal proteins from hemolymph of immunized pupae of Hyalophora cecropia. *Eur. J. Biochem.* **1980**, *106*, 7–16. [CrossRef]
9. Hultmark, D.; Engström, Å.; Bennich, H.; Kapur, R.; Boman, H.G. Insect immunity: Isolation and structure of cecropin D and four minor antibacterial components from Cecropia pupae. *Eur. J. Biochem.* **1982**, *127*, 207–217. [CrossRef]
10. Steiner, H.; Hultmark, D.; Engström, Å.; Bennich, H.; Boman, H.G. Sequence and specificity of two antibacterial proteins involved in insect immunity. *Nature* **1981**, *292*, 246. [CrossRef]
11. Mylonakis, E.; Podsiadlowski, L.; Muhammed, M.; Vilcinskas, A. Diversity, evolution and medical applications of insect antimicrobial peptides. *Philos. Trans. R. Soc. Lond. B Biol. Sci.* **2016**, *371*, 20150290. [CrossRef]
12. Misof, B.; Liu, S.; Meusemann, K.; Peters, R.S.; Donath, A.; Mayer, C.; Frandsen, P.B.; Ware, J.; Flouri, T.; Beutel, R.G.; et al. Phylogenomics resolves the timing and pattern of insect evolution. *Science* **2014**, *346*, 763–767. [CrossRef]
13. Zhao, C.; Liaw, L.; Lee, I.H.; Lehrer, R.I. cDNA cloning of three cecropin-like antimicrobial peptides (Styelins) from the tunicate, Styela clava. *FEBS Lett.* **1997**, *412*, 144–148. [CrossRef]
14. Lee, J.-Y.; Boman, A.; Sun, C.X.; Andersson, M.; Jörnvall, H.; Mutt, V.; Boman, H.G. Antibacterial peptides from pig intestine: Isolation of a mammalian cecropin. *Proc. Natl. Acad. Sci. USA* **1989**, *86*, 9159–9162. [CrossRef] [PubMed]
15. Andersson, M.; Boman, A.; Boman, H.G. Ascaris nematodes from pig and human make three anti-bacterial peptides: Isolation of cecropin P1 and two ASABF peptides. *Cell. Mol. Life Sci.* **2003**, *60*, 599–606. [CrossRef] [PubMed]
16. Pütsep, K.; Normark, S.; Boman, H.G. The origin of cecropins; implications from synthetic peptides derived from ribosomal protein L1. *FEBS Lett.* **1999**, *451*, 249–252. [CrossRef]
17. Tamang, D.G.; Saier, M.H., Jr. The cecropin superfamily of toxic peptides. *J. Mol. Microbiol. Biotechnol.* **2006**, *11*, 94–103. [CrossRef]
18. Tarr, D.E.K. Distribution and characteristics of ABFs, cecropins, nemapores, and lysozymes in nematodes. *Dev. Comp. Immunol.* **2012**, *36*, 502–520. [CrossRef]

19. Segovia, L.J.T.; Ramirez, G.A.T.; Arias, D.C.H.; Duran, J.D.R.; Bedoya, J.P.; Osorio, J.C.C. Identification and characterization of novel cecropins from the *Oxysternon conspicillatum* neotropic dung beetle. *PLoS ONE* **2017**, *12*, e0187914. [CrossRef]

20. Saito, A.; Ueda, K.; Imamura, M.; Atsumi, S.; Tabunoki, H.; Miura, N.; Watanabe, A.; Kitami, M.; Sato, R. Purification and cDNA cloning of a cecropin from the longicorn beetle, *Acalolepta luxuriosa*. *Comp. Biochem. Physiol. B Biochem. Mol. Biol.* **2005**, *142*, 317–323. [CrossRef]

21. Memarpoor-Yazdi, M.; Zare-Zardini, H.; Asoodeh, A. A novel antimicrobial peptide derived from the insect Paederus dermatitis. *Int. J. Pept. Res. Ther.* **2013**, *19*, 99–108. [CrossRef]

22. Wu, J.; Mu, L.; Zhuang, L.; Han, Y.; Liu, T.; Li, J.; Yang, Y.; Yang, H.; Wei, L. A cecropin-like antimicrobial peptide with anti-inflammatory activity from the black fly salivary glands. *Parasites Vectors* **2015**, *8*, 561. [CrossRef] [PubMed]

23. Vizioli, J.; Bulet, P.; Charlet, M.; Lowenberger, C.; Blass, C.; Müller, H.M.; Dimopoulos, G.; Hoffmann, J.; Kafatos, F.C.; Richman, A. Cloning and analysis of a cecropin gene from the malaria vector mosquito, *Anopheles gambiae*. *Insect Mol. Biol.* **2000**, *9*, 75–84. [CrossRef] [PubMed]

24. Lowenberger, C.; Charlet, M.; Vizioli, J.; Kamal, S.; Richman, A.; Christensen, B.M.; Bulet, P. Antimicrobial activity spectrum, cDNA cloning, and mRNA expression of a newly isolated member of the cecropin family from the mosquito vector *Aedes aegypti*. *J. Biol. Chem.* **1999**, *274*, 20092–20097. [CrossRef] [PubMed]

25. Jayamani, E.; Rajamuthiah, R.; Larkins-Ford, J.; Fuchs, B.B.; Conery, A.L.; Vilcinskas, A.; Ausubel, F.M.; Mylonakis, E. Insect-derived cecropins display activity against *Acinetobacter baumannii* in a whole-animal high-throughput *Caenorhabditis elegans* model. *Antimicrob. Agents Chemother.* **2015**, *59*, 1728–1737. [CrossRef]

26. Wei, L.; Yang, Y.; Zhou, Y.; Li, M.; Yang, H.; Mu, L.; Qian, Q.; Wu, J.; Xu, W. Anti-inflammatory activities of *Aedes aegypti* cecropins and their protection against murine endotoxin shock. *Parasites Vectors* **2018**, *11*, 470. [CrossRef]

27. Sun, D.; Eccleston, E.D.; Fallon, A.M. Peptide sequence of an antibiotic cecropin from the vector mosquito, *Aedes albopictus*. *Biochem. Biophys. Res. Commun.* **1998**, *249*, 410–415. [CrossRef]

28. Kaushal, A.; Gupta, K.; Shah, R.; van Hoek, M.L. Antimicrobial activity of mosquito cecropin peptides against *Francisella*. *Dev. Comp. Immunol.* **2016**, *63*, 171–180. [CrossRef]

29. Xu, X.; Yang, H.; Ma, D.; Wu, J.; Wang, Y.; Song, Y.; Wang, X.; Lu, Y.; Yang, J.; Lai, R. Toward an understanding of the molecular mechanism for successful blood feeding by coupling proteomics analysis with pharmacological testing of horsefly salivary glands. *Mol. Cell. Proteomics* **2008**, *7*, 582–590. [CrossRef]

30. Park, S.I.; Yoe, S.M. A novel cecropin-like peptide from black soldier fly, *Hermetia illucens*: Isolation, structural and functional characterization. *Entomol. Res.* **2017**, *47*, 115–124. [CrossRef]

31. Ekengren, S.; Hultmark, D. *Drosophila* cecropin as an antifungal agent. *Insect Biochem. Mol. Biol.* **1999**, *29*, 965–972. [CrossRef]

32. Samakovlis, C.; Kimbrell, D.A.; Kylsten, P.; Engström, Å.; Hultmark, D. The immune response in Drosophila: Pattern of cecropin expression and biological activity. *EMBO J.* **1990**, *9*, 2969–2976. [CrossRef]

33. Lu, X.; Shen, J.; Jin, X.; Ma, Y.; Huang, Y.; Mei, H.; Chu, F.; Zhu, J. Bactericidal activity of *Musca domestica* cecropin (Mdc) on multidrug-resistant clinical isolate of *Escherichia coli*. *Appl. Microbiol. Biotechnol.* **2012**, *95*, 939–945. [CrossRef] [PubMed]

34. Lu, X.M.; Jin, X.B.; Zhu, J.Y.; Mei, H.F.; Ma, Y.; Chu, F.J.; Wang, Y.; Li, X.B. Expression of the antimicrobial peptide cecropin fused with human lysozyme in *Escherichia coli*. *Appl. Microbiol. Biotechnol.* **2010**, *87*, 2169–2176. [CrossRef] [PubMed]

35. Gui, S.; Li, R.; Feng, Y.; Wang, S. Transmission electron microscopic morphological study and flow cytometric viability assessment of *Acinetobacter baumannii* susceptible to *Musca domestica* cecropin. *Sci. World J.* **2014**, *2014*, 657536. [CrossRef] [PubMed]

36. Boulanger, N.; Brun, R.; Ehret-Sabatier, L.; Kunz, C.; Bulet, P. Immunopeptides in the defense reactions of *Glossina morsitans* to bacterial and *Trypanosoma brucei* brucei infections. *Insect Biochem. Mol. Biol.* **2002**, *32*, 369–375. [CrossRef]

37. Boulanger, N.; Munks, R.J.; Hamilton, J.V.; Vovelle, F.; Brun, R.; Lehane, M.J.; Bulet, P. Epithelial innate immunity. A novel antimicrobial peptide with antiparasitic activity in the blood-sucking insect *Stomoxys calcitrans*. *J. Biol. Chem.* **2002**, *277*, 49921–49926. [CrossRef]

38. Okada, M.; Natori, S. Purification and characterization of an antibacterial protein from haemolymph of *Sarcophaga peregrina* (flesh-fly) larvae. *Biochem. J.* **1983**, *211*, 727–734. [CrossRef]

39. Okada, M.; Natori, S. Mode of action of a bactericidal protein induced in the haemolymph of *Sarcophaga peregrina* (flesh-fly) larvae. *Biochem. J.* **1984**, *222*, 119–124. [CrossRef]

40. Okada, M.; Natori, S. Primary structure of sarcotoxin I, an antibacterial protein induced in the hemolymph of *Sarcophaga peregrina* (flesh fly) larvae. *J. Biol. Chem.* **1985**, *260*, 7174–7177.

41. Pöppel, A.K.; Vogel, H.; Wiesner, J.; Vilcinskas, A. Antimicrobial peptides expressed in medicinal maggots of the blow fly *Lucilia sericata* show combinatorial activity against bacteria. *Antimicrob. Agents Chemother.* **2015**, *59*, 2508–2514. [CrossRef]

42. De Lucca, A.J.; Bland, J.M.; Jacks, T.J.; Grimm, C.; Walsh, T.J. Fungicidal and binding properties of the natural peptides cecropin B and dermaseptin. *Med. Mycol.* **1998**, *36*, 291–298. [CrossRef] [PubMed]

43. Wachinger, M.; Kleinschmidt, A.; Winder, D.; von Pechmann, N.; Ludvigsen, A.; Neumann, M.; Holle, R.; Salmons, B.; Erfle, V.; Brack-Werner, R. Antimicrobial peptides melittin and cecropin inhibit replication of human immunodeficiency virus 1 by suppressing viral gene expression. *J. Gen. Virol.* **1998**, *79*, 731–740. [CrossRef] [PubMed]

44. Andrä, J.; Berninghausen, O.; Leippe, M. Cecropins, antibacterial peptides from insects and mammals, are potently fungicidal against *Candida albicans*. *Med. Microbiol. Immunol.* **2001**, *189*, 169–173. [CrossRef] [PubMed]

45. Lee, E.; Shin, A.; Kim, Y. Anti-inflammatory activities of cecropin a and its mechanism of action. *Arch. Insect Biochem. Physiol.* **2015**, *88*, 31–44. [CrossRef] [PubMed]

46. Moore, A.J.; Beazley, W.D.; Bibby, M.C.; Devine, D.A. Antimicrobial activity of cecropins. *J. Antimicrob. Chemother.* **1996**, *37*, 1077–1089. [CrossRef]

47. Liu, X.; Guo, C.; Huang, Y.; Zhang, X.; Chen, Y. Inhibition of porcine reproductive and respiratory syndrome virus by Cecropin D in vitro. *Infect. Genet. Evol.* **2015**, *34*, 7–16. [CrossRef]

48. Qu, Z.; Steiner, H.; Engström, A.; Bennich, H.; Boman, H.G. Insect immunity: Isolation and structure of cecropins B and D from pupae of the Chinese oak silk moth, *Antheraea pernyi*. *Eur. J. Biochem.* **1982**, *127*, 219–224. [CrossRef]

49. Wang, J.; Ma, K.; Ruan, M.; Wang, Y.; Li, Y.; Fu, Y.V.; Song, Y.; Sun, H.; Wang, J. A novel cecropin B-derived peptide with antibacterial and potential anti-inflammatory properties. *PeerJ* **2018**, *6*, e5369. [CrossRef]

50. Fang, S.L.; Wang, L.; Fang, Q.; Chen, C.; Zhao, X.S.; Qian, C.; Wei, G.Q.; Zhu, B.J.; Liu, C.L. Characterization and functional study of a Cecropin-like peptide from the Chinese oak silkworm, *Antheraea pernyi*. *Arch. Insect Biochem. Physiol.* **2017**, *94*, e21368. [CrossRef]

51. Yang, W.; Cheng, T.; Ye, M.; Deng, X.; Yi, H.; Huang, Y.; Tan, X.; Han, D.; Wang, B.; Xiang, Z.; et al. Functional divergence among silkworm antimicrobial peptide paralogs by the activities of recombinant proteins and the induced expression profiles. *PLoS ONE* **2011**, *6*, e18109. [CrossRef]

52. Lu, D.; Geng, T.; Hou, C.; Huang, Y.; Qin, G.; Guo, X. *Bombyx mori* cecropin A has a high antifungal activity to entomopathogenic fungus *Beauveria bassiana*. *Gene* **2016**, *583*, 29–35. [CrossRef] [PubMed]

53. Romoli, O.; Mukherjee, S.; Mohid, S.A.; Dutta, A.; Montali, A.; Franzolin, E.; Brady, D.; Zito, F.; Bergantino, E.; Rampazzo, C.; et al. Enhanced Silkworm Cecropin B Antimicrobial Activity against *Pseudomonas aeruginosa* from Single Amino Acid Variation. *ACS Infect. Dis.* **2019**, *5*, 1200–1213. [CrossRef] [PubMed]

54. Mukherjee, K.; Mraheil, M.A.; Silva, S.; Müller, D.; Cemic, F.; Hemberger, J.; Hain, T.; Vilcinskas, A.; Chakraborty, T. Anti-Listeria activities of Galleria mellonella hemolymph proteins. *Appl. Environ. Microbiol.* **2011**, *77*, 4237–4240. [CrossRef] [PubMed]

55. Oñate-Garzón, J.; Manrique-Moreno, M.; Trier, S.; Leidy, C.; Torres, R.; Patiño, E. Antimicrobial activity and interactions of cationic peptides derived from *Galleria mellonella* cecropin D-like peptide with model membranes. *J. Antibiot. (Tokyo)* **2017**, *70*, 238–245. [CrossRef]

56. Kim, S.R.; Hong, M.Y.; Park, S.W.; Choi, K.H.; Yun, E.Y.; Goo, T.W.; Kang, S.W.; Suh, H.J.; Kim, I.; Hwang, J.S. Characterization and cDNA cloning of a cecropin-like antimicrobial peptide, papiliocin, from the swallowtail butterfly, *Papilio xuthus*. *Mol. Cells* **2010**, *29*, 419–423. [CrossRef]

57. Kim, J.K.; Lee, E.; Shin, S.; Jeong, K.W.; Lee, J.Y.; Bae, S.Y.; Kim, S.H.; Lee, J.; Kim, S.R.; Lee, D.G.; et al. Structure and function of papiliocin with antimicrobial and anti-inflammatory activities isolated from the swallowtail butterfly, *Papilio xuthus*. *J. Biol. Chem.* **2011**, *286*, 41296–41311. [CrossRef]

58. Lee, E.; Kim, J.-K.; Jeon, D.; Jeong, K.-W.; Shin, A.; Kim, Y. Functional roles of aromatic residues and helices of papiliocin in its antimicrobial and anti-inflammatory activities. *Sci. Rep.* **2015**, *5*, 12048. [CrossRef]

59. Choi, C.-S.; Yoe, S.-M.; Kim, E.-S.; Chae, K.-S.; Kim, H.R. Purification and Characterization of Antibacterial Peptides, Spodopsin Ia and Ib Induced in the Larval Haemolymph of the Common Cutworm, Spodoptera litura. *Korean J. Biol. Sci.* **1997**, *1*, 457–462.

60. Choi, C.S.; Lee, I.H.; Kim, E.; Kim, S.I.; Kim, H.R. Antibacterial properties and partial cDNA sequences of cecropin-like antibacterial peptides from the common cutworm, *Spodoptera litura*. *Comp. Biochem. Physiol. C Toxicol. Pharmacol.* **2000**, *125*, 287–297. [CrossRef]

61. Wang, L.; Li, Z.; Du, C.; Chen, W.; Pang, Y. Characterization and expression of a cecropin-like gene from *Helicoverpa armigera*. *Comp. Biochem. Physiol. B Biochem. Mol. Biol.* **2007**, *148*, 417–425. [CrossRef]

62. Lockey, T.D.; Ourth, D.D. Formation of pores in *Escherichia coli* cell membranes by a cecropin isolated from hemolymph of *Heliothis virescens* larvae. *Eur. J. Biochem.* **1996**, *236*, 263–271. [CrossRef] [PubMed]

63. Kim, C.R.; Lee, Y.H.; Bang, I.S.; Kim, E.S.; Kang, C.S.; Yun, C.Y.; Lee, I.H. cDNA cloning and antibacterial activities of cecropin D-like peptides from *Agrius convolvuli*. *Arch. Insect Biochem. Physiol.* **2000**, *45*, 149–155. [CrossRef]

64. Park, S.I.; An, H.S.; Chang, B.S.; Yoe, S.M. Expression, cDNA cloning, and characterization of the antibacterial peptide cecropin D from *Agrius convolvuli*. *Anim. Cells Syst.* **2013**, *17*, 23–30. [CrossRef]

65. Bang, I.S.; Son, S.Y.; Yoe, S.M. Hinnavin I, an antibacterial peptide from cabbage butterfly, *Artogeia rapae*. *Mol. Cells* **1997**, *7*, 509–513.

66. Yoe, S.M.; Kang, C.S.; Han, S.S.; Bang, I.S. Characterization and cDNA cloning of hinnavin II, a cecropin family antibacterial peptide from the cabbage butterfly, *Artogeia rapae*. *Comp. Biochem. Physiol. B Biochem. Mol. Biol.* **2006**, *144*, 199–205. [CrossRef]

67. Duwadi, D.; Shrestha, A.; Yilma, B.; Kozlovski, I.; Sa-Eed, M.; Dahal, N.; Jukosky, J. Identification and screening of potent antimicrobial peptides in arthropod genomes. *Peptides* **2018**, *103*, 26–30. [CrossRef]

68. Kylsten, P.; Samakovlis, C.; Hultmark, D. The cecropin locus in Drosophila; a compact gene cluster involved in the response to infection. *EMBO J.* **1990**, *9*, 217–224. [CrossRef]

69. Tryselius, Y.; Samakovlis, C.; Kimbrell, D.A.; Hultmark, D. CecC, a cecropin gene expressed during metamorphosis in *Drosophila pupae*. *Eur. J. Biochem.* **1992**, *204*, 395–399. [CrossRef]

70. Sackton, T.B.; Lazzaro, B.P.; Clark, A.G. Rapid expansion of immune-related gene families in the house fly, *Musca domestica*. *Mol. Biol. Evol.* **2017**, *34*, 857–872. [CrossRef]

71. Gudmundsson, G.H.; Lidholm, D.A.; Asling, B.; Gan, R.; Boman, H.G. The cecropin locus. Cloning and expression of a gene cluster encoding three antibacterial peptides in *Hyalophora cecropia*. *J. Biol. Chem.* **1991**, *266*, 11510–11517.

72. Ponnuvel, K.M.; Subhasri, N.; Sirigineedi, S.; Murthy, G.N.; Vijayaprakash, N.B. Molecular evolution of the cecropin multigene family in silkworm *Bombyx mori*. *Bioinformation* **2010**, *5*, 97–103. [CrossRef]

73. Ntwasa, M.; Goto, A.; Kurata, S. Coleopteran antimicrobial peptides: Prospects for clinical applications. *Int. J. Microbiol.* **2012**, *2012*, 101989. [CrossRef]

74. Zou, Z.; Evans, J.D.; Lu, Z.; Zhao, P.; Williams, M.; Sumathipala, N.; Hetru, C.; Hultmark, D.; Jiang, H. Comparative genomic analysis of the *Tribolium immune* system. *Genome Biol.* **2007**, *8*, R177. [CrossRef] [PubMed]

75. Quesada, H.; Ramos-Onsins, S.E.; Aguadé, M. Birth-and-death evolution of the *Cecropin multigene* family in Drosophila. *J. Mol. Evol.* **2005**, *60*, 1–11. [CrossRef] [PubMed]

76. Tassanakajon, A.; Somboonwiwat, K.; Amparyup, P. Sequence diversity and evolution of antimicrobial peptides in invertebrates. *Dev. Comp. Immunol.* **2015**, *48*, 324–341. [CrossRef] [PubMed]

77. Ramos-Onsins, S.; Aguade, M. Molecular evolution of the cecropin multigene family in *Drosophila*: Functional genes vs. pseudogenes. *Genetics* **1998**, *150*, 157–171.

78. Unckless, R.L.; Lazzaro, B.P. The potential for adaptive maintenance of diversity in insect antimicrobial peptides. *Phil. Tran. R. Soc. B* **2016**, *371*, 20150291. [CrossRef]

79. Tennessen, J.A. Molecular evolution of animal antimicrobial peptides: Widespread moderate positive selection. *J. Evol. Biol.* **2005**, *18*, 1387–1394. [CrossRef]

80. Wu, Y.-L.; Xia, L.-J.; Li, J.-Y.; Zhang, F.-C. CecropinXJ inhibits the proliferation of human gastric cancer BGC823 cells and induces cell death in vitro and in vivo. *Int. J. Oncol.* **2015**, *46*, 2181–2193. [CrossRef]

81. Price, M.N.; Dehal, P.S.; Arkin, A.P. FastTree 2–approximately maximum-likelihood trees for large alignments. *PLoS ONE* **2010**, *5*, e9490. [CrossRef]

82. Uvell, H.; Engström, Y. A multilayered defense against infection: Combinatorial control of insect immune genes. *Trends Genet.* **2007**, *23*, 342–349. [CrossRef] [PubMed]

83. Romoli, O.; Saviane, A.; Bozzato, A.; D'Antona, P.; Tettamanti, G.; Squartini, A.; Cappellozza, S.; Sandrelli, F. Differential sensitivity to infections and antimicrobial peptide-mediated immune response in four silkworm strains with different geographical origin. *Sci. Rep.* **2017**, *7*, 1048. [CrossRef] [PubMed]

84. Sato, H.; Feix, J.B. Peptide–membrane interactions and mechanisms of membrane destruction by amphipathic α-helical antimicrobial peptides. *Biochim. Biophys. Acta.* **2006**, *1758*, 1245–1256. [CrossRef] [PubMed]

85. Efimova, S.S.; Schagina, L.V.; Ostroumova, O.S. Channel-forming activity of cecropins in lipid bilayers: Effect of agents modifying the membrane dipole potential. *Langmuir* **2014**, *30*, 7884–7892. [CrossRef]

86. Efimova, S.S.; Medvedev, R.Y.; Chulkov, E.G.; Schagina, L.V.; Ostroumova, O.S. Regulation of the Pore-Forming Activity of Cecropin A by Local Anesthetics. *Cell Tiss. Biol.* **2018**, *12*, 331–341. [CrossRef]

87. Yeaman, M.R.; Yount, N.Y. Mechanisms of antimicrobial peptide action and resistance. *Pharmacol. Rev.* **2003**, *55*, 27–55. [CrossRef]

88. Andreu, D.; Merrifield, R.B.; Steiner, H.; Boman, H.G. N-terminal analogs of cecropin A: Synthesis, antibacterial activity, and conformational properties. *Biochemistry* **1985**, *24*, 1683–1688. [CrossRef]

89. Christensen, B.; Fink, J.; Merrifield, R.B.; Mauzerall, D. Channel-forming properties of cecropins and related model compounds incorporated into planar lipid membranes. *Proc. Natl. Acad. Sci. USA* **1988**, *85*, 5072–5076. [CrossRef]

90. Durell, S.R.; Raghunathan, G.; Guy, H.R. Modeling the ion channel structure of cecropin. *Biophys. J.* **1992**, *63*, 1623–1631. [CrossRef]

91. Tacconelli, E.; Magrini, N.; Kahlmeter, G.; Singh, N. *Global Priority List of Antibiotic-Resistant Bacteria to Guide Research, Discovery, and Development of New Antibiotics*; World Health Organization: Geneva, Switzerland, 2017; Volume 27.

92. Jaynes, J.M.; Burton, C.A.; Barr, S.B.; Jeffers, G.W.; Julian, G.R.; White, K.L.; Enright, F.M.; Klei, T.R.; Laine, R.A. In vitro cytocidal effect of novel lytic peptides on *Plasmodium falciparum* and *Trypanosoma cruzi*. *FASEB J.* **1988**, *2*, 2878–2883. [CrossRef]

93. De Lucca, A.J.; Bland, J.M.; Vigo, C.B.; Jacks, T.J.; Peter, J.; Walsh, T.J. D-cecropin B: Proteolytic resistance, lethality for pathogenic fungi and binding properties. *Med. Mycol.* **2000**, *38*, 301–308. [CrossRef] [PubMed]

94. Boman, H.G.; Wade, D.; Boman, I.A.; Wåhlin, B.; Merrifield, R.B. Antibacterial and antimalarial properties of peptides that are cecropin-melittin hybrids. *FEBS Lett.* **1989**, *259*, 103–106. [CrossRef]

95. Saugar, J.M.; Rodríguez-Hernández, M.J.; de la Torre, B.G.; Pachón-Ibañez, M.E.; Fernández-Reyes, M.; Andreu, D.; Pachón, J.; Rivas, L. Activity of cecropin A-melittin hybrid peptides against colistin-resistant clinical strains of *Acinetobacter baumannii*: Molecular basis for the differential mechanisms of action. *Antimicrob. Agents Chemother.* **2006**, *50*, 1251–1256. [CrossRef] [PubMed]

96. Garbacz, K.; Kamysz, W.; Piechowicz, L. Activity of antimicrobial peptides, alone or combined with conventional antibiotics, against Staphylococcus aureus isolated from the airways of cystic fibrosis patients. *Virulence* **2017**, *8*, 94–100. [CrossRef] [PubMed]

97. Geitani, R.; Moubareck, C.A.; Touqui, L.; Sarkis, D.K. Cationic antimicrobial peptides: Alternatives and/or adjuvants to antibiotics active against methicillin-resistant *Staphylococcus aureus* and multidrug-resistant *Pseudomonas aeruginosa*. *BMC Microbiol.* **2019**, *19*, 54. [CrossRef] [PubMed]

98. Ji, S.; Li, W.; Zhang, L.; Zhang, Y.; Cao, B. Cecropin A–melittin mutant with improved proteolytic stability and enhanced antimicrobial activity against bacteria and fungi associated with gastroenteritis in vitro. *Biochem. Biophys. Res. Commun.* **2014**, *451*, 650–655. [CrossRef]

99. Chicharro, C.; Granata, C.; Lozano, R.; Andreu, D.; Rivas, L. N-terminal fatty acid substitution increases the leishmanicidal activity of CA (1-7) M (2-9), a cecropin-melittin hybrid peptide. *Antimicrob. Agents Chemother.* **2001**, *45*, 2441–2449. [CrossRef]

100. Shin, S.Y.; Lee, M.K.; Kim, K.L.; Hahm, K.S. Structure-antitumor and hemolytic activity relationships of synthetic peptides derived from cecropin A-magainin 2 and cecropin A-melittin hybrid peptides. *J. Pept. Res.* **1997**, *50*, 279–285. [CrossRef]

101. Oh, D.; Shin, S.Y.; Lee, S.; Kang, J.H.; Kim, S.D.; Ryu, P.D.; Hahm, K.S.; Kim, Y. Role of the hinge region and the tryptophan residue in the synthetic antimicrobial peptides, cecropin A (1–8)–magainin 2 (1–12) and its analogues, on their antibiotic activities and structures. *Biochemistry* **2000**, *39*, 11855–11864. [CrossRef]

102. Jeong, K.-W.; Shin, S.-Y.; Kim, J.-K.; Kim, Y.-M. Antibacterial activity and synergism of the hybrid antimicrobial peptide, CAMA-syn. *Bull. Korean Chem. Soc.* **2009**, *30*, 1839–1844.

103. Ryu, S.; Choi, S.Y.; Acharya, S.; Chun, Y.J.; Gurley, C.; Park, Y.; Armstrong, C.A.; Song, P.I.; Kim, B.J. Antimicrobial and anti-inflammatory effects of Cecropin A(1–8)-Magainin2(1–12) hybrid peptide analog p5 against Malassezia furfur infection in human keratinocytes. *J. Invest. Dermatol.* **2011**, *131*, 1677–1683. [CrossRef] [PubMed]

104. Lee, D.G.; Park, Y.; Jin, I.; Hahm, K.S.; Lee, H.H.; Moon, Y.H.; Woo, E.R. Structure-antiviral activity relationships of cecropin A-magainin 2 hybrid peptide and its analogues. *J. Pept. Sci.* **2004**, *10*, 298–303. [CrossRef] [PubMed]

105. Lee, J.K.; Seo, C.H.; Luchian, T.; Park, Y. Antimicrobial Peptide CMA3 Derived from the CA-MA Hybrid Peptide: Antibacterial and Anti-inflammatory Activities with Low Cytotoxicity and Mechanism of Action in *Escherichia coli*. *Antimicrob. Agents Chemother.* **2015**, *60*, 495–506. [CrossRef] [PubMed]

106. Wei, X.-B.; Wu, R.-J.; Si, D.-Y.; Liao, X.-D.; Zhang, L.-L.; Zhang, R.-J. Novel hybrid peptide cecropin A (1–8)-LL37 (17–30) with potential antibacterial activity. *Int. J. Mol. Sci.* **2016**, *17*, 983. [CrossRef]

107. Liu, D.; Liu, J.; Li, J.; Xia, L.; Yang, J.; Sun, S.; Ma, J.; Zhang, F. A potential food biopreservative, CecXJ-37N, non-covalently intercalates into the nucleotides of bacterial genomic DNA beyond membrane attack. *Food Chem.* **2017**, *217*, 576–584. [CrossRef]

108. Wei, L.; Huang, C.; Yang, H.; Li, M.; Yang, J.; Qiao, X.; Mu, L.; Xiong, F.; Wu, J.; Xu, W. A potent anti-inflammatory peptide from the salivary glands of horsefly. *Parasites Vectors.* **2015**, *8*, 556. [CrossRef]

109. Heredia, N.; García, S. Animals as sources of food-borne pathogens: A review. *Anim. Nutr.* **2018**, *4*, 250–255. [CrossRef]

110. Whaley, S.G.; Berkow, E.L.; Rybak, J.M.; Nishimoto, A.T.; Barker, K.S.; Rogers, P.D. Azole antifungal resistance in Candida albicans and emerging non-albicans Candida species. *Front. Microbiol.* **2017**, *7*, 2173. [CrossRef]

111. Chen, L.; Deng, H.; Cui, H.; Fang, J.; Zuo, Z.; Deng, J.; Li, Y.; Wang, X.; Zhao, L. Inflammatory responses and inflammation-associated diseases in organs. *Oncotarget* **2017**, *9*, 7204–7218. [CrossRef]

112. Kell, D.B.; Pretorius, E. On the translocation of bacteria and their lipopolysaccharides between blood and peripheral locations in chronic, inflammatory diseases: The central roles of LPS and LPS-induced cell death. *Integr. Biol. (Camb.)* **2015**, *7*, 1339–1377. [CrossRef]

113. Kim, J.; Jacob, B.; Jang, M.; Kwak, C.; Lee, Y.; Son, K.; Lee, S.; Jung, I.D.; Jeong, M.S.; Kwon, S.H.; et al. Development of a novel short 12-meric papiliocin-derived peptide that is effective against Gram-negative sepsis. *Sci. Rep.* **2019**, *9*, 3817. [CrossRef]

114. Giacometti, A.; Cirioni, O.; Ghiselli, R.; Viticchi, C.; Mocchegiani, F.; Riva, A.; Saba, V.; Scalise, G. Effect of mono-dose intraperitoneal cecropins in experimental septic shock. *Crit. Care Med.* **2001**, *29*, 1666–1669. [CrossRef]

115. Zhang, L.; Gui, S.; Liang, Z.; Liu, A.; Chen, Z.; Tang, Y.; Xiao, M.; Chu, F.; Liu, W.; Jin, X.; et al. Musca domestica Cecropin (Mdc) Alleviates Salmonella typhimurium-Induced Colonic Mucosal Barrier Impairment: Associating With Inflammatory and Oxidative Stress Response, Tight Junction as Well as Intestinal Flora. *Front. Microbiol.* **2019**, *10*, 522. [CrossRef] [PubMed]

116. Chen, H.M.; Wang, W.; Smith, D.; Chan, S.C. Effects of the anti-bacterial peptide cecropin B and its analogs, cecropins B-1 and B-2, on liposomes, bacteria, and cancer cells. *Biochim. Biophys. Acta* **1997**, *1336*, 171–179. [CrossRef]

117. Moore, A.J.; Devine, D.A.; Bibby, M.C. Preliminary experimental anticancer activity of cecropins. *Pept. Res.* **1994**, *7*, 265–269.

118. Anghel, R.; Jitaru, D.; Bădescu, L.; Bădescu, M.; Ciocoiu, M. The cytotoxic effect of magainin II on the MDA-MB-231 and M14K tumour cell lines. *Biomed. Res. Int.* **2013**, *2013*, 831709. [CrossRef]

119. Chan, S.-C.; Hui, L.; Chen, H.M. Enhancement of the cytolytic effect of anti-bacterial cecropin by the microvilli of cancer cells. *Anticancer Res.* **1998**, *18*, 4467–4474.

120. Jin, X.; Mei, H.; Li, X.; Ma, Y.; Zeng, A.H.; Wang, Y.; Lu, X.; Chu, F.; Wu, Q.; Zhu, J. Apoptosis-inducing activity of the antimicrobial peptide cecropin of Musca domestica in human hepatocellular carcinoma cell line BEL-7402 and the possible mechanism. *Acta Biochim. Biophys. Sin.* **2010**, *42*, 259–265. [CrossRef]

121. Suttmann, H.; Retz, M.; Paulsen, F.; Harder, J.; Zwergel, U.; Kamradt, J.; Wullich, B.; Unteregger, G.; Stöckle, M.; Lehmann, J. Antimicrobial peptides of the Cecropin-family show potent antitumor activity against bladder cancer cells. *BMC Urol.* **2008**, *8*, 5. [CrossRef]

122. Hilchie, A.L.; Hoskin, D.W.; Coombs, M.R.P. Anticancer Activities of Natural and Synthetic Peptides. In *Antimicrobial Peptides. Advances in Experimental Medicine and Biology*; Matsuzaki, K., Ed.; Springer: Singapore, 2019; Volume 1117, pp. 131–147. [CrossRef]

123. Xia, L.J.; Wu, Y.L.; Ma, J.; Zhang, F.C. Therapeutic effects of antimicrobial peptide on malignant ascites in a mouse model. *Mol. Med. Rep.* **2018**, *17*, 6245–6252. [CrossRef]

124. Economou, V.; Gousia, P. Agriculture and food animals as a source of antimicrobial-resistant bacteria. *Infect. Drug Resist.* **2015**, *8*, 49–61. [CrossRef] [PubMed]

125. Jozefiak, A.; Engberg, R.M. Insect proteins as a potential source of antimicrobial peptides in livestock production. A review. *J. Anim. Feed Sci.* **2017**, *26*, 87–99. [CrossRef]

126. Chiou, P.P.; Chen, M.J.; Lin, C.-M.; Khoo, J.; Larson, J.; Holt, R.; Leong, J.A.; Thorgarrd, G.; Chen, T.T. Production of homozygous transgenic rainbow trout with enhanced disease resistance. *Mar. Biotechnol.* **2014**, *16*, 299–308. [CrossRef] [PubMed]

127. Bundó, M.; Montesinos, L.; Izquierdo, E.; Campo, S.; Mieulet, D.; Guiderdoni, E.; Rossignol, M.; Badosa, E.; Montesinos, E.; San Segundo, B.; et al. Production of cecropin A antimicrobial peptide in rice seed endosperm. *BMC Plant Biol.* **2014**, *14*, 102. [CrossRef] [PubMed]

128. Rajesh, K.S.; Zareena, S.H.; Kumar, M.S.A. Assessment of salivary calcium, phosphate, magnesium, pH, and flow rate in healthy subjects, periodontitis, and dental caries. *Contemp. Clin. Dent.* **2015**, *6*, 461–465. [CrossRef] [PubMed]

129. Joris, L.; Dab, I.; Quinton, P.M. Elemental composition of human airway surface fluid in healthy and diseased airways. *Am. Rev. Respir. Dis.* **1993**, *148*, 1633–1637. [CrossRef]

130. Shrestha, A.; Duwadi, D.; Jukosky, J.; Fiering, S.N. Cecropin-like antimicrobial peptide protects mice from lethal *E. coli* infection. *PLoS ONE* **2019**, *14*, e0220344. [CrossRef]

131. Lewies, A.; Du Plessis, L.H.; Wentzel, J.F. Antimicrobial Peptides: The Achilles' Heel of Antibiotic Resistance? *Probiotics Antimicrob. Proteins* **2019**, *11*, 370–381. [CrossRef]

132. Rodríguez-Rojas, A.; Makarova, O.; Rolff, J. Antimicrobials, stress and mutagenesis. *PLoS Pathog.* **2014**, *10*, e1004445. [CrossRef]

133. Grassi, L.; Maisetta, G.; Esin, S.; Batoni, G. Combination strategies to enhance the efficacy of antimicrobial peptides against bacterial biofilms. *Front. Microbiol.* **2017**, *8*, 2409. [CrossRef]

134. Dosler, S.; Karaaslan, E. Inhibition and destruction of *Pseudomonas aeruginosa* biofilms by antibiotics and antimicrobial peptides. *Peptides* **2014**, *62*, 32–37. [CrossRef] [PubMed]

135. Flemming, H.-C.; Wingender, J.; Szewzyk, U.; Steinberg, P.; Rice, S.A.; Kjelleberg, S. Biofilms: An emergent form of bacterial life. *Nat. Rev. Microbiol.* **2016**, *14*, 563–575. [CrossRef] [PubMed]

136. Dostert, M.; Belanger, C.R.; Hancock, R.E.W. Design and assessment of anti-biofilm peptides: Steps toward clinical application. *J. Innate Immun.* **2019**, *11*, 193–204. [CrossRef] [PubMed]

137. Riool, M.; de Breij, A.; Drijfhout, J.W.; Nibbering, P.H.; Zaat, S.A.J. Antimicrobial peptides in biomedical device manufacturing. *Front. Chem.* **2017**, *5*, 63. [CrossRef]

138. Mataraci, E.; Dosler, S. In vitro activities of antibiotics and antimicrobial cationic peptides alone and in combination against methicillin-resistant *Staphylococcus aureus* biofilms. *Antimicrob. Agents Chemother.* **2012**, *56*, 6366–6371. [CrossRef]

139. Cole, M.A.; Scott, T.F.; Mello, C.M. Bactericidal Hydrogels via Surface Functionalization with Cecropin A. *ACS Biomater. Sci. Eng.* **2016**, *2*, 1894–1904. [CrossRef]

140. Querido, M.M.; Felgueiras, H.P.; Rai, A.; Costa, F.; Monteiro, C.; Borges, I.; Oliveira, D.; Ferreira, L.; Martins, M.C. Cecropin–Melittin Functionalized Polyurethane Surfaces Prevent *Staphylococcus epidermidis* Adhesion without Inducing Platelet Adhesion and Activation. *Adv. Mater. Interfaces* **2018**, *5*, 1801390. [CrossRef]

141. Bai, L.; Zhu, L.; Min, S.; Liu, L.; Cai, Y.; Yao, J. Surface modification and properties of *Bombyx mori* silk fibroin films by antimicrobial peptide. *Appl. Surf. Sci.* **2008**, *254*, 2988–2995. [CrossRef]

142. Saviane, A.; Romoli, O.; Bozzato, A.; Freddi, G.; Cappelletti, C.; Rosini, E.; Cappellozza, S.; Tettamanti, G.; Sandrelli, F. Intrinsic antimicrobial properties of silk spun by genetically modified silkworm strains. *Transgenic Res.* **2018**, *27*, 87–101. [CrossRef]

143. Biswaro, L.S.; da Costa Sousa, M.G.; Rezende, T.; Dias, S.C.; Franco, O.L. Antimicrobial peptides and nanotechnology, recent advances and challenges. *Front. Microbiol.* **2018**, *9*, 855. [CrossRef]

144. Rai, A.; Pinto, S.; Velho, T.R.; Ferreira, A.F.; Moita, C.; Trivedi, U.; Evangelista, M.; Comune, M.; Rumbaugh, K.P.; Simões, P.N.; et al. One-step synthesis of high-density peptide-conjugated gold nanoparticles with antimicrobial efficacy in a systemic infection model. *Biomaterials* **2016**, *85*, 99–110. [CrossRef] [PubMed]
145. Bédard, F.; Biron, E. Recent progress in the chemical synthesis of class II and S-glycosylated bacteriocins. *Front. Microbiol.* **2018**, *9*, 1048. [CrossRef] [PubMed]
146. Thorpe, K.E.; Joski, P.; Johnston, K.J. Antibiotic-resistant infection treatment costs have doubled since 2002, now exceeding $2 billion annually. *Health Affairs* **2018**, *37*, 662–669. [CrossRef] [PubMed]

International Journal of
Molecular Sciences

Article

Whey-Derived Peptides Interactions with ACE by Molecular Docking as a Potential Predictive Tool of atural ACE Inhibitors

Yara Chamata [1], Kimberly A. Watson [2] and Paula Jauregi [1,*]

[1] Harry Nursten Building, Department of Food and Nutritional Sciences, University of Reading, Reading RG6 6AP, UK; y.chamata@pgr.reading.ac.uk
[2] Harborne Building, School of Biological Sciences, University of Reading, Reading RG6 6AP, UK; k.a.watson@reading.ac.uk
* Correspondence: p.jauregi@reading.ac.uk

Received: 18 December 2019; Accepted: 27 January 2020; Published: 29 January 2020

Abstract: Several milk/whey derived peptides possess high in vitro angiotensin I-converting enzyme (ACE) inhibitory activity. However, in some cases, poor correlation between the in vitro ACE inhibitory activity and the in vivo antihypertensive activity has been observed. The aim of this study is to gain insight into the structure-activity relationship of peptide sequences present in whey/milk protein hydrolysates with high ACE inhibitory activity, which could lead to a better understanding and prediction of their in vivo antihypertensive activity. The potential interactions between peptides produced from whey proteins, previously reported as high ACE inhibitors such as IPP, LIVTQ, IIAE, LVYPFP, and human ACE were assessed using a molecular docking approach. The results show that peptides IIAE, LIVTQ, and LVYPFP formed strong H bonds with the amino acids Gln 259, His 331, and Thr 358 in the active site of the human ACE. Interestingly, the same residues were found to form strong hydrogen bonds with the ACE inhibitory drug Sampatrilat. Furthermore, peptides IIAE and LVYPFP interacted with the amino acid residues Gln 259 and His 331, respectively, also in common with other ACE-inhibitory drugs such as Captopril, Lisinopril and Elanapril. Additionally, IIAE interacted with the amino acid residue Asp 140 in common with Lisinopril, and LIVTQ interacted with Ala 332 in common with both Lisinopril and Elanapril. The peptides produced naturally from whey by enzymatic hydrolysis interacted with residues of the human ACE in common with potent ACE-inhibitory drugs which suggests that these natural peptides may be potent ACE inhibitors.

Keywords: ACE-inhibitory activity; whey peptides; molecular docking; hypertension

---

1. Introduction

The number of people with unhealthy living habits who have developed cardiovascular disease (CVD) has increased in recent years. The WHO reported that an estimated 17.9 million people lose their lives as a result of cardiovascular disease every year [1]. CVDs have become the leading cause of death globally [2]. High blood pressure (hypertension) is one of the most important well-defined risk factors for CVD [3], therefore, cardiovascular diseases can be prevented with blood-pressure lowering treatment. Hypertension is regulated by the renin-angiotensin system (RAS), through modulating the angiotensin-converting enzyme ACE, bradykinin and other factors [4–6].

ACE (dipeptidyl carboxypeptidase, EC 3.4.15.1) is a zinc metallopeptidase, found in male genital, vascular endothelial, neuro-epithelial, and absorptive epithelial cells [7–9], and displays both endopeptidase and exopeptidase activities, acting on a wide range of substrates [10]. ACE is a key enzyme for regulating blood pressure in the renin-angiotensin system. Renin cleaves the N-terminal segment of angiotensinogen from the biologically inert AT-1. ACE then hydrolyzes AT-1 by cleaving the carboxyl terminal His-Leu dipeptide from the inactive AT-1 to the active angiotensin II (AT-2), a potent vasoconstrictor responsible for the development of hypertension [5,6,11,12]. ACE also indirectly influences the kallikrein–kinin system, by promoting the inactivation and degradation of the catalytic function of bradykinin, a vasodilator involved in blood pressure control [11–13]. By repressing AT-2 production and restraining bradykinin degradation, ACE inhibitory peptides control the increase of blood pressure [13].

Consequently, ACE-inhibiting natural products have been vigorously investigated during the last decades, due to their potential in lowering blood pressure during hypertension. Among various types of bioactive peptides, ACE-inhibitory peptides from food sources have been most extensively studied for their potential use as natural alternatives to drugs for reducing blood pressure through binding and inhibiting ACE, and thus preventing and managing hypertension [14,15]. Food-derived peptides are believed to represent a healthier and more natural alternative source for chronic treatment of hypertension. Moreover, and although the inhibitory capacity of food-derived peptides is lower than that of chemically-designed antihypertensive drugs, such as Captopril, Sampatrilat, Lisinopril, and Enalapril, it is thought that food-derived peptides are safer than pharmaceutical drugs due to their lack of some drug-associated adverse side effects such as angioedema, skin rashes, and dry cough [6,16]. However, considering the lack of consensus in their physiological antihypertensive effects in different human populations, the role of food peptides in regulating blood pressure is still a subject of ongoing debate [17–19].

Although different animal and plant proteins have been used in the development of functional foods providing antihypertensive activity, milk is the main source of antihypertensive ACE-inhibitory peptides reported to date [20]. Milk is made up of 3.5% proteins of which 80% are caseins, classified as α-, β- and k-caseins, and 20% whey proteins. Whey contains α-lactalbumin, β- lactoglobulin and other minor proteins. Upon the degradation of milk proteins, peptide fragments with many biological effects that can be different from those of the parent protein, are released. Several bioactive peptides in milk proteins have been identified [21], and they serve an array of biological activities, including angiotensin-converting enzyme (ACE) inhibition, antimicrobial, antioxidative functions, dipeptidyl peptidase IV (DPP-IV) inhibition, opioid agonist and antagonist activities, immunomodulation, and mineral binding [22]. Several milk/whey derived peptides possess high in vitro ACE inhibitory activity; particularly, hydrolysates of whey proteins, caseinates, fractions-enriched in individual milk proteins, and whole milk proteins have been reported to be a good source of ACE-inhibitory peptides [14]. Ile-Pro-Pro (IPP) has been identified as the most potent ACE inhibitor from milk protein, and it is derived from casein [23]. The antihypertensive activity of this tripeptide has been demonstrated in several animal studies and human trials [24]. However, in some cases, poor correlation between the in vitro ACE inhibitory activity of milk-derived peptides and the in vivo antihypertensive activity has been observed. This can be partly due to digestion which renders less active peptide sequences and/or due to their low bioavailability [25]. Also, antihypertensive activity may be exerted by mechanisms other than ACE inhibition [26,27], e.g., specific ACE inhibitors were demonstrated to increase the risk of microscopic colitis in a recent study, suggesting that milk-derived peptides may exert their antihypertensive activity through the microbiome [28].

The activity of these peptides depends on their inherent amino acid composition and sequence [29,30]. Shorter peptides up to 12 amino acids with hydrophobic and positively charged amino acids at the carboxyl end are more likely to interact with ACE [31]. In terms of favorable structure–function relationship for high ACE-inhibitory activity, dipeptides including bulky and hydrophobic amino acids are more potent whereas tripeptides having aromatic amino acids at the C-terminus end, positively charged amino acids in the middle and hydrophobic amino acids at the N-terminus end, are most potent [31]. Kobayashi et al. (2008) investigated the effects of aromatic amino acids in the third position of the tripeptides on ACE-inhibitory activity. They found that the difference in the ACE-inhibitory activity between the bioactive peptides (IKW, LKW, IKY, and LKF) resulted from the aromatic amino acids W, Y, and F. The highest inhibitory activity was presented by

LKW, with the largest amino acid in the C-terminal. Accordingly, ACE-inhibitory activity is affected by the size of the amino acid, as well as its hydrophobicity [32]. In the same study, Kobayashi et al. (2008) examined the effects of the charged amino acid in the second position and they reported that in order to obtain a high inhibitory activity, it is essential to have a positively charged residue next to an aromatic residue. They also highlighted that the tripeptide sequence consisting of either I or L + positively charged amino acids + aromatic amino acids is likely to have high ACE-inhibitory activity. The charged amino acid takes part in binding by ACE while the bulky aromatic amino acid prevents the access between substrates and the active site of ACE [32]. Some studies have indicated that tri-peptides show higher ACE-inhibitory activity, and the C terminus end of the tri-peptides substantially affects binding to ACE. Hydrophobic amino acid residues or Proline residues at the carboxyl end are important for ACE inhibition and inhibitors containing these residues are resistant to digestion [33].

ACE inhibitors are commonly discovered using classic investigation techniques which include hydrolysis of proteins with different proteolytic enzymes, isolation and purification of peptides using chromatographic systems, and synthesis of corresponding peptides for the confirmation of activity and structure [8]. Although some structure-activity relationships have been established for food protein derived peptides, they are still quite generic and thus could not be solely used to predict the ACE inhibitory activity of peptide sequences [34]. In order to avoid some challenges of the classical approach, such as having to apply cumbersome purification processes to isolate active peptides, the computer-based approach is considered a useful and effective method to identify novel peptides [35]. A number of docking algorithms are being used in multiple studies to predict potent ACE inhibitory peptides encrypted in food proteins [36–42], and more specifically in milk proteins [43,44], whereby attempts to understand the interactions between receptor and ligand are being attempted [45]. Molecular docking enables the investigation of the specific interactions between certain peptide sequences and specific binding site residues in ACE, which could help to provide a better prediction of bioactivity in vivo through a molecular understanding of the structure-function relationship. Such an approach can be a powerful tool that can be used in pre-screening potentially bioactive peptides, prior to their testing in vivo.

Herein we investigate the potential interactions between whey protein derived peptides with high ACE inhibitory activity and human ACE, utilising a molecular docking approach. This study follows our previous work [30], where the peptides were produced by enzymatic hydrolysis of whey and were further fractionated for their chemical and activity characterization. Moreover, the interactions between these peptides and ACE are compared with those of the ACE inhibitory drugs Sampatrilat, Captopril, Lisinopril, and Enalapril.

2. Results

2.1. Molecular Homology between Human ACE and Rabbit ACE

According to the EMBOSS NEEDLE results (Figure 1), there is 93.4% of similarity between human ACE and rabbit ACE. As shown in Figure 1, the structural comparison of these two enzymes indicates that there is a close homology between the human ACE and the rabbit ACE, and that the active sites between human ACE and rabbit ACE are very similar. The rabbit ACE is generally used for the in vitro testing of ACE inhibition, hence it can be assumed that similar results will be obtained with human ACE. There have not been any previous studies that reported the homology between the human ACE and the rabbit ACE. In a study by Soubrier et al. (1988), amino-terminal sequence analysis was conducted between amino-terminal amino acid sequences of human ACE and other mammalians (rabbit, calf, pig, and mouse), and a high degree of similarity was found between human ACE and these mammalians [46].

```
ACE_HUMAN     1 MGAASGRRGPGLLLPLP-----LLLLLPPQPALALDPGLQPGNFSADEAG    45
                ||||.|||||.||.|.|       ||||.||..||.|||||.||:|:|||||
ACE_RABIT     1 MGAAPGRRGPRLLRPPPPLLLLLLLLRPPPAALTLDPGLLPGDFAADEAG    50

ACE_HUMAN    46 AQLFAQSYNSSAEQVLFQSVAASWAHDTNITAENARRQEEAALLSQEFAE    95
                |:|||.|||||||||||:|.||||||||||||||||||||.|||||||||
ACE_RABIT    51 ARLFASSYNSSAEQVLFRSTAASWAHDTNITAENARRQEEEALLSQEFAE   100

ACE_HUMAN    96 AWGQKAKELYEPIWQNFTDPQLRRIIGAVRTLGSANLPLAKRQQYNALLS   145
                |||:||||||:|:||||||||:||||||||||||||.||||||||||:|||
ACE_RABIT   101 AWGKKAKELYDPVWQNFTDPELRRIIGAVRTLGPANLPLAKRQQYNSLLS   150

ACE_HUMAN   146 NMSRIYSTAKVCLPNKTATCWSLDPDLTNILASSRSYAMLLFAWEGWHNA   195
                |||:||||.|||.||||||:||||||||||.|||||||||||||||||||||
ACE_RABIT   151 NMSQIYSTGKVCFPNKTASCWSLDPDLMNILASSRSYAMLLFAWEGWHNA   200

ACE_HUMAN   196 AGIPLKPLYEDFTALSNEAYKQDGFTDTGAYWRSWYNSPTFEDDLEHLYQ   245
                .|||||||::|||||||||:||||:|||||||||:||||||||:|||||:|||.:|.
ACE_RABIT   201 VGIPLKPLYQEFTALSNEAYRQDGFSDTGAYWRSWYDSPTFEEDLERIYH   250

ACE_HUMAN   246 QLEPLYLNLHAFVRRALHRRYGDRYINLRGPIPAHLLGDMWAQSWENIYD   295
                |||||||||||:|||.|||||||||||||||||||||||||:||||||:|||
ACE_RABIT   251 QLEPLYLNLHAYVRRVLHRRYGDRYINLRGPIPAHLLGNMWAQSWESIYD   300

ACE_HUMAN   296 MVVPFPDKPNLDVTSTMLQQGWNATHMFRVAEEFFTSLELSPMPPEFWEG   345
                |||||||||||||||||:|:|||||||||||||||||||.|.||||||..
ACE_RABIT   301 MVVPFPDKPNLDVTSTMVQKGWNATHMFRVAEEFFTSLGLLPMPPEFWAE   350

ACE_HUMAN   346 SMLEKPADGREVVCHASAWDFYNRKDFRIKQCTRVTMDQLSTVHHEMGHI   395
                ||||||.|||||||||||||||||||||||||||:|||||||||||||:
ACE_RABIT   351 SMLEKPEDGREVVCHASAWDFYNRKDFRIKQCTQVTMDQLSTVHHEMGHV   400
```

Figure 1. EMBOSS NEEDLE multiple sequence alignment results. Colour coding is as follows: Yellow indicates identical residues at the active site, green indicates similar residues at the active site, and red indicates that a part of the residue is similar. (I) residues are identical; (.) conserved change; (:) part of the residue is similar but not that conserved.

2.2. Molecular Docking

Molecular docking was conducted to elucidate the potential molecular interactions between the whey-protein derived peptide sequences and specific amino acids at the binding site of human ACE. The peptide sequences were docked into the binding site of the human ACE, using the X ray crystallographic structure of the human ACE receptor (PDB code 6F9V). The extracted co-crystallized ligand, Sampatrilat, [47] was first re-docked into the prepared protein to be used for docking in order to validate the docking procedure. The RMSD between the docked conformation, as generated by the program PyMol, and the native co-crystallized ligand conformation was 0.1 Å, which was well within the 2 Å grid spacing used in the docking procedure, demonstrating that the docking method to be used was valid and reliable. Additionally, the interactions between the docked ligand and the prepared target receptor mimicked those observed in the crystal structure of the same protein.

Hydrogen bonds are a significant factor that contribute to the specificity and stability of protein-ligand interactions. Figures 2–5 and Tables 1–4 show the hydrogen bond interactions associated with each ligand and the surrounding ACE residues. IPP formed 3 hydrogen bonds with the ACE residues: one with Asp 354 and two with Gln 355 (Figure 2, Table 1). IIAE formed three hydrogen bonds with residues Thr 144, Gln 259, and Thr 358 (Figure 3, Table 2). LIVTQ formed three hydrogen bonds: one with Ala 332, one with Gln 355, and one with Thr 358 (Figure 4, Table 3). As for the ligand LVYPFP, five hydrogen bonds were formed with residues Asp 255, Ser 260, His 331, Arg 350, and Thr 358 (Figure 5, Table 4). It is interesting to note that several peptides had same H bonds in common: Thr 358 formed H bonds with three of the peptides, IIAE, LIVTQ, LVYPFP; Gln 355 with IPP and LIVTQ. Additionally, all except one of the aminoacids in the active site of the ACE were polar (charged and non charged) and some of these charged aminoacids were also involved in salt bridge (electrostatic) interactions (Tables 1–4). IPP formed one salt-bridge interaction with residue Arg 350 (Figure 2, Table 1), whereas ligands IIAE and LIVTQ formed only one salt bridge interaction with residues Asp 140, and Asp 255, respectively (Figures 3 and 4, Tables 2 and 3). As for LVYPFP, two salt-bridge interactions were formed with residues Glu 262, and His 331 (Figure 5, Table 4). Both, LVYPFP and LIVTQ interacted with Asp 255 via H-bonding and a salt bridge, respectively; and, both IPP and LVYPFP interacted with Arg 350 via a salt bridge and H-bonding, respectively.

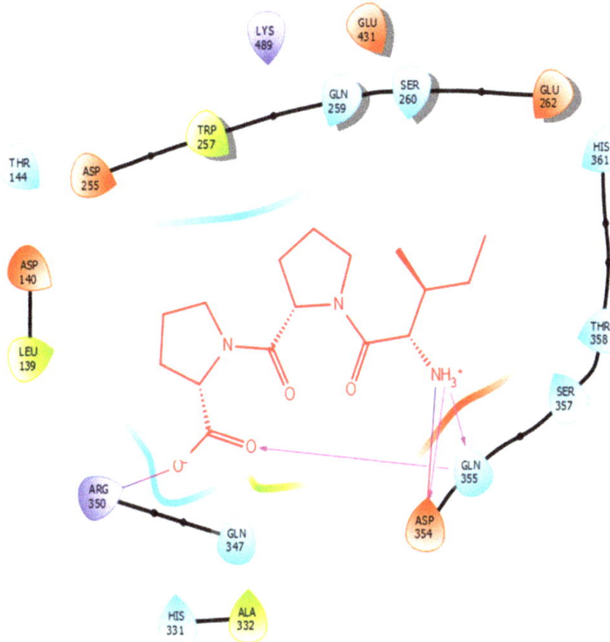

Figure 2. Docking results of the peptide IPP in the active site of human angiotensin I-converting enzyme (ACE). IPP is represented in red, interactions of human ACE residues with the peptide are indicated by arrows of different colours, with purple representing hydrogen bond interactions and blue arrows representing salt bridge interactions. The figure was generated using the software Maestro.

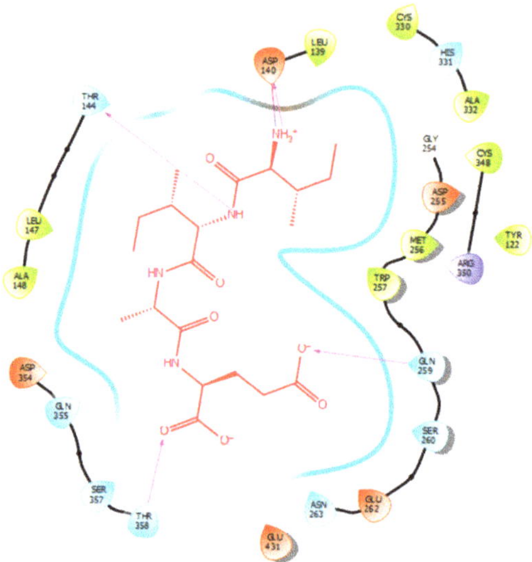

Figure 3. Docking results of the peptide IIAE in the active site human ACE. IIAE is represented in red, the interactions of human ACE residues with the peptide are indicated by arrows of different colours with purple representing hydrogen bond interactions, and blue arrows representing salt bridge interactions. The Figure was obtained using the software Maestro.

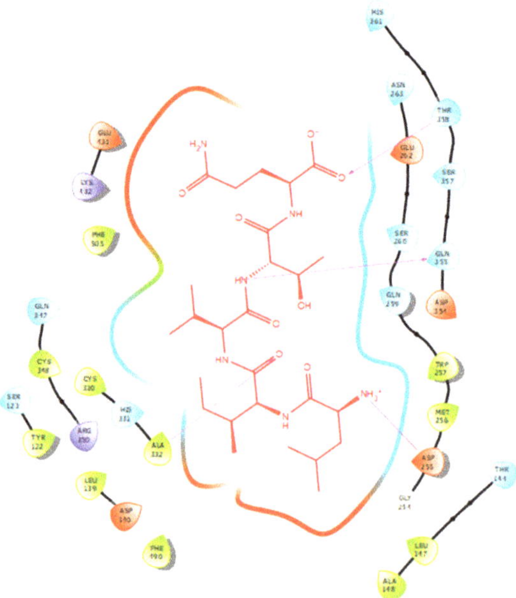

Figure 4. Docking results of the peptide LIVTQ in the human ACE active site. LIVTQ is represented in red, the interactions of human ACE residues with the peptide are indicated by arrows of different colours with purple representing hydrogen bond interactions, and blue arrows representing salt bridge interactions. The software Maestro was used for the generation of this figure.

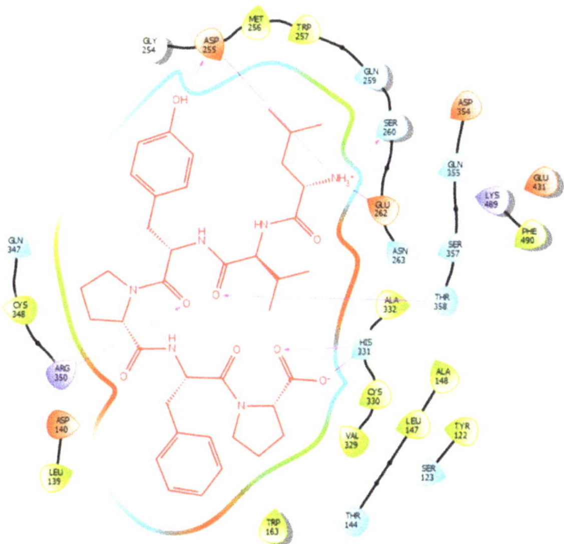

Figure 5. Docking results of the peptide LVYPFP in the human ACE active site. LVYPFP is represented in red, the interactions of human ACE residues with the peptide are indicated by arrows of different colours with purple representing hydrogen bond interactions, and blue arrows representing salt bridge interactions. The software Maestro was used for the generation of this figure.

Table 1. IPP docking results.

Protein 69FV	Ligand IPP		
Residue	Atom Name	Interaction Type	Distance (Å)
NH_2 Arg 350	O^- (Pro)	Salt bridge	2.86
OD2 Asp 354	NH^{3+} (Ile)	Hydrogen bond	1.86
OE1 Gln 355	NH^{3+} (Ile)	Hydrogen bond	1.81
NE2 Gln 355	O- (Pro)	Hydrogen bond	2.04

Table 2. IIAE docking results.

Protein 69FV	Ligand IIAE		
Residue	Atom Name	Interaction Type	Distance (Å)
OD2 Asp 140	NH (Ile)	Salt bridge	2.57
OG1 Thr 144	N (Ile)	Hydrogen bond	2.15
NE2 Gln 259	O (Glu)	Hydrogen bond	2.2
OG1 Thr 358	O (Glu)	Hydrogen bond	1.88

Table 3. LIVTQ docking results.

Protein 69FV	Ligand LIVTQ		
Residue	Atom Name	Interaction Type	Distance (Å)
OD2 Asp 255	NH3+ (Leu)	Salt bridge	4.79
N Ala 332	O (Ile)	Hydrogen bond	2.56
OE1 Gln 355	N (Thr)	Hydrogen bond	1.85
OG1 Thr 358	O- (Gln)	Hydrogen bond	1.88

Table 4. LVYPFP docking results.

Protein 69FV	Ligand LVYPFP		
Residue	**Atom Name**	**Interaction Type**	**Distance (Å)**
OD2 Asp 255	OH (Tyr)	Hydrogen bond	2.12
OG Ser 260	NH3+ (Leu)	Hydrogen bond	1.89
OE2 Glu 262	NH3+ (Leu)	Salt bridge	3.58
ND1 His 331	O (Pro)	Hydrogen bond	1.92
ND1 His 331	O (Pro)	Salt bridge	2.71
NH2 Arg 350	O (Tyr)	Hydrogen bond	2.61
OG1 Thr 358	O (Valine)	Hydrogen bond	1.75

3. Discussion

Hydrogen bonds interactions were demonstrated to play a crucial role in stabilizing the docked ligand complexes [48]. The distance of hydrogen bond interactions between the whey derived peptides and ACE amino acid residues typically were short (< 3.0Å; Tables 1–4), indicating that the peptides' binding affinity to ACE was strong [49]. In addition, these peptides formed a number of favorable salt bridge interactions with ACE residues, indicating that the ligands can pack tightly into the binding site and effectively inhibit ACE. Furthermore, it is interesting to note that hydrophobic amino acid residues such as proline, leucine, and isoleucine were mainly involved in establishing strong interactions with ACE, which goes in accordance with what is reported in SAR (Tables 1–4) [33].

Sampatrilat ((S, S, S)-*N*-{1-[2-carboxy-3-(N-mesyllysylamino) propyl]-1-cyclopentylcarbonyl} tyr-osine) (Figure 6) is a potent dual inhibitor of ACE and neutral endopeptidase. In the treatment of chronic heart failure, Sampatrilat could potentially provide a greater benefit than traditional ACE inhibitors [50,51]. In a recent study investigating the binding of Sampatrilat to the active site of ACE, the amino acid residues involved in the interactions with Sampatrilat were reported [47]. Interestingly, IIAE, LIVTQ, and LVYPFP interacted with three of these previously identified amino acid residues: IIAE interacted with residue Gln 259, LVYPFP interacted with residue His 331 and IIAE, LIVTQ, and LVYPFP interacted with residue Thr 358. Furthermore, previous studies stated that the ACE-inhibitory drugs Captopril, Lisinopril, and Enalapril interact with ACE amino acid residues Gln281, His353, Glu384, Lys511, His 513, and Tyr520 [52–54]. Apart from the amino acid residues in common, Lisinopril was reported to interact with Ala 354, Tyr 523, and Glu 162, and Enalapril to interact with Ala 354 and Tyr 523 [52,54]. According to the docking results, the peptides IIAE and LIVTQ interacted also with two of these residues: IIAE interacted with residue Asp 140 in common with Lisinopril, and LIVTQ interacted with Ala 332 in common with Lisinopril and Enalapril. (Amino acids residues are reported according to the Sampatrilat (PDB code 6F9V) amino acid sequence numbering, please see Table A1).

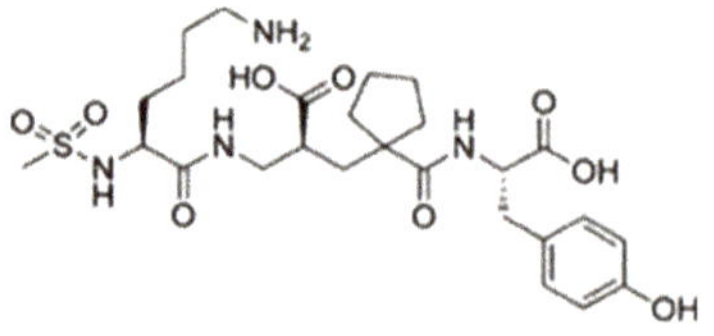

Figure 6. Chemical structure of Sampatrilat [47].

Overall, the docking results together with comparisons with the ACE inhibitory drugs provide strong evidence for the ACE inhibitory activity of IIAE, LIVTQ, and LVYPFP. In our previous work [31], IIAE, LIVTQ, and LVYPFP were identified as major peptides within fractions of high ACE inhibitory activity. Additionally, based on known structure-activity relationships, it was assumed that these were the main contributors to the ACE inhibitory activity measured. The docking results herein corroborate these assumptions and suggest that most probably these are potent ACE inhibitors that will contribute to the ACE inhibitory and antihypertensive activity in vivo. Further work will be needed, using pure synthesized peptides, to confirm ACE inhibition and activity in vivo.

4. Materials and Methods

4.1. Whey-Protein Derived Peptides

In our previous work where we characterized angiotensin-converting enzyme (ACE) inhibitory peptides produced by enzymatic hydrolysis of whey proteins [31], peptide sequences were identified as major peptides in fractions from the enzymatic hydrolysates CDP (casein-derived peptides) and β-lactoglobulin. The well-known antihypertensive peptide IPP, along with some other novel peptide sequences that have structural similarities with reported ACE inhibitory peptides, such as Leu-Val-Tyr-Pro-Phe-Pro (LVYPFP), Leu-Ile-Val-Thr-Gln (LIVTQ), and Ile-Ile-Ala-Glu (IIAE) were characterized and identified by a combination of chemical characterization (LC/MS; MS/MS) and SAR data. Their ACE inhibitory activity is summarized in Table 5; the IC_{50} is defined as the peptide concentration required to reduce the ACE activity by half.

Table 5. IC_{50} (µg/mL) value of the ACE-inhibitory peptide sequences.

Peptide	Protein Source	IC50 (µg/mL)	Reference
IPP	k-Casein	1.23	[23]
IIAE	β-Lg	128 *	[31]
LVYPFP	Casein	97	[55]
LIVTQ	β-Lg	113	[31]

* IC_{50} of β-Lg hydrolysate containing this peptide as one of the major peptides.

4.2. Homology between Human ACE and Rabbit ACE

EMBL-EBI (https://www.ebi.ac.uk/) was queried for human and rabbit ACE amino acid sequences, together with known three-dimensional protein structures. Reviewed sequences were selected and the protein sequence files were downloaded. The accession codes for the human ACE and the rabbit ACE used in this work are P12821 and P12822, respectively. The two sequences were then uploaded to Emboss Needle (https://www.ebi.ac.uk/Tools/psa/emboss_needle/) for multiple sequence alignment and comparison.

4.3. Molecular Docking

4.3.1. Docking Validation

In order to validate the accuracy and the reliability of the docking procedure to be used in this study, the original ligand (extracted from the coordinate files and taken from the Protein Data Bank; PDB code 6F9V was docked into the corresponding crystal structure of the receptor, using the automated docking procedure in the program Surflex-Dock (SFXC) [56], as provided by SYBYL-X2.1. The docked ligand mode and orientation from the docking procedure were compared to that found in the actual crystal structure of the complex using Pymol and PDBeFold [57,58]. Following the docking procedure, the root mean square deviation (RMSD) between the docked ligand and the ligand, as found in the crystal structure, was calculated. The success of the docking process depended on whether the value of RMSD between the real and best-scored docked conformations were within the 2 Å grid spacing, used in the docking procedure [59], and whether the molecular interactions were replicated. In this case, Sampatrilat was docked into the human ACE receptor as validation of the docking procedure.

4.3.2. Docking Procedure

Whey protein-derived peptides Ile-Pro-Pro (IPP), Leu-Ile-Val-Thr-Gln (LIVTQ), Ile-Ile-Ala-Glu (IIAE), and Leu-Val-Tyr-Pro-Phe-Pro (LVYPFP) were used as ligands in separate docking runs. Docking was performed using the docking algorithm Surflex-Dock, as provided in Sybyl-X 2.1. The X-ray crystallographic structure of sampatrilat-Asp in complex with Angiotensin-I-converting enzyme (PDB code 6F9V, 1.69 Å resolution) retrieved from the protein data bank (PDB) was chosen as the target protein for the docking studies, based on its high resolution structure co-crystallized with sampatrilat-Asp [47].

The Biopolymer Structure Preparation Tool, with the implemented default settings provided in the SYBYL programme suite, was used to prepare the protein structure for docking; hydrogens were added to the protein structure in idealised geometries, backbone and sidechains were repaired, residues were protonated, sidechain amides and sidechain bumps were fixed, stage minimization was performed, and all water and any ligand molecules were removed.

The three-dimensional (3D) structure of each ligand was constructed, using the "Build Protein" tool, as provided in Sybyl-X. Once constructed, charges were assigned to each atom of each molecule, using Merck Molecular Force Field (MMFF94) charges. Localized energy minimizations were then performed, and the final structure for each ligand in its lowest energy conformation was used for subsequent docking experiments. The resulting 3D coordinate files were converted to a MOL2 format for subsequent use in Surflex-Dock experiments, as provided in the SYBYL-X 2.0 software suite.

Surflex-Dock is a search algorithm that utilizes an empirically derived scoring function whose parameters are based on protein-ligand complexes of known affinities and structures. This method employs a "protomol", which is an idealized active site, as a target to generate presumed poses of molecules or molecular fragments. The protomol is employed as a mimic of the ideal interactions made by a perfect ligand to the active site of the protein. This molecular-similarity based alignment allows for optimization of potentially favorable molecular interactions, such as those defined by van der Waals forces and hydrogen bonds. In the present work, the protomol was defined by optimizing the threshold and bloat values to 0.5 and 0, respectively, to create a protomol that adequately described the binding pocket of interest. The extent of the protomol and its degree of coverage of an active site are controlled by these two parameters: the threshold value indicates the amount of buried-ness for the primary volume used to generate the protomol, and the bloat parameter determines the number of Ångstroms by which the search grid beyond that primary volume should be expanded. It is generally better to err on the side of a small protomol than on a protomol that is too large [60]. All parameters within the docking suite were left as the default values as established by the software [61,62]. Each peptide was then individually docked into the protomol site, using the "Docking Suite" application, as provided in the SYBYL programme suite. The docking results were visualised using the programme Maestro.

Molecular interactions, for the docking results, are reported according to the Sampatrilat (PDB code 6F9V) amino acid sequence numbering; for comparisons between different sequence numbering in studies referred here (See Appendix A (Table A1)). The software Maestro was used for the identification and characterisation of hydrogen bonds and salt-bridge interactions established between residues at the ACE active site and the peptides.

5. Conclusions

For the first time, reported herein, potential interactions between the naturally produced peptides from whey and ACE have been investigated, using a molecular docking approach. Peptides, IPP, IIAE, LIVTQ, and LVYPFP formed strong H bonds and salt bridge interactions with residues in the active site of human ACE. Moreover, a comparison with commercial ACE inhibitory drugs showed that the natural peptides interacted similarly to the drugs mimicking the same interactions with ACE active site residues. This study provides strong evidence for the ACE inhibitory activity of milk derived peptides, which have not been tested in vivo before. The results of this study, of novel milk derived whey peptides, could lead to the production of novel ACE inhibitors.

Author Contributions: Conceptualization, K.A.W. and P.J.; methodology, K.A.W. and P.J.; software, K.A.W.; validation, Y.C., K.A.W. and P.J.; formal analysis, Y.C., K.A.W. and P.J.; investigation, Y.C. and P.J.; resources, P.J.; data curation, Y.C.; writing—original draft preparation, Y.C.; writing—review and editing, K.A.W. and P.J.; visualization, Y.C.; supervision, K.A.W. and P.J.; project administration, P.J.; funding acquisition, P.J. All authors have read and agreed to the published version of the manuscript.

Funding: This research received no external funding.

Conflicts of Interest: The authors declare no conflict of interest.

Appendix A

Table A1. Comparisons between Captopril (PDB code 1O86) and Sampatrilat (PDB code 6F9V) sequence numbering.

Captopril (PDB Code 1O86) Amino Acid Sequence Numbering [52]	Sampatrilat (PDB Code 6F9V) Amino Acid Sequence Numbering [47]
Glu 162	Asp 140
Gln 281	Gln 259
His 353	His 331
Ala 354	Ala 332

References

1. World Health Organization. Cardiovascular Diseases (CVDs): Fact Sheet No.317. 2015. Available online: http://www.who.int/mediacentre/factsheets/fs317/en/ (accessed on 20 October 2016).
2. Celermajer, D.S.; Chow, C.K.; Marijon, E.; Anstey, N.M.; Woo, K.S. Cardiovascular disease in the developing world. *J. Am. Coll. Cardiol.* **2012**, *60*, 1207–1216. [CrossRef] [PubMed]
3. Cannon, C.P. Cardiovascular disease and modifiable cardiometabolic risk factors. *Clin. Cornerstone* **2008**, *9*, 24–41. [CrossRef]
4. Aluko, R.E. Antihypertensive Peptides from Food Proteins. *Annu. Rev. Food Sci. Technol.* **2015**, *6*, 235–262. [CrossRef] [PubMed]
5. Chen, Z.Y.; Peng, C.; Jiao, R.; Wong, Y.M.; Yang, N.; Huang, Y. Anti-hypertensive Nutraceuticals and Functional Foods. *J. Agric. Food Chem.* **2009**, *57*, 4485–4499. [CrossRef]
6. Donkor, O.N.; Henriksson, A.; Singh, T.K.; Vasiljevic, T.; Shah, N.P. ACE-inhibitory Activity of Probiotic Yoghurt. *Int. Dairy J.* **2007**, *17*, 1321–1331. [CrossRef]
7. Acharya, K.R.; Sturrock, E.D.; Riordan, J.F.; Ehlers, M.R. Ace revisited: A new target for structure-based drug design. *Nat. Rev. Drug Discov.* **2003**, *2*, 891. [CrossRef]
8. Li, G.H.; Le, G.W.; Shi, Y.H.; Shrestha, S. Angiotensin I–converting enzyme inhibitory peptides derived from food proteins and their physiological and pharmacological effects. *Nutr. Res.* **2004**, *24*, 469–486. [CrossRef]
9. Wei, L.; Alhenc-Gelas, F.; Corvol, P.; Clauser, E. The two homologous domains of human angiotensin I-converting enzyme are both catalytically active. *J. Biol. Chem.* **1991**, *266*, 9002–9008.
10. Sturrock, E.D.; Natesh, R.; Van Rooyen, J.M.; Acharya, K.R. Structure of angiotensin I-converting enzyme. *Cell. Mol. Life Sci.* **2004**, *61*, 2677–2686. [CrossRef]
11. Natesh, R.; Schwager, S.L.; Sturrock, E.D.; Acharya, K.R. Crystal structure of the human angiotensin-converting enzyme–lisinopril complex. *Nature* **2003**, *421*, 551. [CrossRef]
12. Tzakos, A.G.; Galanis, A.S.; Spyroulias, G.A.; Cordopatis, P.; Manessi-Zoupa, E.; Gerothanassis, I.P. Structure–function discrimination of the N-and C-catalytic domains of human angiotensin-converting enzyme: Implications for Cl–activation and peptide hydrolysis mechanisms. *Protein Eng.* **2003**, *16*, 993–1003. [CrossRef]
13. Korhonen, H.; Pihlanto, A. Technological options for the production of health-promoting proteins and peptides derived from milk and colostrum. *Curr. Pharm. Des.* **2007**, *13*, 829–843. [CrossRef]
14. FitzGerald, R.J.; Murray, B.A.; Walsh, D.J. Hypotensive peptides from milk proteins. *J. Nutr.* **2004**, *134*, 980S–988S. [CrossRef]
15. Miguel, M.; López-Fandino, R.; Ramos, M.; Aleixandre, A. Short-term effect of egg-white hydrolysate products on the arterial blood pressure of hypertensive rats. *Br. J. Nutr.* **2005**, *94*, 731–737. [CrossRef]

16. Beltrami, L.; Zingale, L.C.; Carugo, S.; Cicardi, M. Angiotensin-converting enzyme inhibitor-related angioedema: How to deal with it. *Expert Opin. Drug Saf.* **2006**, *5*, 643–649. [CrossRef]
17. Cicero, A.F.G.; Gerocarni, B.; Laghi, L.; Borghi, C. Blood pressure lowering effect of lactotripeptides assumed as functional foods: A meta-analysis of current available clinical trials. *J. Hum. Hypertens.* **2011**, *25*, 425. [CrossRef]
18. Fekete, A.A.; Givens, D.I.; Lovegrove, J.A. The impact of milk proteins and peptides on blood pressure and vascular function: A review of evidence from human intervention studies. *Nutr. Res. Rev.* **2013**, *26*, 177–190. [CrossRef]
19. Geleijnse, J.M.; Engberink, M.F. Lactopeptides and human blood pressure. *Curr. Opin. Lipidol.* **2010**, *21*, 58–63. [CrossRef]
20. Martínez-Maqueda, D.; Miralles, B.; Recio, I.; Hernández-Ledesma, B. Antihypertensive peptides from food proteins: A review. *Food Funct.* **2012**, *3*, 350–361. [CrossRef]
21. Meisel, H. Overview on milk protein-derived peptides. *Int. Dairy J.* **1998**, *8*, 363–373. [CrossRef]
22. Kitts, D.D.; Weiler, K. Bioactive proteins and peptides from food sources. Applications of bioprocesses used in isolation and recovery. *Curr. Pharm. Des.* **2003**, *9*, 1309–1323. [CrossRef] [PubMed]
23. Nakamura, Y.; Yamamoto, N.; Sakai, K.; Okubo, A.; Yamazaki, S.; Takano, T. Purification and characterization of angiotensin I-converting enzyme inhibitors from sour milk. *J. Dairy Sci.* **1995**, *78*, 777–783. [CrossRef]
24. Ehlers, P.I.; Nurmi, L.; Turpeinen, A.M.; Korpela, R.; Vapaatalo, H. Casein-derived tripeptide Ile–Pro–Pro improves angiotensin-(1–7)-and bradykinin-induced rat mesenteric artery relaxation. *Life Sci.* **2011**, *88*, 206–211. [CrossRef] [PubMed]
25. Sánchez-Rivera, L.; Martínez-Maqueda, D.; Cruz-Huerta, E.; Miralles, B.; Recio, I. Peptidomics for discovery, bioavailability and monitoring of dairy bioactive peptides. *Food Res. Int.* **2014**, *63*, 170–181. [CrossRef]
26. Majumder, K.; Wu, J. Molecular targets of antihypertensive peptides: Understanding the mechanisms of action based on the pathophysiology of hypertension. *Int. J. Mol. Sci.* **2014**, *16*, 256–283. [CrossRef] [PubMed]
27. Udenigwe, C.C.; Mohan, A. Mechanisms of food protein-derived antihypertensive peptides other than ACE inhibition. *J. Funct. Foods* **2014**, *8*, 45–52. [CrossRef]
28. Masclee, G.M.; Coloma, P.M.; Kuipers, E.J.; Sturkenboom, M.C. Increased risk of microscopic colitis with use of proton pump inhibitors and non-steroidal anti-inflammatory drugs. *Am. J. Gastroenterol.* **2015**, *110*, 749–759. [CrossRef]
29. Dallas, D.C.; Murray, N.M.; Gan, J. Proteolytic systems in milk: Perspectives on the evolutionary function within the mammary gland and the infant. *J. Mammary Gland Biol. Neoplasia* **2015**, *20*, 133–147. [CrossRef]
30. Meisel, H.; FitzGerald, R.J. Biofunctional peptides from milk proteins: Mineral binding and cytomodulatory effects. *Curr. Pharm. Des.* **2003**, *9*, 1289–1296.
31. Welderufael, F.T.; Gibson, T.; Methven, L.; Jauregi, P. Chemical characterisation and determination of sensory attributes of hydrolysates produced by enzymatic hydrolysis of whey proteins following a novel integrative process. *Food Chem.* **2012**, *134*, 1947–1958. [CrossRef]
32. Kobayashi, Y.; Yamauchi, T.; Katsuda, T.; Yamaji, H.; Katoh, S. Angiotensin-I converting enzyme (ACE) inhibitory mechanism of tripeptides containing aromatic residues. *J. Biosci. Bioeng.* **2008**, *106*, 310–312. [CrossRef] [PubMed]
33. Pan, D.; Guo, H.; Zhao, B.; Cao, J. The molecular mechanisms of interactions between bioactive peptides and angiotensin-converting enzyme. *Bioorganic Med. Chem. Lett.* **2011**, *21*, 3898–3904. [CrossRef] [PubMed]
34. Udenigwe, C.C. Bioinformatics approaches, prospects and challenges of food bioactive peptide research. *Trends Food Sci. Technol.* **2014**, *36*, 137–143. [CrossRef]
35. Tu, M.; Feng, L.; Wang, Z.; Qiao, M.; Shahidi, F.; Lu, W.; Du, M. Sequence analysis and molecular docking of antithrombotic peptides from casein hydrolysate by trypsin digestion. *J. Funct. Foods* **2017**, *32*, 313–323. [CrossRef]
36. García-Mora, P.; Martín-Martínez, M.; Bonache, M.A.; González-Múniz, R.; Peñas, E.; Frias, J.; Martinez-Villaluenga, C. Identification, functional gastrointestinal stability and molecular docking studies of lentil peptides with dual antioxidant and angiotensin I converting enzyme inhibitory activities. *Food Chem.* **2017**, *221*, 464–472. [CrossRef]
37. Guo, M.; Chen, X.; Wu, Y.; Zhang, L.; Huang, W.; Yuan, Y.; Wei, D. Angiotensin I-converting enzyme inhibitory peptides from Sipuncula (Phascolosoma esculenta): Purification, identification, molecular docking and antihypertensive effects on spontaneously hypertensive rats. *Process Biochem.* **2017**, *63*, 84–95. [CrossRef]
38. Sangsawad, P.; Choowongkomon, K.; Kitts, D.D.; Chen, X.M.; Li-Chan, E.C.; Yongsawatdigul, J. Transepithelial transport and structural changes of chicken angiotensin I-converting enzyme (ACE) inhibitory peptides through Caco-2 cell monolayers. *J. Funct. Foods* **2018**, *45*, 401–408. [CrossRef]

39. Shi, L.; Wu, T.; Sheng, N.; Yang, L.; Wang, Q.; Liu, R.; Wu, H. Characterization of angiotensin-I converting enzyme inhibiting peptide from Venerupis philippinarum with nano-liquid chromatography in combination with orbitrap mass spectrum detection and molecular docking. *J. Ocean Univ. China* **2017**, *16*, 473–478. [CrossRef]

40. Wang, X.; Chen, H.; Fu, X.; Li, S.; Wei, J. A novel antioxidant and ACE inhibitory peptide from rice bran protein: Biochemical characterization and molecular docking study. *Lwt-Food Sci. Technol.* **2017**, *75*, 93–99. [CrossRef]

41. Wu, Q.; Du, J.; Jia, J.; Kuang, C. Production of ACE inhibitory peptides from sweet sorghum grain protein using alcalase: Hydrolysis kinetic, purification and molecular docking study. *Food Chem.* **2016**, *199*, 140–149. [CrossRef]

42. Yu, Z.; Chen, Y.; Zhao, W.; Li, J.; Liu, J.; Chen, F. Identification and molecular docking study of novel angiotensin-converting enzyme inhibitory peptides from Salmo salar using in silico methods. *J. Sci. Food Agric.* **2018**. [CrossRef] [PubMed]

43. Ashok, N.R.; Aparna, H. Empirical and bioinformatic characterization of buffalo (Bubalus bubalis) colostrum whey peptides & their angiotensin I-converting enzyme inhibition. *Food Chem.* **2017**, *228*, 582–594. [PubMed]

44. Lin, K.; Zhang, L.W.; Han, X.; Cheng, D.Y. Novel angiotensin I-converting enzyme inhibitory peptides from protease hydrolysates of Qula casein: Quantitative structure-activity relationship modeling and molecular docking study. *J. Funct. Foods* **2017**, *32*, 266–277. [CrossRef]

45. Nongonierma, A.B.; FitzGerald, R.J. Strategies for the discovery and identification of food protein-derived biologically active peptides. *Trends Food Sci. Technol.* **2017**. [CrossRef]

46. Soubrier, F.; Alhenc-Gelas, F.; Hubert, C.; Allegrini, J.; John, M.; Tregear, G.; Corvol, P. Two putative active centers in human angiotensin I-converting enzyme revealed by molecular cloning. *Proc. Natl. Acad. Sci. USA* **1998**, *85*, 9386–9390. [CrossRef] [PubMed]

47. Cozier, G.E.; Schwager, S.L.; Sharma, R.K.; Chibale, K.; Sturrock, E.D.; Acharya, K.R. Crystal structures of sampatrilat and sampatrilat-Asp in complex with human ACE–a molecular basis for domain selectivity. *Febs J.* **2018**, *285*, 1477–1490. [CrossRef]

48. Tu, M.; Wang, C.; Chen, C.; Zhang, R.; Liu, H.; Lu, W.; Du, M. Identification of a novel ACE-inhibitory peptide from casein and evaluation of the inhibitory mechanisms. *Food Chem.* **2018**, *256*, 98–104. [CrossRef]

49. Ling, Y.; Sun, L.P.; Zhuang, Y.L. Preparation and identification of novel inhibitory Angiotensin-I-converting enzyme peptides from tilapia skin gelatin hydrolysates: Inhibition kinetics and molecular docking. *Food Funct.* **2018**, *9*, 5251–5259. [CrossRef]

50. Venn, R.F.; Barnard, G.; Kaye, B.; Macrae, P.V.; Saunders, K.C. Clinical analysis of sampatrilat, a combined renal endopeptidase and angiotensin-converting enzyme inhibitor: II: Assay in the plasma and urine of human volunteers by dissociation enhanced lanthanide fluorescence immunoassay (DELFIA). *J. Pharm. Biomed. Anal.* **1998**, *16*, 883–892. [CrossRef]

51. Wallis, E.J.; Ramsay, L.E.; Hettiarachchi, J. Combined inhibition of neutral endopeptidase and angiotensin-converting enzyme by sampatrilat in essential hypertension. *Clin. Pharmacol. Ther.* **1998**, *64*, 439–449. [CrossRef]

52. Natesh, R.; Schwager, S.L.; Evans, H.R.; Sturrock, E.D.; Acharya, K.R. Structural details on the binding of antihypertensive drugs captopril and enalaprilat to human testicular angiotensin I-converting enzyme. *Biochemistry* **2004**, *43*, 8718–8724. [CrossRef] [PubMed]

53. Vercruysse, L.; Van Camp, J.; Morel, N.; Rougé, P.; Herregods, G.; Smagghe, G. Ala-Val-Phe and Val-Phe: ACE inhibitory peptides derived from insect protein with antihypertensive activity in spontaneously hypertensive rats. *Peptides* **2010**, *31*, 482–488. [CrossRef] [PubMed]

54. Wang, Z.L.; Zhang, S.S.; Wei, W.A.N.G.; Feng, F.Q.; Shan, W.G. A novel angiotensin I converting enzyme inhibitory peptide from the milk casein: Virtual screening and docking studies. *Agric. Sci. China* **2011**, *10*, 463–467. [CrossRef]

55. Gonzalez-Gonzalez, C.; Gibson, T.; Jauregi, P. Novel probiotic-fermented milk with angiotensin I-converting enzyme inhibitory peptides produced by Bifidobacterium bifidum MF 20/5. *Int. J. Food Microbiol.* **2013**, *167*, 131–137. [CrossRef]

56. Jain, R.K. Molecular regulation of vessel maturation. *Nat. Med.* **2003**, *9*, 685. [CrossRef]

57. Schrodinger LLC. *Version 1.8, The PyMOL Molecular Graphics System*; Technical Report; Schrödinger LLC: New York, NY, USA, 2015.

58. PDBeFold - Structure Similarity, Embl-Ebi. 2019. Available online: http://www.ebi.ac.uk/msd-srv/ssm/cgi-bin/ssmserver (accessed on 2 December 2019).
59. Wang, R.; Lu, Y.; Wang, S. Comparative evaluation of 11 scoring functions for molecular docking. *J. Med. Chem.* **2003**, *46*, 2287–2303. [CrossRef]
60. Sharma, R.; Dhingra, N.; Patil, S. CoMFA, CoMSIA, HQSAR and molecular docking analysis of ionone-based chalcone derivatives as antiprostate cancer activity. *Indian J. Pharm. Sci.* **2016**, *78*, 54. [CrossRef]
61. Ai, Y.; Wang, S.T.; Sun, P.H.; Song, F.J. Combined 3D-QSAR modeling and molecular docking studies on Pyrrole-Indolin-2-ones as Aurora A Kinase inhibitors. *Int. J. Mol. Sci.* **2011**, *12*, 1605–1624. [CrossRef]
62. Lan, P.; Chen, W.N.; Chen, W.M. Molecular modeling studies on imidazo [4, 5-b] pyridine derivatives as Aurora A kinase inhibitors using 3D-QSAR and docking approaches. *Eur. J. Med. Chem.* **2011**, *46*, 77–94. [CrossRef]

 International Journal of
Molecular Sciences

Article

BIOPEP-UWM Database of Bioactive Peptides: Current Opportunities

Piotr Minkiewicz *, Anna Iwaniak and Małgorzata Darewicz

Chair of Food Biochemistry, University of Warmia and Mazury in Olsztyn, Plac Cieszyński 1,
10-726 Olsztyn-Kortowo, Poland; ami@uwm.edu.pl (A.I.); darewicz@uwm.edu.pl (M.D.)
* Correspondence: minkiew@uwm.edu.pl; Tel.: +48-89-523-37-15

Received: 25 October 2019; Accepted: 25 November 2019; Published: 27 November 2019

Abstract: The BIOPEP-UWM™ database of bioactive peptides (formerly BIOPEP) has recently become a popular tool in the research on bioactive peptides, especially on these derived from foods and being constituents of diets that prevent development of chronic diseases. The database is continuously updated and modified. The addition of new peptides and the introduction of new information about the existing ones (e.g., chemical codes and references to other databases) is in progress. New opportunities include the possibility of annotating peptides containing D-enantiomers of amino acids, batch processing option, converting amino acid sequences into SMILES code, new quantitative parameters characterizing the presence of bioactive fragments in protein sequences, and finding proteinases that release particular peptides.

Keywords: bioactive peptides; database; proteolysis; SMILES code; foods; nutrition; chronic diseases; nutraceuticals

---

1. Introduction

The BIOPEP-UWM database is freely-accessible without registration at the following website: http://www.uwm.edu.pl/biochemia/index.php/pl/biopep. Recently, bioinformatic databases and software represent basic tools in the research on biologically active peptides, e.g., those derived from food. Their role was described in several reviews [1–7]. The BIOPEP-UWM™ (formerly BIOPEP) database of bioactive peptides is one of these tools. It has been available on the internet since 2003. Its previous versions have been described in publications by Minkiewicz et al. [8] and Iwaniak et al. [9]. The database has recently been widely used in food and nutrition science as a source of information about peptides being in the focus of interest as putative components of functional foods involved in the prevention of chronic diseases [5,7,10]. Over 350 articles are available that describe results that had been obtained, verified, or interpreted with the help of the BIOPEP-UWM database of bioactive peptides (excluding these contributed by database curators). Links to the BIOPEP-UWM™ database are recently available via such websites as MetaComBio [11], LabWorm, and OmicX. Information about peptides from the database is integrated into the SpirPep [12] and FeptideDB [13] databases.

The BIOPEP-UWM™ database is continuously updated and modified. Several new options have been introduced since the publication of the last article describing it [9]. The aim of the present publication is to provide information helpful in work with the current version of the database and associated tools, including the use of new options introduced in the last three years.

2. Database Organization

The scheme of organization of the BIOPEP-UWM homepage is presented in Figure 1. The screenshot of the homepage is available in Supplementary Figure S1. Apart from a database of bioactive peptides described in this article, the BIOPEP-UWM contains databases of proteins, allergenic proteins, and their epitopes [14] as well as sensory peptides and amino acids [9]. The homepage also has a tab that allows users to submit new peptide sequences (not annotated yet in the database) or new activities (not annotated) of the existing peptides (See Supplementary Figure S2), and also a new BIOPEP-UWM news tab (not indicated in Figure 1).

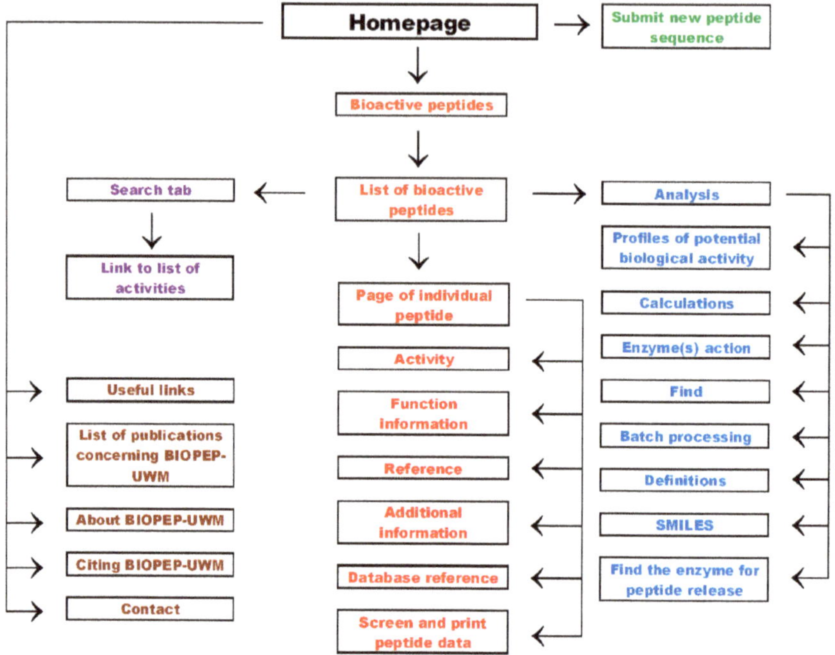

Figure 1. Scheme of organization of the BIOPEP-UWM database of bioactive peptides.

The "bioactive peptides" tab links with the list of bioactive peptides (Supplementary Figure S3). Access to more detailed information about a particular peptide sequence is available via the "peptide data" tab attributed to each peptide. The page with a peptide list contains links to associated tools enabling the processing of peptide and protein sequences (via the "analysis" tab). Scrolling down using the bar left from the table (Supplementary Figure S3) opens the window, which allows the input of queries, enabling a search.

3. Enlarging the Number of Peptides in the Database by BIOPEP-UWM™ Users

The BIOPEP-UWM database is a curated database. Although it is regularly enriched with the new peptides, it is rather impossible to insert all bioactive peptides that are continuously being found in the literature. Thus, the "submit new peptides" option (see the BIOPEP-UWM homepage; Supplementary Figure S1) enables users to send us a peptide sequence not found in our database so far. The peptide sequence to be added to BIOPEP-UWM has to be provided in a one-letter code by pasting it to the window that appears after clicking the "submit new peptides" tab. All peptides sent this way are verified by our curators and can be uploaded to the database on condition that the sender had provided e-mail and reference data (i.e., details of an article the peptide was published in). Providing the senders' address enables generating an automatic e-mail confirming that the peptide of interest was successfully submitted by the user to the BIOPEP-UWM database. Publication details are needed to verify the information sent. The lack of the sender's e-mail as well as reference data on peptide to be inserted to BIOPEP-UWM (mandatory fields for successful submission) makes the submitted information incomplete and may temporarily eliminate the sequence from the process of uploading it to our database.

4. Peptide Information

The current layout of peptide information in the BIOPEP-UWM database has earlier been used in the database of sensory peptides and amino acids [9]. Its implementation into the database of bioactive peptides is still in progress. Information about an example peptide with a GHS sequence (BIOPEP-UWM ID 9473) [15] is presented in Table 1. The screenshot of a peptide page is presented in Figure S4.

Table 1. Content of a page of a representative peptide.

ID	9473		
Name	ACE inhibitor		
Sequence	GHS		
InChIKey	LPCKHUXOGVNZRS-YUMQZZPRSA-N		
Function	Inhibitor of Angiotensin-Converting Enzyme (ACE) (EC 3.4.15.1) (MEROPS ID: M02-001)		
Number of Amino Acid Residues	3	**Activity Code**	ah
Activity	ACE inhibitor		
Chemical Mass	299.2740	**Monoisotopic Mass**	299.1110
IC$_{50}$	0.00 µM		
Bibliographic Data			
Authors	He R., Malomo S. A., Alashi A., Girgih A. T., Ju X., Aluko R. E.		
Title	Glycinyl-histidinyl-serine (GHS), a novel rapeseed protein-derived peptide, has a blood pressure-lowering effect in spontaneously hypertensive rats. J. Agric. Food Chem., 61, 8396-8402, 2013		
Year	2013	**Source**	Journal
Additional Information			
BIOPEP-UWM database of bioactive peptides SMILES: NCC(=O)N[C@@H](Cc1c[nH]cn1)C(=O)N[C@@]([H])(CO)C(=O)O InChI=1S/C11H17N5O5/c12-2-9(18)15-7(1-6-3-13-5-14-6)10(19)16-8(4-17)11(20)21/h3,5,7-8,17H,1-2,4,12H2, (H,13,14)(H,15,18)(H,16,19)(H,20,21)/t7-,8-/m0/s1 InChIKey: LPCKHUXOGVNZRS-YUMQZZPRSA-N Inhibitor of Renin (EC 3.4.23.15) (MEROPS ID: A01.007) according to the BIOPEP-UWM database of bioactive peptides (ID 9472)			
Database Reference			
AHTPDB: ID 1053, 2949 BioPepDB: ID biopep00354 BIOPEP-UWM database of bioactive peptides: ID 9472 SATPdb: ID satpdb13065			

The ID number is the first piece of information displayed on a peptide page. A peptide with a single activity annotated in the BIOPEP-UWM possesses one ID number. A peptide annotated as multifunctional possesses more ID numbers. The representative GHS peptide is annotated in the BIOPEP-UWM database twice, i.e., as an inhibitor of renin (EC 3.4.23.15) and angiotensin-converting enzyme (EC 3.4.15.1) [15] (ID 9472 and 9473, respectively). Database ID may serve as an unambiguous identifier of a compound, e.g., peptide. Examples of using ID numbers from peptide databases (e.g., BIOPEP-UWM) as peptide identifiers may be found in, e.g., recent publications of Skrzypczak et al. [16] and Khazaei et al. [17].

Name is the second piece of information on the page of an individual compound. Peptide names are often identical to their activity (e.g., ACE inhibitor). Some well-known peptides possess their own names, e.g., soybean lunasin (BIOPEP-UWM ID 9525 and 9526), the role of which has been reviewed by Hsieh et al. [18].

Peptide sequences are annotated in the BIOPEP-UWM database of bioactive peptides using a standard one-letter code describing 20 protein amino acids and their D-enantiomers (a recently added option). A peptide with the FhL sequence (L-Phe-D-His-L-Leu) [19] (BIOPEP-UWM ID 9475) may serve as an example of the peptide containing D-amino acid residue. The database offers an opportunity to annotate C-terminal amidation using the "~" symbol [8]. This symbol is, however, not universal. The EROP-Moscow database [20] uses the "z" symbol for the same purpose. There is also an opportunity (not exploited to date) to annotate phosphoserine using "B" and "b" symbols for L- and D-enantiomer, respectively. InChIKey is an unambiguous chemical identifier [21]. It always contains 27 characters and is sufficient for search via both search engines of chemical databases and common search engines such as Google^{TM}. InChIKey is used as a name in the case of some compounds annotated in the PubChem database [22].

Information about the biological activity is inserted as activity (short version), activity code (abbreviation of activity), and function (more detailed version). The current list of activities of peptides found in the BIOPEP-UWM database of bioactive peptides is provided in Table 2. The list of bioactivities has been rearranged as compared to this published in 2008 [8] to remove redundancy (e.g., remove synonymous or extremely rare activities). On the other hand, several new activities, especially these concerning inhibition of enzymes, have recently been added. Annotation of bioactive peptides as compounds interacting with individual enzymes is preferred by users of the BIOPEP-UWM database, as in the case of, e.g., renin inhibitors [23–25]. Information concerning the role particular enzymes play in metabolic pathways has recently become available in specialized databases [26].

The peptide entry page also provides the chemical (average) and monoisotopic molecular mass of the peptide and a reference describing its given activity.

Completion of the contents of "additional information" and "database references" tabs is in progress. The "additional information" tab includes peptide structure written using chemical codes called SMILES [27]—the most popular chemical code, and InChI—recommended by IUPAC [21]. These codes represent a typical language of cheminformatics (i.e., chemical informatics) [26]. Cheminformatics is considered as an emerging method in food science [28,29]. SMILES and InChI codes, as well as InChIKeys, are used as input data for the search of molecules in chemical databases [26,30]. The supplement to our previous review [31] may provide insights on how much information about peptide bioactivity is presented in chemical databases. InChIKey is sufficient to search via common search engines such as Google^{TM}. This option enables, e.g., finding peptides annotated in the BIOPEP-UWM database. There are many types of software that enable predicting the physicochemical and biological properties of chemical compounds and using, e.g., SMILES. This code may be converted into more than one hundred formats used in chemical informatics, for instance, by OpenBabel software [32]. Examples of using programs which require chemical codes as input data for in silico analysis and prediction of properties of food peptides have been recently presented by Ortiz-Martinez et al. [33], Mojica et al. [34], and Yu et al. [35]. Amino acid sequences are converted into SMILES code using applications available in the BIOPEP-UWM database via the "analysis" tab. Conversion of SMILES representations into InChI and InChIKeys is performed using OpenBabel or MarvinSketch software.

A peptide with a C-terminal amide group cannot be found in protein sequences. Precursors of these peptides, containing C-terminal glycine residues, are thus added. The mechanism of amidation includes the substitution of C-terminal glycine with an amide group [36]. A peptide with ID 2580 may serve as an example of this type of annotation. It is a precursor of antibacterial peptide [37] annotated as ID 2579. Information about amidation is provided in the "additional information" tab of a peptide, being a precursor of the amidated form (in the above example, peptide annotated as ID 2580).

The "additional information" tab also contains brief information about activities of the peptide taken from the BIOPEP-UWM database of bioactive peptides and other databases as well as information about peptide taste from the BIOPEP-UWM database of sensory peptides and amino acids [9].

Information about food resources and products, different values of IC_{50} are also included in the "additional information" tab for some of the peptides.

Table 2. List of activities of peptides annotated in the BIOPEP-UWM database of bioactive peptides.

Activity	Description [1]
ACE inhibitor [2]	Inhibitors of angiotensin-converting enzyme (ACE) (EC 3.4.15.1) (MEROPS ID: M02-001)
activating ubiquitin-mediated proteolysis	Peptides activating proteolysis mediated by ubiquitin
alpha-amylase inhibitor [2]	Inhibitors of α-amylase (EC 3.2.1.1)
alpha-glucosidase inhibitor [2]	Inhibitors of α-glucosidase (EC 3.2.1.20)
anorectic	Peptides causing a decrease in food intake and suppression of appetite.
antiamnestic	Inhibitors of prolyl oligopeptidase (EC 3.4.21.26) (MEROPS ID: S09.001). The enzyme catalyzes degradation of neuropeptides, e.g., involved in processes associated with memory.
antibacterial	Peptides revealing any action against bacteria
anticancer	Peptides revealing any action against cancers
antifungal	Peptides revealing any action against fungi
anti-inflammatory	Peptides reducing inflammation or swelling
antioxidative	Peptides inhibiting oxidation
antithrombotic	Inhibitors of blood coagulation. Inhibitors of thrombin (EC 3.4.21.5) (MEROPS ID: S01.217) are attributed to this activity.
antiviral	Peptides revealing any action against viruses. Inhibitors of viral enzymes are included.
bacterial permease ligand	Ligands of bacterial permeases
binding [2]	Peptides binding any biomolecules. Mineral binding peptides are also attributed to this activity.
CaMKII inhibitor [2]	Inhibitors of Ca^{2+}/calmodulin-dependent protein kinase (CaMKII) (EC 2.7.11.17)
CaMPDE inhibitor [2]	Inhibitors of 3′,5′-cyclic-nucleotide phosphodiesterase (Calmodulin-dependent phosphodiesterase 1—CaMPDE) (EC 3.1.4.17)
chemotactic	Peptides inducing chemotaxis, i.e. movement in response to a chemical stimulus
celiac toxic	Peptides toxic to people suffering from celiac disease
contracting	Peptides stimulating muscle contraction
dipeptidyl peptidase III inhibitor [2]	Inhibitors of dipeptidyl peptidase III (EC 3.4.14.4) (MEROPS ID M49.001)
dipeptidyl peptidase IV inhibitor [2]	Inhibitors of dipeptidyl peptidase IV (EC 3.4.14.5) (MEROPS ID S09.003)
embryotoxic	Peptides toxic to animal embryos
hemolytic	Peptides destroying red blood cells
heparin binding [2]	Heparin binding peptides
HMG-CoA reductase inhibitor [2]	Inhibitors of 3-hydroxy-3-methyl-glutaryl-coenzyme A reductase (HMG-CoA reductase) (EC 1.1.1.34)

Table 2. *Cont.*

Activity	Description [1]
hypotensive	Peptides causing blood pressure decrease
immunomodulating	Peptides modulating activity of the immune system
immunostimulating	Peptides stimulating activity of the immune system
inhibitor [2]	Peptides inhibiting various biological processes. Information about processes is provided on the pages of individual peptides.
membrane-active [2]	Peptides affecting transmembrane transport
natriuretic	Peptides inducing the excretion of sodium by kidneys (natriuresis)
neuropeptide	Peptides affecting activity of the nervous system
opioid	Ligands of opioid receptors
opioid agonist	Agonists of opioid receptors
opioid antagonist	Antagonists of opioid receptors
orphan receptor GPR14 agonist	Agonists of orphan receptor GPR14
Protein Kinase C inhibitor [2]	Inhibitors of protein kinase C (EC 2.7.11.13)
regulating	Peptides regulating various biological processes. Information about processes is provided on the pages of individual peptides.
renin inhibitor [2]	Inhibitors of renin (EC 3.4.23.15) (MEROPS ID A01.007)
stimulating	Peptides stimulating various biological processes. Information about processes is provided on the pages of individual peptides.
toxic [2]	Toxic peptides
vasoconstrictor	Peptides causing blood pressure increase

[1] More information concerning enzymes inhibited by peptides is available in the following databases: ExplorEnz [38], BRENDA [39], ChEMBL [40], and MEROPS [41]. Information about associations between abnormal enzyme activity and diseases may be found in the OpenTargets database [42]. [2] Activities absent in the version described in our publication from 2008 [8].

The "database reference" summarizes databases providing information about a given peptide (for example, see Table 1). The list of databases most commonly cited in the above tab is presented in Table 3. The list has been significantly enriched since the publication of our previous article describing the database [9]. ID numbers of peptides are also provided in particular databases. Some databases (such as ACToR [43] or ChemIDPlus [44]) use CAS registry numbers as compound identifiers. The databases are available via the MetaComBio website [11] or the "useful links" tab on the BIOPEP-UWM website. The list of databases cited has been significantly enlarged since 2016 (Table 3).

The last tab "screen and print peptide data" summarizes all data concerning a given peptide. Supplementary Table S1 is copied directly from the above tab. In the supplement to our previous publication [9], we have pointed out the opportunity for providing links to this tab from other resources. Examples of such links are available in the supplement to our review concerning taste-affecting peptides [31]. Here we offer the opportunity to construct links to peptide pages ("activity" tabs). The data of example peptide (ID 9473) can be found at the following address: http://www.uwm.edu.pl/ biochemia/biopep/peptide_data_page1.php?zm_ID=9473. ID at the end of the address (ID = 9473) may be replaced by another one to generate a link to another peptide data. An analogous link to a representative sensory peptide is as follows: http://www.uwm.edu.pl/biochemia/biopep/sensory_data_ page1.php?zm_ID=2.

Table 3. Databases cited on the "Database reference" page and other bioinformatic tools mentioned in the publication.

Database Name	Website [1]	Reference
ACToR [2]	https://actor.epa.gov/actor/home.xhtml	[43]
AHTPDB [2]	https://webs.iiitd.edu.in/raghava/ahtpdb/	[45]
APD	http://aps.unmc.edu/AP/main.html	[46]
BindingDB [2]	http://www.bindingdb.org/bind/index.jsp	[47]
BioPepDB	http://bis.zju.edu.cn/biopepdbr/	[48]
BitterDB [2]	http://bitterdb.agri.huji.ac.il/dbbitter.php	[49]
Brainpeps	http://brainpeps.ugent.be/	[50]
BRENDA [1]	https://www.brenda-enzymes.org/	[39]
CAMP	http://www.camp.bicnirrh.res.in/	[51]
CancerPPD	http://crdd.osdd.net/raghava/cancerppd/index.php	[52]
ChEBI [2]	https://www.ebi.ac.uk/chebi/	[53]
ChEMBL [2]	https://www.ebi.ac.uk/chembl/	[40]
ChemIDplus [2]	https://chem.nlm.nih.gov/chemidplus/chemidlite.jsp	[44]
ChemSpider [2]	http://www.chemspider.com/Default.aspx	[54]
CompTox	https://comptox.epa.gov/dashboard	[55]
CutDB	http://cutdb.burnham.org	[56]
DBAASP	https://dbaasp.org/	[57]
Dendrimer Builder	http://dendrimerbuilder.gdb.tools/	*
DrugBank [2]	https://www.drugbank.ca/	[58]
EROP-Moscow [2]	http://erop.inbi.ras.ru/	[20]
ExplorEnz	https://www.enzyme-database.org/index.php	[38]
FeptideDB	http://www4g.biotec.or.th/FeptideDB/index.php	[13]
FooDB [2]	http://foodb.ca/	*
Hemolytik	http://crdd.osdd.net/raghava/hemolytik/	[59]
HMDB [2]	http://www.hmdb.ca/	[60]
J-Global	https://jglobal.jst.go.jp/en/	*
KEGG [2]	https://www.genome.jp/kegg/	[61]
LabWorm	https://labworm.com/	*
MarvinSketch	https://chemaxon.com/products	*
MBPDB	http://mbpdb.nws.oregonstate.edu/	[62]
MEROPS [2]	https://www.ebi.ac.uk/merops/	[41]
MetaboLights	https://www.ebi.ac.uk/metabolights/index	[63]
MetaComBio [2]	http://www.uwm.edu.pl/metachemibio/index.php/about-metacombio	[11]
MilkAMP	http://milkampdb.org/home.php	[64]
NANPDB	http://african-compounds.org/nanpdb/	[65]
NeuroPep	http://isyslab.info/NeuroPep/	[66]
omicX	https://omictools.com/	*
OpenBabel [2]	http://openbabel.org/wiki/Main_Page	[32]
OpenTargets	https://www.targetvalidation.org/	[42]
PepBank [2]	http://pepbank.mgh.harvard.edu/	[67]
PeptideDB	http://www.peptides.be/	[68]
ProPepper	https://propepper.net/	[69]
PubChem [2]	https://pubchem.ncbi.nlm.nih.gov/	[22]
SATPdb [2]	http://crdd.osdd.net/raghava/satpdb/links.php	[70]
SpirPep	http://spirpepapp.sbi.kmutt.ac.th/SpirPep/Home	[12]
SureChEMBL	https://www.surechembl.org/search/	[71]
SwissSidechain	https://swisssidechain.ch/	[72]
ZINC [2]	http://zinc.docking.org/	[73]

[1] Accessed in July and August 2019. [2] Tools cited in our previous publication [9]. * No reference available.

5. Search Options

Search options are summarized in Table 4 and Supplementary Figure S5.

Table 4. List of search options available in the BIOPEP-UWM database of bioactive peptides. Options described in this table have been announced in [30].

Search Option	Output	
	Version without Exact Search	**Version with Exact Search** [1]
ID	Peptide with given ID	
Name	List of all peptides with the name containing the given word (words)	Peptide with the given name (may appear more than once if it is annotated with more activities)
Activity	Complete list of peptides with all activities named using the given word (e.g., inhibitor)	List of all peptides with the given activity
Mass	List of all peptides having molecular masses within the given range (e.g., 500–600)	
Reference	List of all peptides described in articles published by the given author (or authors with the same second name)	
Sequence	List of all peptides with sequences containing the given fragment	Peptide with the given sequence (may appear more than once if it is annotated with more activities). [2]
Number of amino acid residues	List of all peptides containing the given number of amino acid residues (e.g., 3)	
InChIKey [1]	Peptide with the given InChIKey. Peptide exhibiting more than one activity annotated in the BIOPEP-UWM will appear more than once [2]	

[1] New search options. [2] These options give equivalent search results.

Search options available in the BIOPEP-UWM database of bioactive peptides fall into the following major categories: text-based (ID, name, activity, reference, and InChIKey), structure-based (sequence-based), and property-based (number of amino acid residues and molecular mass). They are typical of peptide databases [4]. The use of an ID number as a query is the first search option. A single ID number corresponds to a single peptide with one defined activity. Search by name or by activity offers two possibilities to the user: finding all names or all activities including the chosen word or text fragment or exact search (see Supplementary Figure S5). The first opportunity leads to finding more peptides that fulfill the search criterion. Using the word "hemorphin-7" as a query, we can find four peptides (ID 2570, 2973, 3079, and 9001) without using the exact search option and only one (ID 3079) using the exact search option (search performed on 30 August 2019).

The search menu contains a link to the list of activities (Supplementary Figure S5), which serve for a query choice. In contrast to Table 2, the bioactivities are listed in the chronological (not alphabetical) order. Again, it is possible to use the exact search option. Using the word "inhibitor" as a query without using the exact search option has given a list of 1552 peptides as an output (30 August 2019). The list contains all inhibitors of enzymes (e.g., ACE, dipeptidyl peptidase IV, and dipeptidyl peptidase III). The exact search option with the same query found only 67 peptides with the activity annotated as "inhibitor" (see Table 2).

InChIKey is the most typical identifier of compounds (e.g., peptides) in chemical databases (e.g., PubChem [22]; ChemSpider [54], and ChEMBL [40]). Although it is a unique identifier of any chemical compound, it does not provide information about its structure [21]. InChIKeys in the BIOPEP-UWM database correspond to linear peptides with all chirality centers defined, acidic and basic groups electrically neutral, and cysteine residues reduced (if any in the peptide sequence). Incomplete InChIKey used as a query may result in finding more peptides. For instance, a "DYKIIFRCSA-N" fragment occurs in three InChIKeys corresponding to the celiac toxic peptide with the sequence PSQQQP (ID 2578), ACE inhibitor GPAGAPGAA (ID 3363), and antibacterial peptide ALCSEK (ID 4011). These peptides have no common fragments (subsequences). The use of incomplete InChIKey with the exact search option will fail to produce any results.

The sequence-based search is the most common and most intuitive option used to find peptide information in the database [4]. The BIOPEP-UWM database offers an opportunity to find all longer sequences containing a query fragment and to find a given sequence (exact). The first opportunity allows user to find peptides containing a defined continuous motif, e.g., attributed to the given function [74,75]. This search option also follows the fragmentomics concept [76]. It assumes that shorter (functional) bioactive subsequences present in a sequence may be crucial for the biological activity of the entire peptide molecule (peptide). Examples of peptides inscribing into this concept may be found in the BIOPEP-UWM (e.g., hemorphins or ACE inhibitors from caseins) and in other peptide databases such as EROP-Moscow [20], PepBank [67], SATPdb [70] or AHTPDB [45]. The exact search option is sufficient to check the bioactivity of peptides identified among protein hydrolysis products. An example of such an experiment has recently been described by Martini et al. [77] and Garcia-Vaquero et al. [78].

In the case of the property-based search (involving the number of amino acid residues or molecular mass range), choosing the exact search option does not change the output. We generally recommend using the exact search option for the sequence-based search.

6. Analysis

The "analysis" page includes the following tabs: "profiles of potential biological activity", "calculations", "enzyme(s) action", "find", "batch processing", "definitions", "SMILES", and "find the enzyme for peptide release" (Supplementary Figure S6).

The profile of a potential biological activity is defined as the type and location of bioactive fragments in a protein or a peptide chain [79]. This idea is based on the assumption that the same bioactive fragment, especially a short one (2–3 amino acid residues), cannot be attributed to a given protein, but may be present in many sequences (many form the so-called common subsequences) [75,79]. The concept of profiles of the potential activity of peptide fragments is consistent with the fragmentomic approach proposed by Zamyatnin [76] (see above). The profiles of potential biological activity of proteins can be obtained using the asterisk by default. Examples of published profiles of the potential activity of peptide or protein fragments may be found in publications of Bauchart et al. [80], Huang et al. [81], Tapal et al. [82], Khazaei et al. [17], and Jakubczyk et al. [83]. The profile may also be constructed for the specific bioactivity (bioactivity of interest) when selecting the activity instead of an asterisk from a toolbar. The menu to be used for the construction of potential biological activity profiles is shown in Supplementary Figures S7–S9. The profile of a potential biological activity of a protein or a peptide sequence is presented as a table including the following columns: ID, name of peptide, activity, number of repetitions of a particular bioactive fragment in a query sequence, sequence of the bioactive fragment, and location of the bioactive fragment in a query sequence. An example of the above profile is presented in Supplementary Table S2.

The "calculations" tab enables calculating two quantitative parameters that characterize proteins as potential precursors of bioactive peptides: the frequency of bioactive fragments occurrence in a protein sequence (A) and a potential biological activity of protein fragments (B). Equations 1 and 2 enabling calculation of the above parameters are provided in Table 5. The menu of the "calculations" tab is shown in Supplementary Figure S10. An example of the output is presented in Supplementary Table S3. The frequency of bioactive fragments occurrence in a protein sequence (A) is calculated for all bioactive peptides present in the query sequence (using the asterisk) or for one specific peptide (by choosing the bioactivity from a toolbar). Potential biological activity of protein fragments (B) may be calculated only if peptide IC_{50} or EC_{50} is available. The program skips peptides without known IC_{50} or EC_{50} value. For instance, Supplementary Table S3 provides B values for ACE and DPPIV inhibitors only. In the case of other activities, B values have not been calculated due to the lack of IC_{50} or EC_{50} attributed to particular peptides. Articles published by Udenigwe et al. [84] and Lin et al. [85] contain representative results of calculations of quantitative parameters characterizing food proteins as potential precursors of bioactive peptides.

Table 5. Quantitative parameters characterizing proteins as potential precursors of bioactive peptides, available in the BIOPEP-UWM database.

Equation No.	Parameter	Reference
1. [1]	The frequency of bioactive fragments occurrence in a protein sequence (A) $A = a/N$ a—the number of fragments with a given activity, N—the number of amino acid residues	[86]
2. [1]	Potential biological activity of protein fragments (B) [μM^{-1}] $B = [\Sigma(a_i/EC_{50i})]/N$ or $B = [\Sigma(a_i/IC_{50i})]/N$ a_i—the number of repetitions of i-th bioactive fragment in a protein sequence, EC_{50i}—the concentration of i-th bioactive peptide corresponding to its half-maximal activity [μM], IC_{50i}—the concentration of i-th bioactive peptide corresponding to half-maximal inhibition [μM], N—the number of amino acid residues	[86]
3. [2]	The frequency of release of fragments with a given activity by selected enzymes (A_E) $A_E = d/N$ d—the number of peptides with a given activity (e.g., ACE inhibitors) released by a given enzyme (e.g., trypsin) N—the number of amino acid residues in protein	[87]

Table 5. *Cont.*

Equation No.	Parameter	Reference
4. [2]	The relative frequency of release of fragments with a given activity by selected enzymes (W) $W = A_E/A$ A_E—the frequency of release of fragments with a given activity by selected enzymes (from Equation (3)) A—the frequency of bioactive fragments occurrence in a protein sequence (from Equation (1))	[87]
5. [2]	Activity of fragments potentially released by proteolytic enzyme (enzymes) (B_E) $B_E = [\Sigma(d_j/EC_{50j})]/N$ or $B_E = [\Sigma(d_j/IC_{50j})]/N$ d_j—the number of repetitions of j-th bioactive fragment released by a given enzyme (enzymes) from a protein sequence, EC_{50j}—the concentration of j-th bioactive peptide corresponding to its half-maximal activity [μM], IC_{50j}—the concentration of j-th bioactive peptide corresponding to half-maximal inhibition [μM], N—the number of amino acid residues in a protein chain	*
6. [2]	Relative activity of fragments potentially released by proteolytic enzyme (enzymes) (V) $V = B_E/B$ B_E—activity of fragments potentially released by proteolytic enzyme (enzymes) (from Equation (5)) B—potential biological activity of protein fragments (from Equation (2))	*
7. [2]	Theoretical degree of hydrolysis (DH_T) $DH_T = d/D \times 100\%$ d—number of hydrolyzed peptide bonds in a protein/peptide chain D—total number of peptide bonds in a protein/peptide chain	[88]
8. [3]	The number of repetitions of the bioactive fragment in all sequences of the protein/peptide set analyzed (a_T) $a_T = a_1 + a_2 + \ldots + a_L$ a_1–a_L—the number of repetitions of a given bioactive fragment in particular sequences in the dataset submitted for analysis L—the number of sequences in the protein/peptide set analyzed	*
9. [3]	The number of repetitions of a given fragment in all sequences of the selected protein/peptide fraction (a_S) $a_S = a_T/L$ a_T—the number of repetitions of the bioactive fragment in all sequences of the protein/peptide set analyzed (from Equation (8)) L—the number of sequences in the protein/peptide set analyzed	*
10. [3]	The mean frequency of the occurrence of a single fragment in a sequence of protein/peptide classified to a given group (A_S) $A_S = a_T/N_T$ a_T—the number of repetitions of the bioactive fragment in all sequences of the protein/peptide set analyzed N_T—the total number of amino acid residues in all protein/peptide sequences belonging to the set (from Equation 10)	*
11. [4]	The total number of amino acid residues in all protein/peptide sequences belonging to the set (N_T) $N_T = N_1 + N_2 + \ldots + N_L$ N—the number of amino acid residues in a single protein/peptide chain L—the number of protein/peptide chains in the set	*
12. [3]	The number of cases of release of the bioactive fragment from all sequences of the protein/peptide set analyzed (a_{TE}) $a_{TE} = a_{1E} + a_{2E} + \ldots + a_{LE}$ a_{1E}–a_{LE}—the number of cases of release of the bioactive fragment from particular sequences of the protein/peptide set analyzed L—the number of protein/peptide chains in the set	*

Table 5. *Cont.*

Equation No.	Parameter	Reference
13. [3]	Mean number of cases of predicted release of a single fragment by a selected enzyme from the chain of protein/peptide belonging to the set analyzed (a_{SE}) $a_{SE} = a_{TE}/L$ a_{TE}—the number of cases of release of the bioactive fragment from all sequences of the protein/peptide set analyzed L—the number of protein/peptide chains in the set	*
14. [3]	Predicted frequency of release of a single peptide by proteolytic enzyme from the set of protein/peptide sequences analyzed (A_{SE}) $A_{TE} = a_{TE}/N_T$ a_T—the number of cases of release of the bioactive fragment from all sequences of the protein/peptide set analyzed N_T—the total number of amino acid residues in all protein/peptide sequences belonging to the set (from Equation (10))	*

[1] available via the "profiles" tab and "batch processing" tab. [2] available via the "enzyme (s) action" tab and "Batch processing" tab. [3] available via the "batch processing" tab only. [4] not displayed among the results. Shown only to explain the calculation of other parameters. * New parameters described for the first time in this publication. Some of them have been announced in [4].

The "Enzyme(s) action" tab allows simulating proteolysis catalyzed by endopeptidases. The scheme of steps required to obtain the peptides potentially released by a given enzyme (or enzymes) is presented in Figure 2. Screenshots of menus of particular tabs are presented in Supplementary Figures S11–S16. The menu also enables enzyme choice (Supplementary Figures S12 and S13). It allows the simulation of proteolysis using one to three enzymes. Example information about a single enzyme (plasmin; EC 3.4.21.7; MEROPS ID: S01.233) is presented in Supplementary Figure S14. The enzyme is annotated using a connection ID, indicating a single peptide bond hydrolyzed by the enzyme and enzyme ID. One enzyme may cover few connection IDs (in the case of plasmin—two). Enzyme specificity is described using two terms: a recognition sequence understood as a fragment of an amino acid sequence recognized by the proteolytic enzyme and a cutting sequence understood as an amino acid residue preceding or following the bond hydrolyzed by protease [8]. The recognition sequence may contain a single amino acid residue (e.g., for plasmin) or a longer fragment such as for a ginger protease—zingipain (EC 3.4.22.67; MEROPS ID: C01.017). Annotations "C-terminus" and "N-terminus" indicate bonds formed by a carboxyl and amine group of an amino acid residue, respectively, hydrolyzed by the enzyme. Data concerning particular enzymes contain references: databases such as MEROPS [41] and CutDB [56] or publications (Bastian and Brown [89] for plasmin and Huang et al. [90] for zingipain). Apart from the addition of new enzymes, the specificity has recently been modified for some of the existing ones. The modification included the addition of new recognition sequences and cutting sequences (possessing connection IDs within the range 141–184), according to data presented in the so-called specificity matrices in the MEROPS database. These matrices are continuously updated to follow newly appearing information about new sites susceptible to proteolysis in protein sequences [41]. Proteolysis simulation is simplified. It assumes that all bonds theoretically susceptible to a given proteinase are hydrolyzed. In real experiments, the proteolysis is often incomplete. This finding may explain false-positive results, i.e., lack of expected peptides. False-negative results may be explained by incomplete knowledge about proteolytic specificity, i.e., the situation when some bonds susceptible to the proteolytic enzyme are considered resistant. The addition of new recognition and cutting sequences to the enzyme data aims to minimize the occurrence of false-negative results.

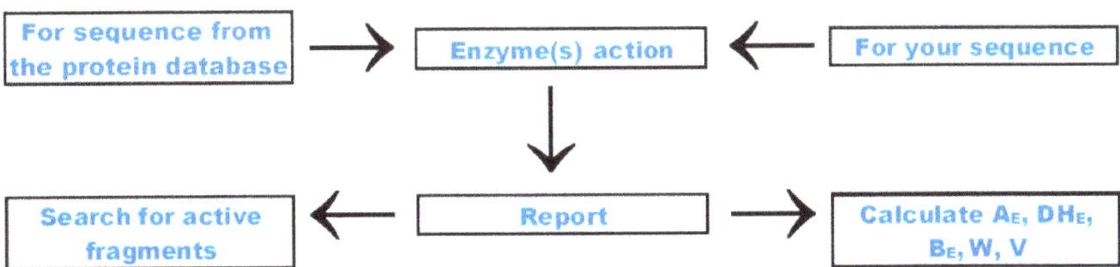

Figure 2. Scheme of the "enzyme(s) action" tab. Option (see Figure 1) "search for enzyme with given specificity" is not included in the Figure. A screenshot of the menu of this tab is presented in Supplementary Figure S11.

Results of simulated proteolysis of an example peptide can be found in Supplementary Figure S15. Displayed results of the initial step of simulation include sequences of peptides being products of proteolysis and their location in the precursor sequence. The next step may include the search for bioactive peptides among products of simulated proteolysis or calculation of quantitative parameters characterizing the proteolysis (Figure 2 and Figure S15). The parameters available via the "enzyme(s) action" tab are calculated according to Equations (3)–(7) from Table 5. Representative results of the search for active peptides among simulated proteolysis products and calculation of quantitative parameters are presented in the Supplement (Tables S4 and S5, respectively). Calculation of parameters B_E and V involves EC_{50} or IC_{50} values. If they are not available, peptides are not taken into account. Simulation of proteolysis using the BIOPEP-UWM database has recently been described by, e.g., Lin et al. [85], Yu D. et al. [91], and Kandemir-Cavas et al. [92]. Data concerning proteolysis simulation may be interpreted together with protein structures [93].

A new tab named "search for enzymes with given specificity" enables the search for information about an enzyme using recognition sequence, cutting sequence, and choice between C- and N-terminus (bond formed by carboxyl or amine group of amino acid residue, respectively). Results include the list of enzymes with a given specificity. An example of a query and result produced using the above option may be found in Supplementary Figure S16 and Supplementary Table S6. For most of the enzymes, the recognition sequence contains only one amino acid residue.

The content of the new "find" tab enables quickly finding of some information in protein and bioactive peptide databases. Particular tabs enable display of a full list of protein sequences annotated in the BIOPEP-UWM database, a full list of peptides revealing a given activity, and a list of all proteins or peptides containing the query sequence (see Supplementary Figure S17). The last option enables finding all proteins or peptides containing a given bioactive fragment or a recognition sequence available for the proteolytic enzyme. An example result of a search for a VPP sequence in the database of bioactive peptides is presented in Table S7 (Supplement). Results cover links to peptide or protein data, ID number, name, and sequence.

Another new "batch processing" option serves for the simultaneous processing of a set of few sequences of proteins or peptides being potential precursors of bioactive peptides. The total length of all sequences forming the query set may be up to c.a. 1500 amino acid residues. The scheme of activities available via this option is presented in Figure 3. The screenshot of the input window is available in Supplementary Figure S18. The FASTA format [94] is used to input a set of sequences. The "batch processing" option enables performing any action available via the tabs: "profiles of potential biological activity", "calculations", and "enzyme(s) action". Moreover, there are new parameters characterizing the occurrence and possibility of enzymatic release of an individual peptide from few precursor sequences (a_T, a_S, A_S, a_{TE}, a_{SE}, A_{TE}) calculated according to Equations (8)–(10) and (12)–(14) in Table 5. Distribution of particular fragments in the set of sequences may be in the focus of scientific interests when using in silico methodologies [76,77,95]. Analysis may cover all possible or selected options. Supplementary Figure S18 shows a set of sequences ready for an analysis concerning bioactive peptides (excluding options concerning data from the database of allergenic proteins). The batch analysis is performed in two steps (Figure 3). The first step may be performed for all activities (default option) or a selected one. The second step may be performed after the first one had been completed. The parameters may be calculated for all bioactive fragments found in the set of sequences or for manually selected peptides. Results of the first and the second step of batch analysis are presented in Supplementary Tables S8 and S9, respectively.

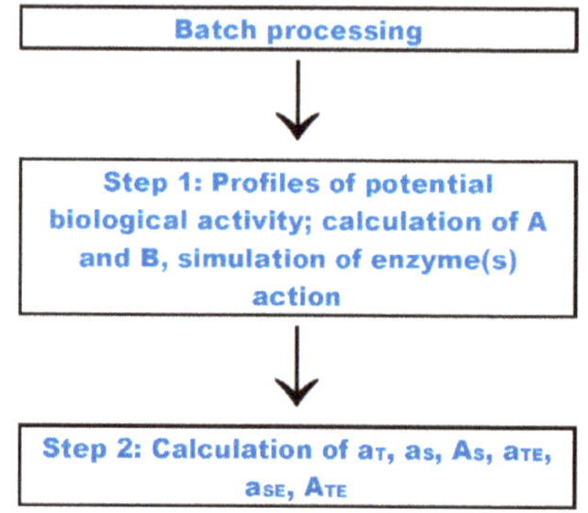

Figure 3. Scheme of the "batch processing" tab action.

The "definitions" tab summarizes terms and definitions used in the BIOPEP-UWM database including equations used to calculate quantitative parameters, as shown in Table 5.

The "SMILES" tab, introduced in 2018, enables translating amino acid sequences (written using standard one-letter code) into the chemical language "SMILES". SMILES representations are built according to a simplified algorithm described by Siani et al. [96]. SMILES codes of particular amino acid residues are written using the same layout as used in the SwissSidechain database [72] and source codes of the CycloPs program [97] (program temporarily unavailable). The procedure was tested and verified according to recommendations proposed in our previous publication [98]. The MarvinSketch 17.28 software (ChemAxon, Budapest, Hungary) was used to test and verify SMILES strings of peptides. The application utilizes the sequences of peptides built from 20 protein amino acids, their D-enantiomers, L- and D-phosphoserine (Symbols B and b, respectively), and C-terminal amide group. It is easy and fast in use and can process linear peptides only. Disulfide bonds and other modifications may be inserted using molecule editors (e.g., MarvinSketch, Dendrimer Builder program provided by the University of Bern, Switzerland, and molecule editor of the NANPDB [65] database) which may serve as alternatives to our application. The first one may be used to construct any molecules from building blocks drawn or imported as SMILES strings, the second, to build representations of branched peptides containing some non-protein amino acids, whereas the third to encode pyrrolysine and selenocysteine apart from 20 most common protein amino acids. Our application converts amino acid sequences into the so-called aromatic SMILES. Some search engines do not utilize this version [30]. The aromatic version of the SMILES string may be converted into an alternative, so-called Kekule version using, e.g., the molecule editor of the PubChem database [99] or MarvinSketch software. Screenshots of the "SMILES" tab window with query and result are given in the Supplementary Figures S19 and S20, respectively. Two types of SMILES representations of the example peptide may be found in Supplementary Table S10.

The way of understanding the output information when using the new tab entitled "find the enzyme for peptide release" is summarized in Figure 4. The screenshot of the menu of this tab and representative results are shown in the Supplementary Figure S21 and Supplementary Table S11, respectively. The input includes peptide sequences provided in FASTA format and the precursor (protein or peptide) sequence. The output includes a list of all enzymes with the specificity sufficient to catalyze particular proteolytic events. A proteolytic event is understood as a case of cleavage of an individual peptide bond. This term has been introduced in the CutDB database [56]. Release of a peptide from the precursor sequence requires two proteolytic events: cleavage of bond preceding N- and following C-terminus (indicated in Supplementary Table S11 as N and C, respectively). If a given peptide appears in the precursor sequence more than once, then the particular events attributed to this peptide are indicated as 1N, 1C, 2N, 2C, and so forth. An example peptide with the AP sequence occurs in the precursor sequence RWAFAPGFAPGHIP twice (positions 5–6, 9–10). Its release is associated with four proteolytic events: 1N—cleavage of the bond between the residues 4 and 5, 1C—cleavage of the bond between the residues 6 and 7, 2N—cleavage of the bond between the residues 8 and 9, and 2C—cleavage of the bond between the residues 10 and 11. Displayed results concerning enzyme catalyzing the particular proteolytic events cover the following data: name, EC number, enzyme ID in the BIOPEP-UWM database, connection ID, cutting sequence, and recognition sequence. The cutting sequences are described using the symbols "+" and "−" assigned to the amino acid symbols. Symbol "+" means that the amino acid residue follows the cleaved bond, i.e., this bond is formed by the amine group. For example, the symbol "A+" means that the cleaved bond is formed by the amine group of alanine. The symbol "−" means that the amino acid residue is located before the cleaved bond, i.e., this bond is formed by the carboxyl group of the amino acid. For instance, the symbol "W-" means a bond formed by the carboxyl group of tryptophan. Enzymes releasing N- and C-terminus are summarized separately. This solution may be justified by the fact that peptides may be released by more than one enzyme (N- and C- terminus are not released by the same enzyme). This process can be exemplified by protein digestion in the human gastrointestinal tract [100].

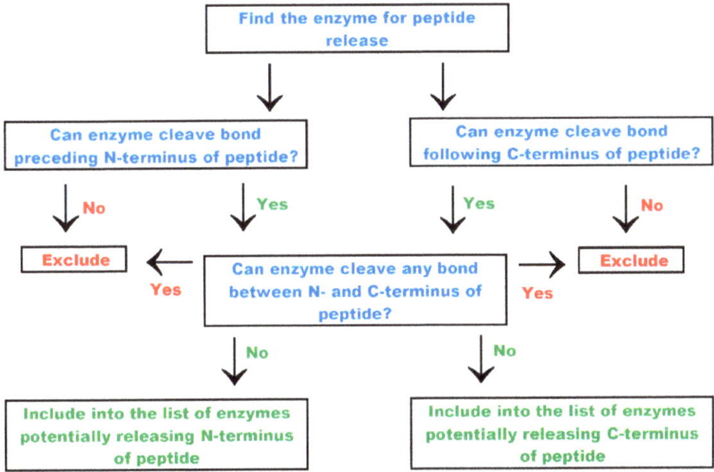

Figure 4. Scheme of "find enzyme for peptide release" tab action.

7. Useful Links and Other Tabs

The BIOPEP-UWM plays the role of a metaserver enabling access to databases and software useful in research concerning peptides and proteins. The linked tools available via the "useful links" tab (Figure 1; Supplementary Figure S1) are divided into categories according to Minkiewicz et al. [101]. These categories are summarized in Table 6.

Table 6. Categories of bioinformatic tools available via the "useful links" tab.

Category	Description
Bioactive peptide databases	Databases of biologically active peptides including general databases (covering several activities) or databases of particular activities (e.g., antimicrobial)
Bioactivity prediction	Software predicting biological activity of peptides, especially interactions with proteins, e.g., enzymes
Immunology of proteins and peptides	Databases of allergens and epitopes, software for predicting allergenicity and occurrence of epitopes as well as other software from the area of immunology
Literature data mining	Software supporting search for biomedical data (e.g., concerning proteins and peptides) in literature
Miscellaneous	Databases and software not belonging to other categories. Chemical databases and metabases are attributed to this category.
Motifs	Programs enabling constructing sequence motifs and finding them in protein or peptide sequences
Physicochemical properties	Software used to predict and exploit the physicochemical properties of peptides
Prediction of post-translational modifications	Software used to predict the location of post-translational modifications (phosphorylation, glycosylation) in protein and peptide sequences
Programs supporting peptide design	Software supporting design of peptides with desired biological properties
Protein resources	Databases and software concerning proteins but not peptides, including databases of protein sequences and structures
Proteolysis	Databases annotating proteolytic enzymes, software for proteolysis simulation
Proteomic tools	Tools supporting proteomics research including mass spectrometry
Sequence alignments	Software for constructing protein and peptide sequence alignments and for searching in protein sequence databases
Structure prediction and visualization	Software for modeling secondary and tertiary structures of proteins and peptides

Other tabs available from the BIOPEP-UWM main page are as follows: List of publications of our group concerning the BIOPEP-UWM database, brief summary concerning the database ("about BIOPEP-UWM" tab), publications concerning particular parts of the BIOPEP-UWM database recommended to be cited by users, and contact data of database curators.

8. Final Remarks

This paper presents the current status of the BIOPEP-UWM^{TM} database including changes introduced within the period of 2016–2019. Apart from the addition of new peptides (562 items added since submission of our last publication describing the database of sensory peptides and amino acids [9]), information about the existing ones has been completed (especially chemical codes and database references). We also added several new options that are summarized in the Table 7.

Table 7. New options in the BIOPEP-UWM database and modifications of existing ones, not described in the previous publications [8,9].

Option	Description
Peptide annotation	Possibility of annotation of peptides containing D-amino acids
Search options [1]	Search on the basis of InChIKey; addition of "exact match" search as user's choice, designed especially for sequence search
List of peptide activities	List of peptide activities rearranged and enriched
Proteolytic enzyme annotation	Updated list of bonds susceptible to proteolytic enzyme action
New search options	Search on the basis of InChIKey; addition of "exact match" search as user's choice
"SMILES" tab [1]	Application converting amino acid sequences into the SMILES code
New options available via the "enzyme(s) action" tab	New quantitative parameters describing possibility of release of bioactive peptides by proteolytic enzymes—Equations (5)–(7) in Table 5, option enabling finding enzyme with a given specificity among proteinases annotated in the database
"find the enzymes for peptide release" tab	Option which enables finding proteolytic enzymes liberating of N- and C-termini of bioactive peptides
"find" tab	Shortcut to the list of peptides with a given activity
Batch processing	Option which enables finding profiles of potential biological activity of fragments, calculating quantitative parameters that characterize protein or peptide, and simulating proteolysis for a set of sequences
Quantitative parameters characterizing occurrence and possibility of release of bioactive peptide from a set of sequences	Parameters calculated via the "batch processing" option—Equations (8)–(10) and (12)–(14) in the Table 5
The "BIOPEP-UWM news" tab	Tab designed to provide important news concerning the database

[1] Application serving for conversion amino acid sequences into SMILES code has been announced in [4].

The content of this publication is not restricted to description of new changes in the database and associated tools during the last three years. We try to provide a complete description including both old and new options.

The next modifications would be aimed at removing the weak points of the database and associated applications. We would like to ask users to submit new peptides (via the current version of the "submit new peptide" tab) and any remarks helpful in improving the bioinformatic tool described in this paper.

Supplementary Materials: Supplementary materials can be found at http://www.mdpi.com/1422-0067/20/23/5978/s1.

Author Contributions: P.M., A.I., and M.D. are curators of the BIOPEP-UWM database. P.M., M.D., and A.I. designed new options and applications associated with the BIOPEP-UWM database. P.M., A.I., and M.D. have written the manuscript. Funding acquisition—M.D. and A.I.

Funding: The project was financially supported by the Minister of Science and Higher Education in the range of the program entitled "Regional Initiative of Excellence" for the years 2019–2022, Project No. 010/RID/2018/19, amount of funding 12,000,000 PLN and University of Warmia and Mazury, grant number 17.610.014-300.

Acknowledgments: Authors thank Krzysztof Sieniawski and Mariusz Falkowski (Enter Krzysztof Sieniawski, Olsztyn, Poland) for IT support; and also Monika Hrynkiewicz, Marta Turło, Agnieszka Skwarek, Monika Pliszka, and Piotr Starowicz, for adding new data to the BIOPEP-UWM database; furthermore Iwona Szerszunowicz and Kamila Licka for pointing out some weak points of the database and associated software; and finally ChemAxon (Budapest, Hungary) for academic license for MarvinSketch program.

Conflicts of Interest: The authors declare no conflict of interests.

Abbreviations

ACE	Angiotensin-converting enzyme (EC 3.4.15.1)
ACToR	Aggregated Computational Toxicology Online Resource
AHTPDB	Antihypertensive Peptide Database
APD	Antimicrobial peptide database
BioPepDB	Bioactive Peptide Database
BRENDA	Braunschweig Enzyme Database
CAMKII	Ca^{2+}/calmodulin-dependent protein kinase (EC 2.7.11.17)
CAMP	Collection of antimicrobial peptides
CaMPDE	Calmodulin-dependent phosphodiesterase 1 (EC 3.1.4.17)
CancerPPD	Anticancer protein and peptide database
CAS	Chemical Abstract Service provided by American Chemical Society
CID	Compound Identifier (in PubChem database)
DB	database
DBAASP	Database of Antimicrobial Activity and Structure of Peptides
EBI	European Bioinformatics Institute
EC_{50}	Concentration corresponding to half-maximal activity
EMBL	European Molecular Biology Laboratory
EROP	Endogenous Regulatory Oligopeptide knowledgebase
FeptideDB	Food Peptide Database
GPR14	Abbreviation of urotensin II receptor
HMDB	Human Metabolome Database
IC_{50}	Concentration corresponding to half-maximal inhibition
InChI	International Chemical Identifier
HMG-CoA	3-hydroxy-3-methyl-glutaryl-coenzyme A (PubChem CID: 445127; CAS registry No 1553-55-5)
InChIKey	Key of International Chemical Identifier
IUPAC	International Union of Pure and Applied Chemistry
KEGG	Kyoto Encyclopedia of Genes and Genomes
MBPDB	Milk Bioactive Peptide Database
MetaComBio	Meta Compound Bioactivity
MilkAMP	Milk antimicrobial peptide database
SATPdb	Structurally Annotated Therapeutic Peptide database
SMILES	Simplified Molecular Input Line Entry System or Simplified Molecular Input Line Entry Specification
UWM	University of Warmia and Mazury

References

1. Holton, T.A.; Vijayakumar, V.; Khaldi, N. Bioinformatics: Current perspectives and future directions for food and nutritional research facilitated by a food-wiki database. *Trends Food Sci. Technol.* **2013**, *34*, 5–17. [CrossRef]
2. Udenigwe, C.C. Bioinformatics approaches, prospects and challenges of food bioactive peptide research. *Trends Food Sci. Technol.* **2014**, *36*, 134–143. [CrossRef]
3. Iwaniak, A.; Minkiewicz, P.; Darewicz, M.; Protasiewicz, M.; Mogut, D. Chemometrics and cheminformatics in the analysis of biologically active peptides from food sources. *J. Funct. Foods* **2015**, *16*, 334–351. [CrossRef]
4. Iwaniak, A.; Darewicz, M.; Mogut, D.; Minkiewicz, P. Elucidation of the role of in silico methodologies in approaches to studying bioactive peptides derived from foods. *J. Funct. Foods* **2019**, *61*, 103486. [CrossRef]
5. Agyei, D.; Tsopmo, A.; Udenigwe, C.C. Bioinformatic and peptidomic approaches to the discovery and analysis of food-derived bioactive peptides. *Anal. Bioanal. Chem.* **2018**, *410*, 3463–3472. [CrossRef] [PubMed]
6. Kalmykova, S.D.; Arapidi, G.P.; Urban, A.S.; Osetrova, M.S.; Gordeeva, V.D.; Ivanov, V.T.; Govorun, V.M. In silico analysis of peptide potential biological functions. *Russ. J. Bioorg. Chem.* **2018**, *44*, 367–385. [CrossRef]

7. Tu, M.; Cheng, S.; Lu, W.; Du, M. Advancement and prospects of bioinformatics analysis for studying bioactive peptides from food-derived protein: Sequence, structure, and functions. *Trends Anal. Chem.* **2018**, *105*, 7–17. [CrossRef]

8. Minkiewicz, P.; Dziuba, J.; Iwaniak, A.; Dziuba, M.; Darewicz, M. BIOPEP database and other programs for processing bioactive peptide sequences. *J. AOAC Int.* **2008**, *91*, 965–980.

9. Iwaniak, A.; Minkiewicz, P.; Darewicz, M.; Sieniawski, K.; Starowicz, P. BIOPEP database of sensory peptides and amino acids. *Food Res. Int.* **2016**, *85*, 155–161. [CrossRef]

10. Piovesana, S.; Capriotti, A.L.; Cavaliere, C.; La Barbera, G.; Montone, C.M.; Chiozzi, R.Z.; Laganà, A. Recent trends and analytical challenges in plant bioactive peptide separation, identification and validation. *Anal. Bioanal. Chem.* **2018**, *410*, 3425–3444. [CrossRef]

11. Minkiewicz, P.; Iwaniak, A.; Darewicz, M. Using internet databases for food science organic chemistry students to discover chemical compound information. *J. Chem. Educ.* **2015**, *92*, 874–876. [CrossRef]

12. Anekthanakul, K.; Hongsthong, A.; Senachak, J.; Ruengjitchatchawalya, M. SpirPep: An in silico digestion-based platform to assist bioactive peptides discovery from a genome-wide database. *BMC Bioinform.* **2018**, *19*, 149. [CrossRef] [PubMed]

13. Panyayai, T.; Ngamphiw, C.; Tongsima, S.; Mhuantong, W.; Limsripraphan, W.; Choowongkomon, K.; Sawatdichaikul, O. FeptideDB: A web application for new bioactive peptides from food protein. *Heliyon* **2019**, *5*, e02076. [CrossRef] [PubMed]

14. Dziuba, M.; Minkiewicz, P.; Dąbek, M. Peptides, specific proteolysis products as molecular markers of allergenic proteins—In Silico studies. *Acta Sci. Pol. Technol. Aliment.* **2013**, *12*, 101–112. [PubMed]

15. He, R.; Malomo, S.A.; Alashi, A.; Girgih, A.T.; Ju, X.; Aluko, R.E. Glycinyl-histidinyl-serine (GHS), a novel rapeseed protein-derived peptide has blood pressure-lowering effect in spontaneously hypertensive rats. *J. Agric. Food Chem.* **2013**, *61*, 8396–8402. [CrossRef] [PubMed]

16. Skrzypczak, K.; Fornal, E.; Waśko, A.; Gustaw, W. Effects of probiotic fermentation of selected milk and whey protein preparations on bioactive peptides and technological properties. *Ital. J. Food Sci.* **2019**, *31*, 437–450.

17. Khazaei, H.; Subedi, M.; Nickerson, M.; Martínez-Villaluenga, C.; Frias, J.; Vandenberg, A. Seed protein of lentils: Current status, progress, and food applications. *Foods* **2019**, *8*, 391. [CrossRef]

18. Hsieh, C.-C.; Martínez-Villaluenga, C.; de Lumen, B.O.; Hernández-Ledesma, B. Updating the research on the chemopreventive and therapeutic role of the peptide lunasin. *J. Sci. Food Agric.* **2018**, *98*, 2070–2079. [CrossRef]

19. Savitha, M.N.; Siddesha, J.M.; Suvilesh, K.N.; Yariswamy, M.; Vivek, H.K.; D'Souza, C.J.; Umashankar, M.; Vishwanath, B.S. Active-site directed peptide L-Phe-D-His-L-Leu inhibits angiotensin converting enzyme activity and dexamethasone-induced hypertension in rats. *Peptides* **2019**, *112*, 34–42. [CrossRef]

20. Zamyatnin, A.A.; Borchikov, A.S.; Vladimirov, M.G.; Voronina, O.L. The EROP-Moscow oligopeptide database. *Nucleic Acids Res.* **2006**, *34*, D261–D266. [CrossRef]

21. Heller, S.R.; McNaught, A.; Pletnev, I.; Stein, S.; Tchekhovskoi, D. InChI, the IUPAC International Chemical Identifier. *J. Cheminform.* **2015**, *7*, 23. [CrossRef] [PubMed]

22. Kim, S.; Chen, J.; Cheng, T.; Gindulyte, A.; He, J.; He, S.; Li, Q.; Shoemaker, B.A.; Thiessen, P.A.; Yu, B.; et al. PubChem 2019 update: Improved access to chemical data. *Nucleic Acids Res.* **2019**, *47*, D1102–D1109. [CrossRef] [PubMed]

23. Ashok, A.; Aparna, B.H.S. Discovery, synthesis, and In vitro evaluation of a novel bioactive peptide for ACE and DPP-IV inhibitory activity. *Eur. J. Med. Chem.* **2019**, *180*, 99–110. [CrossRef] [PubMed]

24. Gallego, M.; Mora, L.; Toldrá, F. The relevance of dipeptides and tripeptides in the bioactivity and taste of dry-cured ham. *Food Prod. Process. Nutr.* **2019**, *1*, 2. [CrossRef]

25. Pinciroli, M.; Aphalo, P.; Nardo, A.E.; Añón, M.C.; Quiroga, A.V. Broken rice as a potential functional ingredient with inhibitory activity of renin and angiotensin-converting enzyme (ACE). *Plant Foods Hum. Nutr.* **2019**, *74*, 405–413. [CrossRef]

26. Minkiewicz, P.; Darewicz, M.; Iwaniak, A.; Bucholska, J.; Starowicz, P.; Czyrko, E. Internet databases of the properties, enzymatic reactions, and metabolism of small molecules-search options and applications in food science. *Int. J. Mol. Sci.* **2016**, *17*, 2039. [CrossRef]

27. Weininger, D. SMILES, a chemical language and information system. 1. Introduction to methodology and encoding rules. *J. Chem. Inf. Comput. Sci.* **1988**, *28*, 31–36. [CrossRef]

28. Peña-Castillo, A.; Méndez-Lucio, O.; Owen, J.R.; Martínez-Mayorga, K.; Medina-Franco, J.L. Chemoinformatics in food science. In *Applied Chemoinformatics: Achievements and Future Opportunities*; Engel, T., Gasteiger, J., Eds.; Wiley-VCH Verlag GmbH & Co. KGaA: Weinheim, Germany, 2018; pp. 501–525.

29. Scotti, L.; Júnior, F.J.; Ishiki, H.M.; Ribeiro, F.F.; Duarte, M.C.; Santana, G.S.; Oliveira, T.B.; Diniz, M.D.; Quintans-Júnior, L.J.; Scotti, M.T. Computer-aided drug design studies in food chemistry. In *Natural and Artificial Flavoring Agents and Food Dyes*; Grumezescu, A.M., Holban, A.M., Eds.; Elsevier: Amsterdam, The Netherlands, 2018; pp. 261–297.

30. Minkiewicz, P.; Turło, M.; Iwaniak, A.; Darewicz, M. Free accessible databases as a source of information about food components and other compounds with anticancer activity–brief review. *Molecules* **2019**, *24*, 789. [CrossRef]

31. Iwaniak, A.; Minkiewicz, P.; Darewicz, M.; Hrynkiewicz, M. Food protein-originating peptides as tastants-Physiological, technological, sensory, and bioinformatic approaches. *Food Res. Int.* **2016**, *89*, 27–38. [CrossRef]

32. O'Boyle, N.M.; Banck, M.; James, C.A.; Morley, C.; Vandermeersch, T.; Hutchison, G.R. Open Babel: An open chemical toolbox. *J. Cheminform.* **2011**, *3*, 33. [CrossRef]

33. Ortiz-Martinez, M.; Gonzalez de Mejia, E.; García-Lara, S.; Aguilar, O.; Lopez-Castillo, L.M.; Otero-Pappatheodorou, J.T. Antiproliferative effect of peptide fractions isolated from a quality protein maize, a white hybrid maize, and their derived peptides on hepatocarcinoma human HepG2 cells. *J. Funct. Foods* **2017**, *34*, 36–48. [CrossRef]

34. Mojica, L.; Luna-Vital, D.A.; Gonzalez de Mejia, E. Black bean peptides inhibit glucose uptake in Caco-2 adenocarcinoma cells by blocking the expression and translocation pathway of glucose transporters. *Toxicol. Rep.* **2018**, *5*, 552–560. [CrossRef] [PubMed]

35. Yu, Z.; Fan, Y.; Zhao, W.; Ding, L.; Li, J.; Liu, L. Novel angiotensin-converting enzyme inhibitory peptides derived from Oncorhynchus mykiss nebulin: Virtual screening and in silico molecular docking study. *J. Food Sci.* **2018**, *83*, 2375–2383. [CrossRef] [PubMed]

36. Bradbury, A.F.; Smyth, D.G. Peptide amidation. *Trends Biochem. Sci.* **1991**, *16*, 112–115. [CrossRef]

37. Lee, K.H.; Hong, S.Y.; Oh, J.E.; Kwon, M.Y.; Yoon, J.H.; Lee, J.H.; Lee, B.L.; Moon, H.M. Identification and characterization of the antimicrobial peptide corresponding to C-terminal beta-sheet domain of tenecin 1, an antibacterial protein of larvae of *Tenebrio molitor*. *Biochem. J.* **1998**, *334*, 99–105. [CrossRef]

38. McDonald, A.G.; Boyce, S.; Tipton, K.F. ExplorEnz: The primary source of the IUBMB enzyme list. *Nucleic Acids Res.* **2009**, *37*, D593–D597. [CrossRef]

39. Jeske, L.; Placzek, S.; Schomburg, I.; Chang, A.; Schomburg, D. BRENDA in 2019: A European ELIXIR core data resource. *Nucleic Acids Res.* **2019**, *47*, D542–D549. [CrossRef]

40. Mendez, D.; Gaulton, A.; Bento, A.P.; Chambers, J.; De Veij, M.; Félix, E.; Magariños, M.P.; Mosquera, J.F.; Mutowo, P.; Nowotka, M.; et al. ChEMBL: Towards direct deposition of bioassay data. *Nucleic Acids Res.* **2019**, *47*, D930–D940. [CrossRef]

41. Rawlings, N.D.; Barrett, A.J.; Thomas, P.D.; Huang, X.; Bateman, A.; Finn, R.D. The MEROPS database of proteolytic enzymes, their substrates and inhibitors in 2017 and a comparison with peptidases in the PANTHER database. *Nucleic Acids Res.* **2018**, *46*, D624–D632. [CrossRef]

42. Carvalho-Silva, D.; Pierleoni, A.; Pignatelli, M.; Ong, C.; Fumis, L.; Karamanis, N.; Carmona, M.; Faulconbridge, A.; Hercules, A.; McAuley, E.; et al. Open Targets Platform: New developments and updates two years on. *Nucleic Acids Res.* **2019**, *47*, D1056–D1065. [CrossRef]

43. Judson, R.S.; Martin, M.T.; Egeghy, P.; Gangwal, S.; Reif, D.M.; Kothiya, P.; Wolf, M.; Cathey, T.; Transue, T.; Smith, D.; et al. Aggregating data for computational toxicology applications: The U.S. Environmental Protection Agency (EPA) Aggregated Computational Toxicology Resource (ACToR) system. *Int. J. Mol. Sci.* **2012**, *13*, 1805–1831. [CrossRef] [PubMed]

44. Tomasulo, P. ChemIDplus-super source for chemical and drug information. *Med. Ref. Serv. Quart.* **2002**, *21*, 53–59. [CrossRef] [PubMed]

45. Kumar, R.; Chaudhary, K.; Sharma, M.; Nagpal, G.; Chauhan, J.S.; Singh, S.; Gautam, A.; Raghava, G.P. AHTPDB: A comprehensive platform for analysis and presentation of antihypertensive peptides. *Nucleic Acids Res.* **2015**, *43*, D956–D962. [CrossRef] [PubMed]

46. Wang, G.; Li, X.; Wang, Z. APD3: The antimicrobial peptide database as a tool for research and education. *Nucleic Acids Res.* **2016**, *44*, D1087–D1093. [CrossRef] [PubMed]

47. Gilson, M.K.; Liu, T.; Baitaluk, M.; Nicola, G.; Hwang, L.; Chong, J. BindingDB in 2015: A public database for medicinal chemistry, computational chemistry and systems pharmacology. *Nucleic Acids Res.* **2016**, *44*, D1045–D1053. [CrossRef] [PubMed]

48. Li, Q.; Zhang, C.; Chen, H.; Xue, J.; Guo, X.; Liang, M.; Chen, M. BioPepDB: An integrated data platform for food-derived bioactive peptides. *Int. J. Food Sci. Nutr.* **2018**, *69*, 963–968. [CrossRef]

49. Dagan-Wiener, A.; Di Pizio, A.; Nissim, I.; Bahia, M.S.; Dubovski, N.; Margulis, E.; Niv, M.Y. BitterDB: Taste ligands and receptors database in 2019. *Nucleic Acids Res.* **2019**, *47*, D1179–D1185. [CrossRef]

50. Van Dorpe, S.; Bronselaer, A.; Nielandt, J.; Stalmans, S.; Wynendaele, E.; Audenaert, K.; Van De Wiele, C.; Burvenich, C.; Peremans, K.; Hsuchou, H.; et al. Brainpeps: The blood-brain barrier peptide database. *Brain Struct. Funct.* **2012**, *217*, 687–718. [CrossRef]

51. Waghu, F.H.; Barai, R.S.; Gurung, P.; Idicula-Thomas, S. CAMP$_{R3}$: A database on sequences, structures and signatures of antimicrobial peptides. *Nucleic Acids Res.* **2016**, *44*, D1094–D1097. [CrossRef]

52. Tyagi, A.; Tuknait, A.; Anand, P.; Gupta, S.; Sharma, M.; Mathur, D.; Joshi, A.; Singh, S.; Gautam, A.; Raghava, G.P. CancerPPD: A database of anticancer peptides and proteins. *Nucleic Acids Res.* **2015**, *43*, D837–D843. [CrossRef]

53. Hastings, J.; Owen, G.; Dekker, A.; Ennis, M.; Kale, N.; Muthukrishnan, V.; Turner, S.; Swainston, N.; Mendes, P.; Steinbeck, C. ChEBI in 2016: Improved services and an expanding collection of metabolites. *Nucleic Acids Res.* **2016**, *44*, D1214–D1219. [CrossRef] [PubMed]

54. Williams, A.; Tkachenko, V. The Royal Society of Chemistry and the delivery of chemistry data repositories for the community. *J. Comput. Aided Mol. Des.* **2014**, *28*, 1023–1030. [CrossRef] [PubMed]

55. Williams, A.J.; Grulke, C.M.; Edwards, J.; McEachran, A.D.; Mansouri, K.; Baker, N.C.; Patlewicz, G.; Shah, I.; Wambaugh, J.F.; Judson, R.S.; et al. The CompTox Chemistry Dashboard: A community data resource for environmental chemistry. *J. Cheminform.* **2017**, *9*, 61. [CrossRef] [PubMed]

56. Igarashi, Y.; Eroshkin, A.; Gramatikova, S.; Gramatikoff, K.; Zhang, Y.; Smith, J.W.; Osterman, A.L.; Godzik, A. CutDB: A proteolytic event database. *Nucleic Acids Res.* **2007**, *35*, D546–D549. [CrossRef] [PubMed]

57. Pirtskhalava, M.; Gabrielian, A.; Cruz, P.; Griggs, H.L.; Squires, R.B.; Hurt, D.E.; Grigolava, M.; Chubinidze, M.; Gogoladze, G.; Vishnepolsky, B.; et al. DBAASP v.2: An enhanced database of structure and antimicrobial/cytotoxic activity of natural and synthetic peptides. *Nucleic Acids Res.* **2016**, *44*, D1104–D1112. [CrossRef] [PubMed]

58. Wishart, D.S.; Feunang, Y.D.; Guo, A.C.; Lo, E.J.; Marcu, A.; Grant, J.R.; Sajed, T.; Johnson, D.; Li, C.; Sayeeda, Z.; et al. DrugBank 5.0: A major update to the DrugBank database for 2018. *Nucleic Acids Res.* **2018**, *46*, D1074–D1082. [CrossRef] [PubMed]

59. Gautam, A.; Chaudhary, K.; Singh, S.; Joshi, A.; Anand, P.; Tuknait, A.; Mathur, D.; Varshney, G.C.; Raghava, G.P. Hemolytik: A database of experimentally determined hemolytic and non-hemolytic peptides. *Nucleic Acids Res.* **2014**, *42*, D444–D449. [CrossRef]

60. Wishart, D.S.; Feunang, Y.D.; Marcu, A.; Guo, A.C.; Liang, K.; Vázquez-Fresno, R.; Sajed, T.; Johnson, D.; Li, C.; Karu, N.; et al. HMDB 4.0: The human metabolome database for 2018. *Nucleic Acids Res.* **2018**, *46*, D608–D617. [CrossRef]

61. Kanehisa, M.; Sato, Y.; Furumichi, M.; Morishima, K.; Tanabe, M. New approach for understanding genome variations in KEGG. *Nucleic Acids Res.* **2019**, *47*, D590–D595. [CrossRef]

62. Nielsen, S.D.; Beverly, R.L.; Qu, Y.; Dallas, D.C. Milk bioactive peptide database: A comprehensive database of milk protein-derived bioactive peptides and novel visualization. *Food Chem.* **2017**, *232*, 673–682. [CrossRef]

63. Haug, K.; Salek, R.M.; Conesa, P.; Hastings, J.; de Matos, P.; Rijnbeek, M.; Mahendraker, T.; Williams, M.; Neumann, S.; Rocca-Serra, P.; et al. MetaboLights—An open-access general-purpose repository for metabolomics studies and associated meta-data. *Nucleic Acids Res.* **2013**, *41*, D781–D786. [CrossRef] [PubMed]

64. Théolier, J.; Fliss, I.; Jean, J.; Hammami, R. MilkAMP: A comprehensive database of antimicrobial peptides of dairy origin. *Dairy Sci. Technol.* **2014**, *94*, 181–193. [CrossRef]

65. Ntie-Kang, F.; Telukunta, K.K.; Döring, K.; Simoben, C.V.; Moumbock, A.F.A.; Malange, Y.I.; Njume, L.E.; Yong, J.N.; Sippl, W.; Günther, S. NANPDB: A resource for natural products from northern African sources. *J. Nat. Prod.* **2017**, *80*, 2067–2076. [CrossRef] [PubMed]

66. Wang, Y.; Wang, M.; Yin, S.; Jang, R.; Wang, J.; Xue, Z.; Xu, T. NeuroPep: A comprehensive resource of neuropeptides. *Database* **2015**, *2015*, bav038. [CrossRef] [PubMed]

67. Shtatland, T.; Guettler, D.; Kossodo, M.; Pivovarov, M.; Weissleder, R. PepBank—A database of peptides based on sequence text mining and public peptide data sources. *BMC Bioinform.* **2007**, *8*, 280. [CrossRef] [PubMed]

68. Liu, F.; Baggerman, G.; Schoofs, L.; Wets, G. The construction of a bioactive peptide database in *Metazoa*. *J. Proteome Res.* **2008**, *7*, 4119–4131. [CrossRef]

69. Juhász, A.; Haraszi, R.; Maulis, C. ProPepper: A curated database for identification and analysis of peptide and immune-responsive epitope composition of cereal grain protein families. *Database* **2015**, *2015*, bav100. [CrossRef]

70. Singh, S.; Chaudhary, K.; Dhanda, S.K.; Bhalla, S.; Usmani, S.S.; Gautam, A.; Tuknait, A.; Agrawal, P.; Mathur, D.; Raghava, G.P. SATPdb: A database of structurally annotated therapeutic peptides. *Nucleic Acids Res.* **2016**, *44*, D1119–D1126. [CrossRef]

71. Papadatos, G.; Davies, M.; Dedman, N.; Chambers, J.; Gaulton, A.; Siddle, J. SureChEMBL: A large-scale, chemically annotated patent document database. *Nucleic Acids Res.* **2016**, *44*, D1220–D1228. [CrossRef]

72. Gfeller, D.; Michielin, O.; Zoete, V. SwissSidechain: A molecular and structural database of non-natural sidechains. *Nucleic Acids Res.* **2013**, *41*, D327–D332. [CrossRef]

73. Sterling, T.; Irwin, J.J. ZINC 15—ligand discovery for everyone. *J. Chem. Inf. Model.* **2015**, *55*, 2324–2337. [CrossRef] [PubMed]

74. Liu, Z.P.; Wu, L.Y.; Wang, Y.; Zhang, X.S.; Chen, L. Bridging protein local structures and protein functions. *Amino Acids* **2008**, *35*, 627–650. [CrossRef] [PubMed]

75. Minkiewicz, P.; Darewicz, M.; Iwaniak, A.; Sokołowska, J.; Starowicz, P.; Bucholska, J.; Hrynkiewicz, M. Common amino acid subsequences in a universal proteome-relevance for food science. *Int. J. Mol. Sci.* **2015**, *16*, 20748–20773. [CrossRef] [PubMed]

76. Zamyatnin, A.A. Fragmentomics of natural peptide structures. *Biochemistry (Moscow)* **2009**, *74*, 1575–1585. [CrossRef] [PubMed]

77. Martini, S.; Conte, A.; Tagliazucchi, D. Comparative peptidomic profile and bioactivities of cooked beef, pork, chicken and turkey meat after In vitro gastro-intestinal digestion. *J. Proteom.* **2019**, *208*, 103500. [CrossRef] [PubMed]

78. Garcia-Vaquero, M.; Mora, L.; Hayes, M. In vitro and in silico approaches to generating and identifying angiotensin-converting enzyme I inhibitory peptides from green macroalga *Ulva lactuca*. *Marine Drugs* **2019**, *17*, 204. [CrossRef]

79. Dziuba, J.; Minkiewicz, P.; Nałęcz, D.; Iwaniak, A. Database of biologically active peptide sequences. *Nahrung* **1999**, *43*, 190–195. [CrossRef]

80. Bauchart, C.; Morzel, M.; Chambon, C.; Mirand, P.P.; Reynès, C.; Buffière, C.; Rémond, D. Peptides reproducibly released by in vivo digestion of beef meat and trout flesh in pigs. *Br. J. Nutr.* **2007**, *98*, 1187–1195. [CrossRef]

81. Huang, B.-B.; Lin, H.-C.; Chang, Y.-W. Analysis of proteins and potential bioactive peptides from tilapia (*Oreochromis* spp.) processing co-products using proteomic techniques coupled with BIOPEP database. *J. Funct. Foods* **2015**, *19*, 629–640. [CrossRef]

82. Tapal, A.; Vegarud, G.E.; Sreedhara, A.; Tiku, P.K. Nutraceutical protein isolate from pigeon pea (*Cajanus cajan*) milling waste by-product: Functional aspects and digestibility. *Food Funct.* **2019**, *10*, 2710–2719. [CrossRef]

83. Jakubczyk, A.; Karaś, M.; Złotek, U.; Szymanowska, U.; Baraniak, B.; Bochnak, J. Peptides obtained from fermented faba bean seeds (*Vicia faba*) as potential inhibitors of an enzyme involved in the pathogenesis of metabolic syndrome. *LWT Food Sci. Technol.* **2019**, *105*, 306–313. [CrossRef]

84. Udenigwe, C.C.; Gong, M.; Wu, S. In silico analysis of the large and small subunits of cereal RuBisCO as precursors of cryptic bioactive peptides. *Process Biochem.* **2013**, *48*, 1794–1799. [CrossRef]

85. Lin, K.; Zhang, L.W.; Han, X.; Xin, L.; Meng, Z.X.; Gong, P.M.; Cheng, D.Y. Yak milk casein as potential precursor of angiotensin I-converting enzyme inhibitory peptides based on in silico proteolysis. *Food Chem.* **2018**, *254*, 340–347. [CrossRef] [PubMed]

86. Dziuba, J.; Iwaniak, A.; Minkiewicz, P. Computer-aided characteristics of proteins as potential precursors of bioactive peptides. *Polimery* **2003**, *48*, 50–53. [CrossRef]

87. Minkiewicz, P.; Dziuba, J.; Michalska, J. Bovine meat proteins as potential precursors of biologically active peptides—A computational study based on the BIOPEP database. *Food Sci. Technol. Int.* **2011**, *17*, 39–45. [CrossRef] [PubMed]

88. Nielsen, P.M.; Petersen, D.; Dambmann, C. Improved method for determining food protein degree of hydrolysis. *J. Food Sci.* **2001**, *65*, 642–646. [CrossRef]

89. Bastian, E.D.; Brown, R.J. Plasmin in milk and dairy products: An update. *Int. Dairy J.* **1996**, *6*, 435–457. [CrossRef]

90. Huang, X.W.; Chen, L.J.; Luo, Y.B.; Guo, H.Y.; Ren, F.Z. Purification, characterization, and milk coagulating properties of ginger proteases. *J. Dairy Sci.* **2011**, *94*, 2259–2269. [CrossRef]

91. Yu, D.; Wang, C.; Song, Y.; Zhu, J.; Zhang, X. Discovery of novel angiotensin-converting enzyme inhibitory peptides from *Todarodes pacificus* and their inhibitory mechanism: In silico and In vitro studies. *Int. J. Mol. Sci.* **2019**, *20*, 4159. [CrossRef]

92. Kandemir-Cavas, C.; Pérez-Sanchez, H.; Mert-Ozupek, N.; Cavas, L. In silico analysis of bioactive peptides in invasive sea grass *Halophila stipulacea*. *Cells* **2019**, *8*, 557. [CrossRef]

93. Dziuba, J.; Niklewicz, M.; Iwaniak, A.; Darewicz, M.; Minkiewicz, P. Structural properties of proteolytic-accessible bioactive fragments of selected animal proteins. *Polimery* **2005**, *50*, 424–428. [CrossRef]

94. Pearson, W.R. Flexible sequence similarity searching with the FASTA3 program package. *Methods Mol. Biol.* **2000**, *132*, 185–219. [PubMed]

95. Nardo, A.E.; Añón, M.C.; Parisi, G. Large-scale mapping of bioactive peptides in structural and sequence space. *PLoS ONE* **2018**, *13*, e0191063. [CrossRef] [PubMed]

96. Siani, M.A.; Weininger, D.; Blaney, J.M. CHUCKLES: A method for representing and searching peptide and peptoid sequences on both monomer and atomic levels. *J. Chem. Inf. Comput. Sci.* **1994**, *34*, 588–593. [CrossRef] [PubMed]

97. Duffy, F.J.; Verniere, M.; Devocelle, M.; Bernard, E.; Shields, D.C.; Chubb, A.J. CycloPs: Generating virtual libraries of cyclized and constrained peptides including nonnatural amino acids. *J. Chem. Inf. Model.* **2011**, *51*, 829–836. [CrossRef] [PubMed]

98. Minkiewicz, P.; Iwaniak, A.; Darewicz, M. Annotation of peptide structures using SMILES and other chemical codes–practical solutions. *Molecules* **2017**, *22*, 2075. [CrossRef]

99. Hähnke, V.D.; Kim, S.; Bolton, E.E. PubChem chemical structure standardization. *J. Cheminform.* **2018**, *10*, 36. [CrossRef]

100. Brodkorb, A.; Egger, L.; Alminger, M.; Alvito, P.; Assunção, R.; Ballance, S.; Bohn, T.; Bourlieu-Lacanal, C.; Boutrou, R.; Carrière, F.; et al. INFOGEST static In vitro simulation of gastrointestinal food digestion. *Nat. Protoc.* **2019**, *14*, 991–1014. [CrossRef]

101. Minkiewicz, P.; Dziuba, J.; Darewicz, M.; Iwaniak, A.; Michalska, J. Online programs and databases of peptides and proteolytic enzymes—A brief update for 2007–2008. *Food Technol. Biotechnol.* **2009**, *47*, 345–355.

International Journal of
Molecular Sciences

Article

Systematical Analysis of the Protein Targets of Lactoferricin B and Histatin-5 Using Yeast Proteome Microarrays

Pramod Shah [1,2], **Wei-Sheng Wu** [3] **and Chien-Sheng Chen** [1,2,4,*]

1   Graduate Institute of Systems Biology and Bioinformatics, National Central University, Jhongli 32001, Taiwan
2   Department of Biomedical Science and Engineering, National Central University, Jhongli 32001, Taiwan
3   Department of Electrical Engineering, National Cheng Kung University, Tainan City 701, Taiwan
4   Department of Food Safety/Hygiene and Risk Management, College of Medicine,
    National Cheng Kung University, Tainan City 701, Taiwan
*   Correspondence: cchen103@gmail.com; Tel.: +886-6-2353535 (ext. 5964)

Received: 18 July 2019; Accepted: 23 August 2019; Published: 28 August 2019

Abstract: Antimicrobial peptides (AMPs) have potential antifungal activities; however, their intracellular protein targets are poorly reported. Proteome microarray is an effective tool with high-throughput and rapid platform that systematically identifies the protein targets. In this study, we have used yeast proteome microarrays for systematical identification of the yeast protein targets of Lactoferricin B (Lfcin B) and Histatin-5. A total of 140 and 137 protein targets were identified from the triplicate yeast proteome microarray assays for Lfcin B and Histatin-5, respectively. The Gene Ontology (GO) enrichment analysis showed that Lfcin B targeted more enrichment categories than Histatin-5 did in all GO biological processes, molecular functions, and cellular components. This might be one of the reasons that Lfcin B has a lower minimum inhibitory concentration (MIC) than Histatin-5. Moreover, pairwise essential proteins that have lethal effects on yeast were analyzed through synthetic lethality. A total of 11 synthetic lethal pairs were identified within the protein targets of Lfcin B. However, only three synthetic lethal pairs were identified within the protein targets of Histatin-5. The higher number of synthetic lethal pairs identified within the protein targets of Lfcin B might also be the reason for Lfcin B to have lower MIC than Histatin-5. Furthermore, two synthetic lethal pairs were identified between the unique protein targets of Lfcin B and Histatin-5. Both the identified synthetic lethal pairs proteins are part of the Spt-Ada-Gcn5 acetyltransferase (SAGA) protein complex that regulates gene expression via histone modification. Identification of synthetic lethal pairs between Lfcin B and Histatin-5 and their involvement in the same protein complex indicated synergistic combination between Lfcin B and Histatin-5. This hypothesis was experimentally confirmed by growth inhibition assay.

Keywords: Lactoferricin B (Lfcin B); Histatin-5; antimicrobial peptides (AMPs); antifungal activity; proteome microarray; synergy

---

1. Introduction

Treatment of cancer and bacterial infection, as well as immunocompromised patients, has often resulted in fungal infection. An evolutionarily close relationship with humans has limited the development of antifungal drugs. Moreover, the development of antifungal resistance has worsened the condition, causing major health threats [1,2]. All these conditions have led to the urgent need for alternative antifungal agents.

Antimicrobial peptides (AMPs) are key components of the innate immune system that protects the host from invading microorganisms: bacteria, fungi, and viruses. AMPs are short peptides (mostly cationic peptides) with a wide range of antimicrobial activities as well as an adjunct role in immunomodulation and wound-healing [3,4]. The broad-spectrum activity, selective targeting to microorganisms, highly sensitive, multiple modes of the mechanism of action and little toxic effect on human cells makes AMPs a potent alternative to conventional antibiotics [5,6]. Antifungal AMPs are reported to target different fungal cell components [7,8], mostly cell wall and cell membrane, causing leakage of ions and ATPs [9]. AMPs also exert intracellular activities and generation of reactive oxygen species (ROS) that lead to apoptosis/necrosis and cell death [10,11]. Despite multiple mechanisms of antifungal AMPs, very few targets have been reported.

Lactoferricin B (Lfcin B) is a 25-residue peptide (derived from bovine lactoferrin; residues 17–41) with a net positive charge of +8. Lfcin B has twisted antiparallel β-sheet structure containing hydrophilic and hydrophobic residues on the alternative strand [12]. Lfcin B is also detected in the human gut as a result of bovine milk consumption [13]. Lfcin B has antimicrobial activity against bacteria, fungi, cancer cells, and others. The minimum inhibitory concentration (MIC) of Lfcin B against *Escherichia coli* (*E. coli*) wild type and *Saccharomyces cerevisiae* is 15.6–31.2 μg/mL and 0.68 μg/mL, respectively [14,15]. Lfcin B has been reported to exert intracellular activity against yeast [14,16]; however, the intracellular binding targets are unknown. Moreover, the hydrophobic residue is crucial for the translocation of Lfcin B through the cell membrane [17]. Proteome microarray is a high-throughput detection platform for the identification of proteome interactions [18]. In this study, we have systematically identified all the yeast-binding proteins of Lfcin B by using yeast proteome microarrays [19,20].

In addition to Lfcin B, we also used yeast proteome microarrays to identify the yeast protein targets of another most potent antifungal AMP, Histatin-5. Histatin-5 is a 24-residue peptide derived from Histatin-3 (family: Histatins) present in human saliva. Histatin-5 has a net charge of +5 and exerts its potential mode of action by targeting intracellular molecules [21,22]. Histatin-5 has no defined structure in water but acquires α-helix structure in dimethyl sulfoxide and aqueous trifluoroethanol [23]. Presence of Histatin-5 in the mouth provides initial defense against pathogens and avoid their entry to the human gut [24]. The MIC reported for Histatin-5 against *Saccharomyces cerevisiae* is 128 μg/mL [25]. Despite its potential role, only a few targets of Histatin-5 have been reported [26–30]. Thus, comprehensive identification of Histatin-5 protein targets is needed.

The identified yeast protein targets of Lfcin B and Histatin-5 were subjected to bioinformatics analysis to identify functional enrichment in gene ontology (GO) and analysis of synthetic lethal pairs. Synthetic lethality is the well-studied pairwise genetic mutation combination that occurs between two non-lethal genes' mutation, where their simultaneous loss causes cell death while the individual gene deletion has no impact on cell viability [31]. Synthetic lethality approach has been recently applied in chemotherapy for the treatment of cancer [32,33] but not fully exploited in case of pathogenic infections in the lack of systematical identification of synthetic lethal pairs. Synthetic lethality is well studied in yeast and other organisms (fruit fly, worm, mouse, and human). Of the entire genes identified from whole-genome sequence in *Saccharomyces cerevisiae* (~6000 genes) only ~1000 are essential genes (known by single-gene deletion mutant) whereas the other non-essential genes, when disrupted in any two combinations (synthetic lethality), cause over 170,000 synthetic lethal pairs [34]. These observations provide the essentiality of every gene in an organism [35]. The canonical explanation governing the mechanism of synthetic lethality is based on the deletion of two genes (synthetic lethal partner) in three ways: parallel pathway, same pathway and reversible steps in a pathway. First, the deletion of two genes working on the parallel pathways that perform the same essential function is redundant in nature (mutually compensatory) [31,36]. Second, deletion of two genes working on the same pathway can be explained by three possible phenomena: (1) partial redundancy where the loss of one gene causes only the partial degradation that might be tolerable but not the loss of both genes; (2) internal redundancy of two steps within a pathway; and (3) double defect in essential protein complex formed by many proteins [34,37,38]. Third, deletion of two genes that target the reversible forward and backward steps of the non-essential pathway [38]. Furthermore, the synthetic lethality approach was applied to determine the effect of pairwise protein essentiality among the identified protein targets of Lfcin B and Histatin-5.

In our previous studies, we have systematically explored the entire *E. coli* protein targets of Lfcin B by used *E. coli* proteome microarrays [39,40]. To analyze the similarity and differences in the pattern of targets of Lfcin B in different species, i.e., yeast and *E. coli*, we have compared the enrichment results of Lfcin B obtained from the protein targets of yeast and the protein targets of *E. coli*.

2. Results

2.1. Yeast Proteome Microarrays Assay

To systematically identify the protein targets of Lfcin B and Histatin-5, high-throughput yeast proteome microarrays were employed. Yeast proteome microarray analysis aids in the parallel identification of protein targets of Lfcin B and Histatin-5 from the entire proteome of yeast. The overall schematic diagram of this study is depicted in Figure 1. Biotinylated Lfcin B and Histstin-5 were individually probed on yeast proteome microarrays and then further probed with DyLight 650-labeled streptavidin and DyLight 550-labeled anti-Glutathione-S-transferase (anti-GST) antibody. DyLight 650-labeled streptavidin was used to detect the biotinylated AMPs that bound to the proteins in yeast proteome microarrays. Because individually purified yeast proteins contain GST tags, anti-GST antibody labeled with DyLight 550 was probed to represent their relative protein amounts on the yeast proteome microarray. Each protein on yeast proteome microarray was printed in duplicate and yeast proteome microarrays assay were conducted in triplicate for Lfcin B and Histatin-5, individually.

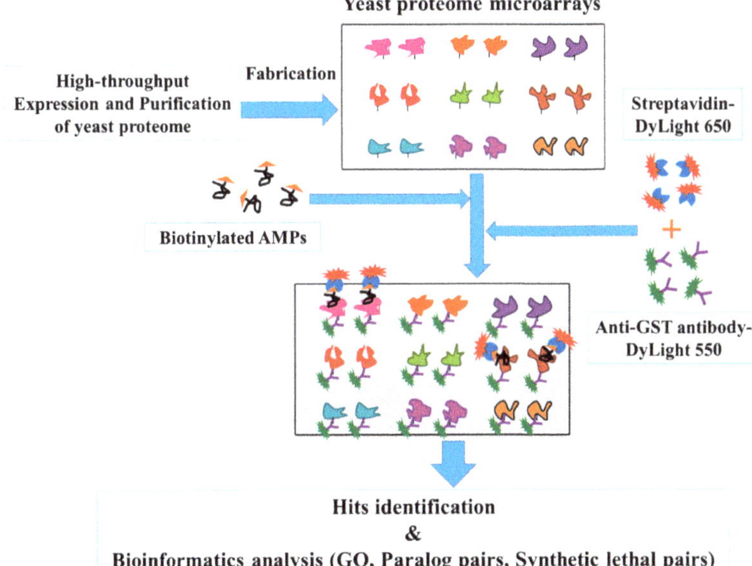

Figure 1. The schematic diagram of this study. Yeast proteome microarrays were fabricated from the ~5800 yeast proteins which were individually expressed and purified from yeast. To identify the protein targets, the fabricated yeast proteome microarrays were probed with biotinylated Lfcin B and Histatin-5, individually. Followed by the probing of DyLight 650 labeled streptavidin and DyLight 550 labeled anti-GST antibody on the yeast proteome microarrays to detect the signal of biotinylated AMPs bound to their specific binding partners and the amount of relative protein on proteome microarray, respectively. The identified hits of Lfcin B and Histatin-5 were systematically analyzed for GO and synthetic lethal pairs by using several online bioinformatics databases (DAVID tools and Synthetic Lethality).

The data obtained from the triplicate yeast proteome microarrays were analyzed together to identify the protein targets of Lfcin B and Histatin-5. After median scaling normalization, hits of Lfcin B and Histatin-5 from yeast proteome microarrays were selected by the local cutoff value that is greater than mean plus two standard deviations and ratio higher than 0.5 of AMP to the anti-GST antibody signal. The total number of protein targets identified for Lfcin B and Histatin-5 from yeast proteome microarrays were 140 and 137, respectively. These hits were all validated by eyeballing their images. The representative images of yeast proteome microarrays assay of Lfcin B (Figure 2A) and the enlarged protein images of representative protein targets of Lfcin B that appeared in the triplicate yeast proteome microarrays assay of Lfcin B are depicted in Figure 2B. Similarly, the representative images of yeast proteome microarrays assay of Histatin-5 on yeast proteome microarrays (Figure 2C) and the enlarged protein images of representative protein targets that appeared in the triplicate microarray assays of Histatin-5 on yeast proteome microarrays are depicted in Figure 2D. The identified protein targets of Lfcin B and Histatin-5 were analyzed for common and unique targets. Common hits are the protein targets present both in Lfcin B and Histatin-5 whereas unique hits are the protein targets present only in Lfcin B or histatin-5. That means the unique protein targets of Lfcin B are only present in Lfcin B and not in Histatin-5. The Venn diagram (Figure 3) shows that Lfcin B and Histatin-5 have common protein targets of 77. The unique protein targets of Lfcin B and Histatin-5 are 63 and 60, respectively. The total unique and common protein targets of Lfcin B and Histatin-5 from yeast proteome microarrays are displayed in Supplementary Table S1.

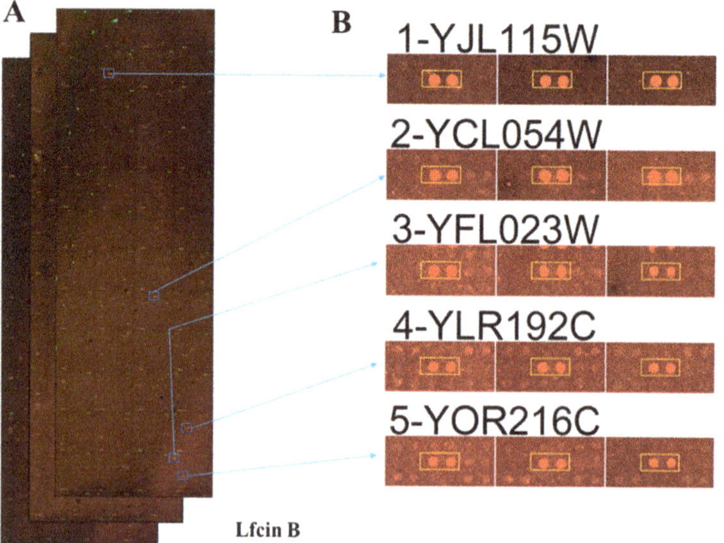

Figure 2. *Cont.*

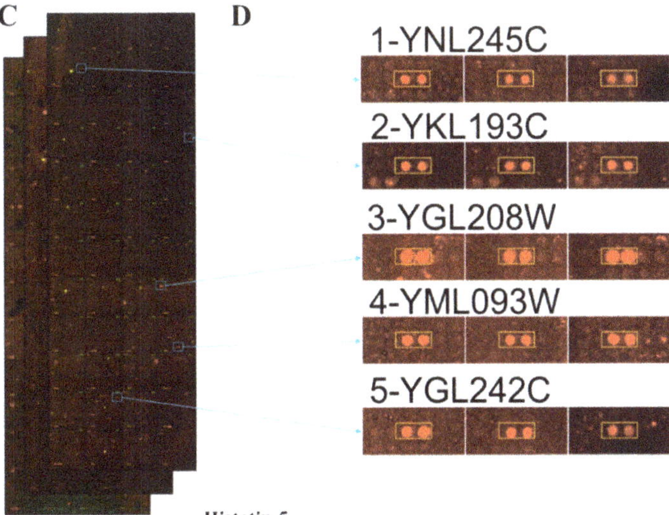

Figure 2. The representative yeast proteome microarray images and the representative hits of Lfcin B and Histatin-5 on yeast proteome microarrays. Images show the microarray and the enlarge hit image of Lfcin B and Histatin-5. Red and green color denotes the signal from DyLight 650 labeled streptavidin and DyLight 550 labeled anti-GST antibody, respectively. Red spots represent the signal of Lfcin B and Histatin-5 on yeast proteome microarrays. (**A**) Triplicate microarray images probed with biotinylated Lfcin B. The position of the representative 5 hits image is shown on the microarray. (**B**) Representative enlarged hits images of Lfcin B. (**C**) Triplicate microarray images probed with biotinylated Histatin-5. (**D**) Representative enlarged hits images of Histatin-5.

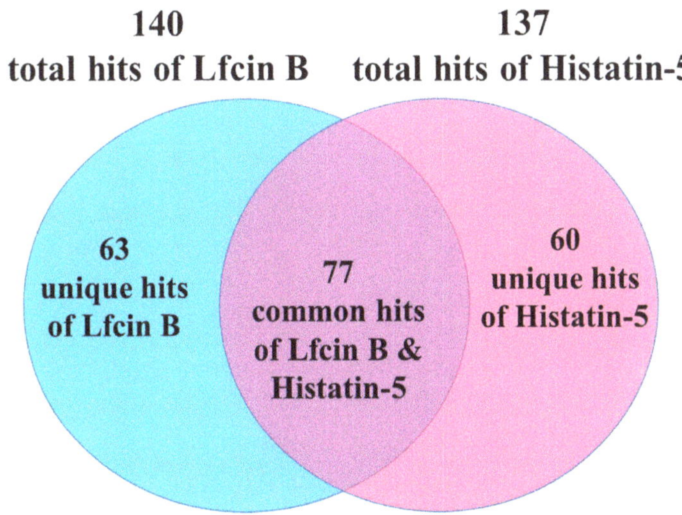

Figure 3. Unique and common hits of Lfcin B and Histatin-5 identified from yeast proteome microarrays. The protein targets of Lfcin B and Histatin-5 identified from yeast proteome microarrays are 140 and 137, respectively. These identified hits of Lfcin B and Histatin-5 were categorized as unique and common hits. Unique protein targets that are present only in Lfcin B are 63 hits whereas the unique protein targets that are present only in Histatin-5 are 60 hits. 77 protein targets were common to both Lfcin B and Histatin-5.

2.2. Enrichment Analysis in GO Biological Process for the Protein Hits of Lfcin B and Histatin-5

To know the over-representation proteins with similar function among the protein targets of Lfcin B and Histatin-5, GO enrichment analysis was performed. Using Database for Annotation, Visualization and Integrated Discovery (DAVID) online database [41], we obtained GO enrichment results in biological processes for the protein hits of Lfcin B and Histatin-5. The results display significant over-representation for the protein targets of Lfcin B and Histatin-5 (p-value cutoff of 0.05) in several biological processes (Figure 4). Interestingly, Lfcin B showed enrichment in most of the displayed categories; this depicts the involvement of Lfcin B to a broader targets range than Histatin-5. This might be the reason that Lfcin B has a lower MIC than Histatin-5 does. Lfcin B depicted the most obvious enrichment in "macromolecular complex subunit organization" that regulates macromolecule aggregation or disaggregation to form or alter protein complexes. Lfcin B showed enrichment in several unique categories that are present only in Lfcin B and not in Histatin-5. They are several negative regulation processes that stop or reduce several cellular process, such as: "negative regulation of cellular process", "negative regulation of cellular metabolic process", "negative regulation of metabolic process", and "negative regulation of macromolecule metabolic process" as well as regulation of several unique processes that modulates chemical reactions and pathways involved in normal metabolic processes such as "regulation of primary metabolic process", "regulation of cellular metabolic process", "regulation of macromolecule metabolic process", "regulation of catalytic activity", and "regulation of cellular component size" (Figure 4). These results indicate that Lfcin B targets proteins that manipulate the transcriptional response. Moreover, Lfcin B showed unique enrichment in the organization of cytoskeletal structures comprised of actin filaments, such as "actin filament organization", "regulation of actin filament-based process", and "actin filament-based process" as well as "cellular component assembly". This over-representation of Lfcin B protein targets shows the possible mechanism of Lfcin B entry inside the yeast and hamper their cytoskeleton. On the other hand, the unique over-representation for the protein targets of Histatin-5 was only observed in two biological processes, i.e., "cellular catabolic process" and "translational initiation". These two categories are related to the breakdown of substances and formation of an initial translational complex of the ribosome, mRNA, respectively.

Common enrichment for the protein targets of Lfcin B and Histatin-5 shows the conserved targets of these two AMPs on yeast. "Ribonucleoprotein complex biogenesis" is the well-conserved protein targets of Lfcin B and Histatin-5 that are involved in RNA-protein complex formation. Several compound metabolic processes were also conserved such as "cellular aromatic compound metabolic process", "organic cyclic compound metabolic process", "nucleobase-containing compound metabolic process", and "heterocycle metabolic process". "Negative regulation of molecular function" and "cellular component disassembly" is also common enrichment of Lfcin B and Histatin-5 (Figure 4).

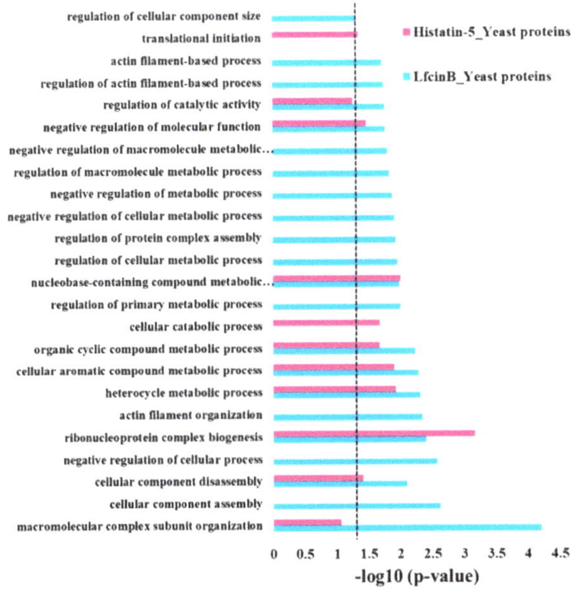

Figure 4. Enrichment in biological process of Lfcin B and Histatin-5 hits obtained from yeast proteome microarrays. The significant enrichment categories in biological process of Lfcin B and Histatin-5 of yeast-binding proteins (dotted line indicates the *p*-value of 0.05).

2.3. Enrichment Analysis in GO Molecular Function and Cellular Component

In addition to GO enrichment in the biological process, we also analyzed GO enrichment in molecular function and cellular component for the protein targets of Lfcin B and Histatin-5 (Figure 5). The analysis of over-representation for the protein targets of Lfcin B in GO molecular function (Figure 5A) showed unique enrichment in "histone binding" (related to DNA binding protein), "basal transcription machinery binding" (related to proteins of basal transcription factors and RNA polymerase core enzyme) indicating the involvement of Lfcin B in several proteins related to DNA and RNA binding that regulate gene expression in yeast. Protein targets of Lfcin B also showed unique over-representation in "protein complex binding" and "ubiquitin-like protein transferase activity". On the other hand, Histatin-5 showed no unique enrichment. However, common enrichment for Lfcin B and Histatin-5 were observed in "ribonucleoprotein complex binding" and "enzyme binding". The enrichment in "ribonucleoprotein complex binding" is similar to the enrichment observed in biological process which indicates the highly conserved targets of Lfcin B and Histatin-5. These results also indicate that the target protein functions of Histatin-5 are also the target protein functions of Lfcin B, and yet Lfcin B targets additional protein functions (Figure 5A).

Figure 5B depicts the GO enrichment in cellular component for the protein targets of Lfcin B and Histatin-5. Unique enrichment of Lfcin B is observed in "cell projection part", "mating projection", "mating projection tip", "cell cortex", "site of polarized growth", and "eisosome filament"—these are related to cell and mating projection as well as several signaling pathways which manipulate the development and communication. Lfcin B specifically showed unique enrichment in "Dom34-Hbs1 complex" that indicates its effect on cotranslational mRNA quality control. On the other hand, Histatin-5 showed no unique enrichment. These results also depicted the wider range of over-representation of protein targets in Lfcin B than Histatin-5. In common enrichment, the enriched categories belong to "intracellular non-membrane-bounded organelle", "intracellular ribonucleoprotein complex", "organelle lumen", and "intracellular organelle lumen". These results showed conserved enrichment of Lfcin B and Histatin-5 in targeting yeast.

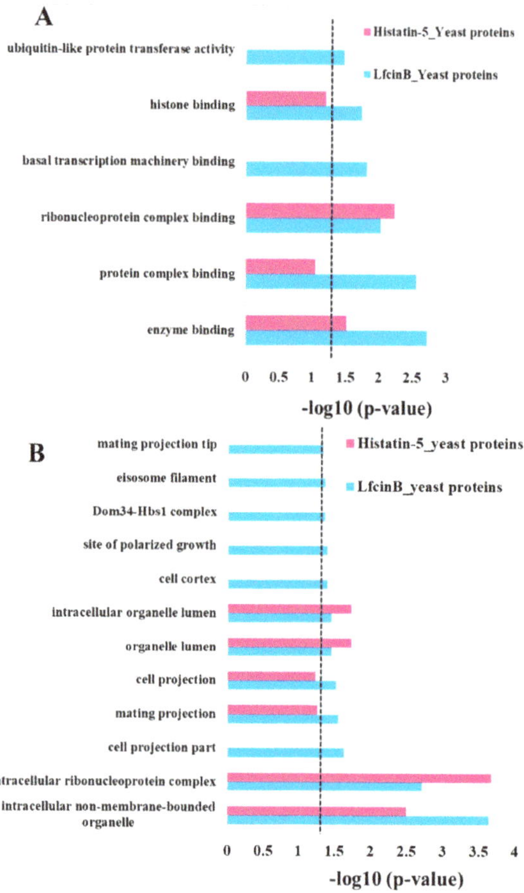

Figure 5. Functional enrichment of Lfcin B and Histatin-5 hits obtained from yeast proteome microarrays. The significant enrichment categories of Lfcin B and Histatin-5 of yeast-binding proteins. (**A**) Enrichment in molecular function (**B**) Enrichment in cellular component (dotted line indicates the *p*-value of 0.05).

2.4. Comparison of Lfcin B Protein Targets of Yeast and E. coli

To analyze the target pattern of Lfcin B in yeast and *E. coli*, we compared the enrichment results of protein targets of Lfcin B obtained from yeast proteome microarrays (in this study) and *E. coli* proteome microarrays (our previous study) [39]. Figure 6 shows the comparison results of GO enrichment in biological process (Figure 6A), molecular function (Figure 6B) and cellular component (Figure 6C) of the protein targets of Lfcin B in yeast and *E. coli*. In biological process, the protein targets of Lfcin B from yeast and *E. coli* showed only 7 common enrichment categories whereas 15 and 14 unique enrichment categories, respectively (Figure 6A). Several unique enrichments result of Lfcin B for yeast and *E. coli* indicated that Lfcin B exerted different target patterns in yeast and *E. coli*. These results also mean that the mechanism by which Lfcin B inhibits yeast is different from *E. coli*. Some targets of Lfcin B are conserved between yeast and *E. coli* and showed common enrichment that were mostly related to metabolic processes, such as: "heterocycle metabolic process", "cellular aromatic compound metabolic process", "organic cyclic compound metabolic process", "regulation of primary metabolic process", "nucleobase-containing compound metabolic process", "regulation of cellular metabolic process", and "regulation of macromolecular metabolic process".

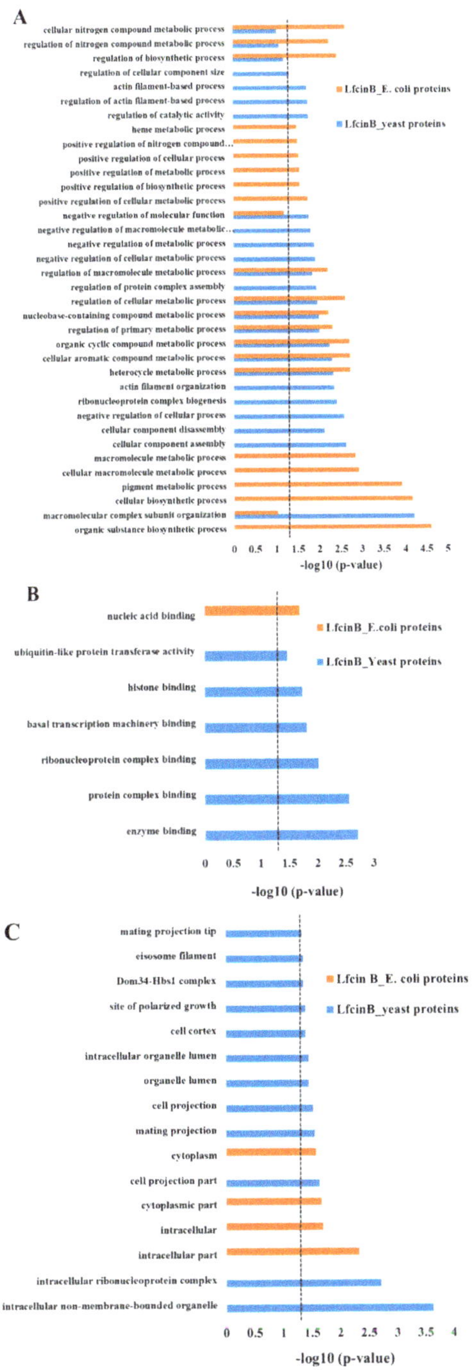

Figure 6. Functional enrichment of Lfcin B hits obtained from *E. coli* proteome microarray and yeast proteome microarrays. The comparison of enrichment categories of Lfcin B target proteins from yeast and *E. coli* proteomes. (**A**) Enrichment in biological process (**B**) Enrichment in molecular function and (**C**) Enrichment in cellular component (dotted line indicates the *p*-value of 0.05).

The difference in the mechanism of actions of Lfcin B in yeast and *E. coli* were also obtained by comparing enrichment results of molecular function and cellular component. The result depicted in Figure 6B shows the enrichment categories in molecular function. Lfcin B showed single unique enrichment in "nucleic acid binding" for the protein targets of *E. coli* whereas for protein targets of Lfcin B from yeast showed significant enrichment in several functions related mostly to protein-binding, such as "enzyme binding", "protein complex binding", "ribonucleoprotein complex binding", "basal transcription machinery binding", "histone binding" and "ubiquitin-like protein transferase activity". It is very interesting to observe that Lfcin B targets most yeast proteins with protein-binding function and most *E. coli* proteins with nucleic acid binding function. In case of cellular component (Figure 6C), *E. coli* protein targets of Lfcin B showed enrichment in the cytoplasm and intracellular whereas yeast protein targets of Lfcin B were enriched in several categories related to cell, ribonucleoprotein, organelle lumen, and mating projection, as well as several signaling pathways.

The above enrichment comparison results of protein targets of Lfcin B from yeast and *E. coli* showed several functional enrichments in yeast than that of *E. coli*, indicating the wider targets of Lfcin B in the case of yeast than *E. coli*. Thus, Lfcin B causes higher effect against yeast than that of *E. coli*.

2.5. Identification of Synthetic Lethal Pairs Targeted by Lfcin B and Histatin-5

The identification of synthetic lethal pairs is critical for deciphering the mechanism of action as they exert lethal effects on growth. Synthetic lethality database [42] was used to identify the synthetic lethal pair for the protein targets of Lfcin B and Histatin-5. Within the protein targets of Lfcin B, we identified 11 synthetic lethal pairs. All the identified synthetic lethal pairs within the protein targets of Lfcin B are shown in Table 1. Among these 11 synthetic lethal pairs of Lfcin B, one paralog pair (SIS2–VHS3) was also identified. Within the protein targets of Histatin-5, we identified a total of three synthetic lethal pairs. Synthetic lethal pairs cause lethal effect on yeast growth, and the higher the number of identified synthetic lethal pairs, the higher is the lethality impact on growth. A total of 11 synthetic lethal pairs identified within the yeast protein targets of Lfcin B would cause intense lethality effect on yeast than the three synthetic lethal pairs identified within the protein targets of Histatin-5. This might also be the potential reason for the previously observed lower MIC of Lfcin B against *Saccharomyces cerevisiae* than that of Histatin-5.

Table 1. Synthetic lethal pairs within the total protein targets of Lfcin B and Histatin-5. * denotes paralog pairs among the identified synthetic lethal pairs.

Synthetic Lethal Pairs within the Total Protein Targets of Lfcin B	Synthetic Lethal Pairs within the Total Protein Targets of Histatin-5
ASF1–RAD50	ASF1–RAD50
SIS2–VHS3 *	ASF1–SPT16
ASF1–ORC2	CDC53–DBP10
RAD6–SPT8	
CDC34–CDC53	
ASF1–SPT16	
RAD50–SLX8	
NPL3–RAD6	
CDC53–DBP10	
CDC34–RAD30	
ABZ1–CDC34	

We further analyzed the synthetic lethal pair between the unique protein targets of Lfcin B and Histatin-5 by using synthetic lethality database. Two synthetic lethal pairs were identified between the unique protein targets of Lfcin B and Histatin-5. The identified synthetic lethal pairs and the function of individual proteins are depicted in Table 2. Also, the individual protein images and their signals from triplicate yeast proteome microarrays were validated by eyeballing and the enlarged protein images are illustrated in Figure 7. The first synthetic lethal pair (SPT8 and HFI1) identified between the unique protein targets of Lfcin B and Histatin-5 are the subunits of the Spt-Ada-Gcn5 acetyltransferase (SAGA) complex. SAGA complex is involved in histone modification that characterizes histone acetyltransferase and histone deubiquitinase [43]. The synthetic lethality caused by Lfcin B and Histatin-5 treatment will hamper the structural integrity as well as the histone modification function of SAGA complex. Moreover, SPT8 and HFI1 subunits of SAGA complex are reported to bind with TATA-binding protein (TBP) and function in delivering TBP to TATA box [44]. Thus, this lethal pair will also effect initiation of transcription processes.

Table 2. Synthetic lethal pairs between the unique hits of Lfcin B and Histatin-5 along with the functions of individual proteins involved in these synthetic lethal pairs.

Synthetic Lethal Pairs between the Unique Protein Targets of Lfcin B and Histatin-5	
SPT8–HFI1	
RAD6–HFI1	
Protein target of Lfcin B	
SPT8	Subunit of SAGA complex; SPT8 bind to TATA box binding protein (TBP) and direct TBP to bind TATA box. TBP binding transcription coregulator involved in histone acetylation, chromatin organization, and both positive and negative regulation of transcription by RNA polymerase II
RAD6	The ubiquitin-conjugating enzyme (E2), involved in ubiquitination of histones and substrates of the N-end rule pathway
Protein target of Histatin-5	
HFI1	Subunit of SAGA complex; required for structural integrity; a histone acetyltransferase-coactivator complex that is involved in global regulation of gene expression through acetylation and transcription functions

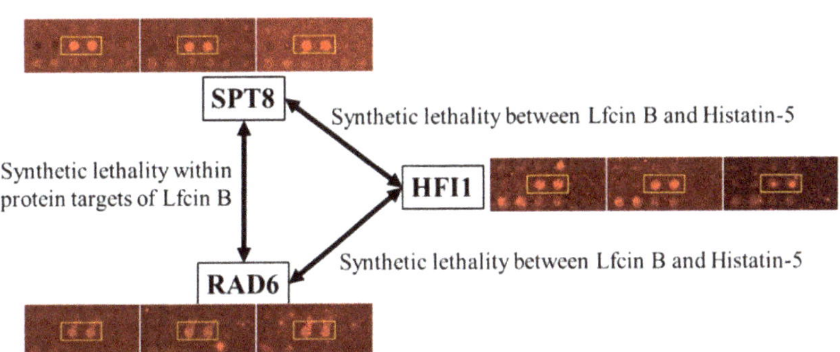

Figure 7. Synthetic lethality and enlarge protein images of SPT8, RAD6, and HFI1 on yeast proteome microarray. The enlarged protein images of two synthetic lethality pairs identified between the unique hits of Lfcin B and Histatin-5 as well as the synthetic lethality pair within the protein targets of Lfcin B.

The second synthetic lethal pair (RAD6 and HFI1) identified between Lfcin B and Histatin-5 also targets the histone modification function. RAD6 (ubiquitin associated enzyme E2), together with BRE1 (ubiquitin enzyme E3) is known to modify histone H2B-lysine at position 123 with ubiquitin [45]. This monoubiquitination of histone H2B is not for H2B degradation, rather the ubiquitin in H2B

(H2B~Ub) server as a signal for various aspects of gene expression, such as initiation and elongation of transcription as well as DNA replication and repair [46]. Ubiquitination in H2B-lysine 123 is reversed by SAGA complex that functions as deubiquitination of H2B~Ub and causes acetylation of histone [47]. H2B~Ub up-regulate histone H3-lysine 4 methylation and down-regulate histone H3-lysine 36 methylation, whereas SAGA deubiquitination of H2B acts in the opposite way and reduce H3-lysine 4 methylation and increase H3-lysine 36 methylation levels. Together, ubiquitination and deubiquitination are involved in transcriptional activation [48]. Lfcin B targeting RAD6 would hamper histone H2B ubiquitination whereas Histatin-5 targeting HFI1 would disrupt SAGA complex and cause functional defect (deubiquitination and acetylation) of SAGA complex to modify histone H2B. Moreover, SPT8 and RAD6 are both the protein targets of Lfcin B and were also reported to have synthetic lethality (Table 1). The mechanism of synthetic lethality is similar to RAD6 and HFI1 (as explained above).

It is known that synthetic lethal pairs identified during treatment with two drugs may have a synergistic effect [49–51], as the two drugs targeting the individual protein of synthetic lethal pair cause extra synthetic lethality. The two synthetic lethal pairs identified between the unique protein targets of Lfcin B and Histatin-5 ensure additional synthetic lethal effects on yeast growth that are not observed with individual treatment of Lfcin B or Histatin-5. Moreover, both these identified synthetic lethal pairs are involved in the structure and function of SAGA protein complex that regulates gene expression by modifying histone. Additional synthetic lethal pairs, as well as the same protein complex target, will cause a significantly higher lethality effect on yeast by the combined treatment of Lfcin B and Histatin-5. Thus, we hypothesized synergistic combination between Lfcin B and Histatin-5 on yeast.

2.6. Validation of Synergistic Combination between Lfcin B and Histatin-5

Based on two pairs of synthetic lethality interactions targeted by unique hits of Lfcin B and Histatin-5, we predicted a synergistic effect between Lfcin B and Histatin-5. This prediction was experimentally validated by the growth inhibition curve in the presence of individuals and a combination of Lfcin B and Histatin-5 (Figure 8). A combination of Lfcin B and Histatin-5 clearly showed drastic inhibition effects on yeast growth curves compared with individual AMPs. To show the result obtained is synergistic, we calculated the expected combinational value of Lfcin B and Histatin-5 from the anticipated optical density (OD) values of the inhibitory effect of the individual Lfcin B and Histatin-5 on yeast growth curves in sequence (i.e., OD of yeast without AMP multiplied by the percentage of the remaining yeast after the treatment of Lfcin B and further multiplied by the percentage of the remaining yeast after the treatment of the Histatin-5). A significant difference was observed between the expected combination value and experimental combination value of Lfcin B and Histatin-5. This result demonstrated the synergistic combination of Lfcin B and Histatin-5 against yeast growth.

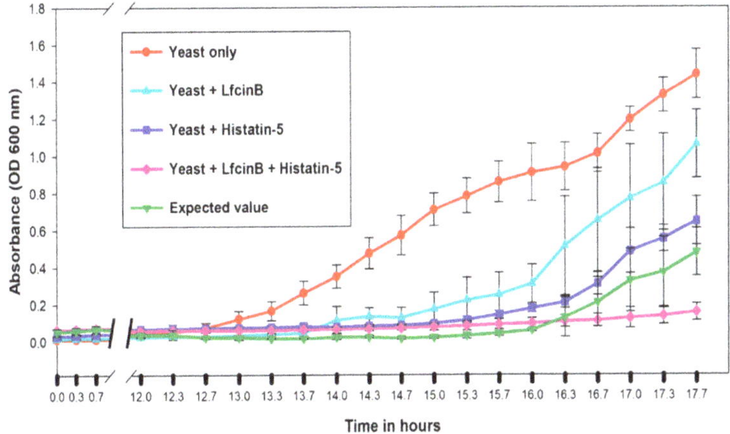

Figure 8. Growth inhibition effect of individual and combination of Lfcin B and Histatin-5 on yeast. Yeast was grown without AMPs and in the presence of individual and combination of Lfcin B (15 µg/mL) and Histatin-5 (20 µg/mL). Significant inhibition in growth was observed with the combination of Lfcin B and Histstin-5. The combinational inhibition of Lfcin B and Histatin was significantly lower than the expected value of Lfcin B and Histatin-5 combination calculated from the individual inhibition of Lfcin B and Histatin-5, concluding the synergy combination between Lfcin B and Histatin-5.

3. Discussion

Lfcin B and Histatin-5 are potent AMPs with antifungal activities [16,52]. Given the lack of entire targets knowledge, the potential mechanisms of antifungal AMPs are not fully explored. In this study, we have systematically identified the entire yeast protein targets of Lfcin B and Histatin-5 by using yeast proteome microarrays. A total of 140 and 137 protein targets of Lfcin B and Histatin-5 were identified from yeast proteome microarrays, respectively. Lfcin B and Histatin-5 can penetrate the yeast cell envelope so we assume that the bioavailability of Lfcin B and Histatin-5 on the yeast proteome microarrays will be similar to yeast in in vivo.

Earlier studies with *Candida albicans* have shown the binding of Histatin-5 to the cell surface proteins SSA1 and SSA2 [53,54]. In our yeast proteome microarray, both SSA1 and SSA2 were absent, thus we did not identify them in our hit list. Histatin-5 is reported to target mitochondria [26,55] and generate and release reactive oxygen species (ROS) that causes cell death [52]. Thus, we looked for the protein targets of Histatin-5 that are involved in mitochondria. We identified seven proteins: AIM21, AIM26, DOA1, YMR26, ETR1, ATP5, and RML2, which are involved in several mitochondrial functions, and six of them (not AIM21) are localized inside the mitochondria. AIM21 is a cytoplasmic protein that helps mitochondrial migration along actin filaments. This result depicted that our finding not only supported the previous finding but also provided the complete targets of Histatin-5 in mitochondria.

Ergosterol is the major fungal membrane sterol essential for fungal cell viability and is absent in humans [56]. Most of the antifungal drugs on the market target ergosterol biosynthesis enzymes [57]. Among the hits of Lfcin B, Lanosterol synthase (ERG7) belonging to ergosterol biosynthesis was identified. Molsidomine, a drug used as a vasodilator for the treatment of angina, was reported to target ERG7 and showed potential antifungal activity [58]. However, no antifungal drug targeting ERG7 is commercially available on the market.

Furthermore, two protein targets belonging to the ergosterol pathway were identified individually in Lfcin B and Histatin-5 by lowering the hit identification cutoff value from 2SD to 1SD. Lfcin B targeted ERG1 and ERG7 whereas Histatin-5 targeted ERG1 and ERG12. These protein targets were individually validated by eyeballing and the enlarged images of these proteins shown in Supplementary Figure S1. ERG1, the common target of Lfcin B and Histatin-5, is the first enzyme in ergosterol biosynthesis and a well-known target of antifungal drugs class, Allylamines (naftifine and terbinafine) [59]. ERG7 is the second enzyme in the ergosterol biosynthesis after ERG1. ERG1, ERG7, and ERG12 are essential genes in the biosynthesis of ergosterol and their individual deletion is lethal for yeast. Deletion of Erg7 results in accumulation of ERG9, which hampers squalene production and amounts to ergosterol biosynthesis. String analysis showed protein–protein interaction between ERG1, ERG7, and ERG12 (data not shown). ERG7 and ERG12 have not been exploited as the target; thus, Lfcin B and Histatin-5 can be a potential target to cope with emerging antifungal drug resistance fungi.

Synthetic lethality approach has recently gained attention for its potential to understand and design medication for cancer [32,33]. Synthetic lethality describes a useful pairwise interaction, where the simultaneous deletion of both the component of the pair cause growth defects but the deletion of an individual component has no effect on growth [31]. We have used a synthetic lethality approach to identify synthetic lethal pairs within the protein targets of Lfcin B and Histain-5. The number of synthetic lethal pairs identified within the protein targets of Lfcin B and Histatin-5 might demonstrate the potential effect it exerts on yeast. Within the protein targets, 11 synthetic lethal pairs in Lfcin B and three synthetic lethal pairs in Histatin-5 were identified. These results showed that Lfcin B exerts a more lethal effect than Histain-5. Interestingly, it is reported that the MIC of Lfcin B is lower than Histatin-5. Thus, our analysis provided the mechanism of lower MIC of Lfcin B against yeast than Histatin-5, based on the number of identified synthetic lethal pairs. Our analysis also identified two synthetic lethal pairs between the unique protein targets of Lfcin B and Histatin-5. Apart from the individual synthetic lethal pairs of Lfcin B and Histatin-5, the combination of Lfcin B and Histatin-5 targeted two additional synthetic lethal pairs. These additional synthetic lethal pairs are the subunits of SAGA protein complex and are also involved in similar functions. The treatment of yeast with Lfcin B and Histatin-5 might cause a greater inhibition effect as they will shut down the structure and function of SAGA complex. Together, Lfcin B and Histatin-5 might exert synergistic combinations. We performed in vivo growth inhibition assays to test our synergistic combination hypothesis between Lfcin B and Histatin-5. The significantly higher inhibition results in the combined treatment of Lfcin B and Histain-5 (experimental data) than the expected combinational value confirmed the synergistic combination between Lfcin B and Histatin-5.

In this study, we have explored the entire biological targets of Lfcin B and Histatin-5 using yeast proteome microarrays. The identified protein hits were analyzed to observe over-representation in different GO enrichment categories. The results showed wider GO enrichment results for Lfcin B than Histatin-5 in all the three categories of the biological process, molecular function, and cellular component. Moreover, 11 synthetic lethal pairs were identified within the protein targets of Lfcin B whereas only three synthetic lethal pairs were identified within the protein targets of Histatin-5. Both these results proved the higher lethal effect of Lfcin B on yeast than Histatin-5. This might result in lower MIC of Lfcin B than Histatin-5 which is in accordance with the previously reported results. Two synthetic lethal pairs were identified between the unique protein targets of Lfcin B than Histatin-5. Thus, we hypothesized a synergistic combination between Lfcin B and Histatin-5. Based on the hypothesis, we designed an inhibition assay to test it and successfully validated our hypothesis. In the future, we will further explore the mechanism of actions of other AMPs with antifungal activities.

4. Materials and Methods

4.1. Expression and Purification of the Entire Yeast Proteome

Yeast entire proteome was expressed and purified using the previously reported protocol [19,20]. Briefly, yeast library consists of ~5800 open reading frames, individually cloned in high-copy URA3 expression vector with Glutathione-S-transferase–poly-histidine (GST–HisX6) tag. These clones use galactose-inducible GAL1 promoter to produce GST-HisX6 fusion proteins. Yeast clones were stored in 96-well format at −80 °C. For protein expression, yeast clones were first grown on SC-URA3-glucose agar plate at 30 °C for 48 h. Colonies were transferred to SC-URA3-glucose medium in 96-deep well plates and incubated at 30 °C for 24 h. Yeast clones were sub-cultured in SC-URA3-raffinose medium in 12 channel reservoirs (a single 96-well plate requires eight 12-channel reservoir plates; in total, the yeast library contains 66 plates of 96 well plates) and incubated at 30 °C for 14–16 h (till Optical Density at 600 nm reached to 0.6–1.0). For protein expression, 2% galactose was added to the cultures in SC-URA3-raffinose medium and further incubated for 4 h. Cells in 12-channel reservoirs were centrifuged at 4000 rpm for 2 min and cells were re-suspended in 800 μL cold water. Each strain in 12-channel reservoir was pooled together in each well of 96-deep well plates. Cells were harvested by centrifuging at 4000 rpm for 2 min and stored at −80 °C ahead of protein purification.

For protein purification, yeast colonies were ruptured, and GST tag proteins were purified by using glutathione (GSH) beads, following standard protocol. Briefly, 200 μL of 0.7 mm zirconia beads (Biospec Products, Inc., Bartles- ville, OK, USA) were loaded in each well of 96-deep well plate with cell pellets. Freshly prepared 400 μL of lysis buffer with protease inhibitors (1 mM phenylmethylsulfonyl fluoride (PMSF), 50 μM calpain Inhibitor I (LLnL), 1 μM MG132 and 50 × dilution of Roche protease inhibitor (Roche Molecular Biochemicals, Basel, Switzerland)) were added in each well and cell were thaw at 4 °C for 15 min. If not stated otherwise, the chemicals used here were purchased from Sigma-Aldrich (Saint Louis, MO, USA) Again, 400 μL of freshly prepared lysis buffer with protease inhibitors was added and the 96-deep well plates were vortexed at 4 °C for 30 min. After centrifugation at 4000 rpm for 15 min, the supernatant was transferred to new 96-deep well plates. Pre-washed 100 μL of GSH Sepharose 4B beads (GE Healthcare, Chicago, IL, USA) was added to each well and plates were sealed tight using 96-well cap mats (Thermo Fisher Scientific, Waltham, MA, USA). The plates were placed vertically on the shaker and shaken gently (80 rpm) at 4 °C for 80 min. Homogeneous mixtures in 96-deep well plates were transferred to 96-well filter plates (Thermo Fisher Scientific, Waltham, MA, USA) with a filter pore size of 20 μm. The contents were washed with wash buffer I and II and gentle spin dry (1000 rpm for 1 min) to remove extra wash buffer. The bottoms of filter plates were sealed and 50 μL of elution buffer containing reduced GSH was added in each well. 96-well filter plates were shaken vigorously at 4 °C for 1 h. The elutes (proteins) were collected in 96-well receiver plates by centrifugation at 4000 rpm for 2 min. To determine the purity and concentration of purified proteins, proteins were randomly selected for SDS-PAGE and Coomassie blue staining was used afterward.

4.2. Fabrication of Yeast Proteome Microarrays

Previously reported protocol was applied to fabricated yeast proteome microarrays [20]. Briefly, the entire purified yeast proteins in 96-well format were transferred to 384-well format by using Liquidator 96 manual pipetting system (Mettler-Toledo Rainin, LLC Oakland, CA, USA). Before printing, the optimal concentrations of landmarks assisted to align blocks on the microarray were determined and tested. In a cold room, the individual proteins and landmarks were printed in duplicate on aldehyde-coated glass slides by using CapitalBio SmartArrayer™ 136 (CapitalBio Corporation, Beijing, China). CapitalBio SmartArrayer is a high-throughput microarray spotter with 48 pins and print 48 proteins at the same time. After printing, the chips were left in the cold room for overnight and finally stored at −80 °C. To monitor the shape, size, and uniformity for each protein spot on a chip, DyLight 550 conjugated anti-GST monoclonal antibody (Rockland Immunochemicals, Gilbertsville, PA, USA) was probed, washed and scanned with LuxScan™ (10K Microarray Scanner; CapitalBio Corporation, Beijing, China).

4.3. Yeast Proteome Chip Assays with Lfcin B and Histatin-5

N-terminal biotin labeled (biotinylated) AMPs (Lfcin B and Histatin-5) were purchased from Kelowna International Scientific Inc. (Taipei, Taiwan). AMPs were aliquot and stored at −80 °C. Below are the amino acid sequences of Lfcin B and Histatin-5 used in this study.

Lfcin B (25 residues):   H2N–FKCRRWQWRMKKLGAPSITCVRRAF–COOH
Histatin-5 (24 residues):   H2N–DSHAKRHHGYKRKFHEKHHSHRGY–COOH

Yeast proteome microarray was first blocked with 3% bovine serum albumin (BSA; Sigma-Aldrich, Saint Louis, MO, USA) in 1X PBS for 1 h at room temperature (RT) with shaking (50 rpm). Chips were washed once with 300 mL PBS-T (0.05% Tween 20) at RT with shaking (50 rpm) for 5 min. Biotinylated AMP (5 μM) was diluted in 1% BSA in 1X PBS and probed individually on yeast proteome microarray with incubation for 1 h at RT with shaking (50 rpm). Chips were washed once with 300 mL PBS-T at RT with shaking (50 rpm) for 5 min. To detect signal from GST tag in yeast proteins and biotinylated AMP bound to yeast proteins, DyLight™ 550 labeled anti-GST antibody (Abcam, Cambridge, UK) and DyLight™ 650 labeled streptavidin (Thermo Fisher Scientific, Waltham, MA, USA) were probed on yeast proteome microarray and incubated with shaking (50 rpm) at RT for 1 h. Finally, the chips were washed for 3 times with 300 mL TBS-T at RT with shaking (50 rpm) for 5 min each. The chips were dried using centrifuge at 1000 rpm for 1 min and scanned with LuxScan.

The chip scan files (TIF format) were opened with GenePix Pro 6.0 software (Axon Instruments, Foster City, CA, USA) and each protein spots on yeast proteome microarray were aligned with their protein names. Binding signals of protein spots on yeast proteome chips were exported in GPR files and later opened with excel files to analyze the results. Median scaling normalization was applied to normalize the signals of Lfcin B, Histatin-5, and anti-GST antibody. After the normalization, the relative binding ability of each AMP to each protein was estimated by the ratio of the fluorescence intensity of each AMP to the anti-GST antibody. The hits were defined as positive only if they met two cutoffs below. First, the signal is higher than the local cutoff, which was defined as one standard deviation (SD) above the signal mean for each spot. Second, the fold change of each AMP signal to anti-GST antibody should be higher than 0.5. Thus, the protein hit list was generated according to the relative binding ability.

4.4. Bioinformatics Analysis of Gene Ontology

The *Saccharomyces* Genome Database (SGD) (https://www.yeastgenome.org/) and Universal Protein Resource (UniProt) database (https://www.uniprot.org/) provide comprehensive biological information on particular species [60,61]. SGD provides genome-wide information only on *Saccharomyces cerevisiae* whereas UniProt covers a large group of organisms from prokaryotes to eukaryotic, including *Saccharomyces cerevisiae*. These databases were used for the identification of recommended name (Standard name) from ordered locus names (Systematic name) as well as a brief description of proteins and related function and location inside the cell.

The DAVID (https://david.ncifcrf.gov/) [41] is an online database that provides several tools such as GO, Pfam domain analysis and other enrichment analysis. DAVID was used for comprehensive analysis of functional annotations of positive hits of Lfcin B and Histatin-5, individually. This investigation was necessary to understand the biological meaning of the identified yeast proteins targets of Lfcin B and Histatin-5. Here, we used the GO terms to identify biological processes, cellular component and molecular function, as well as the statistical analysis, was performed via *p*-value to determine the significant enrichment in each category.

4.5. Bioinformatics Analysis of Synthetic Lethality Pairs

Synthetic Lethality database (http://histone.sce.ntu.edu.sg/SynLethDB/index.php) (SynLethDB) is the open-source database of synthetic lethality gene pairs from different source organisms [42]. SynLethDB was used for the comprehensive identification of synthetic lethality pairs between the protein hits of Lfcin B and Histatin-5.

4.6. Growth Inhibition Assay on Yeast

Yeast (*Saccharomyces cerevisiae* Y258) was incubated with or without AMPs to observe individual and combination inhibition effect of AMP. Yeast was grown on yeast extract peptone dextrose (YPD) agar plate for 48 h at 30 °C. A single colony was transferred into YPD liquid media and incubated with shaking for 24 h at 30 °C. Optical density at 600 nm (OD_{600} nm) was detected for culture grown for

24 h and the culture was diluted to approximately 0.001 in YPD medium. The diluted culture of yeast was added in Nunc^{TM} F 96 well plate (Nalge Nunc International, Rochester, NY, USA) containing an individual and combined number of AMPs in specific wells. The working concentration of Lfcin B and Histatin-5 were 15 µg/mL and 20 µg/mL, respectively. The 96-well plate was incubated at 30 °C in an automated Synergy 2 Multi-Mode Microplate Reader (BioTek Instruments Inc., winooski, VT, USA) and the growth was monitored on a regular interval of 20 min (Shaking for 15 s prior to reading) at OD_{600} nm. Data were collected automatically using Gen5™ reader control and data analysis software (BioTek Instruments Inc., winooski, VT, USA). To show graphic representation of growth inhibition assay, the data were plotted using Sigmaplot. Expected value at given interval $(E_{XY}) = OD_0 *(OD_X/OD_0) *(OD_Y/OD_0)$ [62]. Where OD_0 is the optical density of yeast in absence of AMP, OD_X is the optical density of yeast in the presence of Lfcin B and OD_Y is the optical density of yeast in the presence of Histatin-5 at a given interval of time.

Supplementary Materials: The following are available online at http://www.mdpi.com/1422-0067/20/17/4218/s1, Figure S1: Image of hits belonging to Ergosterol biosynthesis; Lfcin B (ERG7 & ERG1) and Histatin-5 (ERG 12 & ERG1). Table S1: Two standard deviation (2SD) hits of Lfcin B and Histatin-5 from yeast proteome microarrays.

Author Contributions: C.-S.C. developed the idea and supervised the project. P.S. conducted the experiments. P.S. and W.-S.W. analyzed the data. P.S. and C.-S.C. wrote the manuscript.

Funding: This research was financially supported by Ministry of Science and Technology Taiwan, MOST 104-2320-B-008-002-MY3.

Acknowledgments: Authors thank Jagat Rathod for his valuable contribution in editing this manuscript.

Conflicts of Interest: The authors declare no competing interests.

References

1. Scorzoni, L.; de Paula, E.S.A.C.; Marcos, C.M.; Assato, P.A.; de Melo, W.C.; de Oliveira, H.C.; Costa-Orlandi, C.B.; Mendes-Giannini, M.J.; Fusco-Almeida, A.M. Antifungal Therapy: New Advances in the Understanding and Treatment of Mycosis. *Front. Microbiol.* **2017**, *8*, 36. [CrossRef]

2. McCarthy, M.W.; Kontoyiannis, D.P.; Cornely, O.A.; Perfect, J.R.; Walsh, T.J. Novel Agents and Drug Targets to Meet the Challenges of Resistant Fungi. *J. Infect. Dis.* **2017**, *216*, S474–S483. [CrossRef] [PubMed]

3. Kang, H.K.; Kim, C.; Seo, C.H.; Park, Y. The therapeutic applications of antimicrobial peptides (AMPs): A patent review. *J. Microbiol.* **2017**, *55*, 1–12. [CrossRef] [PubMed]

4. Mahlapuu, M.; Hakansson, J.; Ringstad, L.; Bjorn, C. Antimicrobial Peptides: An Emerging Category of Therapeutic Agents. *Front. Cell Infect Microbiol.* **2016**, *6*, 194. [CrossRef] [PubMed]

5. Hancock, R.E.; Patrzykat, A. Clinical development of cationic antimicrobial peptides: From natural to novel antibiotics. *Curr. Drug Targets Infect. Disord.* **2002**, *2*, 79–83. [CrossRef]

6. Shah, P.; Hsiao, F.S.; Ho, Y.H.; Chen, C.S. The proteome targets of intracellular targeting antimicrobial peptides. *Proteomics* **2016**, *16*, 1225–1237. [CrossRef] [PubMed]

7. Matejuk, A.; Leng, Q.; Begum, M.D.; Woodle, M.C.; Scaria, P.; Chou, S.T.; Mixson, A.J. Peptide-based Antifungal Therapies against Emerging Infections. *Drugs Future* **2010**, *35*, 197. [CrossRef]

8. De Lucca, A.J.; Walsh, T.J. Antifungal peptides: Novel therapeutic compounds against emerging pathogens. *Antimicrob. Agents Chemother* **1999**, *43*, 1–11. [CrossRef]

9. Vylkova, S.; Sun, J.N.; Edgerton, M. The role of released ATP in killing *Candida albicans* and other extracellular microbial pathogens by cationic peptides. *Purinergic. Signal* **2007**, *3*, 91–97. [CrossRef]

10. De Brucker, K.; Cammue, B.P.; Thevissen, K. Apoptosis-inducing antifungal peptides and proteins. *Biochem. Soc. Trans.* **2011**, *39*, 1527–1532. [CrossRef]

11. Delattin, N.; Cammue, B.P.; Thevissen, K. Reactive oxygen species-inducing antifungal agents and their activity against fungal biofilms. *Future Med. Chem.* **2014**, *6*, 77–90. [CrossRef] [PubMed]

12. Moore, S.A.; Anderson, B.F.; Groom, C.R.; Haridas, M.; Baker, E.N. Three-dimensional structure of diferric bovine lactoferrin at 2.8 A resolution. *J. Mol. Biol.* **1997**, *274*, 222–236. [CrossRef] [PubMed]

13. Kuwata, H.; Yip, T.T.; Tomita, M.; Hutchens, T.W. Direct evidence of the generation in human stomach of an antimicrobial peptide domain (lactoferricin) from ingested lactoferrin. *Biochim. Biophys. Acta* **1998**, *1429*, 129–141. [CrossRef]

14. Fernandes, K.E.; Carter, D.A. The Antifungal Activity of Lactoferrin and Its Derived Peptides: Mechanisms of Action and Synergy with Drugs against Fungal Pathogens. *Front. Microbiol.* **2017**, *8*, 2. [CrossRef] [PubMed]

15. Bruni, N.; Capucchio, M.T.; Biasibetti, E.; Pessione, E.; Cirrincione, S.; Giraudo, L.; Corona, A.; Dosio, F. Antimicrobial Activity of Lactoferrin-Related Peptides and Applications in Human and Veterinary Medicine. *Molecules* **2016**, *21*, 752. [CrossRef] [PubMed]

16. Wakabayashi, H.; Hiratani, T.; Uchida, K.; Yamaguchi, H. Antifungal Spectrum and Fungicidal Mechanism of an N-Terminal Peptide of Bovine Lactoferrin. *J. Infect Chemother.* **1996**, *1*, 185–189. [CrossRef]

17. Chan, D.I.; Prenner, E.J.; Vogel, H.J. Tryptophan- and arginine-rich antimicrobial peptides: Structures and mechanisms of action. *Biochim. Biophys. Acta* **2006**, *1758*, 1184–1202. [CrossRef]

18. Sutandy, F.X.; Qian, J.; Chen, C.S.; Zhu, H. Overview of protein microarrays. *Curr. Protoc. Protein Sci.* **2013**, *27*. [CrossRef]

19. Zhu, H.; Bilgin, M.; Bangham, R.; Hall, D.; Casamayor, A.; Bertone, P.; Lan, N.; Jansen, R.; Bidlingmaier, S.; Houfek, T.; et al. Global analysis of protein activities using proteome chips. *Science* **2001**, *293*, 2101–2105. [CrossRef]

20. Lu, K.Y.; Tao, S.C.; Yang, T.C.; Ho, Y.H.; Lee, C.H.; Lin, C.C.; Juan, H.F.; Huang, H.C.; Yang, C.Y.; Chen, M.S.; et al. Profiling lipid-protein interactions using nonquenched fluorescent liposomal nanovesicles and proteome microarrays. *Mol. Cell Proteom.* **2012**, *11*, 1177–1190. [CrossRef]

21. Xu, T.; Levitz, S.M.; Diamond, R.D.; Oppenheim, F.G. Anticandidal activity of major human salivary histatins. *Infect Immun.* **1991**, *59*, 2549–2554. [PubMed]

22. Puri, S.; Edgerton, M. How does it kill?: Understanding the *candidacidal* mechanism of salivary histatin 5. *Eukaryot. Cell* **2014**, *13*, 958–964. [CrossRef] [PubMed]

23. Raj, P.A.; Marcus, E.; Sukumaran, D.K. Structure of human salivary histatin 5 in aqueous and nonaqueous solutions. *Biopolymers* **1998**, *45*, 51–67. [CrossRef]

24. Khurshid, Z.; Naseem, M.; Sheikh, Z.; Najeeb, S.; Shahab, S.; Zafar, M.S. Oral antimicrobial peptides: Types and role in the oral cavity. *Saudi. Pharm. J.* **2016**, *24*, 515–524. [CrossRef] [PubMed]

25. Wakiec, R.; Gabriel, I.; Prasad, R.; Becker, J.M.; Payne, J.W.; Milewski, S. Enhanced susceptibility to antifungal oligopeptides in yeast strains overexpressing ABC multidrug efflux pumps. *Antimicrob. Agents Chemother* **2008**, *52*, 4057–4063. [CrossRef] [PubMed]

26. Helmerhorst, E.J.; Breeuwer, P.; van't Hof, W.; Walgreen-Weterings, E.; Oomen, L.C.; Veerman, E.C.; Amerongen, A.V.; Abee, T. The cellular target of histatin 5 on *Candida albicans* is the energized mitochondrion. *J. Biol. Chem.* **1999**, *274*, 7286–7291. [CrossRef]

27. Baev, D.; Li, X.S.; Dong, J.; Keng, P.; Edgerton, M. Human salivary histatin 5 causes disordered volume regulation and cell cycle arrest in *Candida albicans*. *Infect Immun.* **2002**, *70*, 4777–4784. [CrossRef]

28. Baev, D.; Rivetta, A.; Vylkova, S.; Sun, J.N.; Zeng, G.F.; Slayman, C.L.; Edgerton, M. The TRK1 potassium transporter is the critical effector for killing of *Candida albicans* by the cationic protein, Histatin 5. *J. Biol. Chem.* **2004**, *279*, 55060–55072. [CrossRef]

29. Koshlukova, S.E.; Lloyd, T.L.; Araujo, M.W.; Edgerton, M. Salivary histatin 5 induces non-lytic release of ATP from *Candida albicans* leading to cell death. *J. Biol. Chem.* **1999**, *274*, 18872–18879. [CrossRef]

30. Koshlukova, S.E.; Araujo, M.W.; Baev, D.; Edgerton, M. Released ATP is an extracellular cytotoxic mediator in salivary histatin 5-induced killing of *Candida albicans*. *Infect Immun.* **2000**, *68*, 6848–6856. [CrossRef]

31. Tong, A.H.; Lesage, G.; Bader, G.D.; Ding, H.; Xu, H.; Xin, X.; Young, J.; Berriz, G.F.; Brost, R.L.; Chang, M.; et al. Global mapping of the yeast genetic interaction network. *Science* **2004**, *303*, 808–813. [CrossRef] [PubMed]

32. Hartwell, L.H.; Szankasi, P.; Roberts, C.J.; Murray, A.W.; Friend, S.H. Integrating genetic approaches into the discovery of anticancer drugs. *Science* **1997**, *278*, 1064–1068. [CrossRef] [PubMed]

33. Kaelin, W.G., Jr. The concept of synthetic lethality in the context of anticancer therapy. *Nat. Rev. Cancer* **2005**, *5*, 689–698. [CrossRef] [PubMed]

34. Costanzo, M.; Baryshnikova, A.; Bellay, J.; Kim, Y.; Spear, E.D.; Sevier, C.S.; Ding, H.; Koh, J.L.; Toufighi, K.; Mostafavi, S.; et al. The genetic landscape of a cell. *Science* **2010**, *327*, 425–431. [CrossRef] [PubMed]

35. Hillenmeyer, M.E.; Fung, E.; Wildenhain, J.; Pierce, S.E.; Hoon, S.; Lee, W.; Proctor, M.; St Onge, R.P.; Tyers, M.; Koller, D.; et al. The chemical genomic portrait of yeast: Uncovering a phenotype for all genes. *Science* **2008**, *320*, 362–365. [CrossRef]

36. Collins, S.R.; Miller, K.M.; Maas, N.L.; Roguev, A.; Fillingham, J.; Chu, C.S.; Schuldiner, M.; Gebbia, M.; Recht, J.; Shales, M.; et al. Functional dissection of protein complexes involved in yeast chromosome biology using a genetic interaction map. *Nature* **2007**, *446*, 806–810. [CrossRef]

37. Bandyopadhyay, S.; Mehta, M.; Kuo, D.; Sung, M.K.; Chuang, R.; Jaehnig, E.J.; Bodenmiller, B.; Licon, K.; Copeland, W.; Shales, M.; et al. Rewiring of genetic networks in response to DNA damage. *Science* **2010**, *330*, 1385–1389. [CrossRef]

38. Zinovyev, A.; Kuperstein, I.; Barillot, E.; Heyer, W.D. Synthetic lethality between gene defects affecting a single non-essential molecular pathway with reversible steps. *PLoS Comput. Biol.* **2013**, *9*, e1003016. [CrossRef]

39. Tu, Y.H.; Ho, Y.H.; Chuang, Y.C.; Chen, P.C.; Chen, C.S. Identification of lactoferricin B intracellular targets using an *Escherichia coli* proteome chip. *PLoS ONE* **2011**, *6*, e28197. [CrossRef]

40. Ho, Y.H.; Sung, T.C.; Chen, C.S. Lactoferricin B inhibits the phosphorylation of the two-component system response regulators BasR and CreB. *Mol. Cell. Proteom.* **2012**, *11*, M111-014720. [CrossRef]

41. Huang, D.W.; Sherman, B.T.; Lempicki, R.A. Systematic and integrative analysis of large gene lists using DAVID bioinformatics resources. *Nat. Protoc.* **2009**, *4*, 44–57. [CrossRef] [PubMed]

42. Guo, J.; Liu, H.; Zheng, J. SynLethDB: Synthetic lethality database toward discovery of selective and sensitive anticancer drug targets. *Nucleic. Acids Res.* **2016**, *44*, D1011–D1017. [CrossRef] [PubMed]

43. Koutelou, E.; Hirsch, C.L.; Dent, S.Y. Multiple faces of the SAGA complex. *Curr. Opin. Cell Biol.* **2010**, *22*, 374–382. [CrossRef] [PubMed]

44. Sermwittayawong, D.; Tan, S. SAGA binds TBP via its Spt8 subunit in competition with DNA: Implications for TBP recruitment. *EMBO J.* **2006**, *25*, 3791–3800. [CrossRef] [PubMed]

45. Wood, A.; Krogan, N.J.; Dover, J.; Schneider, J.; Heidt, J.; Boateng, M.A.; Dean, K.; Golshani, A.; Zhang, Y.; Greenblatt, J.F.; et al. Bre1, an E3 ubiquitin ligase required for recruitment and substrate selection of Rad6 at a promoter. *Mol. Cell* **2003**, *11*, 267–274. [CrossRef]

46. Weake, V.M.; Workman, J.L. Histone ubiquitination: Triggering gene activity. *Mol. Cell* **2008**, *29*, 653–663. [CrossRef] [PubMed]

47. Samara, N.L.; Datta, A.B.; Berndsen, C.E.; Zhang, X.; Yao, T.; Cohen, R.E.; Wolberger, C. Structural insights into the assembly and function of the SAGA deubiquitinating module. *Science* **2010**, *328*, 1025–1029. [CrossRef] [PubMed]

48. Zhang, Y. Transcriptional regulation by histone ubiquitination and deubiquitination. *Genes Dev.* **2003**, *17*, 2733–2740. [CrossRef]

49. Ocana, A.; Pandiella, A. Novel Synthetic Lethality Approaches for Drug Combinations and Early Drug Development. *Curr. Cancer Drug Targets* **2017**, *17*, 48–52. [CrossRef]

50. Trosset, J.Y.; Carbonell, P. Synergistic Synthetic Biology: Units in Concert. *Front. Bioeng. Biotechnol.* **2013**, *1*, 11. [CrossRef]

51. Campbell, J.; Singh, A.K.; Santa Maria, J.P., Jr.; Kim, Y.; Brown, S.; Swoboda, J.G.; Mylonakis, E.; Wilkinson, B.J.; Walker, S. Synthetic lethal compound combinations reveal a fundamental connection between wall teichoic acid and peptidoglycan biosyntheses in *Staphylococcus aureus*. *ACS Chem. Biol.* **2011**, *6*, 106–116. [CrossRef] [PubMed]

52. Helmerhorst, E.J.; Troxler, R.F.; Oppenheim, F.G. The human salivary peptide histatin 5 exerts its antifungal activity through the formation of reactive oxygen species. *Proc. Natl. Acad. Sci. USA* **2001**, *98*, 14637–14642. [CrossRef] [PubMed]

53. Li, X.S.; Sun, J.N.; Okamoto-Shibayama, K.; Edgerton, M. *Candida albicans* cell wall ssa proteins bind and facilitate import of salivary histatin 5 required for toxicity. *J. Biol. Chem.* **2006**, *281*, 22453–22463. [CrossRef] [PubMed]

54. Sun, J.N.; Li, W.; Jang, W.S.; Nayyar, N.; Sutton, M.D.; Edgerton, M. Uptake of the antifungal cationic peptide Histatin 5 by *Candida albicans* Ssa2p requires binding to non-conventional sites within the ATPase domain. *Mol. Microbiol.* **2008**, *70*, 1246–1260. [CrossRef] [PubMed]

55. Daum, G. Lipids of mitochondria. *Biochim. Biophys. Acta.* **1985**, *822*, 1–42. [CrossRef]

56. Bhattacharya, S.; Esquivel, B.D.; White, T.C. Overexpression or Deletion of Ergosterol Biosynthesis Genes Alters Doubling Time, Response to Stress Agents, and Drug Susceptibility in *Saccharomyces cerevisiae*. *MBio* **2018**, *9*, e01291-18. [CrossRef] [PubMed]

57. Alcazar-Fuoli, L.; Mellado, E. Ergosterol biosynthesis in *Aspergillus fumigatus*: Its relevance as an antifungal target and role in antifungal drug resistance. *Front. Microbiol.* **2012**, *3*, 439. [CrossRef] [PubMed]

58. Lum, P.Y.; Armour, C.D.; Stepaniants, S.B.; Cavet, G.; Wolf, M.K.; Butler, J.S.; Hinshaw, J.C.; Garnier, P.; Prestwich, G.D.; Leonardson, A.; et al. Discovering modes of action for therapeutic compounds using a genome-wide screen of yeast heterozygotes. *Cell* **2004**, *116*, 121–137. [CrossRef]

59. Ryder, N.S. Terbinafine: Mode of action and properties of the squalene epoxidase inhibition. *Br. J. Dermatol.* **1992**, *126* (Suppl. 39), 2–7. [CrossRef]

60. Cherry, J.M.; Hong, E.L.; Amundsen, C.; Balakrishnan, R.; Binkley, G.; Chan, E.T.; Christie, K.R.; Costanzo, M.C.; Dwight, S.S.; Engel, S.R.; et al. *Saccharomyces* Genome Database: The genomics resource of budding yeast. *Nucleic. Acids Res.* **2012**, *40*, D700–D705. [CrossRef]

61. UniProt Consortium, T. UniProt: The universal protein knowledgebase. *Nucleic. Acids Res.* **2018**, *46*, 2699. [CrossRef] [PubMed]

62. Hegreness, M.; Shoresh, N.; Damian, D.; Hartl, D.; Kishony, R. Accelerated evolution of resistance in multidrug environments. *Proc. Natl. Acad. Sci. USA* **2008**, *105*, 13977–13981. [CrossRef] [PubMed]

 International Journal of
Molecular Sciences

Article

Prediction of Bioactive Peptides from *Chlorella sorokiniana* Proteins Using Proteomic Techniques in Combination with Bioinformatics Analyses

Lhumen A. Tejano [1], **Jose P. Peralta** [1], **Encarnacion Emilia S. Yap** [1], **Fenny Crista A. Panjaitan** [2] and **Yu-Wei Chang** [2,*]

[1] Institute of Fish Processing Technology, College of Fisheries and Ocean Sciences, University of the Philippines Visayas, Miagao 5023, Iloilo, Philippines; lhumentejano@gmail.com (L.A.T.); f153mentor@yahoo.com (J.P.P.); esyap@up.edu.ph (E.E.S.Y.)
[2] Department of Food Science, National Taiwan Ocean University, Keelung 202, Taiwan; fennycap@gmail.com
* Correspondence: bweichang@mail.ntou.edu.tw; Tel.: +886-2-2462-2192 (ext. 5152)

Received: 20 February 2019; Accepted: 9 April 2019; Published: 11 April 2019

Abstract: *Chlorella* is one of the most nutritionally important microalgae with high protein content and can be a good source of potential bioactive peptides. In the current study, isolated proteins from *Chlorella sorokiniana* were subjected to in silico analysis to predict potential peptides with biological activities. Molecular characteristics of proteins were analyzed by sodium dodecyl sulfate polyacrylamide gel electrophoresis (SDS-PAGE) and proteomics techniques. A total of eight proteins were identified by proteomics techniques from 10 protein bands of the SDS-PAGE. The predictive result by BIOPEP's profile of bioactive peptides tools suggested that proteins of *C. sorokiniana* have the highest number of dipeptidyl peptidase-IV (DPP IV) inhibitors, with high occurrence of other bioactive peptides such as angiotensin-I converting enzyme (ACE) inhibitor, glucose uptake stimulant, antioxidant, regulating, anti-amnestic and antithrombotic peptides. In silico analysis of enzymatic hydrolysis revealed that pepsin (pH > 2), bromelain and papain were proteases that can release relatively larger quantity of bioactive peptides. In addition, combinations of different enzymes in hydrolysis were observed to dispense higher numbers of bioactive peptides from proteins compared to using individual proteases. Results suggest the potential of protein isolated from *C. sorokiniana* could be a source of high value products with pharmaceutical and nutraceutical application potential.

Keywords: *Chlorella sorokiniana*; in silico; BIOPEP-UWM database; proteomics; bioactive peptides; nano liquid chromatography tandem mass spectrometry (nanoLC–nanoESI MS/MS)

---

1. Introduction

Microalgae are eukaryotic unicellular organisms that grow easily with inexpensive substrates. Therefore, they are considered to be economical and effective raw materials in industry [1]. Many studies have been conducted to utilize microalgae as useful products. Most of them majorly focused on the potential of microalgae for biofuel production due to their lipid content and abundant availability [1,2]. However, due to the increasing population and demand for protein, there is a call to further utilize microalgae as protein sources to shift away from animal proteins.

Microalgae have been known as one of many promising alternative plants for proteins as they offer up to 50% (*w/w*) of protein [3] with a well-balanced amino acid profile required for the nutrition of human beings [1,4]. Several studies have reported the biological activities of various microalgae protein hydrolysates, including immunostimulant and antitumor activities from *C. sorokiniana* [5,6], angiotensin-I converting enzyme (ACE) inhibitory and hypotensive activities from *Chlorella vulgaris* [7,8] and *Nannochloropsis oculata* [9], antioxidant effects from *Navicula incerta* [10] and *Chlorella ellipsoidea* [11], anti-inflammatory effects from *Spirulina maxima* [12] and antibacterial property from *Spirulina platensis* [13]. From *Chlorella sorokiniana*, Lin et al. [14] were able to purify and identify four active peptides with high ACE inhibitory effects. A variety of other compounds were also detected from microalgae which show certain usefulness for human and animals. Morgese et al. [15] previously reported the beneficial effect of omega (ω)-3 and ω-6 polyunsaturated fatty acids (PUFAs) from *Chlorella sorokiniana* on the emotional, cognitive and social behavior in rats. Talero et al. [16] also reviewed bioactive compounds of microalgae that give the chemopreventive effect on chronic inflammation and cancer. Corresponding to those findings, the exploration of biological activities from microalgae has gained significant attention with regard to its health-promoting properties related to bioactive compounds. Therefore, further observations are still needed to identify more bioactive peptides from *C. sorokiniana*, including dipeptidyl peptidase-IV (DPP IV) inhibitory peptides, ACE-inhibitory peptides and antioxidant peptides.

With the advancements in protein analysis, techniques for protein identification and predictive analysis of potential bioactive peptides have been established. Proteomics techniques have been widely used to analyze proteins presented in protein sample [17]. Moreover, mass spectrometry (MS)-based proteomics have been successfully applied in researches to identify protein from complex materials, including protein characterization of chickpea and oat seeds [18], fish authentication [19], species identification of spoilage and pathogenic bacteria [20], and protein characterization of tilapia processing co-products [21]. Once the sequences of protein are obtained, a bioinformatics tool such as the BIOPEP-UWM database is used to predict bioactive peptides composed in protein sequences [22,23]. Bioinformatics, also known as in silico technique, is a computational method used to estimate bioactive peptides from the known protein sequences [24]. It also allows performing the simulation of enzymatic hydrolysis using proteases to predict bioactive peptides theoretically released from the intact protein sequences [25–27].

The application of proteomics coupled with BIOPEP-UWM will be able to deliver a rapid method to identify and characterize proteins. This technique will reduce cost and time regarding the prediction of the potential bioactive peptides. Thus, this study generally aimed to characterize the isolated proteins from *C. sorokiniana* using in-gel digestion and proteomics techniques. Furthermore, the BIOPEP-UWM database tool was used to predict the potential bioactive peptides derived from the identified proteins of *C. sorokiniana*.

2. Results and Discussion

2.1. Identified Proteins of C. sorokiniana

The protein content of *C. sorokiniana* isolates was 65.08 $\pm$ 0.88% with a yield of 4.40% (w/w initial biomass dry basis). A total of the 10 distinct protein bands of *C. sorokiniana* proteins were observed in 12% acrylamide gel (Figure 1). These labeled protein bands (A–J) were used for the in-gel digestion and subsequently analyzed using nanoLC–nanoESI MS/MS. The molecular weights (MWs) of the proteins estimated by SDS-PAGE were 109.02 (A), 72.08 (B), 54.15 (C), 45.72 (D), 38.33 (E), 29.04 (F), 24.96 (G), 21.13 (H), 16.59 (I), and 7.82 (J). Eight protein hits were discovered from the selected bands in the NCBI database namely, chloroplast rubisco activase, 50S ribosomal protein L7/L12 (chloroplast), phosphoglycerate kinase, Fe-superoxide dismutase, heat shock protein 70, ATP synthase subunit beta (chloroplast), elongation factor 2, partial, and V-type H+ ATPase subunit A, partial. The protein hits, accession number from the NCBI database, length of amino acid (AA), and molecular weight of the band reported in NCBI database and estimated by SDS-PAGE were presented in Table 1. The estimated MWs by SDS-PAGE were comparable to the theoretical molecular weights reported in NCBI database except for elongation factor 2, partial and V-type H+ ATPase Subunit A, partial.

Table 1. Identified *C. sorokiniana* proteins by SDS-PAGE and nanoLC–nanoESI MS/MS analysis.

Protein Name	Accession Number (NCBI)	Score	Sequence Coverage (%)	Length (AA)	Molecular Weights from NCBI Database (kDa)	Molecular Weights (kDa) Estimated from SDS–PAGE
Chloroplast Rubisco Activase	AEL29575.1	323	31%	567	45.68	45.72 (D) 109.02 (A) 72.08 (B) 38.33 (E) 29.04 (F) 24.96 (G) 21.13 (H) 16.59 (I) 7.82 (J)
50S Ribosomal Protein L7/L12 (Chloroplast)	YP_009020879.1	72	18%	130	13.58	7.82 (J)
Phosphoglycerate Kinase	AKP17751.1	250	22%	465	49.13	45.72 (D) 109.02 (A) 72.08 (B) 54.15 (C) 38.33 (E) 29.04 (F) 24.96 (G) 21.13 (H) 16.59 (I) 7.82 (J)
Fe-superoxide Dismutase	AHD25899.1	82	8%	236	26.41	24.96 (G)
Heat Shock Protein 70	AKP17750.1	128	29%	652	71.55	72.08 (B) 109.02 (A) 54.15 (C) 38.33 (E)
ATP Synthase Subunit Beta (Chloroplast)	YP_009020893.1	906	49%	481	51.833	54.15 (C)
elongation factor 2, partial	BAE48222.1	120	4%	816	90.54	45.72 (D) 7.82 (J)
V-type H+ ATPase Subunit A, partial	BAE48224.1	61	4%	596	65.47	7.82 (J)

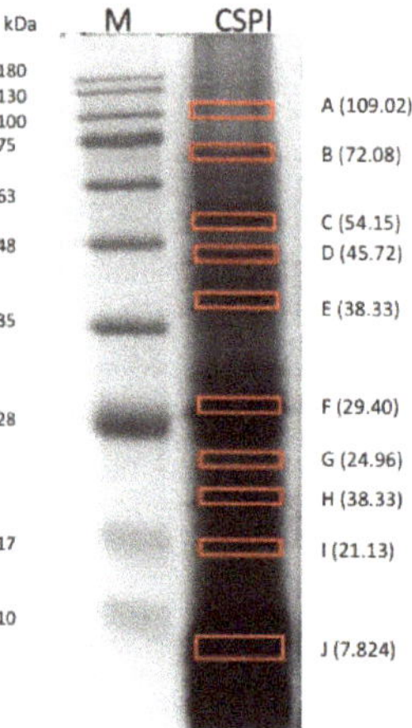

Figure 1. Twelve percent SDS-PAGE of *C. sorokiniana* protein isolates. M: Protein marker; CSPI: *C. sorokiniana* protein isolate.

C. sorokiniana is a freshwater green algae species with high protein content [28]. This species of genus *Chlorella* was originally called as *C. pyrenoidosa* [29]. The NCBI database revealed a total of 20,925 proteins from the *C. sorokiniana*. Most of them are enzymes responsible for various cell functions. According to the nanoLC–nanoESI MS/MS data, eight protein hits from the NCBI database corresponded to the proteins of *C. sorokiniana*. Phosphoglycerate kinase (*Auxenochlorella pyrenoidosa*, NCBI accession number: AKP17751.1) was detected in all selected protein bands in the SDS-PAGE. Watson et al. [30] stated that phosphoglycerate kinase derived from various protein sources are all monomers with MWs around 45 kDa. Moreover, their amino acid composition and catalytic functions are similar [31]. The reference molecular weight for phosphoglycerate kinase from NCBI database was 49.13 kDa, thus phosphoglycerate kinase found in band D (45.72 kDa) was used for further analyses. The possibility of having the same single protein in different bands is high since proteins are denatured and separated in SDS-PAGE. In addition, protein hits discovered by the Mascot database were identified based on the matched tryptic peptides detected by mass spectrometry.

2.2. Identified Tryptic Peptides from C. sorokiniana Proteins

Tryptic peptides derived from the identified proteins of *C. sorokiniana* by in-gel digestion were evaluated by nanoLC–nanoESI MS/MS analysis. Those tryptic peptides identified through mass spectrometry were peptides matching with protein hits in the Mascot database. Tryptic peptides are generated by trypsin through in-gel digestion process which is part of proteomics technique. The result of the MS/MS ion search for the tryptic digests revealed that all the tryptic peptides from the identified proteins were doubly and triply charged. Figures 2 and 3 present the representative spectra of the doubly and triply charged peptides from the identified proteins of *C. sorokiniana* by proteomics analysis.

Figure 2 illustrates the doubly charged tryptic peptide of *C. sorokiniana* of protein band D (Figure 1), with observed signal m/z 1053.01 marked in red box representing a doubly charged peptide (with adjacent signal difference of 0.50, insert A), and nanoLC–nanoESI MS/MS fragmentation spectra of NFNNIEDGFYISPAFLDK found in chloroplast rubisco activase (NCBI accession no. AEL29575.1) represented in insert B. Figure 3 illustrates the triply charged tryptic peptide also found in the same identified protein of the same protein band D. The observed signal was m/z 618.66, also marked in red box demonstrating a triply charged peptide (with adjacent signal difference of 0.33, insert A), and nanoLC–nanoESI MS/MS fragmentation spectra of tryptic peptide LVDAFPGQSIDFFGALR found in chloroplast rubisco activase (NCBI accession no. AEL29575.1) is illustrated in insert B.

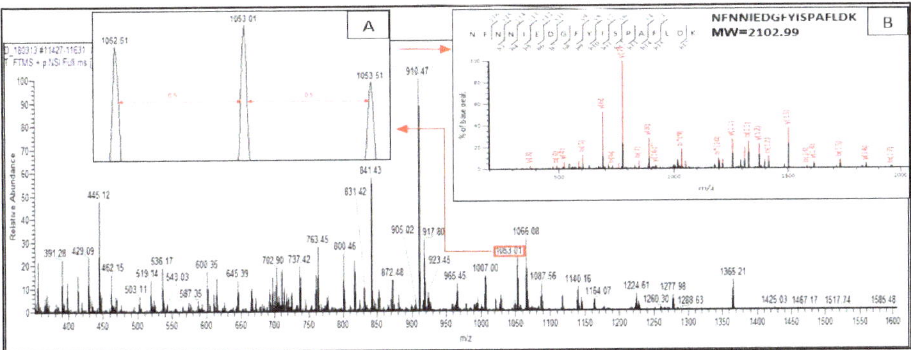

Figure 2. NanoLC–nanoESI MS/MS spectra (m/z region 350–1600) of *C. sorokiniana* protein band D, m/z 1053.401 signal in red box. Insert A presents the identified doubly charged signal by the difference of 0.50 between signals. Insert B illustrates the fragmentation of nanoLC–nanoESI MS/MS spectra of the peptide NFNNIEDGFYISPAFLDK, calculated MW 2102.99 Da.

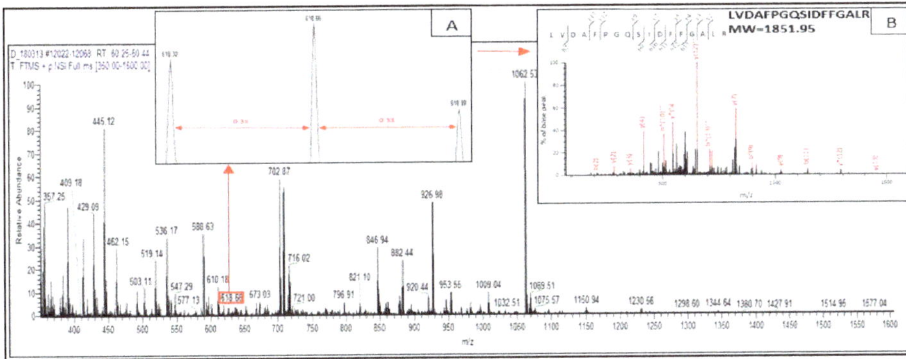

Figure 3. NanoLC–nanoESI MS/MS spectra (m/z region 350–1600) of *C. sorokiniana* protein band D, m/z 618.66 signal in red box. Insert A shows the identified doubly charged signal by the difference of 0.33 between signals. Insert B illustrates the fragmentation of nanoLC–nanoESI MS/MS spectra of the peptide LVDAFPGQSIDFFGALR, calculated MW 1851.95 Da.

In the identification of proteins by proteomics analysis, trypsin is usually used to digest proteins in the gel [32]. Trypsin hydrolyzes protein specifically at the C-terminus of the carboxyl side of the amino acids arginine or lysine, but poorly when lysine and arginine are followed by proline. With this perspective, the tryptic peptides are either doubly or triply charged in ESI since the amino terminal residues are basic which explains the result of the MS/MS ion search [33].

2.3. Potential Bioactive Peptides from Identified Proteins in C. sorokiniana

Potential bioactive peptides presented in identified proteins of *C. sorokiniana* were investigated using the BIOPEP-UWM database. Amino acid sequences of six proteins, namely chloroplast rubisco activase, 50s ribosomal protein l7/l12 (chloroplast), phosphoglycerate kinase, Fe-superoxide dismutase, heat shock protein 70 and ATP synthase subunit beta (chloroplast), were chosen as they are found to be relatively abundant components of *C. sorokiniana* proteins found in SDS-PAGE based on the results. Moreover, they also corresponded to the estimated molecular weights in the NCBI database (Table 1). The profile of the potential bioactive peptides, their biological activities (ACE inhibitory, antioxidant, anti-amnestic, antithrombotic, stimulating, regulating, DPP IV inhibitory), and frequencies are summarized in Table 2. Results revealed that most of the potential bioactive peptides were dipeptides or tripeptides with multiple biological activities. The number of those bioactive peptides was identified based on the amino acid sequences which were predicted to become potential bioactive peptides. The BIOPEP database displays peptides with their bioactivities from inputted protein sequences corresponding to the information in the database.

Table 2. Number of potential bioactive peptides of identified *C. sorokiniana* proteins using BIOPEP's "profiles of potential biological activities" tool.

Protein Name	Number of Bioactive Peptides						
	AC	AO	AA	AT	S	R	DPP
Chloroplast Rubisco Activase	187 (0.329)	18 (0.032)	3 (0.005)	2 (0.004)	14 (0.25)	3 (0.005)	250 (0.441)
50s Ribosomal Protein L7/L12 (Chloroplast)	59 (0.454)	8 (0.062)	1 (0.008)	1 (0.008)	8 (0.062)	1 (0.008)	83 (0.638)
Phosphoglycerate Kinase	224 (0.482)	23 (0.049)	5 (0.011)	4 (0.009)	33 (0.071)	6 (0.013)	297 (0.639)
Fe-Superoxide Dismutase	107 (0.453)	18 (0.076)	-	-	5 (0.021)	-	156 (0.661)
Heat Shock Protein 70	276 (0.423)	30 (0.046)	4 (0.006)	4 (0.006)	41 (0.064)	4 (0.006)	446 (0.684)
ATP Synthase Subunit Beta (Chloroplast)	204 (0.424)	20 (0.042)	5 (0.010)	5 (0.010)	25 (0.052)	6 (0.12)	332 (0.690)

Abbreviation: ACE Inhibitory (AC), antioxidant (AO), anti-amnestic (AA), antithrombotic (AT), stimulating (S), regulating (R), dipeptidyl peptidase-IV (DPP IV) inhibitory (DPP).

Chloroplast rubisco activase (NCBI accession no. AEL29575.1) and phosphoglycerate kinase (NCBI accession no. AKP17751.1) were chosen to illustrate the profile of bioactive peptides within in the protein (Figures 4 and 5) because they appeared to be relatively abundant and were found in almost all picked bands in SDS-PAGE. As shown in Figure 4, the molecular weights of the tryptic peptides corresponded to the identified tryptic peptide at amino acid positions 100–117, 132–144, 158–181, 187–145, 149–213, 255–273, 290–304, and 310–326 (matched tryptic peptides shown in red letters) in the chloroplast rubisco activase amino acid sequence. BIOPEP-UWM analysis results exhibited that potential bioactive peptides encrypted in chloroplast rubisco activase amino acid sequence were mostly DPP IV inhibitors (with 250 peptide fragments, marked with an orange line) and ACE inhibitors (with 187 peptide fragments, marked with a green line). Other bioactive peptides found were 18 antioxidant, 3 anti-amnestic, 2 antithrombotic, 18 stimulant and 3 regulatory peptides. Some bioactive peptides have multiple activities such as VPL, WG, LA, IR, PG, VY, and KP.

On the other hand, Figure 5 shows the molecular weights of the tryptic peptides in phosphoglycerate kinase which corresponds to the theoretical tryptic peptides at amino acid positions 232–244, 258–268, 285–298, 303–313, 383–411, and 436–465 (matched tryptic peptides shown in red letters). There were 297 DPP IV inhibitor, 224 ACE inhibitor, 23 antioxidant, 33 stimulant, 5 anti-amnetic, 4 antithrombotic, and 6 regulatory peptides embedded in the amino acid sequence of phosphoglycerate kinase. Moreover, the profiles of the bioactive peptides of 50s ribosomal protein l7/l12 (chloroplast), Fe-superoxide dismutase, heat shock protein 70 and ATP synthase subunit beta (chloroplast) also show the presence of the above mentioned bioactive peptides in these proteins, except that Fe-superoxide dismutase does not show anti-amnestic, antithrombotic, or regulating peptides. In all the proteins, DPP IV and ACE inhibitors were the most abundant bioactive peptides.

The amino acid composition and sequence of the proteins greatly determines the presence of these bioactive peptides. Results also revealed that most of the DPP IV peptides present in the identified proteins had proline (P), alanine (A), glycine (G), valine (V) and leucine (L) amino acid residues. DPP IV preferably cleaves dipeptides with proline and alanine residues at the N-terminal side of the peptide [34]. It also has relatively lower cleavage rates with serine, glycine, leucine, and valine [35,36]. Moreover, the presence of basic and hydrophobic amino acids at the N-terminal side of the peptides could enhance the cleavage susceptibility of the substrate [34,37]. DPP IV inhibitors were also reported from various protein sources by in silico approach including barley, canola, oat, soybean, wheat, quinoa, chicken egg, bovine milk, bovine meat, pig, tuna, Atlantic salmon, chum salmon, tilapia skin and frame, and palmaria palmate [21,27,38–40]. On the other hand, the abundance of ACE inhibitory peptides in the identified proteins might have also been influenced by the amino acid compositions of the proteins. The presence of amino acid residues such as phenylalanine (F), tyrosine (Y), tryptophan (W), or proline (P) in at the C-terminal side of the peptides have been reported to exhibit high potent ACE inhibitory activity [41–45]. The adjacent amino acid residue of proline can also influence the potency of the ACE inhibitor, which is usually enhanced by hydrophobic amino acids [46]. In silico analysis of different proteins revealed the abundance of ACE inhibitors embedded in various protein sequences [47–49]. In previous studies, some amino acid sequences of ACE inhibitory peptides from *C. sorokiniana* were discovered. Lin et al. [14] reported IC_{50} values of WV, VW, IW, and LW were 307.61, 0.58, 0.50, and 1.11 µM, respectively. Moreover, *C. sorokiniana* protein hydrolysates could reduce systolic and diastolic blood pressure at 20 and 21 mm Hg, respectively. Suetsuna and Chen [8] also mentioned several amino acid sequences generated potential antihypertensive activity through oral administration, such as IVVE (IC_{50}: 315.3 µM), AFL (IC_{50}: 63.8 µM), FAL (IC_{50}: 26.3 µM), AEL (IC_{50}: 57.1 µM), and VVPPA (IC_{50}: 79.5 µM) from *C. vulgaris*; IAE (IC_{50}: 34.7 µM), FAL, AEL, IAPG (IC_{50}: 11.4 µM), and VAF (IC_{50}: 35.8 µM) from *S. platensis*. Those findings showed that ACE inhibitory peptides predicted through in silico analysis obviously possessed potential antihypertensive activity through in vitro and in vivo analysis.

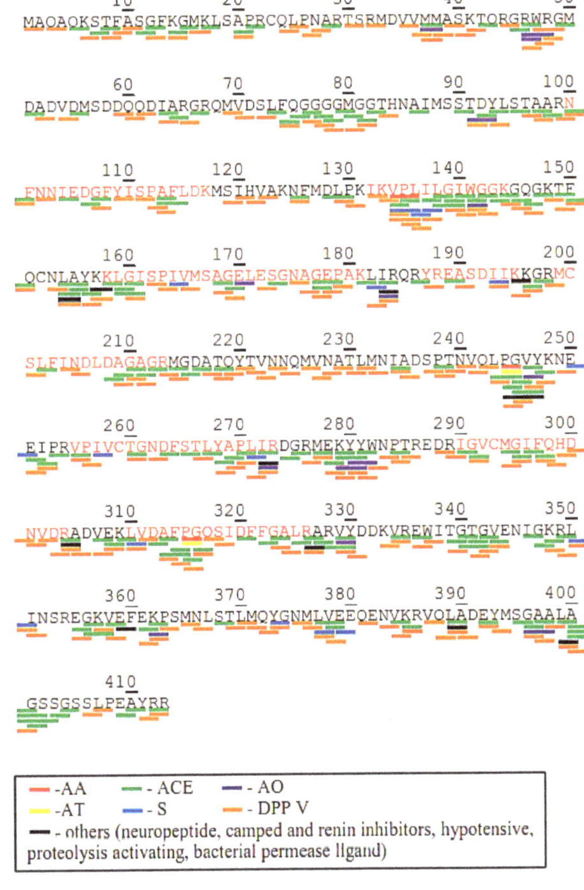

Figure 4. Protein sequence and potential bioactive peptides of chloroplast rubisco activase (AEL29575.1) from *C. sorokiniana*. Abbreviation: ACE Inhibitory (ACE), Antioxidant (AO), Anti-amnestic (AA), Antithrombotic (AT), Stimulating (S), Regulating (R), DPP IV inhibitory (DPP).

For many years now, in silico analysis has been successfully used to predict the potential application of various proteins as a source of bioactive peptides [22]. It provides sufficient information for determining the potential biological activity of proteins which is much faster than conventional methods [21]. The results of the in silico analysis by BIOPEP-UWM suggest the potential of *C. sorokiniana* proteins for pharmaceutical application as demonstrated by its bioactivities. These peptides in the intact proteins are inactive and need to be released in order to perform their functions [50]. The prediction of the potential bioactivities of the proteins after digestion by various proteases can be conducted by the BIOPEP-UWM database tool.

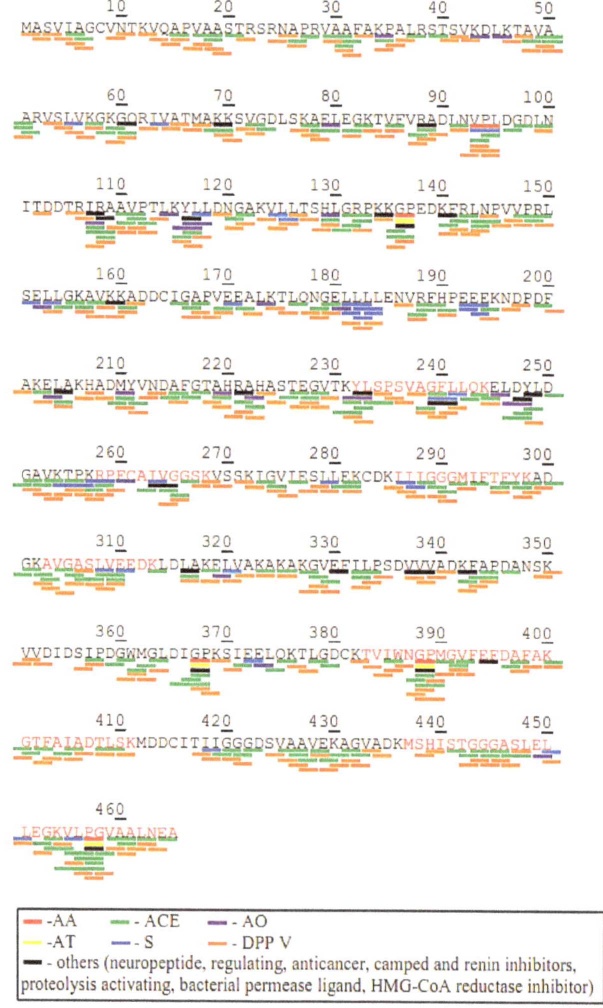

Figure 5. Protein sequence and potential bioactive peptides of phosphoglycerate kinase (AKP17751.1) from *C. sorokiniana*. Abbreviation: ACE Inhibitory (ACE), Antioxidant (AO), Anti-amnestic (AA), Antithrombotic (AT), Stimulating (S), Regulating (R), DPP IV inhibitory (DPP).

2.4. Prediction of Potential Bioactive Peptides after Protease Cleavage using BIOPEP-UWM Tool

Identified proteins such as chloroplast rubisco activase, phosphoglycerate kinase, Fe-superoxide dismutase, heat shock protein 70 and ATP synthase subunit beta (chloroplast) were further analyzed using the "enzyme action" tool in BIOPEP-UWM database; these proteins showed the most numbers of bioactivities from their profiles of bioactive peptides (Table 2). Results of the 15 simulations of enzymatic hydrolysis for each protein sequence are presented in Table 3. The table shows the number of bioactive peptides with specific bioactivities after hydrolysis of the individual proteins by various proteases. The results revealed that DPP IV inhibitory peptides were observed to be dominantly produced from the selected proteins using different proteases. ACE inhibitory peptides were also released in relatively high numbers but lower than DPP IV. This information is in concurrence with the profile of the potential bioactive peptides from the proteins in Table 2. Bromelain, papain, ficin and pepsin (pH > 2) were the individual proteases that released the most diverse and large number peptides with certain biological activities from all the selected proteins. Meanwhile, trypsin had the lowest number of bioactive peptides release after in silico hydrolysis. Trypsin is the most commonly used enzyme in proteomics approach [32], however, in the in silico analysis, it did not release significant numbers of potential bioactive peptides. Nonetheless, based on the results, other single action enzymes could also release relatively high numbers of bioactive peptides. The use of a combination of enzymes in hydrolysis is also offered by the BIOPEP-UWM database. A combination of two to a maximum of three enzymes could be utilized in the hydrolysis simulation of the proteins. The combination of three enzymes (trypsin, α-chymotrypsin, and pepsin) had been identified to produce potential anti-inflammatory peptides in microalgae, such as LDAVNR and MMLDF [12]. Table 3 reveals that the use of combined action of two to three enzymes could actually lead to the release of higher numbers of bioactive peptides from the selected proteins. This implies a greater effectiveness of using the combined action of enzymes in cleaving peptide bonds than the single action enzyme, except for pepsin which has almost the same number of released peptides with the combined enzyme action. Pepsin has been reported from several in vitro studies to produce various bioactive peptides from microalgae hydrolysates such as an ACE inhibitor and antioxidant peptides. Samarakoon et al. [9] mentioned that pepsin generated more potential ACE inhibitory peptides compared to other proteases, such as GMNNLTP (IC$_{50}$: 123 μM) and LEQ (IC$_{50}$: 173 μM). Ko et al. [11] also identified LNGDVW from peptic hydrolysates possessed strong scavenged peroxyl, DPPH and hydroxyl radicals at the IC$_{50}$ values of 0.02, 0.92 and 1.42 mM, respectively. Moreover, peptic hydrolysates from microalgae efficiently generated strong antioxidant activities [51,52].

Furthermore, in comparison to the other three proteins, ATP synthase subunit beta demonstrated higher tendency to release more bioactive peptides using the different proteases. However, these theoretically produced bioactive peptides may not always have a comparable function with the in vitro and in vivo analyses, thus further study of these peptides using in vitro and in vivo studies should be conducted. Nevertheless, the BIOPEP's "enzyme action" tool was able to provide reference information on the possible bioactive peptides that could be released from the selected proteins using various proteases.

Table 3. Number of predicted potential bioactive peptides to be released from identified proteins of *C. sorokiniana* using BIOPEP's "enzyme action" tool.

Protease	Chloroplast Rubisco Activase				Phosphoglycerate Kinase				Heat Shock Protein 70				ATP Synthase Subunit Beta			
	AC	AO	S	DP	AC	AO	S	DP	AC	AO	S	DP	AC	AO	S	DP
Pepsin (EC 3.4.23.1)	31	2	2	48	29	-	5	51	42	11	1	78	44	6	9	66
Thermolysin (EC 3.4.24.27)	12	-	-	14	20	4	-	23	21	-	-	31	20	1	-	25
Bromelain (EC 3.4.22.32)	26	3	1	36	25	8	3	41	36	3	2	50	20	2	4	37
Chymotrypsin A (EC 3.4.21.1)	5	1	1	10	9	1	2	11	8	2	2	15	7	4	2	10
Chymotrypsin C (EC 3.4.21.2)	17	3	1	16	14	-	2	19	20	6	2	35	17	3	4	25
Pancreatic elastase (EC 3.4.21.36)	16	2	-	31	27	5	-	49	21	2	-	34	19	2	-	41
Papain (EC 3.4.22.2)	31	3	1	35	27	6	2	31	40	4	4	50	27	1	4	43
Proteinase K (EC 3.4.21.67)	12	1	-	20	17	1	1	19	23	5	-	41	15	1	-	28
Trypsin (EC 3.4.21.4)	5	-	-	3	3	1	-	3	5	-	-	2	3	-	-	3
Ficin (EC 3.4.22.3)	27	7	2	37	22	8	3	33	26	4	3	35	22	7	1	28
Cathepsin (EC 3.4.21.20)	4	1	1	6	7	1	1	6	6	2	1	7	5	3	1	6
Subtisilin (EC 3.4.21.62)	10	4	2	14	10	3	4	18	9	5	7	24	12	2	6	23
Pepsin + Trypsin	28	4	2	55	33	1	5	56	36	2	10	66	40	6	9	62
Pepsin + Trysin + Chymotrypsin A	31	4	1	40	33	1	5	54	34	1	10	63	38	5	9	61
Pepsin + Trysin + Chymotrypsin C	28	4	1	33	35	1	4	54	41	2	10	70	38	5	9	51

Abbreviation: ACE Inhibitory (AC), Antioxidant (AO), Stimulating (S), DPP IV inhibitory (DP).

3. Materials and Methods

3.1. Materials

The microalgae, *C. sorokiniana* was obtained from the Taiwan Chlorella Manufacturing Co., Ltd. (Taipei, Taiwan), considered as the largest producer of *Chlorella* every year with an average production of 400 tons of dried biomass [1]. All reagents and chemicals used were analytical grade.

3.2. Protein Isolation

The protein isolation process was adapted from the procedure of Parimi et al. [53] with modifications. Briefly, *C. sorokiniana* biomass slurry at 1:16 (*w/v*) ratio was prepared. Sonication for 1 h was done to the slurry for pretreatment and subsequent alkaline protein extraction by solubilization at 11.38 using 2 M NaOH for 35 min with stirring. It was followed by isoelectric precipitation of the supernatant at 4.01 with 1M HCL and stirred for 60 min. Centrifugation at $8750 \times g$ for 35 min was done for the solid-liquid separation during the solubilization and precipitation steps. The protein isolate was lyophilized and stored at $-20\,^{\circ}$C until further use. The modified Lowry method [54] was used to determine the protein content of the isolate.

3.3. SDS-PAGE Analysis

The SDS-PAGE was performed according to a method described by Schägger and Von Jagow [55] 4% stacking gel (*w/v*) and 12% polyacrylamide gel (w/v). 10 milligrams of protein isolate was dissolved in 1 mL of denaturant sample buffer (0.5 M Tris-HCl pH 6.8, glycerol, 10% SDS, *w/v*, 0.5% bromophenol blue, *w/v*, β-mercaptoethanol), and heated at 95 °C. Then, 10 μL of the sample was loaded to the sample wells. Protein separation was carried at 80 V for 30 min followed by 110 V for 90 min for the resolving gel using a Mini Protean II unit (Bio-Rad Laboratories, Hercules, CA, USA). The gel was stained for 40 min with Brilliant Blue (Bio-Rad, Coomassie R250). Destaining of the gel was done three times using water/methanol/acetic acid (7/2/1, *v/v/v*) for 15 min each cycle with shaking using an orbital shaker (Fristek S10, Taichung city, Taiwan). Estimation of the molecular mass of proteins was done using molecular protein mass marker (250 to 10 kDa, Bio-Rad) loaded at 5 uL in the sample well. The gels was scanned with E-Box VX5 (Vilber Lourmat, Paris, France) and the analysis of the captured image was done using Vision Capt software (V16.08a, Vilber Lourmat, Paris, France).

3.4. Proteomics Techniques

3.4.1. In-Gel Tryptic Digestion

The following proteomics technique experiments were carried out in Academia Sinica, Nangang District, Taipei City, Taiwan. Proteomics techniques were adapted from the methods described by Chang et al. [18]. Gel slice and in-gel digestion were performed using the combined modified methods of Rosenfeld et al. [56] and Shevchenko et al. [32]. Briefly, 10 intensive colored protein bands were excised from the SDS-PAGE gel for the in-gel digestion. The gel pieces were destained with 25 mM amonium bicarbonate (ABC)/ 50% acetonitrile (ACN) solution in a microcentrifuge PP tubes. The destained gel pieces were added with 100 μL of 50 mM dithioerythreitol (DTE) / 25 mM ABC and soaked at 37 °C for 1 h. The tubes were centrifuged and the DTE solution was removed. Then, the gel pieces were added with 100 μL of 100 mM iodoacetamide (IAM) / 25 mM ABC and soaked at room temperature in a dark place for 1 h for the alkylation step. The IAM solution was removed after centrifugation. Washing of the gel pieces was done by soaking in 200 μL of 50% ACN / 25 mM ABC for 15 min. The solution was removed after centrifugation and the process was repeated four times. The gel slices were then soaked in 100 μL of 100% ACN for 5 min, repeated twice, and the solution was discarded after centrifugation. The gel slices were dried for 5 min using Speed Vac (Thermo Scientific, Waltham, MA, USA). Trypsin digestion followed by adding Lys-C / 25 mM ABC (enzyme:protein, 1:50) and incubating the mixture for 1 h at 37 °C. Afterwards, the same amount of trypsin was added and incubated for 16 h at 37 °C. Afterwards, the extraction of the tryptic peptides was done with 50 µL of 50% ACN/ 5% trifluoroacetic acid (TFA). The peptide extracts were transferred to new tubes and dried Speed Vac (Thermo Scientific, Waltham, MA, USA). Finally, the peptide extracts were purified using C18 Zip-Tip. The purified peptide extracts were used for the nanoLC–nanoESI MS/MS analysis.

3.4.2. Nanoliquid Chromatography–Nanoelectrospray Ionization Tandem Mass Spectrometry (NanoLC–nanoESI MS/MS) Analysis

Dried tryptic peptide digest was subjected to nanoLC–nanoESI MS/MS analysis using a nanoAcquity system (Waters, Milford, MA, USA) connected to the LTQ Orbitrap Velos hybrid mass spectrometer (Thermo Electron, Bremen, Germany) equipped with a PicoView nanospray interface (New Objective, Woburn, MA). The tryptic peptide mixtures were loaded onto a 75 µm ID, 25 cm length C18 BEH column (Waters, Milford, MA) packed with 1.7 µm particles with a pore width of 130 Å. Separation was performed using a segmented gradient in 60 min from 5 to 35% solvent B (acetonitrile with 0.1% formic acid) at 300 nL/min flow rate and at 35 °C column temperature. Solvent A was 0.1% formic acid in water (*v/v*). The mass spectrometer was operated in the data-dependent mode. In brief, the orbitrap (m/z 350–1600) with the resolution set to 60 K at m/z 400 and automatic gain control (AGC) target at 10^6 was used to obtain the survey full scan MS spectra. The 20 most intense ions were sequentially isolated for collision-induced dissociation (CID) MS/MS fragmentation and detection in the linear ion trap (AGC target at 10,000) with previously selected ions dynamically excluded for 60 s. Ions with singly and unrecognized charge state were also excluded. The LTQ-Orbitrap data were acquired at the Academia Sinica Common Mass Spectrometry Facilities located at the Institute of Biological Chemistry, Academia Sinica, Nangang District, Taipei City, Taiwan.

3.4.3. Tandem MS Data Analysis of Proteins and Peptide Identification

First, the MS raw data was converted to PKL files using de novo sequencing parameter in the ProteinLynx software coupled with Mascot MS/MS ion search (http://140.112.52.63/mascot/cgi/ search_form.pl?FORMVER=2;SEARCH=MIS) [57]. MS/MS data were examined using the National Center for Biotechnology Information (NCBI) database (https://www.ncbi.nlm.nih.gov/) accessed on March 15, 2018 [58] for Viridiplantae (green plants) entries. Search parameters were set to: Carbamidomethyl cysteine as fixed modification; oxidation (M) as variable modification; 10 ppm peptide mass tolerance; 2+, 3+, 4+ peptide charge; ± 0.6 Da MS/MS tolerance; instrument is ESI-TRAP; and the enzyme entry as trypsin with 2 missed cleavages. Peptide masses were acquired as monoisotopic masses.

The Mascot ion score was -10^*Log (P), where P is the probability that the observed match is a random event. Individual ion scores of N 45 indicated identity or extensive homology ($p < 0.05$). Protein scores were derived from ion scores as a non-probabilistic basis for ranking protein hits (Matrix Science, London, United Kingdom). The sequence coverage of protein hits was expressed in percentage (%) indicating the sequence homology of identified tryptic peptides from *C. sorokiniana* to corresponding protein hits based on the Mascot MS/MS ion search results [21].

3.4.4. In Silico Analysis of Bioactive Peptides and Enzyme Cleavages using BIOPEP-UWM Database Tools

Sequences of the identified protein of *C. sorokiniana* proteins from NCBI database were analyzed for bioactive peptides and enzyme cleavages using BIOPEP-UWM database (http://www.uwm.edu. pl/biochemia/index.php/pl/biopep) accessed on March 15, 2018 [23] performed as described by Cheung et al. [25] with modifications. Briefly, the bioactivities, sequences, number and location of the peptides were obtained from the sequences of the identified proteins analyzed using the "profiles of potential bioactivity" tool. Moreover, the sequences of the identified proteins were examined using the "enzyme action" tool to simulate enzymatic hydrolysis. A total of 15 enzymatic hydrolysis simulations (composed of 12 individual proteases, one double enzyme action, and two triple enzyme action) were conducted to each protein sequence. A list of all the potential bioactive peptides was obtained after directing the theoretical peptide sequence data to the "search for active fragments" option. The occurrence of the frequency of the bioactive peptides in the intact proteins was computed as $A = a/N$, where A is occurrence frequency, a is the number of bioactive peptides and N is the total number of amino acid residues in the protein sequence.

4. Conclusions

Proteomics techniques coupled with in silico analysis used in this study showed a rapid method to identify the isolated proteins of *C. sorokiniana*, to predict potential bioactivities and to determine the appropriate proteases that theoretically released more bioactive peptides. Results of the proteomics technique showed the identification of tryptic peptides corresponding to eight proteins from the microalgae. The in silico analysis using BIOPEP-UWM database tools revealed that the combined actions of mixed enzymes and the use of single enzyme action of pepsin (pH > 2) could lead to the production of more diverse and larger numbers of potential bioactive peptides embedded in the protein sequences. According to the results, *C. sorokiniana* proteins are potential sources of bioactive peptides with various bioactivities. Nonetheless, with the use of appropriate extraction methods and purification techniques for certain predicted bioactivities, these proteins could be a good alternative source of high value compounds for pharmaceutical, medical, cosmetics and functional food applications to aid in human health maintenance and enhancement.

Author Contributions: Conceptualization, L.A.T., J.P.P., E.E.S.Y. and Y.-W.C.; methodology, L.A.T.; validation, L.A.T.; formal analysis, L.A.T.; investigation, L.A.T.; resources, L.A.T. and Y.-W.C.; data curation, L.A.T., J.P.P., E.E.S.Y. and Y.-W.C.; writing—original draft preparation, L.A.T.; writing—review and editing, F.C.A.P.; visualization, L.A.T.; supervision, J.P.P. and Y.-W.C.; funding acquisition, J.P.P. and Y.-W.C.

Funding: This research was funded by the Ministry of Science and Technology, Taiwan (MOST: 106-2311-B-019-001).

Acknowledgments: We thank Taiwan Chlorella Manufacturing Co., Ltd. (Taipei, Taiwan) for the donations of *C. sorokiniana* biomass samples used in this research.

Conflicts of Interest: The authors declare no conflict of interest.

References

1. Spolaore, P.; Joannis-Cassan, C.; Duran, E.; Isambert, A. Commercial applications of microalgae. *J. Biosci. Bioeng.* **2006**, *101*, 87–96. [CrossRef] [PubMed]
2. Mubarak, M.; Shaija, A.; Suchithra, T. A review on the extraction of lipid from microalgae for biodiesel production. *Algal Res.* **2015**, *7*, 117–123. [CrossRef]
3. Schwenzfeier, A.; Wierenga, P.A.; Gruppen, H. Isolation and characterization of soluble protein from the green microalgae *Tetraselmis* sp. *Bioresour. Technol.* **2011**, *102*, 9121–9127. [CrossRef] [PubMed]
4. Zhu, Q.; Chen, X.; Wu, J.; Zhou, Y.; Qian, Y.; Fang, M.; Xie, J.; Wei, D. Dipeptidyl peptidase IV inhibitory peptides from *Chlorella vulgaris*: In silico gastrointestinal hydrolysis and molecular mechanism. *Eur. Food Res. Technol.* **2017**, *243*, 1739–1748. [CrossRef]
5. Kralovec, J.; Metera, K.; Kumar, J.; Watson, L.; Girouard, G.; Guan, Y.; Carr, R.; Barrow, C.; Ewart, H. Immunostimulatory principles from *Chlorella pyrenoidosa*—Part 1: Isolation and biological assessment *in vitro*. *Phytomedicine* **2007**, *14*, 57–64. [CrossRef] [PubMed]
6. Reyna-Martinez, R.; Gomez-Flores, R.; López-Chuken, U.; Quintanilla-Licea, R.; Caballero-Hernandez, D.; Rodríguez-Padilla, C.; Beltrán-Rocha, J.C.; Tamez-Guerra, P. Antitumor activity of *Chlorella sorokiniana* and *Scenedesmus* sp. microalgae native of Nuevo León State, México. *PeerJ* **2018**, *6*, e4358. [CrossRef] [PubMed]
7. Sheih, I.-C.; Fang, T.J.; Wu, T.-K. Isolation and characterisation of a novel angiotensin I-converting enzyme (ACE) inhibitory peptide from the algae protein waste. *Food Chem.* **2009**, *115*, 279–284. [CrossRef]
8. Suetsuna, K.; Chen, J.-R. Identification of antihypertensive peptides from peptic digest of two microalgae, *Chlorella vulgaris* and *Spirulina platensis*. *Mar. Biotechnol.* **2001**, *3*, 305–309. [CrossRef] [PubMed]

9. Samarakoon, K.W.; Kwon, O.-N.; Ko, J.-Y.; Lee, J.-H.; Kang, M.-C.; Kim, D.; Lee, J.B.; Lee, J.-S.; Jeon, Y.-J. Purification and identification of novel angiotensin-I converting enzyme (ACE) inhibitory peptides from cultured marine microalgae (*Nannochloropsis oculata*) protein hydrolysate. *J. Appl. Phycol.* **2013**, *25*, 1595–1606. [CrossRef]

10. Kang, K.-H.; Qian, Z.-J.; Ryu, B.; Kim, D.; Kim, S.-K. Protective effects of protein hydrolysate from marine microalgae Navicula incerta on ethanol-induced toxicity in HepG2/CYP2E1 cells. *Food Chem.* **2012**, *132*, 677–685. [CrossRef]

11. Ko, S.-C.; Kim, D.; Jeon, Y.-J. Protective effect of a novel antioxidative peptide purified from a marine *Chlorella ellipsoidea* protein against free radical-induced oxidative stress. *Food Chem. Toxicol.* **2012**, *50*, 2294–2302. [CrossRef] [PubMed]

12. Vo, T.-S.; Ryu, B.; Kim, S.-K. Purification of novel anti-inflammatory peptides from enzymatic hydrolysate of the edible microalgal Spirulina maxima. *J. Funct. Foods* **2013**, *5*, 1336–1346. [CrossRef]

13. Sun, Y.; Chang, R.; Li, Q.; Li, B. Isolation and characterization of an antibacterial peptide from protein hydrolysates of *Spirulina platensis*. *Eur. Food Res. Technol.* **2016**, *242*, 685–692. [CrossRef]

14. Lin, Y.-H.; Chen, G.-W.; Yeh, C.; Song, H.; Tsai, J.-S. Purification and Identification of Angiotensin I-Converting Enzyme Inhibitory Peptides and the Antihypertensive Effect of *Chlorella sorokiniana* Protein Hydrolysates. *Nutrients* **2018**, *10*, 1397. [CrossRef] [PubMed]

15. Morgese, M.; Mhillaj, E.; Francavilla, M.; Bove, M.; Morgano, L.; Tucci, P.; Trabace, L.; Schiavone, S. *Chlorella sorokiniana* extract improves short-term memory in rats. *Molecules* **2016**, *21*, 1311. [CrossRef] [PubMed]

16. Talero, E.; García-Mauriño, S.; Ávila-Román, J.; Rodríguez-Luna, A.; Alcaide, A.; Motilva, V. Bioactive compounds isolated from microalgae in chronic inflammation and cancer. *Mar. Drugs* **2015**, *13*, 6152–6209. [CrossRef] [PubMed]

17. Gevaert, K.; Vandekerckhove, J. Protein identification methods in proteomics. *ELECTROPHORESIS Int. J.* **2000**, *21*, 1145–1154. [CrossRef]

18. Chang, Y.-W.; Alli, I.; Konishi, Y.; Ziomek, E. Characterization of protein fractions from chickpea (*Cicer arietinum* L.) and oat (*Avena sativa* L.) seeds using proteomic techniques. *Food Res. Int.* **2011**, *44*, 3094–3104. [CrossRef]

19. Mazzeo, M.F.; Giulio, B.D.; Guerriero, G.; Ciarcia, G.; Malorni, A.; Russo, G.L.; Siciliano, R.A. Fish authentication by MALDI-TOF mass spectrometry. *J. Agric. Food Chem.* **2008**, *56*, 11071–11076. [CrossRef]

20. Böhme, K.; Fernández-No, I.C.; Barros-Velázquez, J.; Gallardo, J.M.; Cañas, B.; Calo-Mata, P. Rapid species identification of seafood spoilage and pathogenic Gram-positive bacteria by MALDI-TOF mass fingerprinting. *Electrophoresis* **2011**, *32*, 2951–2965. [CrossRef] [PubMed]

21. Huang, B.-B.; Lin, H.-C.; Chang, Y.-W. Analysis of proteins and potential bioactive peptides from tilapia (*Oreochromis* spp.) processing co-products using proteomic techniques coupled with BIOPEP database. *J. Funct. Foods* **2015**, *19*, 629–640. [CrossRef]

22. Jao, C.-L.; Hung, C.-C.; Tung, Y.-S.; Lin, P.-Y.; Chen, M.-C.; Hsu, K.-C. The development of bioactive peptides from dietary proteins as a dipeptidyl peptidase IV inhibitor for the management of type 2 diabetes. *Biomedicine* **2015**, *5*, 14. [CrossRef]

23. Minkiewicz, P.; Dziuba, J.; Iwaniak, A.; Dziuba, M.; Darewicz, M. BIOPEP database and other programs for processing bioactive peptide sequences. *J. AOAC Int.* **2008**, *91*, 965–980.

24. Kumar, C.; Mann, M. Bioinformatics analysis of mass spectrometry-based proteomics data sets. *FEBS Lett.* **2009**, *583*, 1703–1712. [CrossRef]

25. Cheung, I.W.; Nakayama, S.; Hsu, M.N.; Samaranayaka, A.G.; Li-Chan, E.C. Angiotensin-I converting enzyme inhibitory activity of hydrolysates from oat (*Avena sativa*) proteins by *in silico* and *in vitro* analyses. *J. Agric. Food Chem.* **2009**, *57*, 9234–9242. [CrossRef]

26. Gangopadhyay, N.; Wynne, K.; O'Connor, P.; Gallagher, E.; Brunton, N.P.; Rai, D.K.; Hayes, M. *In silico* and *in vitro* analyses of the angiotensin-I converting enzyme inhibitory activity of hydrolysates generated from crude barley (*Hordeum vulgare*) protein concentrates. *Food Chem.* **2016**, *203*, 367–374. [CrossRef]

27. Lafarga, T.; O'Connor, P.; Hayes, M. Identification of novel dipeptidyl peptidase-IV and angiotensin-I-converting enzyme inhibitory peptides from meat proteins using *in silico* analysis. *Peptides* **2014**, *59*, 53–62. [CrossRef]

28. Waghmare, A.G.; Salve, M.K.; LeBlanc, J.G.; Arya, S.S. Concentration and characterization of microalgae proteins from *Chlorella pyrenoidosa*. *Bioresour. Bioprocess.* **2016**, *3*, 16. [CrossRef]

29. Rosenberg, J.N.; Kobayashi, N.; Barnes, A.; Noel, E.A.; Betenbaugh, M.J.; Oyler, G.A. Comparative analyses of three *Chlorella* species in response to light and sugar reveal distinctive lipid accumulation patterns in the microalga *C. sorokiniana*. *PLoS ONE* **2014**, *9*, e92460. [CrossRef]

30. Watson, H.; Walker, N.; Shaw, P.; Bryant, T.; Wendell, P.; Fothergill, L.; Perkins, R.; Conroy, S.; Dobson, M.; Tuite, M. Sequence and structure of yeast phosphoglycerate kinase. *EMBO J.* **1982**, *1*, 1635–1640. [CrossRef]

31. Banks, R.; Blake, C.; Evans, P.; Haser, R.; Rice, D.; Hardy, G.; Merrett, M.; Phillips, A. Sequence, structure and activity of phosphoglycerate kinase: A possible hinge-bending enzyme. *Nature* **1979**, *279*, 773–777. [CrossRef]

32. Shevchenko, A.; Tomas, H.; Havli, J.; Olsen, J.V.; Mann, M. In-gel digestion for mass spectrometric characterization of proteins and proteomes. *Nat. Protoc.* **2006**, *1*, 2856–2860. [CrossRef]

33. Wysocki, V.H.; Resing, K.A.; Zhang, Q.; Cheng, G. Mass spectrometry of peptides and proteins. *Methods* **2005**, *35*, 211–222. [CrossRef]

34. Power, O.; Nongonierma, A.B.; Jakeman, P.; FitzGerald, R.J. Food protein hydrolysates as a source of dipeptidyl peptidase IV inhibitory peptides for the management of type 2 diabetes. *Proc. Nutr. Soc.* **2014**, *73*, 34–46. [CrossRef]

35. Boots, J.-W.P. Protein Hydrolysate Enriched in Peptides Inhibiting DPP IV and Their Use. U.S. Patent No. 8,273,710, 25 September 2012.

36. Lambeir, A.-M.; Durinx, C.; Scharpé, S.; De Meester, I. Dipeptidyl-peptidase IV from bench to bedside: An update on structural properties, functions, and clinical aspects of the enzyme DPP IV. *Crit. Rev. Clin. Lab. Sci.* **2003**, *40*, 209–294. [CrossRef]

37. Liu, R.; Zhou, L.; Zhang, Y.; Sheng, N.-J.; Wang, Z.-K.; Wu, T.-Z.; Wang, X.-Z.; Wu, H. Rapid Identification of Dipeptidyl Peptidase-IV (DPP IV) Inhibitory Peptides from *Ruditapes philippinarum* Hydrolysate. *Molecules* **2017**, *22*, 1714. [CrossRef]

38. Lacroix, I.M.; Li-Chan, E.C. Evaluation of the potential of dietary proteins as precursors of dipeptidyl peptidase (DPP)-IV inhibitors by an *in silico* approach. *J. Funct. Foods* **2012**, *4*, 403–422. [CrossRef]

39. Nongonierma, A.B.; FitzGerald, R.J. An *in silico* model to predict the potential of dietary proteins as sources of dipeptidyl peptidase IV (DPP IV) inhibitory peptides. *Food Chem.* **2014**, *165*, 489–498. [CrossRef]

40. Nongonierma, A.B.; Mooney, C.; Shields, D.C.; FitzGerald, R.J. *In silico* approaches to predict the potential of milk protein-derived peptides as dipeptidyl peptidase IV (DPP IV) inhibitors. *Peptides* **2014**, *57*, 43–51. [CrossRef]

41. Kim, Y.K.; Chung, B.H. A novel angiotensin-I-converting enzyme inhibitory peptide from human αs1-casein. *Biotechnol. Lett.* **1999**, *21*, 575–578. [CrossRef]

42. Kohmura, M.; Nio, N.; Kubo, K.; Minoshima, Y.; Munekata, E.; Ariyoshi, Y. Inhibition of angiotensin-converting enzyme by synthetic peptides of human β-casein. *Agric. Biol. Chem.* **1989**, *53*, 2107–2114. [CrossRef]

43. Li, C.H.; Matsui, T.; Matsumoto, K.; Yamasaki, R.; Kawasaki, T. Latent production of angiotensin I-converting enzyme inhibitors from buckwheat protein. *J. Pept. Sci.* **2002**, *8*, 267–274. [CrossRef]

44. Maruyama, S.; Mitachi, H.; Awaya, J.; Kurono, M.; Tomizuka, N.; Suzuki, H. Angiotensin I-converting enzyme inhibitory activity of the C-terminal hexapeptide of αs1-casein. *Agric. Biol. Chem.* **1987**, *51*, 2557–2561. [CrossRef]

45. Miyoshi, S.; Ishikawa, H.; Kaneko, T.; Fukui, F.; Tanaka, H.; Maruyama, S. Structures and activity of angiotensin-converting enzyme inhibitors in an α-zein hydrolysate. *Agric. Biol. Chem.* **1991**, *55*, 1313–1318. [CrossRef]

46. Li, G.-H.; Le, G.-W.; Shi, Y.-H.; Shrestha, S. Angiotensin I–converting enzyme inhibitory peptides derived from food proteins and their physiological and pharmacological effects. *Nutr. Res.* **2004**, *24*, 469–486. [CrossRef]

47. Abdelhedi, O.; Nasri, R.; Jridi, M.; Mora, L.; Oseguera-Toledo, M.E.; Aristoy, M.-C.; Amara, I.B.; Toldrá, F.; Nasri, M. *In silico* analysis and antihypertensive effect of ACE-inhibitory peptides from smooth-hound viscera protein hydrolysate: Enzyme-peptide interaction study using molecular docking simulation. *Process Biochem.* **2017**, *58*, 145–159. [CrossRef]

48. Bleakley, S.; Hayes, M.; O'Shea, N.; Gallagher, E.; Lafarga, T. Predicted Release and Analysis of Novel ACE-I, Renin, and DPP IV Inhibitory Peptides from Common Oat (*Avena sativa*) Protein Hydrolysates Using *in Silico* Analysis. *Foods* **2017**, *6*, 108. [CrossRef]

49. Muhammad, S.A.; Fatima, N. *In silico* analysis and molecular docking studies of potential angiotensin-converting enzyme inhibitor using quercetin glycosides. *Pharmacogn. Mag.* **2015**, *11*, S123–S126. [CrossRef]

50. Paiva, L.; Lima, E.; Neto, A.I.; Baptista, J. Isolation and characterization of angiotensin I-converting enzyme (ACE) inhibitory peptides from *Ulva rigida* C. Agardh protein hydrolysate. *J. Funct. Foods* **2016**, *26*, 65–76. [CrossRef]

51. Kang, K.-H.; Qian, Z.-J.; Ryu, B.; Kim, S.-K. Characterization of growth and protein contents from microalgae Navicula incerta with the investigation of antioxidant activity of enzymatic hydrolysates. *Food Sci. Biotechnol.* **2011**, *20*, 183–191. [CrossRef]

52. Sheih, I.-C.; Wu, T.-K.; Fang, T.J. Antioxidant properties of a new antioxidative peptide from algae protein waste hydrolysate in different oxidation systems. *Bioresour. Technol.* **2009**, *100*, 3419–3425. [CrossRef]

53. Parimi, N.S.; Singh, M.; Kastner, J.R.; Das, K.C.; Forsberg, L.S.; Azadi, P. Optimization of protein extraction from Spirulina platensis to generate a potential co-product and a biofuel feedstock with reduced nitrogen content. *Front. Energy Res.* **2015**, *3*, 30. [CrossRef]

54. Markwell, M.A.K.; Haas, S.M.; Bieber, L.; Tolbert, N. A modification of the Lowry procedure to simplify protein determination in membrane and lipoprotein samples. *Anal. Biochem.* **1978**, *87*, 206–210. [CrossRef]

55. Schägger, H.; Von Jagow, G. Tricine-sodium dodecyl sulfate-polyacrylamide gel electrophoresis for the separation of proteins in the range from 1 to 100 kDa. *Anal. Biochem.* **1987**, *166*, 368–379. [CrossRef]

56. Rosenfeld, J.; Capdevielle, J.; Guillemot, J.C.; Ferrara, P. In-gel digestion of proteins for internal sequence analysis after one-or two-dimensional gel electrophoresis. *Anal. Biochem.* **1992**, *203*, 173–179. [CrossRef]

57. Perkins, D.N.; Pappin, D.J.; Creasy, D.M.; Cottrell, J.S. Probability-based protein identification by searching sequence databases using mass spectrometry data. *ELECTROPHORESIS Int. J.* **1999**, *20*, 3551–3567. [CrossRef]

58. Pruitt, K.D.; Tatusova, T.; Maglott, D.R. NCBI reference sequences (RefSeq): A curated non-redundant sequence database of genomes, transcripts and proteins. *Nucleic Acids Res.* **2006**, *35*, D61–D65. [CrossRef]

International Journal of
Molecular Sciences

Review

Evolving a Peptide: Library Platforms and Diversification Strategies

Krištof Bozovičar and Tomaž Bratkovič *

Department of Pharmaceutical Biology, Faculty of Pharmacy, University of Ljubljana, Aškerčeva Cesta 7, SI-1000 Ljubljana, Slovenia; kristof.bozovicar@ffa.uni-lj.si
* Correspondence: tomaz.bratkovic@ffa.uni-lj.si; Tel.: +386-1-4769-570

Received: 2 December 2019; Accepted: 25 December 2019; Published: 27 December 2019

Abstract: Peptides are widely used in pharmaceutical industry as active pharmaceutical ingredients, versatile tools in drug discovery, and for drug delivery. They find themselves at the crossroads of small molecules and proteins, possessing favorable tissue penetration and the capability to engage into specific and high-affinity interactions with endogenous receptors. One of the commonly employed approaches in peptide discovery and design is to screen combinatorial libraries, comprising a myriad of peptide variants of either chemical or biological origin. In this review, we focus mainly on recombinant peptide libraries, discussing different platforms for their display or expression, and various diversification strategies for library design. We take a look at well-established technologies as well as new developments and future directions.

Keywords: peptide; combinatorial library; library design; screening; mutagenesis

1. Introduction

Peptides are short polymers composed of 19 L-amino acid and non-chiral glycine residues, linked by amide bonds. The definition is rather vague in terms of chain length, although an arbitrary upper limit of 6000 Da has been assigned to label these molecules peptides, and polymers above that molecular mass are considered proteins [1]. They are ubiquitous in nature and have a role in most physiological processes as host defense (antimicrobial) agents [2], (neuro)hormones [3,4], and toxins [5]. Peptide research has experienced considerable development in the last decades, and over 7000 peptides have been identified in nature [6]. Today, peptides are widely used in drug discovery, drug delivery, food industry, cosmetics, and various other fields.

Peptides and small proteins isolated from natural sources have been used as medicines since the beginning of the 1920s [7], with bovine and pork insulin being the first ones. Transition to synthetic peptide drugs only began in 1950s with synthetic oxytocin and vasopressin entering clinical use subsequently. However, native peptides possess several drawbacks, most notably having poor oral bioavailability and very short plasma half-life, which have tempered enthusiasm for their use, instigating investigators to develop peptides with improved pharmaceutical properties [8]. But it would be one of the biggest breakthroughs in understanding the fundamentals of life itself, the discovery of the genetic code and how it translates to the amino-acid sequence, laying foundations for modern biotechnology, which really pushed the field in a whole new direction in the coming decades. Gene identification and manipulation techniques that developed rapidly from 1973 onwards allowed for producing large quantities of pure gene products [9]. The landmark event was the expression of the first recombinant peptide hormone insulin in *Escherichia coli* and the following approval to commercialize recombinant insulin in 1982.

Peptides are utilized broadly owing to their superiority in specific cellular targeting. They bind cellular receptors with high potency and great selectivity, lowering toxicity potential and occurrence of off-target effects. In addition, peptides in the body are degraded to amino acids, further lowering the risk of toxicity [10]. Chemical synthesis enables peptide fabrication in large quantities, chipping production costs compared to other biologics. More attributes include stability at room temperature and good tissue permeability. Furthermore, physico-chemical traits of peptides (e.g., solubility, hydrophobicity, and charge), metabolic stability, and their residential time in the body can be fine-tuned through chemical modifications. Reiterative chemical modification approach can be honed for development of peptide therapeutics with improved properties [11], including extraordinary target affinity [12].

Areas of the highest concentration of peptide development in medicine are metabolic diseases, oncology, and cardiovascular diseases, not surprisingly, all areas of highest interest to the pharmaceutical industry. By 2018, more than 60 peptide drugs (excluding insulins and other small proteins) have been approved in the US, Europe, and Japan, over 150 were in active clinical development, and an additional 260 were assessed in human clinical trials but did not make it to the market [8]. The peptide therapeutics market was valued at 19,475 million USD in 2015 and it is estimated it will more than double the value by 2024, reaching 45,542 million USD [13].

During the past decade, peptides have also been used in a wide range of applications in other fields. They are found in biosensor applications as biorecognition molecules and are conjugated with transducers or molecular beacons that aid signal detection [14,15]. Additionally, they serve as surfactants or tags promoting solubility of recombinant intrinsic membrane proteins [16–20], increasing their yield, activity, and aiding protein structural studies. Peptides are even replacing enzymes in catalytic reactions [21] and substituting proteins as ligands in affinity chromatography [22,23].

Discovery and design of novel peptides can be guided by various strategies. In this review, we focus mainly on the use of peptide and peptide aptamer [24] (sequences of 5–20 amino acid residues, grafted into loops of a robust protein scaffold) libraries generated through recombinant DNA technology, but discuss chemical peptide libraries as well.

2. Combinatorial Peptide Libraries

Peptides of great number and diversity occur as a natural form of combinatorial chemistry. Conversely, exploiting evolutionary principles in the laboratory by constructing and screening large peptide libraries can yield new lead compounds with desired traits. The discovery of novel binders is a multifaceted process involving scanning of thousands or even millions of potential candidates from combinatorial libraries using in vitro screening analysis, commonly used in target-based drug discovery. Target-based drug discovery (sometimes called "reverse pharmacology") is the opposite of a traditional phenotypic screening strategy. The latter typically leads to the identification of molecules that modify a disease phenotype by acting on previously unidentified target [25]. In contrast, the targets in the target-based approach are well defined. With the molecular target in hand, discovery of novel binders can be facilitated by utilizing crystallographic and biochemical studies, computational modeling, binding kinetics, and mutational analysis to gain insight into how the target and the ligand interact and thus enable efficient structure-activity (SAR) analysis and the development of future generations of binders [26].

Combinatorial peptide libraries can be categorized into two groups—chemical peptide libraries, which are produced via organic synthesis, and biological libraries. Choosing a library platform should be guided by practical manners. Importance of library size, the experience of operators, available equipment, and other technical considerations may well limit the choice [27]. In principle, library-based peptide discovery adheres to the following paradigm: (1) creation of a pooled peptide library, (2) screening of the library against the target molecule and isolation of hits, and (3) hit identification.

Various screening/selection methods are at disposal depending on the peptide library platform. Normally, screening peptide libraries involves incubating the library with a fluorescently labeled soluble target or target-coated magnetic beads, followed by flow cytometry-based systems such as fluorescence activated cell sorting (FACS) [28], or magnetic separation techniques like magnetic-activated cell sorting (MACS) [29], respectively. The former is mostly used for cell-based peptide libraries, although it has also been used for screening chemical library systems such as one-bead-one-compound platform [30] (see below). Hit identification is also dependent on the library type; either iterative deconvolution or positional scanning methods are used for chemical libraries, while sequencing is typically employed for DNA-encoded platforms. In recent years, next-generation sequencing (NGS) methods, capable of massively parallel nucleic acid sequence determination, have transformed the field of screening biological libraries, enabling detection of low abundant clones and quantification of changes in clone copy numbers without performing many rounds of selection. This technology has been paired with various library platforms, immensely enhancing the throughput of these methods [31–38].

2.1. Chemical Peptide Libraries

In chemical approach, the library synthesis is either completed on a solid support (insoluble porous polymeric resin) and then members are cleaved to be screened as free compounds, or the library is synthesized and screened on a solid matrix.

Solid-phase peptide synthesis (SPPS) was first achieved by the Nobel laureate Robert B. Merrifield. The general approach of SPPS is to attach the first amino acid to a solid support through its carboxyl group. Subsequently, each N-protected amino acid is then added in turns. During coupling, the carboxyl group of the incoming amino acid must be activated, which is commonly achieved by using carbodiimides, amino acid halides, uronium (guanidinium N-oxides), or phosphonium salts [39,40]. After each addition, the N-protecting group must be removed before the next amino acid is added. The most used N-protecting group is the fluorenylmethyloxycarbonyl (Fmoc) group [41] which can be orthogonally removed under basic conditions. Since the growing peptide chain is attached to a surface, removing waste products of synthesis is accomplished by simple washing. Combinatorial synthesis is inherently a parallel synthetic process where a single product is obtained in each different reaction flask (or a sealed permeable container ("teabag") submerged in a reaction flask) [42]. Alternatively, a mixture of products can be obtained in a single flask via the "mix-and-split" method (see below) [43]. The whole synthesis process is quantitative, as the reactions are driven to completion by applying reagents in excess at every step. In addition, as the growing peptide chain is attached to a matrix, there is no need for isolation of intermediates. Nevertheless, judicious use of resources should be considered as wasting large amounts of reagents is costly and the surplus effluent produced can be an environmental burden [1], although this is a concern mainly when producing large quantities of product and less so for library generation.

For generating large libraries, the "split-and-mix" is the preferred strategy. This cyclic scheme of SPSS was first demonstrated by Furka et al. [43] in 1991 and involves coupling of individual amino acids to resin beads, mixing the beads together, separating them in equal portions, and then reacting each portion with a different amino acid (Figure 1). Mixing, separating beads, and the reaction step are repeated until the desired peptide length and diversity are achieved. An important virtue of this method is that a single bead contains a single peptide sequence, which is why the libraries produced in this way are termed one-bead-one-compound (OBOC) [44]. The second strategy is called "pre-mix", where all the amino acids to be coupled at each synthetic step are premixed in equimolar ratio and then reacted with a single resin batch. To overcome the problem of different kinetics rate of each amino acid coupling, "smart" mixtures of molar ratios that correspond to different coupling rates were proposed [45].

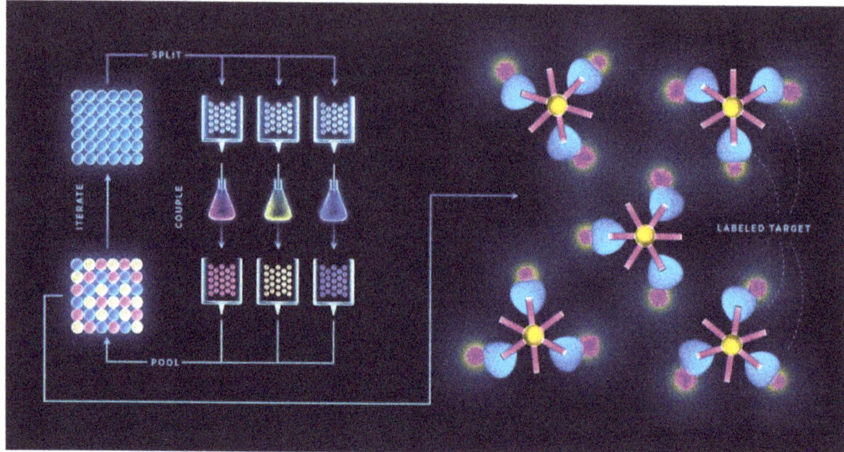

Figure 1. Synthesis of combinatorial peptide libraries by the "split-and-mix" method. Carrier beads are split into aliquots (only 3 are shown for clarity) for coupling individual amino acid residues, pooled, and the process is iterated until the desired length of peptides is achieved. Library diversity increases exponentially with each coupling step. At the end, each bead carries a single peptide sequence; hence, the name "one-bead-one-compound" combinatorial library. Libraries are typically screened by incubating the beads with a fluorophore-labeled target and subsequent fluorescence-activated sorting.

Various deconvolution methods have been developed for screening and identification of hits from chemical peptide libraries; most commonly adopted are iterative deconvolution [46] and positional scanning [47]. Iterative deconvolution is based on dividing the library into non-overlapping subsets containing peptides with defined residues at the specified position(s) (while having the rest of the structure randomized). Each subset is then screened separately. The most active subset of compounds is further divided into new subsets (retaining the identified optimal residue(s) from previous screening round) and retested for activity. This process is continued until the fittest molecule is identified [42]. In positional scanning, sub-libraries individually address each diversity position. This position is defined with a single amino acid residue, while the remaining positions are fully randomized. Positional sub-libraries are assayed in parallel to gather information on the optimal residue for each diversity position, consequently identifying the fittest member [47]. Other hit identification techniques include Edman degradation [48] and mass spectrometry (MS)-based strategies [49,50]. Mass spectrometry is highly sensitive and very fast, and is the method of choice when it comes to generating sequence data in the femtomolar range. Edman degradation, although reliable, only becomes a viable option at picomolar quantities [51]. It is also slow and not easily amenable to multiplexing [52]. In contrast, mass spectrometry has the edge in analyzing mixtures of great diversity, making HPLC separation of peptides unnecessary. However, a complete peptide fragmentation pattern is required in order to unambiguously identify each amino acid sequence [50].

The cutting-edge technology in chemical combinatorial libraries is the so-called DNA-encoded library (DEL) platform. It works by tagging the library members via an adapter module to a double-stranded DNA barcode which enables the unambiguous identification of retained compounds at the end of the selection process [53]. Both the library synthesis and tagging rely on the "split-and-mix" approach. In each synthesis cycle, the process is encoded (recorded) by ligation of a short DNA tag that identifies the amino acid residue added. A pooled library is then assayed by affinity selection and selected binders can be easily identified by PCR amplification and sequencing of tags [54,55]. Some peptides identified using the DEL platform include carbonic anhydrase binders [56], ligands of integrins, and CCR6 [57]. Generation of macrocyclic peptide libraries by DEL has also been described [55,58]. DEL offers several benefits over some other established methods, like phage display (see below), for discovery of peptide ligands. Library sizes (i.e., diversity) are larger, allowing the chemical space to be more deeply sampled in a single screen. Furthermore, DEL platforms are compatible with parallel screening against multiple targets, enable activity-agnostic screening for identification of "silent binders", and consume relatively low amounts of reagents [54]. In addition, not only individual screening hits are identified, but rather, larger families can be detected, in which related building blocks or combinations thereof are enriched [59]. Limitations of the platform include possible interference of the oligonucleotide tag in target–binder interaction that may cause steric hindrance. The barcode can also restrict the extent of possible chemical reactions and influence the properties of binders. Resynthesis after selection must be performed for subsequent target binding validation, which adds to the duration of the assay [54,60].

2.2. Biological Peptide Libraries

A key feature of biological libraries is the linkage of the genotype (i.e., genetic information) with its corresponding phenotype (the encoded peptide/protein). In directed evolution, (random) mutagenesis (gene diversification) and screening for functional gene products are iteratively alternated, and the phenotype:genotype link is crucial for peptide identification via sequence determination of the encoding oligonucleotide. Based on the display type approach, biological libraries can be categorized as either cellular or acellular [61].

2.2.1. Cellular Approach

The main bottleneck to this approach is a transformation step needed for delivering a DNA library into host cells, providing transcriptional and translational machineries for gene expression. One of the most widely used and recognized methods is phage display, first described in 1985 by George P. Smith [62]. Smith shared half of the 2018 Nobel Prize in chemistry (the other half was awarded to Frances H. Arnold) with Gregory P. Winter for the phage display of peptides and antibodies, validating the gargantuan importance of this technology. Filamentous (bacterio)phages (most commonly used vehicles in phage display) are rod-shaped viruses that infect *E. coli*, and are composed of coat (capsid) proteins that encapsulate the phage genome. Surface display is achieved by inserting a peptide-encoding oligonucleotide sequence into one of the genes for a capsid structural protein [63–65]. All 5 filamentous phage coat proteins have been exploited as anchors for display of foreign peptides and proteins. The filamentous phage's minor coat protein p3 is the most widely used and can present 3–5 copies per virion, seconded by the major coat protein p8 with a much larger count, around 2700 copies. The filamentous phage M13 and the closely related fd are most commonly used due to their ease of manipulation and their ability to accommodate fairly large pieces of foreign DNA. Two main display types emerged for displaying libraries on filamentous phage: the polyvalent (one-gene-system) and the monovalent display (two-gene-system). In the polyvalent display, the DNA fragments are inserted into the phage vector, producing fusions with each copy of the chosen coat protein [65]. As opposed to monovalent display, polyvalent systems yield binders that exhibit reduced affinity due to avidity effects, enriching weak (albeit specific) ligands over the course of a multi-step affinity selection process [66]. The monovalent display uses a phagemid vector—essentially a plasmid encoding the foreign peptide-coat protein fusion, and harboring the phage origin of replication and the packaging signal for production of single-stranded DNA copies and assembly into phage particles, respectively. A helper phage (which supplies the rest of phage genes, but has assembly defects) is employed to superinfect the cells harboring the phagemid. Thus, the produced virions display a mixture of recombinant coat proteins, encoded by the phagemid, and the cognate wild-type proteins, encoded by the helper phage [27]. Although two-gene-systems are mainly utilized for displaying larger proteins, both systems have been used to display relatively short peptides [67]. Less commonly, lytic phages, such as T7 or lambda, have been used for phage display of peptides [64,68]. Another interesting phage display-like system allowing tunable display valency is based on virus-like particles (VLPs) of the RNA bacteriophage MS2 [69]. In phage display, target-binding peptides are identified through affinity selection process called panning (Figure 2). This technique is comprised of several steps. First, surface-immobilized target is contacted with phage library, followed by stringent washing to remove unbound and non-specifically bound clones. Specific binders are then eluted, usually by applying pH shock or high salt solutions, and subsequently amplified in host cells. Obtaining binders with high affinity usually requires 3–5 panning rounds. There is a reason why phage display is the method of choice for so many. Phage libraries are more readily affordable compared to using other microbial/cell display vehicles (see below), can easily be amplified by allowing library phage to replicate in a bacterial host. Virions can withstand various selection environments, endure harsh washing and elution conditions, and can be stored at –80 °C for years. Phage library diversities are typically significantly larger compared to chemical peptide libraries and can reach up to 10^{11} clones. Although phage can accommodate a large variety of (poly)peptide structures, amino acid and sequence biases do occur (as with any biological library), leading to inherent library diversity decrease [70]. In filamentous phage, this is due to censorship of charged sequences through the general bacterial secretory (Sec) pathway. Capsid structural proteins fused to charged peptides seem be to less efficiently inserted into the *E. coli* inner membrane, hindering virion assembly [71]. Conversely, lytic phage vehicles have the advantage of not being limited to display of peptides which are efficiently translocated to the periplasm [72]. To display (poly)peptides that fold rapidly in the cytoplasm (and are thus inefficiently transported to periplasm) on *filamentous* phage, the twin-arginine translocase (Tat) export pathway has been exploited instead of the conventional Sec pathway [73,74]. In contrast to Sec translocase complex that mediates export of unfolded proteins through the inner bacterial membrane, Tat only exports fully folded proteins. Another drawback of phage display is the non-quantitative nature of clone screening, although this problem can be elegantly tackled by deep sequencing [31] instead of traditional clone picking. In conventional phage display, the displayed peptides are limited to natural L-amino acids. All-D peptide ligands, showing improved metabolic stability, can be developed using the mirror-image phage display strategy [75]. Cyclization of phage-displayed peptides [76] represents another popular method of augmenting protease resistance and simultaneously increasing affinity (also see Section 3). In addition to in vitro or ex vivo (e.g., against isolated cells), phage library selections can also be performed *in vivo*. The latter approach is used to identify tissue homing peptides for cell-specific targeting of therapeutics [77]. In contrast to phage library pannings to purified targets, selections against intact cells and in vivo screens are inherently more complicated as the library is exposed to many potential decoys, leading to high background [63]. When panning against a cell surface receptor, one typically relies on a cell line ectopically overexpressing the targeted membrane protein. This allows for so called *subtractive* or *negative* selections to be performed against the same cells devoid of the target of interest before contacting the library phages with target-transfected cells in each selection round, thereby effectively reducing background binders. In in vivo selections, phage library is typically injected intravenously or directly into tumor xenografts of an experimental animal. Upon systemic application, library phages are likely not distributed uniformly to all tissues, but are rather retained in the circulation, being primarily exposed to vascular endothelium. Targeting other tissues requires fairly large phage titers; phages can be rescued from tissue of interest, and amplified for iteration of selection. Deep sequencing enables analysis of the entire repertoire of retained phages [78], and should thereby allow the deduction of specific binders even from single-selection-round experiments [79], especially if independent parallel screens are performed and the same peptides are observed enriched in each case. To limit background phages, vascular perfusion can be harnessed.

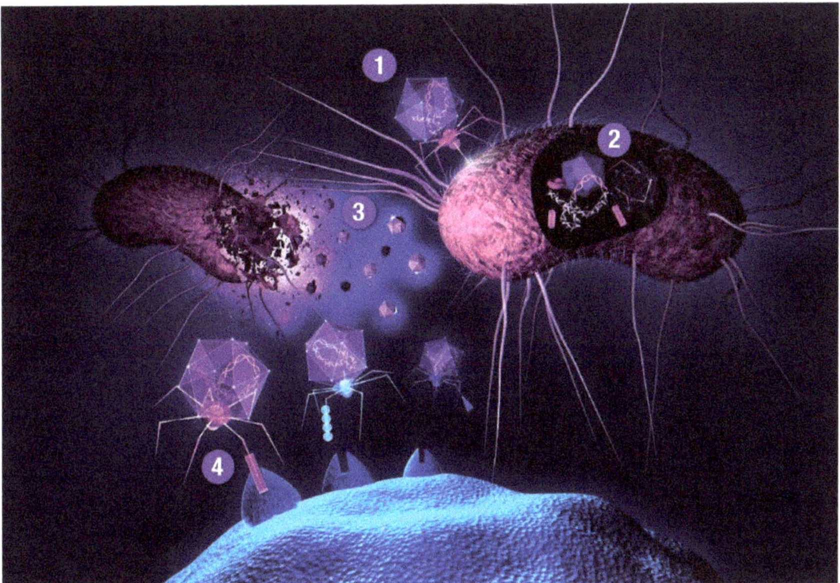

Figure 2. Iterative principle of phage display library screening. Recombinant phage DNA is packed into viral particles in vitro (common for T7- and lambda phage-based systems) for transduction of genetic library into host bacteria (**1**). Alternatively, phage DNA can be electroporated into host cells (typical for filamentous phage-based systems; not shown). Exploiting bacterial transcription and translation mechanisms, progeny virions displaying foreign peptides are amplified and assembled (**2**), and released into growth medium (**3**). Amplified library is isolated and purified, and contacted with an immobilized target (**4**). Non-bound clones are removed by stringent washing, while those retained due to target:displayed peptide interaction are eluted and collected for amplification in host bacteria before being subjected to further selection rounds.

Several peptide display systems based on Gram-negative bacteria were reported. Typically, peptides are grafted in the surface-exposed loops of outer membrane proteins. Fusion partners for peptide library display include the *E. coli* OmpA [80], OmpX [81], and the *Pseudomonas* OprF [82] among others. It is also possible to display peptides on extracellular appendages such as pili [83] and flagella, albeit only the latter has been used for library construction [84,85]. Conventional bacterial surface display libraries are commonly screened using FACS. Library display and selection have also been demonstrated in the periplasm of *E. coli*. In the intra periplasm secretion and selection (PERISS) system, the target molecule gene and peptide library DNA are integrated in tandem into a plasmid, expressing the peptide library in the periplasmic space, while the target molecule is incorporated into the *E. coli* inner membrane. The outer membrane is disrupted with ethylenediaminetetraacetic acid (EDTA) and the cell wall is degraded enzymatically to form spheroplasts. Peptides interacting with the target are collected as binding complexes on magnetic beads through a fusion tag, and identified through PCR amplification and sequencing [86]. Another form of periplasmic display is the "anchored periplasmic expression" (APEx) technique, where a peptide library is expressed and immobilized in the periplasm of *E. coli* via fusion to a lipoprotein targeting motif that enables anchoring to the inner membrane. The outer membrane and the cell wall are then removed, followed by incubation with the fluorescently tagged target molecule, washing, and binder detection by FACS [87,88]. Furthermore, peptide libraries were expressed in the cytoplasm of bacteria via fusions to the C-terminus of the *lac* repressor. The repressor protein binds to *lac* operator sequence on the plasmid encoding the peptide, providing a physical genotype-phenotype link. The library is screened by affinity purification with an immobilized receptor [89]. Screening techniques of the latter are inferior compared to surface display, as the cells must be lysed prior to panning to the immobilized target. Another important technique, the bacterial two-hybrid system, is used for studying protein–protein interaction (PPI). It can be used to screen libraries of peptides to probe and manipulate biological pathways [90]. The premise of all variants of two-hybrid platforms is the identification of target binders via reconstitution of reporter's activity in vivo dependent on the interaction between a pair of mediator proteins. Typically, the target protein ("bait") is fused to the transcriptional activator's DNA-binding domain (interacting specifically with the upstream activating sequence of a reporter gene), while the library (poly)peptide variants ("prey") are fused to the transcriptional activation domain (which recruits the RNA polymerase). Another option is the functional complementation of a split two-domain enzyme adenylate cyclase upon bait-prey interaction, leading to indirect reporter gene activation [91]. Interactions between protein pairs are identified by growing library bacteria either on color indicator or selective media plates [92]. In general, assets of bacterial combinatorial libraries are fast bacterial growth rate, ease of genetic and physical manipulation, and (in case of surface display and periplasmic expression) highly efficient screening protocols based on FACS. On the other hand, display platforms may suffer from interference of the complex bacterial surface with the selective peptide:target recognition. A recently developed peptide library system termed "surface localized antimicrobial display" (SLAY) [93] relies on a "reversed" screening protocol to identify peptides with antimicrobial activity. In SLAY, library plasmids are transformed into bacteria of interest and induced to express the encoded peptides. The peptides with antibacterial properties will lead to bactericidal or bacteriostatic effects, eliminating these clones from the population. Using NGS and in silico techniques, sequences pre- and post-induction are analyzed to identify members with antimicrobial properties.

Phage and bacterial display can be classified as prokaryotic display. There are also vast possibilities of displaying peptides and proteins in eukaryotic systems. Their main advantage is the reliable folding and glycosylation of eukaryotic proteins [94]. One of the most widely used organisms in this category is *Saccharomyces cerevisiae*. Yeast has been utilized as a vessel for numerous types of peptide library screening. Like the bacterial two-hybrid system, the yeast two-hybrid system is used for investigating protein–protein interaction through the activation of reporter genes responding to a reconstituted transcription factor [95,96]. This particular method gave rise to a number of adaptations to study PPIs, such as the *ras* recruitment system [97], split ubiquitin system [98], and the yeast three-hybrid system [99], to name a few. Conversely, yeast surface display is achieved by fusions to a cell surface-anchored protein, followed by screening and selection through magnetic separation or FACS [100]. Foreign peptides are commonly fused to the C-terminus of α-agglutinin Aga2p subunit, a surface protein covalently bound to glucan, which mediates cell to cell adhesion during yeast mating. Aga2p is linked to the Aga1 protein through two disulfide bridges, resulting in a covalent complex on the surface of the yeast cell [101]. Other surface anchor proteins used in this manner are Agα1p, Cwp1p, Cwp2p, Tip1p, Flo1p, Sed1p, YCR89w, and Tir1, a choice that depends on the type of protein/peptide to be displayed [102,103]. Another type of display in yeast is possible via secretory expression [104]. A number of novel ligands have been discovered via screening of peptide or small protein libraries expressed in yeast, both on the surface and intracellularly. Examples of surface display libraries include cysteine knot peptides (knottins) [105] and lanthipeptides [106], whereas head-to-tail cyclized peptide libraries [96,107] (see the SICLOPPS method description below) are expressed intracellularly. The main advantage of yeast display is its eukaryotic protein expression mechanism, which allows for complex post-translational modifications, and quantitative library screening through FACS [108,109]. Disadvantages include smaller library sizes due to low transformation efficiency [110], and lower affinity caused by unintended multivalent binding to oligomeric targets, although this can be surmounted by applying kinetic selections [111]. Yeast also form high-mannose type glycans that render glycoproteins produced in this system unfit for human applications, but this problem can be tackled by humanizing yeast glycosylation pathways [112].

Another platform well-suited for (poly)peptide display is the baculoviral particle (or baculovirus infected cell) system. Baculoviruses, enveloped viruses infecting invertebrates, are distinguished by a large packaging capacity, and (being eukaryotic pathogens) support diverse post-translational modifications [113]. Typically, foreign peptides or proteins are fused with the major envelope glycoprotein, such as the gp64 of the baculovirus *Autographa californica* multiple nuclear polyhedrosis virus (AcMNPV), and the recombinant fusions are embedded in the viral envelope (and the plasma membrane of infected insect cells) along with the wild-type glycoprotein. Other membrane anchoring strategies can be exploited to display a wide range of proteins and peptides at diverse valency range [114,115]. Regardless of the display strategy, libraries on infected insect cells are screened with FACS [116], and libraries displayed on the AcMNPV virions may be selected via conventional affinity panning [117]. In a prominent example, baculovirus-displayed libraries of peptides bound to the class I or II major histocompatibility complex (MHC) have served for identification of T cellular receptor peptide antigen mimetics (mimotopes) [118,119]. Combining the benefits characteristic for prokaryotic platforms (e.g., simple viral propagation as insect cells do not require CO_2 exchange for growth, high transfection rates, and the low-risk biosafety profile [114,120]) with clear eukaryotic advantages (efficient protein folding and a plethora of accessible post-translational modifications), the baculoviral display platform seems underappreciated. On the other hand, downsides of using this technology are expensive growth media [121], similar but still different glycosylation patterns compared to those of mammalian systems [122] and time-consuming cloning in the baculoviral vector, although new systems to avert this lengthy step have been developed (reviewed by Possee and King [123]).

Peptide libraries have also been displayed by using several eukaryotic RNA viruses with inserting short peptides into their native envelope proteins without interfering with the viral infectivity. Examples of eukaryotic retroviral vehicles include the avian leukosis virus [124] and feline leukemia virus [125,126]. Alongside retroviruses, the adeno-associated virus (AAV) has been used as a platform for peptide library display. Here, capsid diversification combined with phenotypic screening was primarily used as a means of achieving re-targeted tropism [34,127–131]. Normally, AAV libraries are produced in human cells, such as HEK 293T or HeLa [132]. AAV library construction starts by inserting encoded capsid gene variants (in place of previously excised wild-type *cap* gene sequence) in a template vector encoding a complete AAV genome. Because the natural vector tropism is disrupted, the rare subpopulation of library peptides are expected to redirect virions to new, formerly inaccessible cell types [133]. Library screening involves passaging the virions over several rounds in the chosen cell type to enrich for capsid variants with higher transduction efficiency and/or specificity compared to the naïve library background [134]. AAV libraries are primarily considered for their modus operandi—they infect humans and can thus be exploited as target-specific gene therapy vectors. AAV does not seem to cause any diseases and is only weakly immunogenic. Many challenges face AAV library platform, most notable being an uncertainty of capsid-genome correlation due to the possibility of co-transfection with different AAV capsid variants, resulting in library members with multiple phenotypes [132].

Peptide libraries have also been displayed on the surface of mammalian cells. Examples include fusions to the chemokine receptor CCR5 for antibody mimotope selection [135] and cystine-dense peptides for difficult to drug targets [36]. On the other hand, peptide libraries have been deployed for intracellular screens in mammalian cells [136–139]. Besides the benefit of authentic post-translational modifications, the most conspicuous advantage of mammalian display libraries is the ability to screen against targets and to investigate PPIs in their native environment. On the other hand, obvious downsides of working with mammalian cells are high cultivation costs and laborious cultivation technologies.

The "split-intein circular ligation of peptides and proteins" (SICLOPPS) is a method for cyclic peptide library generation. It takes advantage of intein splicing to generate peptide libraries in cells [140]. Inteins are self-excising protein domains that process to link their consecutive sequences with a peptide bond, while liberating the intervening portions (exteins) as head-to-tail cyclized peptides [141]. Library peptides to be cyclized are usually 6 amino acids long randomized exteins, flanked by C- and N-terminal intein domains with splice site-adjacent cysteine residues. Propagation of plasmids harboring SICLOPPS expression cassettes in cells resolves the problem of genotype:phenotype linkage. As the libraries are assembled in cells, this approach is particularly well-suited for functional assays; both function and affinity (to a large repertoire of accessible intracellular targets) for each binder can be assessed. Other benefits include simplicity, speed, and ease, and its applicability to different organisms [107,142,143] make this high-throughput approach the platform of choice. On the other hand, the main bottleneck of SICLOPPS is the limited library size of $10^{7\text{-}9}$ (determined by transfection efficiency) [140,144,145].

2.2.2. Acellular Approach

In contrast to the cellular approach, the acellular libraries are not propagated in living cells. These systems are based on in vitro transcription and translation, and one of the main advantages of this approach is the potential to generate high diversity libraries with up to 10^{15} individual entities by averting cell transformation step. Also convenient is the ability to apply stringent screening conditions that would be incompatible with cell viability.

In ribosome display, peptide:ribosome:mRNA complexes are generated, physically associating nascent peptide to their encoding mRNA. This is achieved by eliminating stop codons, consequently stalling both the nascent peptide and its encoding mRNA on the ribosome. The method consists of the following steps: DNA library preparation, in vitro transcription and translation, affinity selection and mRNA recovery, followed by reverse transcription and PCR amplification for sequencing or further selection rounds [146]. Many types of peptide libraries have been screened through this platform, including tumor targeting peptides [147], peptides that bind to monoclonal antibodies [148,149], streptavidin ligands [150], and metal binding peptides [151], to name a few. Besides its simplicity and the obvious ability to create high diversity libraries, another key advantage of the technology is that the diversity of the libraries can be easily manipulated by introducing new mutations at any selection step, and is therefore particularly suited for directed evolution projects [152]. With the advances such as PURE (protein synthesis using recombinant elements), ribosome display has also become very stable, as it is composed only of reconstituted ribosome and translation factors, eliminating degradation of mRNA and nascent (poly)peptides by endonucleases and proteases from the cell extract [153]. Another detriment that is being resolved is the thermal stability of this platform; usually, ribosome display is performed at near freezing temperatures. Fusing the library peptides to the Cv RNA-associating protein (Cvap) and adding the Cv RNA motif to the 5' mRNA end renders the peptide-Cvap:mRNA:ribosome complexes stable at room temperature, even for prolonged time [151,154]. However, the fragile noncovalent phenotype:genotype conjugation still requires mild selection conditions.

Another technique, mRNA display, is similar to ribosome display. The essence of this technology is the employment of puromycin, an antimicrobial product that inhibits translation by mimicking the 3'-end of an aminoacyl-tRNA, conjugated to the 3' terminus of mRNA [155]. When the ribosome reaches the 3' end of the template, the newly synthesized peptide is transferred onto the puromycin, achieving genotype–phenotype linkage with its encoding mRNA. Later, affinity panning is the preferred practice for binder identification. As in ribosome display, after each selection round, mRNA is reverse transcribed to cDNA and PCR amplified for the purpose of sequencing or further selection rounds. This platform has been extensively used and there are numerous examples of discovered novel peptide binders. These include cyclic peptide therapeutics targeting GPCR signaling [12], streptavidin binding peptides [156], and peptide vaccines [157]. Compared to ribosome display, this platform is superior and enables stringent selection conditions [158]. A major advantage over ribosome display is obviously the absence of the ribosome, a huge 2,000,000 Da ribonucleoprotein complex, which can interfere with the selection process [159]. Despite its unique ability in addressing a repertoire of different biological problems, mRNA display has limitations like any other display technology. One of the concerns involves interactions of covalently bound mRNA with the displayed peptide or the target molecule. The interference of flexible mRNA is particularly problematic when dealing with proteins that nonspecifically bind nucleic acids [158]. Furthermore, the highly negatively charged mRNA fusion moiety may interfere with positively charged target molecules [160].

A major problem in both ribosome and mRNA display is mRNA lability. This was the driving motivation behind the development of the cDNA display method. Here, mRNA is ligated to a looped DNA linker, harboring a primer region for reverse transcription and a restrictase cleavage site, and carrying a loop-attached biotin group and a 3′ terminal puromycin moiety. Thus, the translated mRNA-encoded peptide is transferred to the DNA linker, and the complexes are captured on streptavidin beads for reverse transcription. Finally, the library is released from the solid support by restrictase treatment [161]. This refined platform has been used in screening for cysteine-rich peptides with antagonistic activity at interleukin-6-receptor [162], peptide antagonists of growth hormone secretagogue receptor [163], amino group binders [164], binders of vacuolating toxin [165], and peptide agonist and antagonist of angiotensin II type-1 receptor [166]. The problem of RNA instability instigated the design of another DNA-based approach called CIS display. This method harnesses the so-called cis activity—the capacity of a DNA replication initiator protein (RepA), fused to the library peptides, to bind exclusively to the cognate template DNA [167]. CIS display has been applied for identification of binders of antibodies [167], protease-resistant peptides [168], and ligands of human vascular endothelial growth factor receptor [169].

Another acellular platform dubbed "in vitro compartmentalization" (IVC) mimics natural encapsulation of living cells by entrapping a DNA library and a transcription/translation reaction mixture in water-in-oil emulsion droplets. These "microreactors" serve as boundary analogs to a cell membrane, assuring an effective genotype–phenotype linkage; on average, each droplet contains a maximum of one library member. The in vitro transcription/translation paraphernalia within the droplets processes the genes, and the desired phenotype is selected using a suitable strategy [170]. For example, the DNA can be linked to a substrate which is converted into a marker product by the encoded enzyme, rendering it detectable by FACS [171]. It has been reported, that 1 mL of such an emulsion can hold as many as 10^{10} droplets [170]. Adaptations of this technique (e.g., STABLE [172], SNAP [173]) have been summarized in a recent review [174]. Although IVC is a technology especially handy for in vitro enzyme evolution [175], it has also been applied for screening peptide libraries [176,177]. Possessing benefits of the above acellular platforms, this technology is superior as reactions can be controlled inside the droplets, by adding reaction constituents in a step-wise fashion at defined time-points (e.g., after in vitro translation) [178]. IVC is also suitable for quantitative screening using FACS [179]. There are technical limitations to this platform, and the one that stands out is the formation of double emulsion droplets with the consequence of severing the genotype–phenotype link, although this could be solved by employing a high-throughput screening platform using droplet-based microfluidics [180]. IVC libraries are also considerably smaller than mRNA-display libraries [181].

3. Incorporating Unnatural Amino Acids and Constraints into (Library) Peptides

The methods described above have become invaluable for the discovery or refinement of peptide binders. Still, low-molecular-weight leads are generally preferred over peptides in drug discovery, owing to peptides' inadequate properties. For example, libraries of natural peptides are limited to the 20 proteinogenic amino acids and are normally restricted to linear peptides that have high flexibility and poor pharmacokinetic properties. The natural amino acid limit makes it difficult (if not impossible) to introduce predefined pharmacophores into peptide libraries. Moreover, the exploration of "chemical space" is limited by the restricted set of residues, many of which share the same or similar side-chain chemical groups. Marked flexibility of the peptide backbone imposes high entropic cost upon peptide ligand binding to the target molecule, resulting in low affinity interactions [182,183]. Linear peptides can also fall prey to enzymatic cleavage by exoproteases [184]. Potent and proteolytically stable binders can be designed by introducing modifications into peptides, such as incorporation of non-proteinogenic residues [185], cyclization [186,187], or the inclusion of stabilizing moieties (e.g., the parallel β-sheet scaffolds) [188]. Another option is the introduction of chemical post-translational modifications (cPTM) that can vastly expand the diversity of peptide libraries, for example, by phosphorylation [189], conjugation to glycans [190,191], attachment of a fluorescent reporter [192], or ligation to a synthetic peptide fragment [193]. By introducing constrained topologies, cyclic [194] or bicyclic [195] variants can be generated [187]. However, cPTMs are not quantitative. These moieties are also not genetically encoded, and thus cannot be identified via sequencing, although this problem can be overcome with the construction of chemically identical peptide libraries using different codons ("silent barcodes"), signifying a specific cPTM [196].

Genetic code "reprogramming" refers to reassignment of arbitrary codons from proteinogenic to non-proteinogenic amino acids, allowing ribosomal synthesis of non-canonical peptides. This strategy is compatible with acellular in vitro display techniques. By omitting aminoacyl-tRNA synthetases (ARSs), library diversity can be increased by utilizing a reconstituted translation system such as PURE [153]. tRNAs can be charged with nonstandard or artificial amino acids employing natural or modified ARSs [197]. Promiscuous tRNA acylation ribozymes, termed "flexizymes", were created as a surrogate for ARSs. Flexizymes recognize activated carboxylates and allow extensive genetic code reprogramming [198]. Combining flexizymes with a custom in vitro translation system, a new technology, dubbed FIT (flexible in vitro translation), was developed. Adding mRNA display to the mixture, the so-called RaPID (random non-standard peptide integrated discovery) system was born, facilitating the discovery of potent nonstandard peptides for therapeutic and diagnostic use [38].

4. Peptide Library Design and Construction

Completely randomizing even relatively short peptides would require a library size surpassing the capacities of most platforms. Sampling the complete mutational space for peptides exceeding 8–9 residues is therefore practically impossible, and gene diversification strategies only allow for generation and subsequent interrogation of a limited subset of the entire theoretical peptide population. Peptide maturation can be depicted as an ascent in a simplified fitness landscape (Figure 3) in which the x-y coordinates denote the otherwise multidimensional genotype, and the z-axis represent the peptide's "phenotypic" traits, e.g., target affinity. Ascending towards peak activity with mutational steps is the goal of directed evolution. Beneficial mutations accumulate over several generations upon selection pressure, resulting in improved phenotype [199].

In general, library generation can be performed either through focused or random mutagenesis. The latter is usually used in the absence of structure–function relationship knowledge. In focused mutagenesis, residues previously found to be essential for peptide activity are retained (or favored over the rest of the building block set), while the others are (fully or partially) randomized. Of course, the odds that a library contains improved peptide variants are higher for those produced by focused mutagenesis. A plethora of mutagenesis methods can be used for gene diversification in library generation and we will briefly discuss them below.

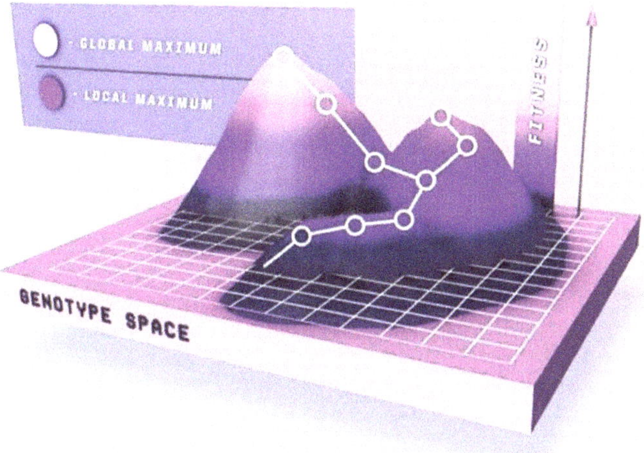

Figure 3. Maturation of a peptide depicted as ascent on a simplified fitness landscape. After each selection round, mutations are introduced into the enriched combinatorial library, and the next generation of peptides is screened for improved affinity and/or activity.

4.1. Random Mutagenesis

Random mutagenesis based on physical and/or chemical mutagens is sufficient for traditional genome screening (gene inactivation), but it is not suitable for directed evolution due to limited mutational spectrum [200,201]. For library generation purposes, random mutagenesis can be performed in vivo in bacterial mutator strains that contain defective proofreading and repair enzymes (mutS, mutT, and mutD) [202–204]. Another approach in *E. coli* relies on mutagenesis plasmids (MP), which carry multiple genes for proteins affecting DNA proofreading, mismatch repair, translesion synthesis, base selection, and base excision repair, thereby enabling broad mutagenic spectra. MPs support mutation rates 322,000-fold over basal levels and are suitable for platforms based on bacterial and phage-mediated directed evolution [205]. Unfortunately, beside the library gene, mutator strains and MPs also induce deleterious mutations in host genome. In eukaryotes, this was overcome by the development of orthogonal in vivo DNA replication apparatus, which in essence utilizes plasmid–polymerase pairs, limiting mutagenesis to a cytoplasm-only event [206]. Related phenomena are also known to occur in nature (e.g., the *Bordetella bronchiseptica* bacteriophage error-prone retroelement, which selectively introduces mutations into the gene encoding the major tropism determinant (Mtd) protein on the phage tail fibers [207]) and can be exploited for creating libraries [208].

One of the most established methods for in vitro random mutagenesis is the error-prone PCR (epPCR), first described in 1989 [209]. It works by harnessing the natural error rate of low-fidelity DNA polymerases, generating point mutations during PCR amplification. However, even the faulty *Taq* DNA polymerase is not erroneous enough to be useful for constructing combinatorial libraries under standard amplification conditions. The fidelity of the reaction can be further reduced by altering the amount of bivalent cations Mn^{2+} and Mg^{2+}, introduction of biased concentrations of deoxyribonucleoside triphosphates (dNTPs) [210], using mutagenic dNTP analogues [211], or adjusting elongation time and the number of cycles [212]. Random mutations can also be induced by utilizing 3′-5′ proofreading-deficient polymerases [213–215].

Despite its popularity, epPCR suffers from limited mutational spectrum as it inclines to transitions (A↔G or T↔C). Thus, epPCR-generated libraries are abundant in synonymous and conservative nonsynonymous mutations as a result of codon redundancy [199]. Ideally, all four transitions (AT→GC and GC→AT) and eight transversions (AT→TA, AT→CG, GC→CG, and GC→TA) would occur at equal ratios, with the desired probability, and without insertions or deletions [216]. This problem has been addressed by the sequence saturation mutagenesis (SeSaM) [217] method, which utilizes deoxyinosine, a promiscuous base-pairing nucleotide that is enzymatically inserted throughout the target gene and later changed for canonical nucleotides using standard PCR amplification of the mutated template gene. SeSaM was later improved with the introduction of SeSaM-Tv-II [218], which generates sequence space unobtainable via conventional epPCR by increasing the number of transversions. It employs a novel polymerase with increased processivity, allowing efficient read through consecutive base-pair mismatches. EpPCR has been successfully adopted for library generation in various platforms, including phage [219], *E. coli* [220], and ribosome [221] display.

Alternatively, mutagenesis can be achieved by performing isothermal rolling circle amplification (RCA) under error-prone conditions. Using a wild-type sequence as a template, this method is able to generate a random DNA mutant library, which can be directly transformed into *E. coli* without subcloning [222]. RCA was advanced further, coupling it with Kunkel mutagenesis [223] (see below). Termed "selective RCA" (sRCA), it operates by producing plasmids in *ung*⁻ (uracil-DNA-glycosylase deficient) *dut*⁻ (dUTP diphosphatase deficient) *E. coli* strain to introduce non-specific uridylation (dT→dU). After PCR with mutagenic primers, abasic sites are created by the uracil-DNA glycosylase in the uracil-containing template. Only mutagenized products are amplified by RCA, excluding non-mutated background sequences [224].

Although epPCR generates high mutational rates, the sequence space remains mostly untapped [225]. DNA shuffling is touted to be superior to epPCR and oligonucleotide-directed mutagenesis because it does not suffer from the possibility of introducing neutral or non-essential mutations from repeated rounds of mutagenesis [226]. DNA shuffling was the first in vitro recombination method and it involves random fragmentation of a pool of closely related dsDNA sequences and subsequent reassembly of fragments by PCR [227]. Such template switching generates a myriad of new sequences and improves library diversity by mimicking natural sexual recombination [228]. Meyer et al. [225] developed an approach where DNase I creates double-stranded breaks at the regions of interest, followed by denaturation and reannealing at homologous regions. Hybridized fragments then serve as templates and are subjected to repeated PCR rounds to form a whole array of new sequences. Improved methods were developed, eliminating the lengthy DNA fragmentation step. In the "staggered extension process" (StEP), polynucleotide sequences can be diversified through severely-abbreviated annealing/polymerase-catalyzed extension. In each cycle, growing fragments switch between different templates and anneal to them based on sequence complementarity. They then extend further and the cycle is repeated until full-length mosaic sequences are formed [229]. Another ingenious method for creating random customized peptide libraries by Fujishima et al. [230] works by shuffling short DNA blocks with dinucleotide overhangs, enabling efficient and seamless library assembly through a simple ligation process.

Currently, recombination methods are shifting from in vitro to *in vivo*. Taking advantage of the high occurrence of homologous DNA recombination events in *S. cerevisiae*, the "mutagenic organized recombination process by homologous in vivo grouping" (MORPHING) method was developed. MORPHING is a "one-pot" random mutagenesis method allowing construction of libraries with various degrees of diversity. Short DNA segments are produced by epPCR, and subsequently assembled with conserved overlapping gene fragments and the linearized plasmid by in vivo recombination upon transformation into yeast cells [231]. Another technique for assembling linear DNA fragments with homologous ends in *E. coli* is called "in vivo assembly" (IVA). IVA uses PCR amplification with primers designed to substitute, delete, or insert portions of DNA, and to simultaneously append homologous sequences at amplicon ends. Finally, it exploits recA-independent homologous recombination *in vivo*, greatly simplifying complex cloning operations. Thus, multiple simultaneous modifications (insertions, deletions, point mutations, and/or site-saturation mutagenesis) are confined to a single PCR reaction, and multi-fragment assembly (library construction) proceeds in bacteria following transformation [232].

4.2. Focused Mutagenesis

Effectively exploring the sequence landscape requires structural and biochemical data (from previous random mutagenesis studies), which can be leveraged to constrain genetic variation to distinct positions of the (poly)peptide, such as regions of the peptide aptamer scaffold which can endure substitutions/insertions/deletions without affecting their overall protein fold, or those peptide residues considered not absolutely essential for specific property of interest (and whose mutation might further augment peptide's activity). Random mutagenesis results in stochastic point mutations at codons corresponding to such residues, but systematically interrogating the entire set of residues at a specific position requires a focused mutagenesis strategy. Focused libraries are typically smaller and more effective, as they only address the residues presumed to bestow the peptide with the property of interest [233].

4.2.1. Enzyme-Based Approaches

Building a library of recombinant DNA constructs is a widely adopted practice accessible to virtually all laboratories, due to the ease of oligonucleotide synthesis and availability of commercial restriction enzymes and DNA ligases. The so-called oligonucleotide-directed mutagenesis enables point or multiple mutations to the target DNA sequence [234]. Normally, a mutagenic primer is designed and synthesized, subsequently elongated by Klenow fragment of DNA polymerase I, ligated into a vector by T4 DNA ligase and finally transformed into a competent *E. coli* strain. This process is long and includes multiple subcloning and ssDNA rescuing steps [235]. Several kits for site-specific mutagenesis based on mutagenic primers are commercially available. One of the systems works by applying a pair of forward and reverse complementary oligonucleotides with designed mutations. The primers are perfectly complementary to the template at 5′ and 3′ ends, but carry a changed central nucleotide sequence. A high-fidelity *Pfu* DNA polymerase is used to amplify the entire plasmid harboring the gene to be mutated, followed by the removal of the template by *Dpn*I (an endonuclease specific for methylated DNA) [236]. There are numerous adaptations of this method (reviewed by Tee and Wong [216]).

An approach termed "Kunkel mutagenesis" is commonly used for constructing libraries displayed on filamentous phage [237,238], because its genome is circular and single-stranded. In Kunkel mutagenesis, mutations are introduced with a mutagenic primer that is complementary to the circular ssDNA template. The template is propagated in an *ung⁻ dut⁻ E. coli* strain. This enzyme handicap results in the template DNA containing uracil bases in place of thymine. The template is recovered and hybridized with the primer and extended by polymerase, followed by transformation into *ung⁺ dut⁺* host cells [223]. Upon transformation, uridylated DNA template is biologically inactivated through the action of uracil glycosylase [239] of the *ung⁺ dut⁺* host, granting a strong selection advantage to the mutated strand(s) over the template.

Overlap extension PCR is another focused mutagenesis approach. First, two DNA fragments with homologous ends (and harboring desired mutation(s)) are amplified in separate PCR reactions by using 5′ complementary oligonucleotides. In a subsequent reaction, the fragments are combined; now, the overlapping 3′ ends from one of the strands of each fragment anneal and serve as "mega" primers for extension of the complementary strands. Finally, the construct is amplified with the two flanking primers [240]. Based on this strategy, the SLIM (site-directed ligase-independent mutagenesis) method, compatible with all three types of sequence modifications (insertion, deletion, and substitution), employs an inverse PCR amplification of the plasmid-embedded template by two 5′ adapter-tailed long forward and reverse primers (which include modifications) and two short forward and reverse primers (identical to the long ones but lacking the 5′ adapter sequences) in a single reaction, producing 4 distinct amplicons. Next, the amplicons are heat denatured and reannealed to yield 16 (hetero)duplexes, 4 of which are directly cloneable, forming circular DNA through ligation-independent pathway via complementary 5′ and 3′ single-stranded overhangs. All steps of the SLIM procedure are carried out in a single tube [241].

Gibson assembly is a method of combining up to 15 DNA fragments containing 20–40 bp overlaps in a single isothermal reaction. It utilizes a cocktail of three enzymes; exonuclease, DNA polymerase, and DNA ligase. The exonuclease nibbles back DNA form the 5′ end, enabling annealing of homologous DNA fragments. DNA polymerase then fills in the gaps, followed by the covalent fragment joining by the DNA ligase [242]. Applications of Gibson assembly include site-directed mutagenesis and library construction [243]. A recent adaptation, QuickLib, is a modified Gibson assembly method that has been used to generate a cyclic peptide library [244]. QuickLib uses two primers that share complementary 5′ ends; one long partially degenerate, and the other short non-degenerate, which are then used for full plasmid PCR amplification. Subsequently, a Gibson reaction is performed which circularizes the library of linear plasmids, followed by template elimination by *Dpn*I restriction.

Besides conventional enzymes involved in cumbersome digestion and ligation steps, other enzymes can be utilized for mutagenesis. In nature, lambda exonuclease aids viral DNA recombination. It progressively degrades the 5′-phosphoryl strand of a duplex DNA from 5′ to 3′, producing ssDNA and (mono)nucleotides [245]. To exploit this property, first, a PCR amplification using template ssDNA and phosphorylated primers with overlapping regions is performed. The PCR product is then treated with lambda exonuclease, generating ssDNA fragments that are subsequently annealed via overlap regions. Afterwards, Klenow fragment is employed to create dsDNA. In this manner, site-specific mutagenesis can be performed using primers that contain degenerate bases [246].

One of the most broadly used approaches for characterization of individual amino acid residues of a (poly)peptide with regards to their contribution to binding affinity or activity is the alanine-scanning mutagenesis. As the name implies, the technique is based on systematic substitution of residues with alanine, and assessing ligand's activity in a biochemical assay. Alanine eliminates the influence of all side chain atoms beyond the beta-carbon, thus exploring the role of side chain functional groups at interrogated positions [247]. For example, a conventional single-site alanine-scanning was used to assess the contribution of individual amino acid residues of a Fc fragment binding peptide displayed on filamentous phage [67]. Since this type of approach is laborious, methods have been developed for multiple alanine substitutions in a high-throughput manner [248]. One such approach builds on the codon-based mutagenesis, analyzing multiple positions, applying split-and-mix synthesis to produce degenerate oligonucleotides (one pool for the alanine codon and another for the wild-type codon) [249]. An alternative to alanine-scanning is serine-scanning, which follows the logic that, sometimes, substitutions with the hydrophobic alanine side chains may be more detrimental to the peptide's affinity compared to the slightly larger but hydrophilic serine side chain. Similarly, homolog-scanning (substitutions at individual positions with similar residues) may be employed with the goal of minimizing structural disruption and identifying residues essential for maintaining a function [250].

Another site-directed mutagenesis type is the cassette mutagenesis. It works by replacing a section of genetic information with an alternative, synthetic sequence—a "cassette" [251]. Different from other approaches that target short regions of a gene, this method is convenient for sequences up to 100 bp in size [252,253]. A prerequisite for this method to be practical is that the gene cassette must be flanked by two restriction sites that are complementary and unique with digest sites on the targeted vector. Restriction enzymes excise the targeted fragment from a vector that can then be replaced with DNA sequences carrying desired mutations. If a larger fragment is to be cloned, the "megaprimer" approach is applied by amplification with a series of oligonucleotides [254]. This method can also benefit from using "spiked" synthetic oligonucleotides, allowing randomization at multiple sites [255,256]. Cassette mutagenesis is based on Kunkel mutagenesis, which is time-consuming, so researchers developed an improved version termed ''PFunkel'', a conflation of *Pfu* DNA polymerase and Kunkel mutagenesis, that can be performed in a day's work [257,258]. To overcome the main constraint of site-directed mutagenesis, which is the tedious primer design, rational design techniques can be utilized to introduce desired mutations at precise positions. Researchers can leverage readily available tools such as AAscan, PCRdesign, and MutantChecker to simplify and boost the mutagenesis process [259].

4.2.2. Chemical-Based Mutagenesis

Chemical-based mutations involve various chemical methods to produce desired mutants. To chemically synthesize fully randomized oligonucleotides, a mixture of nucleotides must be applied at each coupling step [260]. A calamitous problem with this strategy is the pronounced bias resulting from the uneven incorporation frequency of the 4 nucleotide building blocks due to their inherent reactivity differences, rendering statistical random mutations inaccessible. Avoiding incorporation of stop codons is practically unattainable and the system is inclined towards amino acid residues encoded by redundant codons [261]. This problem can be tackled by adjusting the mutational frequency with "spiked oligonucleotides" [255], taking into account the differences in reactivity of mononucleotides and the redundant genetic code. The essence of DNA spiking is that non-equimolar ratio of bases at targeted positions are applied during oligonucleotide synthesis, meaning each wild type nucleotide can be custom "doped", achieving either "soft" (high incidence of a certain nucleotide) or "hard" (equal incidence of all four nucleotides) randomization, manually tuning the occurrence of certain amino acids at defined positions in the (poly)peptide chain.

Site-saturation mutagenesis seeks to achieve mutation at a maximal capacity by examining substitutions of a given residue against all possible amino acids. A fully randomized codon NNN (where N = A/C/G/T) gives rise to all possible 64 variant combinations (also known as 64-fold degeneracy) and codes for all 20 amino acids and 3 stop codons. This causes difficulties during library screening and risks enrichment of non-functional clones due to the random introduction of termination codons [262]. Operating with NNK, NNS, and NNB codons (where K = G/T, S = C/G, and B = C/G/T) minimizes the degeneracy in the third position of each codon, consequently lowering codon redundancy and the frequency of terminations [263]. However, such degenerate primers are expensive to synthesize, and using a single degenerate primer to completely eliminate codon redundancy while providing all 20 amino acids is unattainable, due to disproportional representation of certain amino acids [264,265]. Other strategies have to be employed to circumvent these constraints.

To synthesize redundancy-free mutagenic primers, mono [266], di [267], or trinucleotide phosphoramidite [268] solutions (or combinations [269]) can be used. This way, mixtures of oligonucleotides encoding all possible amino acid substitutions within a defined stretch of peptide or a limited number of amino acids (i.e., "tailored" randomization) can be synthesized. This fine-tuning gives complete control over amino acid prevalence at defined positions in the corresponding (poly)peptide sequence, achieving "soft" or "hard" randomization. With this approach, codon redundancy and stop codons are completely eliminated [261]. Another randomization strategy labeled MAX eliminates genetic redundancy by using a collection of 20 primers containing only codons for each amino acid with the highest expression frequency in *E. coli* [270]. These primers are annealed to a template strand with completely randomized codons (NNN or NNK) at the targeted position. Any misannealing is trivial, since only the ligated selection strand is amplified by a subsequent PCR. The produced random cassettes are then enzyme-digested for cloning. Further development of this strategy gave birth to an upgraded version dubbed ProxiMAX in which multiple contiguous codons are randomized in a non-degenerate manner [271]. Here, a donor blunt-end dsDNA with terminal MAX codons and an upstream *MlyI* restriction site is ligated to an acceptor blunt-end dsDNA. The product strands are amplified, analyzed, and combined at desired ratios in the next randomization cycle. After each ligation cycle, endonuclease *MlyI* is applied to remove the donor DNA strand, making only the randomized sequences available for the successive ligation cycle.

Another strategy that has been developed by Tang et al. [264] is cost-effective and uses degenerate codons to eliminate or achieve near-zero redundancy. A mixture of four codons, NDT, VMA, ATG, and TGG (where D = A/G/T, V = A/C/G, M = A/C) with a molar ratio of 12:6:1:1 at each randomized position results in an equal theoretical distribution for each of the 20 amino acids, without occurrences of stop codons. Following a similar rationale, Kille et al. [265] developed the "22c-trick" which uses only three codons per randomized position; NDT, VHG, and TGG (where H = A/C/T), at 12:9:1 molar ratio. The name sprung from the usage of 22 unique codons, achieving near uniform amino acid distribution (i.e., 2/22 for Leu and Val, and 1/22 for each of the remaining 18 amino acids). Other sophisticated primer mixing strategies have been reported [272–274], although picking the best approach is mostly dependent on the size and quality of the library to be prepared, and the lab's operating budget [275].

5. Conclusions

"Design, build, test, repeat" is the core philosophy of synthetic biology. Considering the intricacies of biological systems, every step of this process is potentially affected by multiple obstacles. Researchers are often put in predicaments where their progress is halted by unpredictable issues, and solving them can last even months at a time. These are not sporadic isolated events, but rather frequent and occur in virtually every field in natural sciences, let alone molecular biology. Building a peptide library is no different; it involves miscellaneous techniques and is often laborious. Recent advances are trumping bottlenecks in practically every facet of this technology. Today, new technologies enable accurately modeling peptide structure as long as 40 residues [276], and in silico design and screening using bioinformatics tools like Rosetta [277]. Furthermore, the ability to synthesize oligonucleotides adequate in size to code for potential peptide binders and assemble them in a display format of choice is now available even to labs with tight resources [275]. Alternatively, commercial peptide libraries can be purchased, which is very convenient especially for small operations with limited personnel, equipment, and know-how. Owing to automation, unprecedented parallel screening ability is now a reality, and coupled with high-throughput deep DNA sequencing [31], discovery of large numbers of novel high-affinity binders is a realistic prospect.

So, what predictions can we make for the future of peptide discovery? Surely, the convergence of new molecular strategies coupled with novel high-throughput methods and machine learning [278] will aid future bench researchers with engineering novel peptide binders. Initiating such projects could one day be as simple as running a computer program, and ordering (or synthesizing) a library—an effortless endeavor.

Author Contributions: Conceptualization: K.B and T.B.; Writing—Original Draft: K.B.; Writing—Review and editing: K.B. and T.B. All authors have read and agreed to the published version of the manuscript.

Funding: This work was funded by Slovenian research Agency, program P4-0127.

Acknowledgments: The authors are grateful to Stella Ivšek for contributing the artwork for this paper.

Conflicts of Interest: The authors declare no conflict of interest.

Abbreviations

AAV	Adeno-associated virus
AcMNPV	*Autographa californica* multiple nuclear polyhedrosis virus
APEx	Anchored periplasmic expression
ARS	Aminoacyl-tRNA synthetase
cPTM	Chemical post-translational modifications
Cvap	Cv RNA-associating protein
DEL	DNA-encoded library
dNTP	deoxyribonucleoside triphosphate
epPCR	Error-prone PCR

FACS	Fluorescence activated cell sorting
FIT	Flexible in vitro translation
Fmoc	Fluorenylmethyloxycarbonyl
IVA	*In vivo assembly*
IVC	*In vitro* compartmentalization
MACS	Magnetic-activated cell sorting
MHC	Major histocompatibility complex
MORPHING	Mutagenic organized recombination process by homologous in vivo grouping
MP	Mutagenesis plasmid
MS	Mass spectrometry
NGS	Next-generation sequencing
OBOC	One-bead-one-compound
PERISS	intra periplasm secretion and selection system
PPI	Protein-protein interaction
PURE	Protein synthesis using recombinant elements
RaPID	Random non-standard peptide integrated discovery
RCA	Rolling circle amplification
SAR	Structure-activity relationship
SeSaM	Sequence saturation mutagenesis
SICLOPPS	Split-intein circular ligation of peptides and proteins
SLAY	Surface localized antimicrobial display
SLIP	Site-directed ligase-independent mutagenesis
SPPS	Solid-phase peptide synthesis
StEP	staggered extension process
VLP	Virus-like particle

References

1. Guzmán, F.; Barberis, S.; Illanes, A. Peptide synthesis: Chemical or enzymatic. *Electron. J. Biotechnol.* **2007**, *10*, 279–314. [CrossRef]
2. Sun, E.; Belanger, C.R.; Haney, E.F.; Hancock, R.E.W. Host defense (antimicrobial) peptides. In *Peptide Applications in Biomedicine, Biotechnology and Bioengineering*; Woodhead Publishing: Cambridge, UK, 2017; pp. 253–285.
3. Gribble, F.M.; Reimann, F. Function and mechanisms of enteroendocrine cells and gut hormones in metabolism. *Nat. Rev. Endocrinol.* **2019**, *15*, 226–237. [CrossRef] [PubMed]
4. Nusbaum, M.P. Neurotransmission: Peptide transmitters turn 36. *J. Exp. Biol.* **2017**, *220*, 2492–2494. [CrossRef] [PubMed]
5. Clark, G.C.; Casewell, N.R.; Elliott, C.T.; Harvey, A.L.; Jamieson, A.G.; Strong, P.N.; Turner, A.D. Friends or foes? Emerging impacts of biological toxins. *Trends Biochem. Sci.* **2019**, *44*, 365–379. [CrossRef]
6. Fosgerau, K.; Hoffmann, T. Peptide therapeutics: Current status and future directions. *Drug Discov. Today* **2015**, *20*, 122–128. [CrossRef]
7. Slominsky, P.A.; Shadrina, M.I. Peptide pharmaceuticals: Opportunities, prospects, and limitations. *Mol. Genet. Microbiol. Virol.* **2018**, *33*, 8–14. [CrossRef]
8. Lau, J.L.; Dunn, M.K. Therapeutic peptides: Historical perspectives, current development trends, and future directions. *Bioorg. Med. Chem.* **2018**, *26*, 2700–2707. [CrossRef]
9. Buchholz, K.; Collins, J. *Concepts in Biotechnology: History, Science and Business*; John Wiley & Sons: Hoboken, NJ, USA, 2013; Volume 49, ISBN 352769109X.
10. Craik, D.J.; Fairlie, D.P.; Liras, S.; Price, D. The future of peptide-based drugs. *Chem. Biol. Drug Des.* **2013**, *81*, 136–147. [CrossRef]
11. Erak, M.; Bellmann-Sickert, K.; Els-Heindl, S.; Beck-Sickinger, A.G. Peptide chemistry toolbox – Transforming natural peptides into peptide therapeutics. *Bioorg. Med. Chem.* **2018**, *26*, 2759–2765. [CrossRef] [PubMed]
12. Millward, S.W.; Fiacco, S.; Austin, R.J.; Roberts, R.W. Design of cyclic peptides that bind protein surfaces with antibody-like affinity. *ACS Chem. Biol.* **2007**, *2*, 625–634. [CrossRef] [PubMed]

13. Mimmi, S.; Maisano, D.; Quinto, I.; Iaccino, E. Phage display: An overview in context to drug discovery. *Trends Pharmacol. Sci.* **2019**, *40*, 87–91. [CrossRef] [PubMed]

14. Damayanti, N.P.; Buno, K.; Voytik Harbin, S.L.; Irudayaraj, J.M.K. Epigenetic process monitoring in live cultures with peptide biosensors. *ACS Sens.* **2019**, *4*, 562–565. [CrossRef] [PubMed]

15. Care, A.; Bergquist, P.L.; Sunna, A. Solid-binding peptides: Smart tools for nanobiotechnology. *Trends Biotechnol.* **2015**, *33*, 259–268. [CrossRef] [PubMed]

16. Schoch, G.A.; Attias, R.; Belghazi, M.; Dansette, P.M.; Werck-Reichhart, D. Engineering of a water-soluble plant cytochrome P450, CYP73A1, and NMR-based orientation of natural and alternate substrates in the active site. *Plant Physiol.* **2003**, *133*, 1198–1208. [CrossRef] [PubMed]

17. Wang, X.; Corin, K.; Baaske, P.; Wienken, C.J.; Jerabek-Willemsen, M.; Duhr, S.; Braun, D.; Zhang, S. Peptide surfactants for cell-free production of functional G protein-coupled receptors. *Proc. Natl. Acad. Sci. USA* **2011**, *108*, 9049–9054. [CrossRef] [PubMed]

18. Yeh, J.I.; Du, S.; Tortajada, A.; Paulo, J.; Zhang, S. Peptergents: Peptide detergents that improve stability and functionality of a membrane protein, glycerol-3-phosphate dehydrogenase. *Biochemistry* **2005**, *44*, 16912–16919. [CrossRef]

19. Matsumoto, K.; Vaughn, M.; Bruce, B.D.; Koutsopoulos, S.; Zhang, S. Designer peptide surfactants stabilize functional Photosystem-I membrane complex in aqueous solution for extended time. *J. Phys. Chem. B* **2009**, *113*, 75–83. [CrossRef]

20. Bukowska, M.A.; Grütter, M.G. New concepts and aids to facilitate crystallization. *Curr. Opin. Struct. Biol.* **2013**, *23*, 409–416. [CrossRef]

21. Wei, Z.; Maeda, Y.; Matsui, H. Discovery of catalytic peptides for inorganic nanocrystal synthesis by a combinatorial phage display approach. *Angew. Chem. Int. Ed.* **2011**, *50*, 10585–10588. [CrossRef]

22. Kruljec, N.; Bratkovič, T. Alternative affinity ligands for immunoglobulins. *Bioconjug. Chem.* **2017**, *28*, 2009–2030. [CrossRef]

23. Kish, W.S.; Roach, M.K.; Sachi, H.; Naik, A.D.; Menegatti, S.; Carbonell, R.G. Purification of human erythropoietin by affinity chromatography using cyclic peptide ligands. *J. Chromatogr. B Anal. Technol. Biomed. Life Sci.* **2018**, *1085*, 1–12. [CrossRef] [PubMed]

24. Reverdatto, S.; Burz, D.; Shekhtman, A. Peptide aptamers: Development and applications. *Curr. Top. Med. Chem.* **2015**, *15*, 1082–1101. [CrossRef] [PubMed]

25. Kotz, J. Phenotypic screening, take two. *Sci. Exch.* **2012**, *5*, 380. [CrossRef]

26. Croston, G.E. The utility of target-based discovery. *Expert Opin. Drug Discov.* **2017**, *12*, 427–429. [CrossRef]

27. Mersich, C.; Jungbauer, A. Generation of bioactive peptides by biological libraries. *J. Chromatogr. B Anal. Technol. Biomed. Life Sci.* **2008**, *861*, 160–170. [CrossRef]

28. Yoshida, H.; Baik, S.H.; Harayama, S. An effective peptide screening system using recombinant fluorescent bacterial surface display. *Biotechnol. Lett.* **2002**, *24*, 1715–1722. [CrossRef]

29. Stratis-Cullum, D.N. Method for discovery of peptide reagents using a commercial magnetic separation platform and bacterial cell surface display technology. *J. Anal. Bioanal. Tech.* **2015**, *6*, 1. [CrossRef]

30. Erharuyi, O.; Simanski, S.; McEnaney, P.J.; Kodadek, T. Screening one bead one compound libraries against serum using a flow cytometer: Determination of the minimum antibody concentration required for ligand discovery. *Bioorg. Med. Chem. Lett.* **2018**, *28*, 2773–2778. [CrossRef]

31. Matochko, W.L.; Derda, R. Next-generation sequencing of phage-displayed peptide libraries. *Methods Mol. Biol.* **2015**, *1248*, 249–266.

32. Jalali-Yazdi, F.; Huong lai, L.; Takahashi, T.T.; Roberts, R.W. High-throughput measurement of binding kinetics by mRNA display and next-generation sequencing. *Angew. Chem. Int. Ed.* **2016**, *55*, 4007–4010. [CrossRef]

33. Lagoutte, P.; Lugari, A.; Elie, C.; Potisopon, S.; Donnat, S.; Mignon, C.; Mariano, N.; Troesch, A.; Werle, B.; Stadthagen, G. Combination of ribosome display and next generation sequencing as a powerful method for identification of affibody binders against β-lactamase CTX-M15. *New Biotechnol.* **2019**, *50*, 60–69. [CrossRef] [PubMed]

34. Körbelin, J.; Sieber, T.; Michelfelder, S.; Lunding, L.; Spies, E.; Hunger, A.; Alawi, M.; Rapti, K.; Indenbirken, D.; Müller, O.J.; et al. Pulmonary targeting of adeno-associated viral vectors by next-generation sequencing-guided screening of random capsid displayed peptide libraries. *Mol. Ther.* **2016**, *24*, 1050–1061. [CrossRef] [PubMed]

35. Mendes, K.R.; Malone, M.L.; Ndungu, J.M.; Suponitsky-Kroyter, I.; Cavett, V.J.; McEnaney, P.J.; MacConnell, A.B.; Doran, T.D.M.; Ronacher, K.; Stanley, K.; et al. High-throughput identification of DNA-encoded IgG ligands that distinguish active and latent mycobacterium tuberculosis infections. *ACS Chem. Biol.* **2017**, *12*, 234–243. [CrossRef] [PubMed]

36. Crook, Z.R.; Sevilla, G.P.; Friend, D.; Brusniak, M.Y.; Bandaranayake, A.D.; Clarke, M.; Gewe, M.; Mhyre, A.J.; Baker, D.; Strong, R.K.; et al. Mammalian display screening of diverse cystine-dense peptides for difficult to drug targets. *Nat. Commun.* **2017**, *8*, 2244. [CrossRef] [PubMed]

37. Pantazes, R.J.; Reifert, J.; Bozekowski, J.; Ibsen, K.N.; Murray, J.A.; Daugherty, P.S. Identification of disease-specific motifs in the antibody specificity repertoire via next-generation sequencing. *Sci. Rep.* **2016**, *6*, 30312. [CrossRef]

38. Suga, H. Max-Bergmann award lecture:A RaPID way to discover bioactive nonstandard peptides assisted by the flexizyme and FIT systems. *J. Pept. Sci.* **2018**, *24*, e3055. [CrossRef]

39. Wang, Y.C.; Distefano, M.D. Synthesis and screening of peptide libraries with free C-termini. *Curr. Top. Pept. Protein Res.* **2014**, *15*, 1–23.

40. Carpino, L.A.; Beyermann, M.; Wenschuh, H.; Bienert, M. Peptide synthesis via amino acid halides. *Acc. Chem. Res.* **1996**, *29*, 268–274. [CrossRef]

41. Behrendt, R.; White, P.; Offer, J. Advances in Fmoc solid-phase peptide synthesis. *J. Pept. Sci.* **2016**, *22*, 4–27. [CrossRef]

42. Houghten, R.A. General method for the rapid solid-phase synthesis of large numbers of peptides: Specificity of antigen-antibody interaction at the level of individual amino acids. *Proc. Natl. Acad. Sci. USA* **1985**, *82*, 5131–5135. [CrossRef]

43. Furka, Á.; Sebestyén, F.; Asgedom, M.; Dibó, G. General method for rapid synthesis of multicomponent peptide mixtures. *Int. J. Pept. Protein Res.* **1991**, *37*, 487–493. [CrossRef]

44. Lam, K.S.; Salmon, S.E.; Hersh, E.M.; Hruby, V.J.; Kazmierskit, W.M.; Knappt, R.J. A new type of synthetic peptide library for identifying ligand-binding activity. *Nature* **1991**, *354*, 82–84. [CrossRef]

45. Ostresh, J.M.; Winkle, J.H.; Hamashin, V.T.; Houghten, R.A. Peptide libraries: Determination of relative reaction rates of protected amino acids in competitive couplings. *Biopolymers* **1994**, *34*, 1681–1689. [CrossRef]

46. Wilson-Lingardo, L.; Davis, P.W.; Ecker, D.J.; Hébert, N.; Acevedo, O.; Sprankle, K.; Brennan, T.; Schwarcz, L.; Freier, S.M.; Wyatt, J.R. Deconvolution of combinatorial libraries for drug discovery: Experimental comparison of pooling strategies. *J. Med. Chem.* **1996**, *39*, 2720–2726. [CrossRef]

47. Pinilla, C.; Appel, J.R.; Blanc, P.; Houghten, R.A.; Blanc, P.; Houghten, R.A. Rapid identification of high affinity peptide ligands using positional scanning synthetic peptide combinatorial libraries. *Biotechniques* **1993**, *13*, 901–905.

48. Edman, P. Method for determination of the amino acid sequence in peptides. *Acta Chem. Scand* **1950**, *4*, 283–293. [CrossRef]

49. Wang, P.; Arabaci, G.; Pei, D. Rapid sequencing of library-derived peptides by partial edman degradation and mass spectrometry. *J. Comb. Chem.* **2001**, *3*, 251–254. [CrossRef]

50. Vinogradov, A.A.; Gates, Z.P.; Zhang, C.; Quartararo, A.J.; Halloran, K.H.; Pentelute, B.L. Library design-facilitated high-throughput sequencing of synthetic peptide libraries. *ACS Comb. Sci.* **2017**, *19*, 694–701. [CrossRef]

51. Deutzmann, R. Structural characterization of proteins and peptides. *Methods Mol. Med.* **2004**, *94*, 269–297.

52. Grant, G.A.; Crankshaw, M.W.; Gorka, J. Edman sequencing as tool for characterization of synthetic peptides. *Methods Enzymol.* **1997**, *289*, 395–419.

53. Goodnow, R.A.; Dumelin, C.E.; Keefe, A.D. DNA-encoded chemistry: Enabling the deeper sampling of chemical space. *Nat. Rev. Drug Discov.* **2017**, *16*, 131–147. [CrossRef]

54. Ottl, J.; Leder, L.; Schaefer, J.V.; Dumelin, C.E. Encoded library technologies as integrated lead finding platforms for drug discovery. *Molecules* **2019**, *24*, 1629. [CrossRef]

55. Zhu, Z.; Shaginian, A.; Grady, L.S.C.; Davie, C.P.; Lind, K.; Pal, S.; Thansandote, P.; Simpson, G.L. DNA-encoded macrocyclic peptide library. *Methods Mol. Biol.* **2019**, *2001*, 273–284.

56. Daguer, J.P.; Ciobanu, M.; Alvarez, S.; Barluenga, S.; Winssinger, N. DNA-templated combinatorial assembly of small molecule fragments amenable to selection/amplification cycles. *Chem. Sci.* **2011**, *2*, 625–632. [CrossRef]

57. Svensen, N.; Díaz-Mochón, J.J.; Bradley, M. Encoded peptide libraries and the discovery of new cell binding ligands. *Chem. Commun.* **2011**, *47*, 7638–7640. [CrossRef]
58. Stress, C.J.; Sauter, B.; Schneider, L.A.; Sharpe, T.; Gillingham, D. A DNA-encoded chemical library incorporating elements of natural macrocycles. *Angew. Chem. Int. Ed.* **2019**, *58*, 9570–9574. [CrossRef]
59. Faver, J.C.; Riehle, K.; Lancia, D.R.; Milbank, J.B.J.; Kollmann, C.S.; Simmons, N.; Yu, Z.; Matzuk, M.M. Quantitative comparison of enrichment from DNA-encoded chemical library selections. *ACS Comb. Sci.* **2019**, *21*, 75–82. [CrossRef]
60. Zimmermann, G.; Li, Y.; Rieder, U.; Mattarella, M.; Neri, D.; Scheuermann, J. Hit-validation methodologies for ligands isolated from DNA-encoded chemical libraries. *ChemBioChem* **2017**, *18*, 853–857. [CrossRef]
61. Galán, A.; Comor, L.; Horvatić, A.; Kuleš, J.; Guillemin, N.; Mrljak, V.; Bhide, M. Library-based display technologies: Where do we stand? *Mol. Biosyst.* **2016**, *12*, 2342–2358. [CrossRef]
62. Smith, G.P. Filamentous fusion phage: Novel expression vectors that display cloned antigens on the virion surface. *Science* **1985**, *228*, 1315–1317. [CrossRef]
63. Molek, P.; Strukelj, B.; Bratkovic, T. Peptide phage display as a tool for drug discovery: Targeting membrane receptors. *Molecules* **2011**, *16*, 857–887. [CrossRef] [PubMed]
64. Bratkovič, T. Progress in phage display: Evolution of the technique and its applications. *Cell. Mol. Life Sci.* **2010**, *67*, 749–767. [CrossRef] [PubMed]
65. Smith, G.P.; Petrenko, V.A. Phage display. *Chem. Rev.* **1997**, *97*, 391–410. [CrossRef] [PubMed]
66. González-Techera, A.; Umpiérrez-Failache, M.; Cardozo, S.; Obal, G.; Pritsch, O.; Last, J.A.; Gee, S.J.; Hammock, B.D.; González-Sapienza, G. High-throughput method for ranking the affinity of peptide ligands selected from phage display libraries. *Bioconjug. Chem.* **2008**, *19*, 993–1000. [CrossRef]
67. Kruljec, N.; Molek, P.; Hodnik, V.; Anderluh, G.; Bratkovič, T. Development and characterization of peptide ligands of immunoglobulin G Fc region. *Bioconjug. Chem.* **2018**, *29*, 2763–2775. [CrossRef]
68. Gray, B.P.; Brown, K.C. Combinatorial peptide libraries: Mining for cell-binding peptides. *Chem. Rev.* **2014**, *114*, 1020–1081. [CrossRef]
69. Chackerian, B.; Caldeira, J.D.C.; Peabody, J.; Peabody, D.S. Peptide epitope identification by affinity selection on bacteriophage MS2 virus-like particles. *J. Mol. Biol.* **2011**, *409*, 225–237. [CrossRef]
70. He, B.; Tjhung, K.F.; Bennett, N.J.; Chou, Y.; Rau, A.; Huang, J.; Derda, R. Compositional bias in naïve and chemically-modified phage-displayed libraries uncovered by paired-end deep sequencing. *Sci. Rep.* **2018**, *8*, 1214. [CrossRef]
71. Peters, E.A.; Schatz, P.J.; Johnson, S.S.; Dower, W.J. Membrane insertion defects caused by positive charges in the early mature region of protein pIII of filamentous phage fd can be corrected by prlA suppressors. *J. Bacteriol.* **1994**, *176*, 4296–4305. [CrossRef]
72. Krumpe, L.R.H.; Atkinson, A.J.; Smythers, G.W.; Kandel, A.; Schumacher, K.M.; McMahon, J.B.; Makowski, L.; Mori, T. T7 lytic phage-displayed peptide libraries exhibit less sequence bias than M13 filamentous phage-displayed peptide libraries. *Proteomics* **2006**, *6*, 4210–4222. [CrossRef]
73. Paschke, M.; Höhne, W. A twin-arginine translocation (Tat)-mediated phage display system. *Gene* **2005**, *350*, 79–88. [CrossRef]
74. Speck, J.; Arndt, K.M.; Mller, K.M. Efficient phage display of intracellularly folded proteins mediated by the TAT pathway. *Protein Eng. Des. Sel.* **2011**, *24*, 473–484. [CrossRef]
75. Feng, Z.; Xu, B. Inspiration from the mirror: D-amino acid containing peptides in biomedical approaches. *Biomol. Concepts* **2016**, *7*, 179–187. [CrossRef]
76. Deyle, K.; Kong, X.D.; Heinis, C. Phage selection of cyclic peptides for application in research and drug development. *Acc. Chem. Res.* **2017**, *50*, 1866–1874. [CrossRef]
77. Wada, A.; Terashima, T.; Kageyama, S.; Yoshida, T.; Narita, M.; Kawauchi, A.; Kojima, H. Efficient prostate cancer therapy with tissue-specific homing peptides identified by advanced phage display technology. *Mol. Ther. Oncolytics* **2019**, *12*, 138–146. [CrossRef]
78. Vekris, A.; Pilalis, E.; Chatziioannou, A.; Petry, K.G. A computational pipeline for the extraction of actionable biological information from NGS-phage display experiments. *Front. Physiol.* **2019**, *10*, 1160. [CrossRef]
79. Gillespie, J.W.; Yang, L.; De Plano, L.M.; Stackhouse, M.A.; Petrenko, V.A. Evolution of a landscape phage library in a mouse xenograft model of human breast cancer. *Viruses* **2019**, *11*, 988. [CrossRef]
80. Bessette, P.H.; Rice, J.J.; Daugherty, P.S. Rapid isolation of high-affinity protein binding peptides using bacterial display. *Protein Eng. Des. Sel.* **2004**, *17*, 731–739. [CrossRef]

81. Bessette, P.H.; Hu, X.; Soh, H.T.; Daugherty, P.S. Microfluidic library screening for mapping antibody epitopes. *Anal. Chem.* **2007**, *79*, 2174–2178. [CrossRef]

82. Wong, R.S.Y.; Wirtz, R.A.; Hancock, R.E.W. Pseudomonas aeruginosa outer membrane protein OprF as an expression vector for foreign epitopes: The effects of positioning and length on the antigenicity of the epitope. *Gene* **1995**, *158*, 55–60. [CrossRef]

83. Saffar, B.; Yakhchali, B.; Arbabi, M. Development of a bacterial surface display of hexahistidine peptide using CS3 pili for bioaccumulation of heavy metals. *Curr. Microbiol.* **2007**, *55*, 273–277. [CrossRef]

84. Westerlund-Wikström, B. Peptide display on bacterial flagella: Principles and applications. *Int. J. Med. Microbiol.* **2000**, *290*, 223–230. [CrossRef]

85. Dong, J.; Liu, C.; Zhang, J.; Xin, Z.T.; Yang, G.; Gao, B.; Mao, C.Q.; Le Liu, N.; Wang, F.; Shao, N.S.; et al. Selection of novel nickel-binding peptides from flagella displayed secondary peptide library. *Chem. Biol. Drug Des.* **2006**, *68*, 107–112. [CrossRef]

86. Kimura, T. Screening techniques using the periplasmic expression of peptide libraries and target molecules. *J. Bioanal. Biomed.* **2017**, *9*, 263–268. [CrossRef]

87. Harvey, B.R.; Georgiou, G.; Hayhurst, A.; Jeong, K.J.; Iverson, B.L.; Rogers, G.K. Anchored periplasmic expression, a versatile technology for the isolation of high-affinity antibodies from Escherichia coli-expressed libraries. *Proc. Natl. Acad. Sci. USA* **2004**, *101*, 9193–9198. [CrossRef]

88. Guo, M.; Xu, L.M.; Zhou, B.; Yin, J.C.; Ye, X.L.; Ren, G.P.; Li, D.S. Anchored periplasmic expression (APEx)-based bacterial display for rapid and high-throughput screening of B cell epitopes. *Biotechnol. Lett.* **2014**, *36*, 609–616. [CrossRef]

89. Cull, M.G.; Miller, J.F.; Schatz, P.J. Screening for receptor ligands using large libraries of peptides linked to the C terminus of the lac repressor. *Proc. Natl. Acad. Sci. USA* **1992**, *89*, 1865–1869. [CrossRef]

90. Zhu, W.; Williams, R.S.; Kodadek, T. A CDC6 protein-binding peptide selected using a bacterial two-hybrid-like system is a cell cycle inhibitor. *J. Biol. Chem.* **2000**, *275*, 32098–32105. [CrossRef]

91. Karimova, G.; Gauliard, E.; Davi, M.; Ouellette, S.P.; Ladant, D. Protein–protein interaction: Bacterial two-hybrid. *Methods Mol. Biol.* **2017**, *1615*, 159–176.

92. Mehla, J.; Caufield, J.H.; Sakhawalkar, N.; Uetz, P. A comparison of two-hybrid approaches for detecting protein–protein interactions. *Methods Enzymol.* **2017**, *586*, 333–358.

93. Tucker, A.T.; Leonard, S.P.; DuBois, C.D.; Knauf, G.A.; Cunningham, A.L.; Wilke, C.O.; Trent, M.S.; Davies, B.W. Discovery of next-generation antimicrobials through bacterial self-screening of surface-displayed peptide libraries. *Cell* **2018**, *172*, 618–628. [CrossRef] [PubMed]

94. Dell, A.; Galadari, A.; Sastre, F.; Hitchen, P. Similarities and differences in the glycosylation mechanisms in prokaryotes and eukaryotes. *Int. J. Microbiol.* **2010**, *2010*, 148178. [CrossRef] [PubMed]

95. Mehla, J.; Caufield, J.H.; Uetz, P. The yeast two-hybrid system: A tool for mapping protein-protein interactions. *Cold Spring Harb. Protoc.* **2015**, *2015*, 425–430. [CrossRef] [PubMed]

96. Barreto, K.; Aparicio, A.; Bharathikumar, V.M.; DeCoteau, J.F.; Geyer, C.R. Yeast two-hybrid screening of cyclic peptide libraries using a combination of random and PI-deconvolution pooling strategies. *Protein Eng. Des. Sel.* **2012**, *25*, 453–464. [CrossRef] [PubMed]

97. Aronheim, A. The Ras Recruitment System (RRS) for the identification and characterization of protein–protein interactions. *Methods Mol. Biol.* **2018**, *1794*, 61–73. [PubMed]

98. Dirnberger, D.; Messerschmid, M.; Baumeister, R. An optimized split-ubiquitin cDNA-library screening system to identify novel interactors of the human Frizzled 1 receptor. *Nucleic Acids Res.* **2008**, *36*, e37. [CrossRef]

99. Licitra, E.J.; Liu, J.O. A three-hybrid system for detecting small ligand-protein receptor interactions. *Proc. Natl. Acad. Sci. USA* **1996**, *93*, 12817–12821. [CrossRef]

100. Gera, N.; Hussain, M.; Rao, B.M. Protein selection using yeast surface display. *Methods* **2013**, *60*, 15–26. [CrossRef]

101. Ueda, M. Principle of cell surface engineering of yeast. In *Yeast Cell Surface Engineering*; Springer: Singapore, 2019; pp. 3–14.

102. Kondo, A.; Ueda, M. Yeast cell-surface display—Applications of molecular display. *Appl. Microbiol. Biotechnol.* **2004**, *64*, 28–40. [CrossRef]

103. Cherf, G.M.; Cochran, J.R. Applications of yeast surface display for protein engineering. *Methods Mol. Biol.* **2015**, *1319*, 155–175.

104. Mersich, C.; Billes, W.; Pabinger, I.; Jungbauer, A. Peptides derived from a secretory yeast library restore factor VIII activity in the presence of an inhibitory antibody. *Biotechnol. Bioeng.* **2007**, *98*, 12–21. [CrossRef] [PubMed]

105. Moore, S.J.; Cochran, J.R. Engineering knottins as novel binding agents. *Methods Enzymol.* **2012**, *503*, 223–251. [PubMed]

106. Hetrick, K.J.; Walker, M.C.; Van Der Donk, W.A. Development and application of yeast and phage display of diverse lanthipeptides. *ACS Cent. Sci.* **2018**, *4*, 458–467. [CrossRef] [PubMed]

107. Kritzer, J.A.; Hamamichi, S.; McCaffery, J.M.; Santagata, S.; Naumann, T.A.; Caldwell, K.A.; Caldwell, G.A.; Lindquist, S. Rapid selection of cyclic peptides that reduce α-synuclein toxicity in yeast and animal models. *Nat. Chem. Biol.* **2009**, *5*, 655–663. [CrossRef] [PubMed]

108. Andreu, C.; del Olmo, M. Yeast arming systems: Pros and cons of different protein anchors and other elements required for display. *Appl. Microbiol. Biotechnol.* **2018**, *102*, 2543–2561. [CrossRef] [PubMed]

109. Angelini, A.; Chen, T.F.; De Picciotto, S.; Yang, N.J.; Tzeng, A.; Santos, M.S.; Van Deventer, J.A.; Traxlmayr, M.W.; Dane Wittrup, K. Protein engineering and selection using yeast surface display. *Methods Mol. Biol.* **2015**, *1319*, 3–36.

110. Benatuil, L.; Perez, J.M.; Belk, J.; Hsieh, C.M. An improved yeast transformation method for the generation of very large human antibody libraries. *Protein Eng. Des. Sel.* **2010**, *23*, 155–159. [CrossRef]

111. Orcutt, K.D.; Wittrup, K.D. Yeast display and selections. In *Antibody Engineering*; Springer: Berlin/Heidelberg, Germany, 2010; pp. 207–233.

112. Zhang, N.; Liu, L.; Dan Dumitru, C.; Cummings, N.R.H.; Cukan, M.; Jiang, Y.; Li, Y.; Li, F.; Mitchell, T.; Mallem, M.R.; et al. Glycoengineered pichia produced anti-HER2 is comparable to trastuzumab in preclinical study. *MAbs* **2011**, *3*, 289–298. [CrossRef]

113. Grabherr, R.; Ernst, W. Baculovirus for eukaryotic protein display. *Curr. Gene Ther.* **2010**, *10*, 195–200. [CrossRef]

114. Ernst, W.; Schinko, T.; Spenger, A.; Oker-Blom, C.; Grabherr, R. Improving baculovirus transduction of mammalian cells by surface display of a RGD-motif. *J. Biotechnol.* **2006**, *126*, 237–240. [CrossRef]

115. Song, L.; Liu, Y.; Chen, J. Baculoviral capsid display of His-tagged ZnO inorganic binding peptide. *Cytotechnology* **2010**, *62*, 133–141. [CrossRef] [PubMed]

116. Ernst, W.; Grabherr, R.; Wegner, D.; Borth, N.; Grassauer, A.; Katinger, H. Baculovirus surface display: Construction and screening of a eukaryotic epitope library. *Nucleic Acids Res.* **1998**, *26*, 1718–1723. [CrossRef] [PubMed]

117. Grabherr, R.; Ernst, W.; Oker-Blom, C.; Jones, I. Developments in the use of baculovirusesfor the surface display of complex eukaryotic proteins. *Trends Biotechnol.* **2001**, *19*, 231–236. [CrossRef]

118. Wang, Y.; Rubtsov, A.; Heiser, R.; White, J.; Crawford, F.; Marrack, P.; Kappler, J.W. Using a baculovirus display library to identify MHC class I mimotopes. *Proc. Natl. Acad. Sci. USA* **2005**, *102*, 2476–2481. [CrossRef]

119. Crawford, F.; Jordan, K.R.; Stadinski, B.; Wang, Y.; Huseby, E.; Marrack, P.; Slansky, J.E.; Kappler, J.W. Use of baculovirus MHC/peptide display libraries to characterize T-cell receptor ligands. *Immunol. Rev.* **2006**, *210*, 156–170. [CrossRef] [PubMed]

120. Kost, T.A.; Kemp, C.W. Fundamentals of baculovirus expression and applications. *Adv. Exp. Med. Biol.* **2016**, *896*, 187–197. [PubMed]

121. Chan, L.C.L.; Reid, S. Development of serum-free media for lepidopteran insect cell lines. *Methods Mol. Biol.* **2016**, *1350*, 161–196.

122. Geisler, C.; Jarvis, D. Insect cell glycosylation patterns in the context of biopharmaceuticals. In *Post-Translational Modification of Protein Biopharmaceuticals*; Wiley-VCH Verlag GmbH & Co. KGaA: Weinheim, Germany, 2009; pp. 165–191. ISBN 9783527320745.

123. Possee, R.D.; King, L.A. Baculovirus transfer vectors. *Methods Mol. Biol.* **2016**, *1350*, 51–71.

124. Khare, P.D.; Rosales, A.G.; Bailey, K.R.; Russell, S.J.; Federspiel, M.J. Epitope selection from an uncensored peptide library displayed on avian leukosis virus. *Virology* **2003**, *315*, 313–321. [CrossRef]

125. Bupp, K.; Sarangi, A.; Roth, M.J. Probing sequence variation in the receptor-targeting domain of feline leukemia virus envelope proteins with peptide display libraries. *J. Virol.* **2005**, *79*, 1463–1469. [CrossRef]

126. Sarangi, A.; Bupp, K.; Roth, M.J. Identification of a retroviral receptor used by an Envelope protein derived by peptide library screening. *Proc. Natl. Acad. Sci. USA* **2007**, *104*, 11032–11037. [CrossRef] [PubMed]

127. Michelfelder, S.; Kohlschütter, J.; Skorupa, A.; Pfennings, S.; Müller, O.; Kleinschmidt, J.A.; Trepel, M. Successful expansion but not complete restriction of tropism of adeno-associated virus by in vivo biopanning of random virus display peptide libraries. *PLoS ONE* **2009**, *4*, e5122. [CrossRef] [PubMed]

128. Müller, O.J.; Kaul, F.; Weitzman, M.D.; Pasqualini, R.; Arap, W.; Kleinschmidt, J.A.; Trepel, M. Random peptide libraries displayed on adeno-associated virus to select for targeted gene therapy vectors. *Nat. Biotechnol.* **2003**, *21*, 1040–1046. [CrossRef]

129. Waterkamp, D.A.; Müller, O.J.; Ying, Y.; Trepel, M.; Kleinschmidt, J.A. Isolation of targeted AAV2 vectors from novel virus display libraries. *J. Gene Med.* **2006**, *8*, 1307–1319. [CrossRef]

130. Michelfelder, S.; Lee, M.K.; deLima-Hahn, E.; Wilmes, T.; Kaul, F.; Müller, O.; Kleinschmidt, J.A.; Trepel, M. Vectors selected from adeno-associated viral display peptide libraries for leukemia cell-targeted cytotoxic gene therapy. *Exp. Hematol.* **2007**, *35*, 1766–1776. [CrossRef]

131. Varadi, K.; Michelfelder, S.; Korff, T.; Hecker, M.; Trepel, M.; Katus, H.A.; Kleinschmidt, J.A.; Müller, O.J. Novel random peptide libraries displayed on AAV serotype 9 for selection of endothelial cell-directed gene transfer vectors. *Gene Ther.* **2012**, *19*, 800–809. [CrossRef]

132. Körbelin, J.; Hunger, A.; Alawi, M.; Sieber, T.; Binder, M.; Trepel, M. Optimization of design and production strategies for novel adeno-associated viral display peptide libraries. *Gene Ther.* **2017**, *24*, 470–481. [CrossRef]

133. Marsic, D.; Zolotukhin, S. Altering tropism of rAAV by directed evolution. *Methods Mol. Biol.* **2016**, *1382*, 151–173.

134. Büning, H.; Srivastava, A. Capsid modifications for targeting and improving the efficacy of AAV vectors. *Mol. Ther. Methods Clin. Dev.* **2019**, *12*, 248–265. [CrossRef]

135. Wolkowicz, R.; Jager, G.C.; Nolan, G.P. A random peptide library fused to CCR5 for selection of mimetopes expressed on the mammalian cell surface via retroviral vectors. *J. Biol. Chem.* **2005**, *280*, 15195–15201. [CrossRef]

136. Xu, X.; Leo, C.; Jang, Y.; Chan, E.; Padilla, D.; Huang, B.C.B.; Lin, T.; Gururaja, T.; Hitoshi, Y.; Lorens, J.B.; et al. Dominant effector genetics in mammalian cells. *Nat. Genet.* **2001**, *27*, 23–29. [CrossRef] [PubMed]

137. Peelle, B.; Lorens, J.; Li, W.; Bogenberger, J.; Payan, D.G.; Anderson, D.C. Intracellular protein scaffold-mediated display of random peptide libraries for phenotypic screens in mammalian cells. *Chem. Biol.* **2001**, *8*, 521–534. [CrossRef]

138. Peelle, B.; Gururaja, T.L.; Payan, D.G.; Anderson, D.C. Characterization and use of green fluorescent proteins from Renilla mulleri and Ptilosarcus guernyi for the human cell display of functional peptides. *J. Protein Chem.* **2001**, *20*, 507–519. [CrossRef] [PubMed]

139. Kinsella, T.M.; Ohashi, C.T.; Harder, A.G.; Yam, G.C.; Li, W.; Peelle, B.; Pali, E.S.; Bennett, M.K.; Molineaux, S.M.; Anderson, D.A.; et al. Retrovirally delivered random cyclic peptide libraries yield inhibitors of interleukin-4 signaling in human B cells. *J. Biol. Chem.* **2002**, *277*, 37512–37518. [CrossRef]

140. Tavassoli, A. SICLOPPS cyclic peptide libraries in drug discovery. *Curr. Opin. Chem. Biol.* **2017**, *38*, 30–35. [CrossRef]

141. Pavankumar, T. Inteins: Localized distribution, gene regulation, and protein engineering for biological applications. *Microorganisms* **2018**, *6*, 19. [CrossRef]

142. Horswill, A.R.; Savinov, S.N.; Benkovic, S.J. A systematic method for identifying small-molecule modulators of protein-protein interactions. *Proc. Natl. Acad. Sci. USA* **2004**, *101*, 15591–15596. [CrossRef]

143. Mistry, I.N.; Tavassoli, A. Reprogramming the transcriptional response to hypoxia with a chromosomally encoded cyclic peptide HIF-1 inhibitor. *ACS Synth. Biol.* **2017**, *6*, 518–527. [CrossRef]

144. Castillo, F.; Tavassoli, A. Genetic selections with SICLOPPS Libraries: Toward the identification of novel protein-protein interaction inhibitors and chemical tools. *Methods Mol. Biol.* **2019**, *2001*, 317–328.

145. Obexer, R.; Walport, L.J.; Suga, H. Exploring sequence space: Harnessing chemical and biological diversity towards new peptide leads. *Curr. Opin. Chem. Biol.* **2017**, *38*, 52–61. [CrossRef]

146. Li, R.; Kang, G.; Hu, M.; Huang, H. Ribosome display: A potent display technology used for selecting and evolving specific binders with desired properties. *Mol. Biotechnol* **2019**, *61*, 60–71. [CrossRef]

147. Gersuk, G.M.; Corey, M.J.; Corey, E.; Stray, J.E.; Kawasaki, G.H.; Vessella, R.L. High-affinity peptide ligands to prostate-specific antigen identified by polysome selection. *Biochem. Biophys. Res. Commun.* **1997**, *232*, 578–582. [CrossRef] [PubMed]

148. Ohashi, H.; Kanamori, T.; Osada, E.; Akbar, B.K.; Ueda, T. Peptide screening using pure ribosome display. *Methods Mol. Biol.* **2012**, *805*, 251–259. [PubMed]

149. Mattheakis, L.C.; Bhatt, R.R.; Dower, W.J. An in vitro polysome display system for identifying ligands from very large peptide libraries. *Proc. Natl. Acad. Sci. USA* **1994**, *91*, 9022–9026. [CrossRef]

150. Lamla, T.; Erdmann, V.A. Searching sequence space for high-affinity binding peptides using ribosome display. *J. Mol. Biol.* **2003**, *329*, 381–388. [CrossRef]

151. Wada, A.; Sawata, S.Y.; Ito, Y. Ribosome display selection of a metal-binding motif from an artificial peptide library. *Biotechnol. Bioeng.* **2008**, *101*, 1102–1107. [CrossRef]

152. Dreier, B.; Plückthun, A. Ribosome display: A technology for selecting and evolving proteins from large libraries. *Methods Mol. Biol.* **2011**, *687*, 283–306.

153. Kanamori, T.; Fujino, Y.; Ueda, T. PURE ribosome display and its application in antibody technology. *Biochim. Biophys. Acta Proteins Proteom.* **2014**, *1844*, 1925–1932. [CrossRef]

154. Wada, A. Ribosome display technology for selecting peptide and protein ligands. In *Biomedical Applications of Functionalized Nanomaterials: Concepts, Development and Clinical Translation*; Elsevier: Amsterdam, The Netherlands, 2018; pp. 89–104. ISBN 9780323508797.

155. Barendt, P.A.; Ng, D.T.W.; McQuade, C.N.; Sarkar, C.A. Streamlined protocol for mRNA display. *ACS Comb. Sci.* **2013**, *15*, 77–81. [CrossRef]

156. Wilson, D.S.; Keefe, A.D.; Szostak, J.W. The use of mRNA display to select high-affinity protein-binding peptides. *Proc. Natl. Acad. Sci. USA* **2001**, *98*, 3750–3755. [CrossRef]

157. Horiya, S.; Bailey, J.K.; Krauss, I.J. Directed evolution of glycopeptides using mRNA display. *Methods Enzymol.* **2017**, *597*, 83–141.

158. Valencia, C.A.; Zou, J.; Liu, R. In vitro selection of proteins with desired characteristics using mRNA-display. *Methods* **2013**, *60*, 55–69. [CrossRef] [PubMed]

159. Gold, L. mRNA display: Diversity matters during in vitro selection. *Proc. Natl. Acad. Sci. USA* **2001**, *98*, 4825–4826. [CrossRef] [PubMed]

160. Lamboy, J.A.; Tam, P.Y.; Lee, L.S.; Jackson, P.J.; Avrantinis, S.K.; Lee, H.J.; Corn, R.M.; Weiss, G.A. Chemical and genetic wrappers for improved phage and RNA display. *Chembiochem* **2008**, *9*, 2846–2852. [CrossRef]

161. Yamaguchi, J.; Naimuddin, M.; Biyani, M.; Sasaki, T.; Machida, M.; Kubo, T.; Funatsu, T.; Husimi, Y.; Nemoto, N. cDNA display: A novel screening method for functional disulfide-rich peptides by solid-phase synthesis and stabilization of mRNA-protein fusions. *Nucleic Acids Res.* **2009**, *37*, e108. [CrossRef]

162. Nemoto, N.; Tsutsui, C.; Yamaguchi, J.; Ueno, S.; Machida, M.; Kobayashi, T.; Sakai, T. Antagonistic effect of disulfide-rich peptide aptamers selected by cDNA display on interleukin-6-dependent cell proliferation. *Biochem. Biophys. Res. Commun.* **2012**, *421*, 129–133. [CrossRef]

163. Ueno, S.; Yoshida, S.; Mondal, A.; Nishina, K.; Koyama, M.; Sakata, I.; Miura, K.; Hayashi, Y.; Nemoto, N.; Nishigaki, K.; et al. In vitro selection of a peptide antagonist of growth hormone secretagogue receptor using cDNA display. *Proc. Natl. Acad. Sci. USA* **2012**, *109*, 11121–11126. [CrossRef]

164. Mochizuki, Y.; Nishigaki, K.; Nemoto, N. Amino group binding peptide aptamers with double disulphide-bridged loops selected by in vitro selection using cDNA display. *Chem. Commun.* **2014**, *50*, 5608–5610. [CrossRef]

165. Hayakawa, Y.; Matsuno, M.; Tanaka, M.; Wada, A.; Kitamura, K.; Takei, O.; Sasaki, R.; Mizukami, T.; Hasegawa, M. Complementary DNA display selection of high-affinity peptides binding the vacuolating toxin (VacA) of Helicobacter pylori. *J. Pept. Sci.* **2015**, *21*, 710–716. [CrossRef]

166. Doi, N.; Yamakawa, N.; Matsumoto, H.; Yamamoto, Y.; Nagano, T.; Matsumura, N.; Horisawa, K.; Yanagawa, H. DNA display selection of peptide ligands for a full-length human G protein-coupled receptor on CHO-K1 cells. *PLoS ONE* **2012**, *7*, e30084. [CrossRef]

167. Odegrip, R.; Coomber, D.; Eldridge, B.; Hederer, R.; Kuhlman, P.A.; Ullman, C.; FitzGerald, K.; McGregor, D. CIS display: In vitro selection of peptides from libraries of protein-DNA complexes. *Proc. Natl. Acad. Sci. USA* **2004**, *101*, 2806–2810. [CrossRef] [PubMed]

168. Eldridge, B.; Cooley, R.N.; Odegrip, R.; McGregor, D.P.; FitzGerald, K.J.; Ullman, C.G. An in vitro selection strategy for conferring protease resistance to ligand binding peptides. *Protein Eng. Des. Sel.* **2009**, *22*, 691–698. [CrossRef] [PubMed]

169. Patel, S.; Mathonet, P.; Jaulent, A.M.; Ullman, C.G. Selection of a high-affinity WW domain against the extracellular region of VEGF receptor isoform-2 from a combinatorial library using CIS display. *Protein Eng. Des. Sel.* **2013**, *26*, 307–315. [CrossRef] [PubMed]

170. Tawfik, D.S.; Griffiths, A.D. Man-made cell-like compartments for molecular evolution. *Nat. Biotechnol.* **1998**, *16*, 652–656. [CrossRef]

171. Bernath, K.; Hai, M.; Mastrobattista, E.; Griffiths, A.D.; Magdassi, S.; Tawfik, D.S. In vitro compartmentalization by double emulsions: Sorting and gene enrichment by fluorescence activated cell sorting. *Anal. Biochem.* **2004**, *325*, 151–157. [CrossRef]

172. Doi, N.; Yanagawa, H. STABLE: Protein-DNA fusion system for screening of combinatorial protein libraries in vitro. *FEBS Lett.* **1999**, *457*, 227–230. [CrossRef]

173. Kaltenbach, M.; Hollfelder, F. SNAP display: In vitro protein evolution in microdroplets. *Methods Mol. Biol.* **2012**, *805*, 101–111.

174. Contreras-Llano, L.E.; Tan, C. High-throughput screening of biomolecules using cell-free gene expression systems. *Synth. Biol.* **2018**, *3*. [CrossRef]

175. Körfer, G.; Pitzler, C.; Vojcic, L.; Martinez, R.; Schwaneberg, U. In vitro flow cytometry-based screening platform for cellulase engineering. *Sci. Rep.* **2016**, *6*, 26128. [CrossRef]

176. Yonezawa, M.; Doi, N.; Kawahashi, Y.; Higashinakagawa, T.; Yanagawa, H. DNA display for in vitro selection of diverse peptide libraries. *Nucleic Acids Res.* **2003**, *31*, e118. [CrossRef]

177. Sepp, A.; Tawfik, D.S.; Griffiths, A.D. Microbead display by in vitro compartmentalisation: Selection for binding using flow cytometry. *FEBS Lett.* **2002**, *532*, 455–458. [CrossRef]

178. Rothe, A.; Surjadi, R.N.; Power, B.E. Novel proteins in emulsions using in vitro compartmentalization. *Trends Biotechnol.* **2006**, *24*, 587–592. [CrossRef] [PubMed]

179. Nishikawa, T.; Sunami, T.; Matsuura, T.; Yomo, T. Directed evolution of proteins through in vitro protein synthesis in liposomes. *J. Nucleic Acids* **2012**, *2012*, 1–11. [CrossRef] [PubMed]

180. Fallah-Araghi, A.; Baret, J.C.; Ryckelynck, M.; Griffiths, A.D. A completely in vitro ultrahigh-throughput droplet-based microfluidic screening system for protein engineering and directed evolution. *Lab Chip* **2012**, *12*, 882–891. [CrossRef] [PubMed]

181. Seelig, B. MRNA display for the selection and evolution of enzymes from in vitro-translated protein libraries. *Nat. Protoc.* **2011**, *6*, 540–552. [CrossRef] [PubMed]

182. Tjhung, K.F.; Kitov, P.I.; Ng, S.; Kitova, E.N.; Deng, L.; Klassen, J.S.; Derda, R. Silent encoding of chemical post-translational modifications in phage-displayed libraries. *J. Am. Chem. Soc.* **2016**, *138*, 32–35. [CrossRef] [PubMed]

183. Heinis, C.; Winter, G. Encoded libraries of chemically modified peptides. *Curr. Opin. Chem. Biol.* **2015**, *26*, 89–98. [CrossRef]

184. Roxin, Á.; Zheng, G. Flexible or fixed: A comparative review of linear and cyclic cancer-targeting peptides. *Future Med. Chem.* **2012**, *4*, 1601–1618. [CrossRef]

185. Soudy, R.; Gill, A.; Sprules, T.; Lavasanifar, A.; Kaur, K. Proteolytically stable cancer targeting peptides with high affinity for breast cancer cells. *J. Med. Chem.* **2011**, *54*, 7523–7534. [CrossRef]

186. Valentine, J.; Tavassoli, A. Genetically encoded cyclic peptide libraries: From hit to lead and beyond. *Methods Enzymol.* **2018**, *610*, 117–134.

187. Gang, D.; Kim, D.W.; Park, H.S. Cyclic peptides: Promising scaffolds for biopharmaceuticals. *Genes* **2018**, *9*, 557. [CrossRef] [PubMed]

188. Freire, F.; Gellman, S.H. Macrocyclic design strategies for small, stable parallel β-sheet scaffolds. *J. Am. Chem. Soc.* **2009**, *131*, 7970–7972. [CrossRef] [PubMed]

189. Huyer, G.; Kelly, J.; Moffat, J.; Zamboni, R.; Jia, Z.; Gresser, M.J.; Ramachandran, C. Affinity selection from peptide libraries to determine substrate specificity of protein tyrosine phosphatases. *Anal. Biochem.* **1998**, *258*, 19–30. [CrossRef] [PubMed]

190. Moradi, S.V.; Hussein, W.M.; Varamini, P.; Simerska, P.; Toth, I. Glycosylation, an effective synthetic strategy to improve the bioavailability of therapeutic peptides. *Chem. Sci.* **2016**, *7*, 2492–2500. [CrossRef] [PubMed]

191. Ng, S.; Jafari, M.R.; Matochko, W.L.; Derda, R. Quantitative synthesis of genetically encoded glycopeptide libraries displayed on M13 phage. *ACS Chem. Biol.* **2012**, *7*, 1482–1487. [CrossRef] [PubMed]

192. Tian, F.; Tsao, M.L.; Schultz, P.G. A phage display system with unnatural amino acids. *J. Am. Chem. Soc.* **2004**, *126*, 15962–15963. [CrossRef]

193. Dwyer, M.A.; Lu, W.; Dwyer, J.J.; Kossiakoff, A.A. Biosynthetic phage display: A novel protein engineering tool combining chemical and genetic diversity. *Chem. Biol.* **2000**, *7*, 263–274. [CrossRef]

194. Kale, S.S.; Villequey, C.; Kong, X.D.; Zorzi, A.; Deyle, K.; Heinis, C. Cyclization of peptides with two chemical bridges affords large scaffold diversities. *Nat. Chem.* **2018**, *10*, 715–723. [CrossRef]

195. Heinis, C.; Rutherford, T.; Freund, S.; Winter, G. Phage-encoded combinatorial chemical libraries based on bicyclic peptides. *Nat. Chem. Biol.* **2009**, *5*, 502–507. [CrossRef]

196. Derda, R.; Ng, S. Genetically encoded fragment-based discovery. *Curr. Opin. Chem. Biol.* **2019**, *50*, 128–137. [CrossRef]

197. Young, T.S.; Schultz, P.G. Beyond the canonical 20 amino acids: Expanding the genetic lexicon. *J. Biol. Chem.* **2010**, *285*, 11039–11044. [CrossRef] [PubMed]

198. Goto, Y.; Katoh, T.; Suga, H. Flexizymes for genetic code reprogramming. *Nat. Protoc.* **2011**, *6*, 779–790. [CrossRef] [PubMed]

199. Packer, M.S.; Liu, D.R. Methods for the directed evolution of proteins. *Nat. Rev. Genet.* **2015**, *16*, 379–394. [CrossRef] [PubMed]

200. Lai, Y.P.; Huang, J.; Wang, L.F.; Li, J.; Wu, Z.R. A new approach to random mutagenesis in vitro. *Biotechnol. Bioeng.* **2004**, *86*, 622–627. [CrossRef] [PubMed]

201. Myers, R.M.; Lerman, L.S.; Maniatis, T. A general method for saturation mutagenesis of cloned DNA fragments. *Science* **1985**, *229*, 242–247. [CrossRef]

202. Cox, E.C. Bacterial mutator genes and the control of spontaneous mutation. *Annu. Rev. Genet.* **1976**, *10*, 135–156. [CrossRef]

203. Greener, A.; Callahan, M.; Jerpseth, B. An efficient random mutagenesis technique using an *E. coli* mutator strain. *Appl. Biochem. Biotechnol. Part B Mol. Biotechnol.* **1997**, *7*, 189–195.

204. Scheuermann, R.; Tam, S.; Burgers, P.M.J. Identification of the ε-subunit of Escherichia coli DNA polymerase III holoenzyme as the dnaQ gene product: A fidelity subunit for DNA replication. *Proc. Natl. Acad. Sci. USA* **1983**, *80*, 7085–7089. [CrossRef]

205. Badran, A.H.; Liu, D.R. Development of potent in vivo mutagenesis plasmids with broad mutational spectra. *Nat. Commun.* **2015**, *6*, 8425. [CrossRef]

206. Ravikumar, A.; Arrieta, A.; Liu, C.C. An orthogonal DNA replication system in yeast. *Nat. Chem. Biol.* **2014**, *10*, 175–177. [CrossRef]

207. Zhang, X.; Guo, H.; Jin, L.; Czornyj, E.; Hodes, A.; Hui, W.H.; Nieh, A.W.; Miller, J.F.; Hong Zhou, Z. A new topology of the HK97-like fold revealed in Bordetella bacteriophage by cryoEM at 3.5 Å resolution. *Elife* **2013**, *2*, e01299. [CrossRef] [PubMed]

208. Yuan, T.Z.; Overstreet, C.M.; Moody, I.S.; Weiss, G.A. Protein Engineering with Biosynthesized Libraries from Bordetella bronchiseptica Bacteriophage. *PLoS ONE* **2013**, *8*, e55617. [CrossRef] [PubMed]

209. Leung, D.W.; Chen, E.; Goeddel, D.V. A Method for random mutagenesis of a defined DNA segment using a modified polymerase chain reaction. *Technique* **1989**, *1*, 11–15.

210. Lin-Goerke, J.L.; Robbins, D.J.; Burczak, J.D. PCRr-based random mutagenesis using manganese and reduced DNTP concentration. *Biotechniques* **1997**, *23*, 409–412. [CrossRef]

211. Zaccolo, M.; Williams, D.M.; Brown, D.M.; Gherardi, E. An approach to random mutagenesis of DNA using mixtures of triphosphate derivatives of nucleoside analogues. *J. Mol. Biol.* **1996**, *255*, 589–603. [CrossRef]

212. McCullum, E.O.; Williams, B.A.R.; Zhang, J.; Chaput, J.C. Random mutagenesis by error-prone PCR. *Methods Mol. Biol.* **2010**, *634*, 103–109.

213. Mondon, P.; Grand, D.; Souyris, N.; Emond, S.; Bouayadi, K.; Kharrat, H. Mutagen™: A random mutagenesis method providing a complementary diversity generated by human error-prone DNA polymerases. *Methods Mol. Biol.* **2010**, *634*, 373–386.

214. Vanhercke, T.; Ampe, C.; Tirry, L.; Denolf, P. Reducing mutational bias in random protein libraries. *Anal. Biochem.* **2005**, *339*, 9–14. [CrossRef]

215. Ye, J.; Wen, F.; Xu, Y.; Zhao, N.; Long, L.; Sun, H.; Yang, J.; Cooley, J.; Todd Pharr, G.; Webby, R.; et al. Error-prone pcr-based mutagenesis strategy for rapidly generating high-yield influenza vaccine candidates. *Virology* **2015**, *482*, 234–243. [CrossRef]

216. Tee, K.L.; Wong, T.S. Polishing the craft of genetic diversity creation in directed evolution. *Biotechnol. Adv.* **2013**, *31*, 1707–1721. [CrossRef]

217. Wong, T.S. Sequence saturation mutagenesis (SeSaM): A novel method for directed evolution. *Nucleic Acids Res.* **2004**, *32*, e26. [CrossRef] [PubMed]

218. Mundhada, H.; Marienhagen, J.; Scacioc, A.; Schenk, A.; Roccatano, D.; Schwaneberg, U. SeSaM-Tv-II generates a protein sequence space that is unobtainable by epPCR. *ChemBioChem* **2011**, *12*, 1595–1601. [CrossRef] [PubMed]

219. Fang, L.; Xu, Z.; Wang, G.S.; Ji, F.Y.; Mei, C.X.; Liu, J.; Wu, G.M. Directed evolution of an LBP/CD14 inhibitory peptide and its anti-endotoxin activity. *PLoS ONE* **2014**, *9*, e101406. [CrossRef] [PubMed]

220. Selas Castiñeiras, T.; Williams, S.G.; Hitchcock, A.; Cole, J.A.; Smith, D.C.; Overton, T.W. Development of a generic β-lactamase screening system for improved signal peptides for periplasmic targeting of recombinant proteins in Escherichia coli. *Sci. Rep.* **2018**, *8*, 6986. [CrossRef] [PubMed]

221. Zahnd, C.; Spinelli, S.; Luginbühl, B.; Amstutz, P.; Cambillau, C.; Plückthun, A. Directed in vitro evolution and crystallographic analysis of a peptide-binding single chain antibody fragment (scFv) with low picomolar affinity. *J. Biol. Chem.* **2004**, *279*, 18870–18877. [CrossRef] [PubMed]

222. Fujii, R.; Kitaoka, M.; Hayashi, K. One-step random mutagenesis by error-prone rolling circle amplification. *Nucleic Acids Res.* **2004**, *32*, e145. [CrossRef]

223. Kunkel, T.A. Rapid and efficient site-specific mutagenesis without phenotypic selection. *Proc. Natl. Acad. Sci. USA* **1985**, *82*, 488–492. [CrossRef]

224. Huovinen, T.; Brockmann, E.C.; Akter, S.; Perez-Gamarra, S.; Ylä-Pelto, J.; Liu, Y.; Lamminmäki, U. Primer extension mutagenesis powered by selective rolling circle amplification. *PLoS ONE* **2012**, *7*, e31817. [CrossRef]

225. Meyer, A.J.; Ellefson, J.W.; Ellington, A.D. Library generation by gene shuffling. In *Current Protocols in Molecular Biology*; John Wiley & Sons, Inc.: Hoboken, NJ, USA, 2013; Volume 105, pp. 15.12.1–15.12.7.

226. Lim, C.C.; Choong, Y.S.; Lim, T.S. Cognizance of molecular methods for the generation of mutagenic phage display antibody libraries for affnity maturation. *Int. J. Mol. Sci.* **2019**, *20*, 1861. [CrossRef]

227. Stemmer, W.P.C. DNA shuffling by random fragmentation and reassembly: In vitro recombination for molecular evolution. *Proc. Natl. Acad. Sci. USA* **1994**, *91*, 10747–10751. [CrossRef]

228. Reid, A.J. DNA shuffling: Modifying the hand that nature dealt. *Vitr. Cell. Dev. Biol. Plant* **2000**, *36*, 331–337. [CrossRef]

229. Zhao, H.; Giver, L.; Shao, Z.; Affholter, J.A.; Arnold, F.H. Molecular evolution by staggered extension process (StEP) in vitro recombination. *Nat. Biotechnol.* **1998**, *16*, 258–261. [CrossRef] [PubMed]

230. Fujishima, K.; Venter, C.; Wang, K.; Ferreira, R.; Rothschild, L.J. An overhang-based DNA block shuffling method for creating a customized random library. *Sci. Rep.* **2015**, *5*, 9740. [CrossRef]

231. Gonzalez-Perez, D.; Molina-Espeja, P.; Garcia-Ruiz, E.; Alcalde, M. Mutagenic organized recombination process by homologous in vivo grouping (MORPHING) for directed enzyme evolution. *PLoS ONE* **2014**, *9*, e90919. [CrossRef] [PubMed]

232. García-Nafría, J.; Watson, J.F.; Greger, I.H. IVA cloning: A single-tube universal cloning system exploiting bacterial in vivo assembly. *Sci. Rep.* **2016**, *6*, 27459. [CrossRef] [PubMed]

233. Chung, D.H.; Potter, S.C.; Tanomrat, A.C.; Ravikumar, K.M.; Toney, M.D. Site-directed mutant libraries for isolating minimal mutations yielding functional changes. *Protein Eng. Des. Sel.* **2017**, *30*, 347–357. [CrossRef]

234. Zoller, M.J.; Smith, M. Oligonucleotide-Directed Mutagenesis: A Simple Method Using two Oligonucleotide Primers and a Single-Stranded DNA Template. *Methods Enzymol.* **1987**, *154*, 329–350.

235. Walker, K.W. Site-directed mutagenesis. *Encycl. Cell Biol.* **2015**, *1*, 122–127.

236. Rapley, R.; Braman, J.; Papworth, C.; Greener, A. Site-directed mutagenesis using double-stranded plasmid DNA templates. In *The Nucleic Acid Protocols Handbook*; Humana Press: Totowa, NJ, USA, 2003; Volume 9, pp. 835–844.

237. Huang, R.; Fang, P.; Kay, B.K. Improvements to the Kunkel mutagenesis protocol for constructing primary and secondary phage-display libraries. *Methods* **2012**, *58*, 10–17. [CrossRef]

238. Scholle, M.D.; Kehoe, J.W.; Kay, B.K. Efficient construction of a large collection of phage-displayed combinatorial peptide libraries. *Comb. Chem. High Throughput Screen.* **2005**, *8*, 545–551. [CrossRef]

239. Lindahl, T. DNA glycosylases, endonucleases for apurinic/apyrimidinic sites, and base excision-repair. *Prog. Nucleic Acid Res. Mol. Biol.* **1979**, *22*, 135–192. [PubMed]

240. Ho, S.N.; Hunt, H.D.; Horton, R.M.; Pullen, J.K.; Pease, L.R. Site-directed mutagenesis by overlap extension using the polymerase chain reaction. *Gene* **1989**, *77*, 51–59. [CrossRef]

241. Chiu, J.; March, P.E.; Lee, R.; Tillett, D. Site-directed, ligase-independent mutagenesis (SLIM): A single-tube methodology approaching 100% efficiency in 4 h. *Nucleic Acids Res.* **2004**, *32*, e174. [CrossRef] [PubMed]

242. Gibson, D.G.; Young, L.; Chuang, R.Y.; Venter, J.C.; Hutchison, C.A.; Smith, H.O. Enzymatic assembly of DNA molecules up to several hundred kilobases. *Nat. Methods* **2009**, *6*, 343–345. [CrossRef] [PubMed]

243. Thomas, S.; Maynard, N.D.; Gill, J. DNA library construction using Gibson Assembly. *Nat. Methods* **2015**, *12*, 1–2. [CrossRef]

244. Galka, P.; Jamez, E.; Joachim, G.; Soumillion, P. QuickLib, a method for building fully synthetic plasmid libraries by seamless cloning of degenerate oligonucleotides. *PLoS ONE* **2017**, *12*, e0175146. [CrossRef]

245. Mitsis, P.G.; Kwagh, J.G. Characterization of the interaction of lambda exonuclease with the ends of DNA. *Nucleic Acids Res.* **1999**, *27*, 3057–3063. [CrossRef]

246. Lim, B.N.; Choong, Y.S.; Ismail, A.; Glökler, J.; Konthur, Z.; Lim, T.S. Directed evolution of nucleotide-based libraries using lambda exonuclease. *Biotechniques* **2012**, *53*, 357–364. [CrossRef]

247. Weiss, G.A.; Watanabe, C.K.; Zhong, A.; Goddard, A.; Sidhu, S.S. Rapid mapping of protein functional epitopes by combinatorial alanine scanning. *Proc. Natl. Acad. Sci. USA* **2000**, *97*, 8950–8954. [CrossRef]

248. Morrison, K.L.; Weiss, G.A. Combinatorial alanine-scanning. *Curr. Opin. Chem. Biol.* **2001**, *5*, 302–307. [CrossRef]

249. Chatellier, J.; Mazza, A.; Brousseau, R.; Vernet, T. Codon-based combinatorial alanine scanning site-directed mutagenesis: Design, implementation, and polymerase chain reaction screening. *Anal. Biochem.* **1995**, *229*, 282–290. [CrossRef] [PubMed]

250. Pál, G.; Fong, S.-Y.; Kossiakoff, A.A.; Sidhu, S.S. Alternative views of functional protein binding epitopes obtained by combinatorial shotgun scanning mutagenesis. *Protein Sci.* **2005**, *14*, 2405–2413. [CrossRef] [PubMed]

251. Wells, J.A.; Vasser, M.; Powers, D.B. Cassette mutagenesis: An efficient method for generation of multiple mutations at defined sites. *Gene* **1985**, *34*, 315–323. [CrossRef]

252. Kegler-ebo, D.M.; Docktor, C.M.; Dimaio, D. Codon cassette mutagenesis: A general method to insert or replace individual codons by using universal mutagenic cassettes. *Nucleic Acids Res.* **1994**, *22*, 1593–1599. [CrossRef] [PubMed]

253. Arkin, M. In vitro mutagenesis. In *Brenner's Encyclopedia of Genetics*, 2nd ed.; Elsevier: Amsterdam, The Netherlands, 2013; pp. 46–50. ISBN 9780080961569.

254. Lai, R.; Bekessy, A.; Chen, C.C.; Walsh, T.; Barnard, R. Megaprimer mutagenesis using very long primers. *Biotechniques* **2003**, *34*, 52–56. [CrossRef]

255. Cárcamo, E.; Roldán-Salgado, A.; Osuna, J.; Bello-Sanmartin, I.; Yánez, J.A.; Saab-Rincón, G.; Viadiu, H.; Gaytán, P. Spiked genes: A method to introduce random point nucleotide mutations evenly throughout an entire gene using a complete set of spiked oligonucleotides for the assembly. *ACS Omega* **2017**, *2*, 3183–3191. [CrossRef]

256. Hermes, J.D.; Parekh, S.M.; Blacklow, S.C.; Koster, H.; Knowles, J.R. A reliable method for random mutagenesis: The generation of mutant libraries using spiked oligodeoxyribonucleotide primers. *Gene* **1989**, *84*, 143–151. [CrossRef]

257. Firnberg, E.; Ostermeier, M. PFunkel: Efficient, expansive, user-defined mutagenesis. *PLoS ONE* **2012**, *7*, e52031. [CrossRef]

258. Valetti, F.; Gilardi, G. Improvement of biocatalysts for industrial and environmental purposes by saturation mutagenesis. *Biomolecules* **2013**, *3*, 778–811. [CrossRef]

259. Sun, D.; Ostermaier, M.K.; Heydenreich, F.M.; Mayer, D.; Jaussi, R.; Standfuss, J.; Veprintsev, D.B. AAscan, PCRdesign and MutantChecker: A suite of programs for primer design and sequence analysis for high-throughput scanning mutagenesis. *PLoS ONE* **2013**, *8*, e78878. [CrossRef]

260. Derbyshire, K.M.; Salvo, J.J.; Grindley, N.D.F. A simple and efficient procedure for saturation mutagenesis using mixed oligodeoxynucleotides. *Gene* **1986**, *46*, 145–152. [CrossRef]

261. Arunachalam, T.S.; Wichert, C.; Appel, B.; Müller, S. Mixed oligonucleotides for random mutagenesis: Best way of making them. *Org. Biomol. Chem.* **2012**, *10*, 4641. [CrossRef] [PubMed]

262. Siloto, R.M.P.; Weselake, R.J. Site saturation mutagenesis: Methods and applications in protein engineering. *Biocatal. Agric. Biotechnol.* **2012**, *1*, 181–189. [CrossRef]

263. Nov, Y. When second best is good enough: Another probabilistic look at saturation mutagenesis. *Appl. Environ. Microbiol.* **2012**, *78*, 258–262. [CrossRef] [PubMed]

264. Tang, L.; Gao, H.; Zhu, X.; Wang, X.; Zhou, M.; Jiang, R. Construction of "small-intelligent" focused mutagenesis libraries using well-designed combinatorial degenerate primers. *Biotechniques* **2012**, *52*, 149–158. [CrossRef] [PubMed]

265. Kille, S.; Acevedo-Rocha, C.G.; Parra, L.P.; Zhang, Z.G.; Opperman, D.J.; Reetz, M.T.; Acevedo, J.P. Reducing codon redundancy and screening effort of combinatorial protein libraries created by saturation mutagenesis. *ACS Synth. Biol.* **2013**, *2*, 83–92. [CrossRef]

266. Gaytán, P.; Roldán-Salgado, A. Elimination of redundant and stop codons during the chemical synthesis of degenerate oligonucleotides. Combinatorial testing on the chromophore region of the red fluorescent protein mkate. *ACS Synth. Biol.* **2013**, *2*, 453–462. [CrossRef]

267. Neuner, P.; Cortese, R.; Monaci, P. Codon-based mutagenesis using dimer-phosphoramidites. *Nucleic Acids Res.* **1998**, *26*, 1223–1227. [CrossRef]

268. Ono, A.; Matsuda, A.; Zhao, J.; Santi, D.V. The synthesis of blocked triplet-phosphoramidites and their use in mutagenesis. *Nucleic Acids Res.* **1995**, *23*, 4677–4682. [CrossRef]

269. Gaytán, P.; Contreras-Zambrano, C.; Ortiz-Alvarado, M.; Morales-Pablos, A.; Yáñez, J. TrimerDimer: An oligonucleotide-based saturation mutagenesis approach that removes redundant and stop codons. *Nucleic Acids Res.* **2009**, *37*, e125. [CrossRef]

270. Hughes, M.D.; Nagel, D.A.; Santos, A.F.; Sutherland, A.J.; Hine, A.V. Removing the redundancy from randomised gene libraries. *J. Mol. Biol.* **2003**, *331*, 973–979. [CrossRef]

271. Ashraf, M.; Frigotto, L.; Smith, M.E.; Patel, S.; Hughes, M.D.; Poole, A.J.; Hebaishi, H.R.M.; Ullman, C.G.; Hine, A.V. ProxiMAX randomization: A new technology for non-degenerate saturation mutagenesis of contiguous codons. *Biochem. Soc. Trans.* **2013**, *41*, 1189–1194. [CrossRef] [PubMed]

272. Pines, G.; Pines, A.; Garst, A.D.; Zeitoun, R.I.; Lynch, S.A.; Gill, R.T. Codon compression algorithms for saturation mutagenesis. *ACS Synth. Biol.* **2015**, *4*, 604–614. [CrossRef] [PubMed]

273. Tang, L.; Wang, X.; Ru, B.; Sun, H.; Huang, J.; Gao, H. MDC-Analyzer: A novel degenerate primer design tool for the construction of intelligent mutagenesis libraries with contiguous sites. *Biotechniques* **2014**, *56*, 301–310. [CrossRef] [PubMed]

274. Li, A.; Acevedo-Rocha, C.G.; Reetz, M.T. Boosting the efficiency of site-saturation mutagenesis for a difficult-to-randomize gene by a two-step PCR strategy. *Appl. Microbiol. Biotechnol.* **2018**, *102*, 6095–6103. [CrossRef] [PubMed]

275. Acevedo-Rocha, C.G.; Reetz, M.T.; Nov, Y. Economical analysis of saturation mutagenesis experiments. *Sci. Rep.* **2015**, *5*, 10654. [CrossRef] [PubMed]

276. Chevalier, A.; Silva, D.A.; Rocklin, G.J.; Hicks, D.R.; Vergara, R.; Murapa, P.; Bernard, S.M.; Zhang, L.; Lam, K.H.; Yao, G.; et al. Massively parallel de novo protein design for targeted therapeutics. *Nature* **2017**, *550*, 74–79. [CrossRef]

277. Alford, R.F.; Leaver-Fay, A.; Jeliazkov, J.R.; O'Meara, M.J.; DiMaio, F.P.; Park, H.; Shapovalov, M.V.; Renfrew, P.D.; Mulligan, V.K.; Kappel, K.; et al. The Rosetta all-atom energy function for macromolecular modeling and design. *J. Chem. Theory Comput.* **2017**, *13*, 3031–3048. [CrossRef]

278. Yang, K.K.; Wu, Z.; Arnold, F.H. Machine-learning-guided directed evolution for protein engineering. *Nat. Methods* **2019**, *16*, 687–694. [CrossRef]

International Journal of
Molecular Sciences

Article

Accumulation of Innate Amyloid Beta Peptide in Glioblastoma Tumors

Lilia Y. Kucheryavykh [1], Jescelica Ortiz-Rivera [1], Yuriy V. Kucheryavykh [1],
Astrid Zayas-Santiago [2], Amanda Diaz-Garcia [2] and Mikhail Y. Inyushin [2,*]

[1] Department of Biochemistry, School of Medicine, Universidad Central del Caribe, PO Box 60327, Bayamon,
    PR 00960-6032, USA; lilia.kucheryavykh@uccaribe.edu (L.Y.K.); 416jortiz@uccaribe.edu (J.O.-R.);
    yuriy.kucheryavykh@uccaribe.edu (Y.V.K.)
[2] Department of Physiology, School of Medicine, Universidad Central del Caribe, PO Box 60327, Bayamon, PR
    00960-6032, USA; astrid.zayas@uccaribe.edu (A.Z.-S.); 417adiaz@uccaribe.edu (A.D.-G.)
* Correspondence: mikhail.inyushin@uccaribe.edu; Tel.: +1-787-667-6469

Received: 8 April 2019; Accepted: 15 May 2019; Published: 20 May 2019

Abstract: Immunostaining with specific antibodies has shown that innate amyloid beta (Aβ) is accumulated naturally in glioma tumors and nearby blood vessels in a mouse model of glioma. In immunofluorescence images, Aβ peptide coincides with glioma cells, and enzyme-linked immunosorbent assay (ELISA) have shown that Aβ peptide is enriched in the membrane protein fraction of tumor cells. ELISAs have also confirmed that the Aβ(1–40) peptide is enriched in glioma tumor areas relative to healthy brain areas. Thioflavin staining revealed that at least some amyloid is present in glioma tumors in aggregated forms. We may suggest that the presence of aggregated amyloid in glioma tumors together with the presence of Aβ immunofluorescence coinciding with glioma cells and the nearby vasculature imply that the source of Aβ peptides in glioma can be systemic Aβ from blood vessels, but this question remains unresolved and needs additional studies.

Keywords: amyloid; Aβ peptide; glioma; platelets

1. Introduction

As Alzheimer's disease (AD) affects mostly the elderly population [1], gliablastoma (GBM) is the most common primary malignant brain tumor in older people [2]. Recently, statistically independent cohort studies have found an inverse association between cancers in general and AD [3–5]. Specifically, most patients with AD are protected from lung cancers [3], and, vice versa, cancer survivors have a lower risk of AD [6]. However, there is a significant positive correlation between the AD mortality rate and the malignant brain tumor mortality rate [4,7,8]. These correlations suggest that there are common factors in these diseases. Mitochondrial metabolism, in general, and the p53, Pin1, and Wnt cellular signaling pathways, in particular, were proposed as possible linkages in this cancer–AD relationship [9,10]. Interestingly, chemotherapy [6] and radiotherapy [9] also affected this correlation.

On the other hand, the buildup of amyloid precursor protein (APP), the precursor of the AD hallmark amyloid beta (Aβ) peptides, have now been found in pancreatic and breast cancer tumors and the corresponding metastatic lymph nodes [11,12]. Proteolytic cleavage of APP by the α-secretase pathway mediates proliferation and migration in breast cancer, while other pathways were not studied [13]. It was also discovered that plasma levels of Aβ peptides in esophageal cancer, colorectal cancer, hepatic cancer, and lung cancer patients were significantly higher than in normal controls [14]. The question arises, what is the source of these Aβ peptides? Moreover, what is their role?

Aβ peptides can be generated by glioma cells themselves. It was shown that glioma cells in culture produce the 4-kDa Aβ peptide, which co-migrates with synthetic Aβ(1–40) (also known as Aβ40) and is specifically recognized by antibodies raised against terminal domains of the Aβ peptide, and releases them into the medium [15]. The role of Aβ peptides in glioma development was investigated in another study [16]. It was reported that full-length Aβ40 is a dose-dependent inhibitor of angiogenesis and suppresses human U87 glioblastoma subcutaneous xenografts in nude mice. A small peptide sequence of Aβ, Aβ(11–20), was found to be a potent, anti-angiogenic molecule. Systemic delivery of this peptide leads to reductions in glioma proliferation, angiogenesis, and invasiveness [16]. Furthermore, parallel experiments in transgenic mice overexpressing Aβ40 also showed reductions in glioma growth, invasion, and angiogenesis [16–18].

However, besides glioma itself, there is another systemic source of Aβ peptide production in the body [19,20]. Recently, we showed that platelets produce a massive release of Aβ after thrombosis in the brain and skin and that this release is concentrated near blood vessels [21,22]. It has been shown that platelets are hyperactivated in cancer patients and form cancer cell-induced aggregates and micro-thrombi in the vasculature near tumors (reviewed in [23]). A high platelet count is associated with poor survival in a large variety of cancers, while thrombocytopenia or antiplatelet drugs can reduce the short-term risk of cancer, cancer mortality, and metastasis (reviewed in [24]). Platelets affect glioma cells by releasing platelet-derived growth factor (PDGF) [25]. May platelet-generated Aβ also diffuse to glioma cells and accumulate inside these brain tumors?

In our study, we chose specific antibodies against Aβ peptides with low reactivity for the precursor APP to see whether Aβ immunoreactivity is present in glioma tumors and nearby blood vessels in mice. We used an enzyme-linked immunosorbent assay (ELISA) to study Aβ40 content in tumor and "healthy" brain area, while also assessing Aβ40 content in the membrane and cytoplasmic fractions of glioma cells. The presence of aggregated forms of amyloid inside glioma tumors was evaluated as well.

2. Results

2.1. Immunoreactivity against Aβ Peptides Is Present in Glioma Cells in Primary and Secondary Tumors as Well as in Blood Vessels and Erythrocytes in the Near Vicinity, Indicating that the Aβ Level Is Elevated in the Tumor Zone

After glioma implantation into mouse brains using standard methods established in our laboratory [26,27] we allowed 16 days of tumor growth. We then prepared brain slices containing tumors within nearby tissue. Immunostaining with polyclonal (Figure 1A,B, green) antibody against Aβ showed that these peptides are present in glioma cells (white arrows), in nearby broken blood vessels, and in escaped erythrocytes. In addition, astrocytes are marked by red fluorescence (anti-Glial Fibrillary Acidic Protein (anti-GFAP)), and the nuclei are marked blue (4′,6-diamidino-2-phenylindole (DAPI) staining). The same images (Figure 1A,B) are presented as moving confocal images (Figure S1A,B) so that blood vessel details and their relation to glioma cells are more discernable. Inside blood vessel segments marked by Aβ green immunofluorescence, erythrocytes were also specifically marked by Aβ (Figure 1A,B, see also Figure S1A,B) as well as erythrocytes diffused locally near broken blood vessels (Figure S1A,B), as blood vessels near the tumor are usually ruptured [28]. As was shown previously, Aβ peptide in blood plasma binds to practically all erythrocytes and may be a marker for AD [29]. Also, the addition of synthetic Aβ specifically marks erythrocyte membranes [30]. We want to stress once again that Aβ immunofluorescence is present only in blood vessel segments near the glioma tumor and in the tumor itself (Figure 1A,B and Figure S1A,B). Therefore, only the glioma cells in the tumor and nearby blood vessels containing erythrocytes and within the distance 0–200 μm from the ruptured blood vessel are fluorescent.

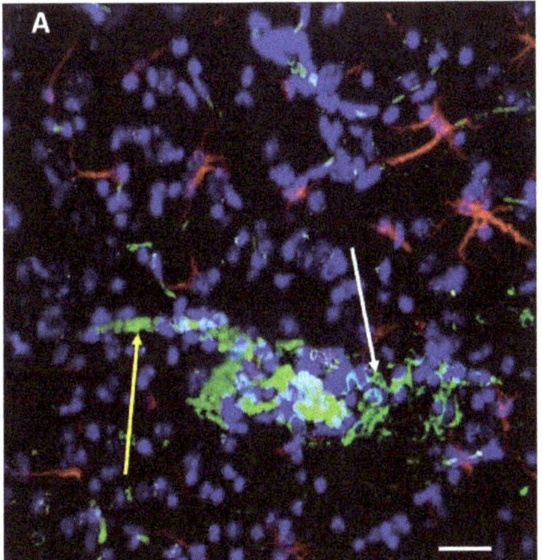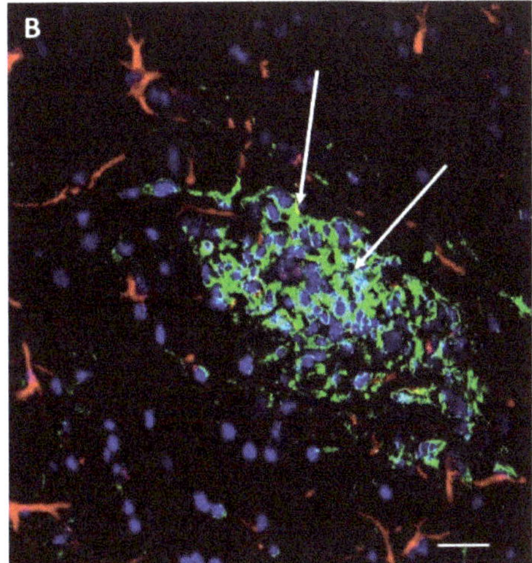

Figure 1. Aβ peptide immunoreactivity (green) in glioma cells and in nearby blood vessels. (**A**) A small glioma tumor near a broken blood vessel. Aβ peptide immunoreactivity (green) visible in glioma cells (white arrow) and in blood vessels. Erythrocytes released from the broken vessel are also marked with Aβ-related immunofluorescence (yellow arrow). (**B**) A larger glioma tumor in which a broken blood vessel passes through the tumor (more clearly visible in the 3D image of this tumor shown in Figure S1B), and white arrows indicate glioma cells marked by green immunofluorescence representing Aβ peptide. For **A** and **B**, astrocytes are indicated by immunoreactivity to GFAP (red) and cell nuclei by DAPI staining (blue). Scale bar, 20 μm. (See also supplemental confocal 3D images of the same tumors in Figure S1A,B, respectively).

We also made ELISA measurements of mouse Aβ40 peptide in the brain sample tissue containing the main tumor versus the "healthy" control from the corresponding cortical zone in the other hemisphere from the same animal 16 days after glioma implantation. Similar amounts of the homogenate were taken for analysis. It was found that the relative amount of Aβ40 in the glioma tissue was $142 \pm 9\%$ larger and statistically different ($p < 0.001$; $t = 4.714$; $df = 4$; $n = 3$) from "healthy" tissue (Figure 2A).

In these experiments, we found that glioma cells exhibit specific Aβ immunofluorescence that clearly marks these cells, but the question arises whether it is inside the cells or somehow attached to the external membrane.

2.2. Aβ40 Is Concentrated in the Membrane Cell Fraction in Glioma Tumor Tissue

To determine more precisely there Aβ is distributed, we separated the cytoplasmic and membrane fractions of proteins from glioma cells from the main tumor extracted from the brain of animals 16 days after implantation. Before processing, blood cells were eliminated from the tumor tissue samples using the Percoll purification method. Membrane and cytoplasmic proteins were isolated, and the total protein content was determined using the Bradford spectrophotometric method to establish a reference point for measuring the amount of Aβ in each fraction. Using ELISA, it was found that the relative amount of Aβ40 in the membrane fraction is significantly greater ($170 \pm 4\%$, $p < 0.001$, $t = 16.23$, $df = 4$, $n = 3$) than in the cytoplasmic fraction (Figure 2B).

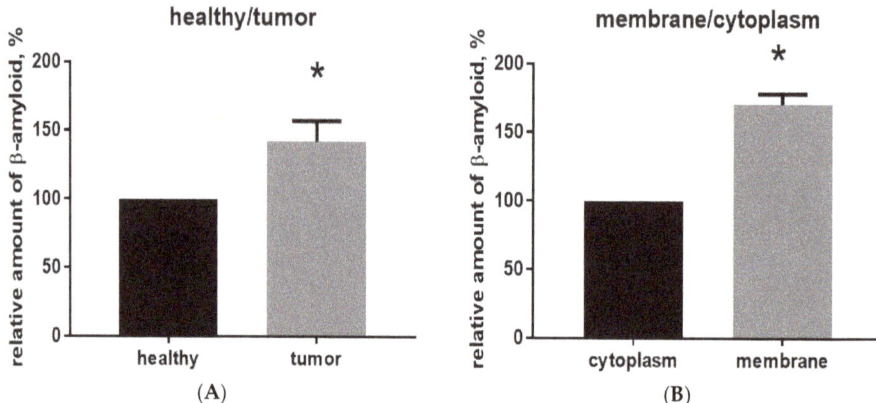

Figure 2. (**A**) The relative amount of Aβ40 in the glioma tissue is elevated. (**B**) Aβ40 in glioma tumor tissue is concentrated in the cell membrane fraction.

2.3. Glioma Tumor Tissue Contains Aggregated Amyloid

To determine whether glioma tumors have aggregated forms of Aβ with cross-β architecture, we used standard thioflavin T and thioflavin S staining of brain slices with glioma from animals with implanted glioma cells. It was previously demonstrated that both thioflavin T and thioflavin S fluorescence originates mainly from dye bound to aggregated forms of amyloids with cross-β-pleated sheet structure, and gives a distinct increase (and a spectral shift in the case of thioflavin T) in fluorescence emission after binding [31,32]. We used IP injection of thioflavin T, while slices containing tumors were additionally stained with thioflavin S. Both dyes specifically marked glioma tumors (Figure 3), in which staining (green for thioflavin T and red for thioflavin S) is obvious only inside the tumor body, while the nearby normal tissue remained unstained.

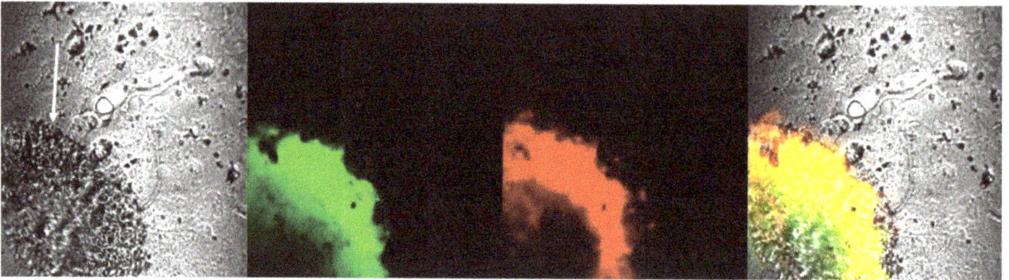

Figure 3. Aggregated amyloid visualized by staining with thioflavin T (green) and thioflavin S (red) inside the glioma tumor body. The white arrow shows the glioma tumor body visible in the brain slice.

3. Discussion

Here we report that antibodies against Aβ with relatively low reactivity against APP [33] show Aβ immunostaining in glioma cells and nearby blood vessels in mice (Figure 1). Using ELISA, we also report that Aβ40 levels are significantly increased in glioma (Figure 2). Glioma tissue from one brain hemisphere contains about two-fold more Aβ than a similar amount of tissue from the "mirror" hemisphere, with Aβ concentrated in the membrane fraction. The question arises whether Aβ is coming from the systemic source—from the blood, and is marking the glioma cell membrane—or is synthetized by glioma cells themselves.

Previous studies support the possibility of systemic source for this Aβ. The results indicating increased Aβ content in blood plasma for different types of cancer have already been reported [14].

Systemic Aβ is generated in large quantities by blood platelets in broken vessels, as we have shown for the thrombotic process [21,22]. Here, broken blood vessels marked by extensive Aβ fluorescence can be seen near tumors in our experiments (Figure 1A,B and Figure S1A,B). It has been shown previously that platelets are hyperactivated in cancer patients and form cancer cell-induced aggregates and micro-thrombi in vasculature near tumors (reviewed in [23]), thus suggesting that the source of Aβ that we have found for the clotting process may also be present here. It seems possible that Aβ released from clots can migrate and somehow mark only glioma cells (Figure 1A,B), but this raises new questions about why Aβ marks glioma cells so specifically.

To bind specifically, Aβ must be recognized by a specific receptor on the external membrane of the glioma cell. A known specific Aβ receptor, such as the PrPC–mGluR5 complex, is associated with proline-rich tyrosine kinase 2 (Pyk2 or PTK2B) [34,35]. This receptor localizes to postsynaptic sites in the brain, but is also overexpressed in all glioblastoma cells, where it controls cell migration [27,36]. Aβ is a known inhibitor of Pyk2 [35]. Thus, its release by platelets may be a part of the intrinsic immunity that is directed against cancerous gliomas. Another suspected molecule related to Aβ binding is PI3K (phosphatidylinositol [PI] type 3 receptor tyrosine kinase). This kinase and its signaling network is also present and hyperactivated in a majority of glioblastoma cells, where it controls membrane microdynamics and cell cycling [37,38]. Its Aβ receptor is unknown, but it complexes with PI3K and most probably is situated on the external membrane [39,40]. It is known that Aβ inhibits PI3K activity as well [41]. We speculate that in this case, Aβ peptides generated by platelets also play a role in the intrinsic immunity directed against cancerous gliomas.

In addition, Aβ may bind to the advanced glycation end products (RAGE) receptor. It is known that this receptor is the binding site for Aβ peptides [42] thus mediating Aβ transport through the blood–brain barrier [43]. Very same RAGE receptor regulates the tumor environment and tumor cell migration, is part of the important microglial activation mechanism and is overexpressed in tumors [44].

On the other hand, it was shown that glioma cells in culture produce Aβ peptides that comigrate with synthetic Aβ40 and are specifically recognized by antibodies raised against the terminal domains of the Aβ peptide and released by these cells into the medium [15]. However, cultured and in vivo astrocytes also produce Aβ peptides is similar amounts [45–47] and astrocytes were not marked by Aβ immunofluorescence in our experiments, probably because these peptides is present in/near the astrocytes in amounts that can be neglected compared with the glioma tumor cells that we have studied here. While derived from the same cell type, glioma cells are clearly marked by Aβ immunofluorescence in our experiments (Figure 1).

It is clear that the question of whether the source of Aβ is inside the glioma cell itself or is a systemic source from blood vessels should be investigated further. Anyway, all our results from these experiments taken together as well as our previous experience with Aβ peptides released during platelet accumulation and aggregation in thrombotic blood vessels [21,22] lead us to the conclusion that most probably Aβ peptides are generated by platelets and somehow bind almost exclusively to glioma cells.

An additional issue is the accuracy of Aβ40 concentrations measurements in brain tissue. In our study of Aβ40 concentrations in tissue, we used relative values, indicating the percentage change from initial values, as the most accurate. It was shown previously that the Invitrogen Aβ40 ELISA Kit is very specific to murine Aβ40, but the data are very sensitive to "noise" (such as the presence of other proteins and lipids), and absolute values can deviate 40–50% [48]. Also, ELISA data may vary considerably, with a variety of collection and storage protocols [49]. Measurement of Aβ by ELISA reveals mainly free peptides, while a significant amount of Aβ peptide remains bound to proteins, lipoproteins, and cell membranes [50].

Our experiments also indicate that there is some thioflavin-positive amyloid inside glioma tumors (Figure 3). While we have shown that Aβ peptides are definitely present in tumor and may constitute a predominant part of this glioma amyloid, the specific type of aggregated amyloid found inside the borders of glioma tumors is unknown. To our opinion, this amyloid is most probably mixed amyloid, as was found for AD [51]. Protein aggregation is sequence specific, not favoring self-assembly over cross-seeding with nonhomologous sequences [52]. However, proteins with aggregation-prone regions may aggregate with each other at elevated concentrations, forming a mixed misfolded amyloid [53]. In this case, one aggregated protein can work as a "seed" for aggregation of other protein types. Previously, different amyloids were found in a variety of tumors. Different carcinomas have amyloid stroma [54,55], and odontogenic tumors are positive for thioflavin T and Congo Red staining and are also immunopositive for the enamel matrix protein ameloblastin [56–58]. Similarly, amyloid was reported in breast cancer tumors but was determined to be a localized amyloid light chain (AL) type (primary amyloidosis caused by ImG light-chain β-sheeting) [59,60]. Localized AL type amyloidosis was also found in myeloma (plasma cell) tumors as well as in kidneys and early-stage non-small-cell lung adenocarcinomas [61]. If the content of amyloid in glioma tumors is mixed, it must be further studied, because tumor-related amyloid could be a new target for anticancer therapy.

4. Materials and Methods

4.1. Ethics Statement

All procedures involving rodents were conducted in accordance with the National Institutes of Health regulations concerning the use and care of experimental animals and approved by the Universidad Central del Caribe Institutional Animal Care and Use Committee. All efforts were made to minimize suffering. In all surgical experiments, animals were anesthetized with isoflurane (4% for induction and 1.75% for maintenance) using a Matrix Quanti-flex VMC Anesthesia Machine for small animals (Midmark Corporation, Dayton, OH, USA). The animals were sacrificed for brain tissue and blood analysis after experiments.

4.2. Glioma Cell Culture

The GL261 glioma cell line derived from C57BL/6 mice was obtained from the NCI (Frederick, MD, USA). All cells were cultured in Dulbecco's Modified Eagle's Medium (DMEM) supplemented with 10% fetal calf serum, 0.2 mM glutamine, and antibiotics (50 U/mL penicillin, 50 µG/mL streptomycin) and maintained in a humidified atmosphere of CO_2/air (5%/95%) at 37 °C. The medium was exchanged with fresh culture medium every 2–3 days.

4.3. Intracranial Implantation of Glioma Cells

All surgery was performed under isoflurane anesthesia, and all efforts were made to minimize suffering. GL261 glioma cells were implanted into the right cerebral hemisphere of 12–16-week-old C57BL/6 mice. Implantation was performed according to the protocol that we described earlier [26]. Briefly, mice were anesthetized with isoflurane, and a midline incision was made on the scalp. At stereotaxic coordinates of bregma, 2 mm lateral, 1 mm caudal, and 3 mm ventral, a small burr hole (0.5 mm diameter) was drilled into the skull. One microliter of cell suspension (2×10^4 cells/µL in phosphate buffer solution (PBS)) was delivered at a depth of 3 mm over 2 min. Sixteen days following injection, the animals were anesthetized with pentobarbital (50 mg/kg) and transcardially perfused with PBS followed by 4% paraformaldehyde (PFA). The brains were removed and post-fixed in 4% PFA/PBS for 24 h at 4 °C, followed by 0.15 M, 0.5 M, and 0.8 M sucrose at 4 °C until fully dehydrated. The brains were then frozen and embedded in Cryo-M-Bed embedding compound (Bright Instrument, Huntingdon, UK) and cut using a Vibratome UltraPro 5000 cryostat (American Instrument, Haverhill, MA, USA).

4.4. Percoll Purification of Blood Cells from Tissue Samples for Membrane Fraction Isolation

To study Aβ distribution inside tumor cells, we first eliminated blood cells from the tumor tissue sample using the Percoll purification method. Tumors and healthy cortex from the contralateral hemisphere were removed from the mouse brains, minced into 1–2-mm pieces with a razor blade, and enzymatically homogenized using a collagenase/hyaluronidase in DMEM (cat. #07912, Stemcell Technologies, WA, USA). Blood cells were separated from the homogenized tissue using Percoll (Sigma-Aldrich, St. Louis, MO, USA) gradients of 30% and 70%. Following this procedure the tissue fraction free from blood cells was collected from the top of the 70% Percoll level and used for further analysis.

4.5. Isolation of Membrane and Cytoplasmic Proteins

A homogenized cell suspension was resuspended and sonicated in 20 mM Tris buffer containing 1 mM ethylenedinitrilotetraacetic acid (EDTA), 1 mM β-mercaptoethanol, and 5% glycerin, pH 8.5 with HCl, 1 μM Na_3VO_4, 0.5 mM phenylmethylsulfonyl fluoride (PMSF), and 10 mM dithiothreitol (DTT). After centrifugation the supernatant was collected and used for further investigations as the cytoplasmic protein fraction. The pellet containing the membranes and the membrane proteins was lysed, and clarified cell lysate was used as the membrane protein fraction.

4.6. Enzyme-Linked Immunosorbent Assay (ELISA) Measurements

A specialized, ready-to-use, mouse-specific, solid-phase sandwich ELISA kit (cat. #KMB3481; Invitrogen, Thermo Fisher Scientific, Waltham, MA, USA) was used for direct measurement of the amount of Aβ40 peptide in the brains of experimental animals in accordance with the manufacturer's documentation. Briefly, the brain samples were homogenized mechanically, and 100 mg of homogenate was then lysed in guanidine solution (5 M guanidine HCl, 50 mM Tris HCl, pH 8.0). In other experiments, the lysate (normalized to total protein content) from membrane and cytoplasmic fractions (see above) were used. A monoclonal antibody against the NH_2-terminus of mouse Aβ40 peptide was coated onto the wells of the microtiter strips provided in the kit. Samples, including standards of known Aβ40 content for calibration purposes as well as experimental specimens, were pipetted into the wells. After washing, the rabbit antibody specific to the COOH-terminus of Aβ40 was added and detected with horseradish peroxidase-labeled anti-rabbit antibody. The optical density values at 450 nm were determined using a Wallac 1420 Victor 2 Microplate Reader (PerkinElmer Inc., Waltham, MA, USA). The calculated mean reading from the healthy hemisphere (normalized cytoplasmic fraction) was defined as 100%, while other readings were presented as the percentage of this value.

4.7. Immunohistochemistry and Confocal Microscopy

Immunostaining was performed using a protocol previously established in our laboratory [22,62]. Frozen 30-μm sections were generated from brain cortex containing the tumor(s). The sections were blocked with 5% normal goat serum/5% normal horse serum (Vector Laboratories, Burlingame, CA, USA) in 0.10 M phosphate buffer solution (PBS: NaCl, 137 mM; KCl, 2.70 mM; Na_2HPO_4, 10.14 mM; KH_2PO_4, 1.77 mM) containing 0.3% Triton X-100 and 0.05% phenylhydrazine for 60 min for permeabilization and then processed separately using two different antibodies against Aβ. For that purpose, slices were incubated with a rabbit polyclonal antibody to Aβ (Abcam, Cambridge, MA, USA, cat. #ab2539) diluted 1:400 in 0.03% Triton X-100, 1% dimethyl sulfoxide (DMSO), 2% bovine serum albumin (BSA), 5% normal horse serum, and 5% normal goat serum in 0.1 M PBS. Anti-GFAP–Cy3 (1:200) was added, and the slice left overnight at 4 °C. After three washes with permeabilization solution for 10 min, the secondary antibodies (fluorescein-labeled goat anti-rabbit IgG) were added at a dilution of 1:200 with shaking for 2 h at room temperature and protected from light. The slices were then washed three times with PBS for 10 min and once with distilled water before being transferred onto a glass slide containing Fluoroshield mounting medium (Sigma-Aldrich, St. Louis, MO, USA, cat. #F6057) with DAPI. Negative controls were routinely performed by removal of primary antibody in each staining experiment to validate the immunohistochemical staining quality and results.

For thioflavin (Th) staining we used: (1) ThT staining, in which mice were injected IP with 10 μL/g of 3 mM solution of ThT in PBS. After 5 min, the animals were euthanized, and the brains were harvested and kept in fixative without light. (2) ThS staining, in which brain slices (30 μm) containing tumors were allowed to completely air dry prior to staining, then stained with a drop of 3 mM ThS in PBS (previously filtered through a 0.2-μm filter) for 5 min, then washed twice with distilled water and dried again. The coverslip was mounted with a drop of Vaseline on the slice. DAPI and Cy3 excitation/emission filters were used to visualize ThT and ThS fluorescence, respectively.

Images were acquired using an Olympus Fluoview FV1000 scanning inverted confocal microscope system equipped with a 20×, 40×, or 60×/1.43 oil objective (Olympus, Melville, NY, USA). The images were analyzed using ImageJ software (http://imagej.nih.gov/ij) with the Open Microscopy Environment Bio-Formats library and plugin, allowing for the opening of Olympus files (http://www.openmicroscopy.org/site/support/bio-formats5.4/). The data were evaluated using custom colorization.

4.8. Statistics and Measurements

Using GraphPad Prism 7.03 (GraphPad Software, Inc., La Jolla, CA, USA) for calculations, an unpaired t-test was employed to estimate statistical differences. Values were determined to be significantly different if the two-tailed p-value was <0.05.

5. Conclusions

- Aggregated amyloid is present inside glioma tumor borders;
- Aβ peptide immunofluorescence is present in glioma tumors, marking glioma cells and nearby ruptured blood vessels.

Supplementary Materials: Supplementary materials can be found at http://www.mdpi.com/1422-0067/20/10/2482/s1.

Author Contributions: Conceptualization, methodology, original draft preparation, review and editing, and formal analysis were performed by M.Y.I. and L.Y.K.; experimental investigation, formal analysis, visualization, data curation, and review and editing were performed by L.Y.K., A.Z.-S., A.D.-G., J.-O.-R., and Y.V.K.

Funding: This research was supported by NIH grants SC1GM122691 to L.Y.K. and SC2GM111149 to M.Y.I. The funding sources had no role in study design; data collection, analysis, or interpretation; or the decision to submit this article.

Acknowledgments: We want to thank personnel of Animal Resources Center in Universidad Central del Caribe for their kind help.

Conflicts of Interest: The authors declare that they have no conflicts of interest. The funders had no role in the design of the study; in the collection, analyses, or interpretation of data; in the writing of the manuscript; or in the decision to publish the results.

Abbreviations

ELISA	enzyme-linked immunosorbent assay
ThS	Thioflavin S
ThT	Thioflavin T
PMSF	Phenylmethylsulfonyl fluoride
DTT	Dithiothreitol
DMEM	Dulbecco's Modified Eagle's Medium
Aβ	Amyloid beta peptide

References

1. Guerreiro, R.; Bras, J. The age factor in Alzheimer's disease. *Genome Med.* **2015**, *7*, 106. [CrossRef] [PubMed]
2. Young, J.S.; Chmura, S.J.; Wainwright, D.A.; Yamini, B.; Peters, K.B.; Lukas, R.V. Management of glioblastoma in elderly patients. *J. Neurol. Sci.* **2017**, *380*, 250–255. [CrossRef]
3. Ou, S.M.; Lee, Y.J.; Hu, Y.W.; Liu, C.J.; Chen, T.J.; Fuh, J.L.; Wang, S.J. Does Alzheimer's disease protect against cancers? A nationwide population-based study. *Neuroepidemiology* **2013**, *40*, 42–49. [CrossRef] [PubMed]

4. Musicco, M.; Adorni, F.; Di Santo, S.; Prinelli, F.; Pettenati, C.; Caltagirone, C.; Palmer, K.; Russo, A. Inverse occurrence of cancer and Alzheimer disease: A population-based incidence study. *Neurology* **2013**, *81*, 322–328. [CrossRef]

5. Yarchoan, M.; James, B.D.; Shah, R.C.; Arvanitakis, Z.; Wilson, R.S.; Schneider, J.; Bennett, D.A.; Arnold, S.E. Association of Cancer History with Alzheimer's Disease Dementia and Neuropathology. *J. Alzheimers Dis.* **2017**, *56*, 699–706. [CrossRef]

6. Driver, J.A.; Beiser, A.; Au, R.; Kreger, B.E.; Splansky, G.L.; Kurth, T.; Kiel, D.P.; Lu, K.P.; Seshadri, S.; Wolf, P.A. Inverse association between cancer and Alzheimer's disease: Results from the Framingham Heart Study. *BMJ* **2012**, *344*, e1442. [CrossRef] [PubMed]

7. Lehrer, S. Glioblastoma and dementia may share a common cause. *Med. Hypotheses* **2010**, *75*, 67–68. [CrossRef]

8. Lehrer, S. Glioma and Alzheimer's Disease. *J. Alzheimers Dis. Rep.* **2018**, *2*, 213–218. [CrossRef]

9. Behrens, M.I.; Lendon, C.; Roe, C.M. A common biological mechanism in cancer and Alzheimer's disease? *Curr. Alzheimer Res.* **2009**, *6*, 196–204. [CrossRef]

10. Sánchez-Valle, J.; Tejero, H.; Ibáñez, K.; Portero, J.L.; Krallinger, M.; Al-Shahrour, F.; Tabarés-Seisdedos, R.; Baudot, A.; Valencia, A. A molecular hypothesis to explain direct and inverse co-morbidities between Alzheimer's Disease, Glioblastoma and Lung cancer. *Sci. Rep.* **2017**, *7*, 4474. [CrossRef]

11. Hansel, D.E.; Rahman, A.; Wehner, S.; Herzog, V.; Yeo, C.J.; Maitra, A. Increased expression and processing of the Alzheimer amyloid precursor protein in pancreatic cancer may influence cellular proliferation. *Cancer Res.* **2003**, *63*, 7032–7037. [PubMed]

12. Tsang, J.Y.S.; Lee, M.A.; Ni, Y.B.; Chan, S.K.; Cheung, S.Y.; Chan, W.W.; Lau, K.F.; Tse, G.M.K. Amyloid Precursor Protein Is Associated with Aggressive Behavior in Nonluminal Breast Cancers. *Oncologist* **2018**, *23*, 1273–1281. [CrossRef] [PubMed]

13. Tsang, J.Y.S.; Lee, M.A.; Chan, T.H.; Li, J.; Ni, Y.B.; Shao, Y.; Chan, S.K.; Cheungc, S.Y.; Lau, K.F.; Tse, G.M.K. Proteolytic cleavage of amyloid precursor protein by ADAM10 mediates proliferation and migration in breast cancer. *EBioMedicine* **2018**, *38*, 89–99. [CrossRef] [PubMed]

14. Jin, W.S.; Bu, X.L.; Liu, Y.H.; Shen, L.L.; Zhuang, Z.Q.; Jiao, S.S.; Zhu, C.; Wang, Q.H.; Zhou, H.D.; Zhang, T.; et al. Plasma Amyloid-Beta Levels in Patients with Different Types of Cancer. *Neurotox. Res.* **2017**, *31*, 283–288. [CrossRef] [PubMed]

15. Morato, E.; Mayor, F., Jr. Production of the Alzheimer's beta-amyloid peptide by C6 glioma cells. *FEBS Lett.* **1993**, *336*, 275–278. [CrossRef]

16. Murphy, S.F.; Banasiak, M.; Yee, G.-T.; Wotoczek-Obadia, M.; Tran, Y.; Prak, A.; Albright, R.; Mullan, M.; Paris, D.; Brem, S. A synthetic fragment of beta-amyloid peptide suppresses glioma proliferation, angiogenesis, and invasiveness in vivo and in vitro. *Neuro-Oncol.* **2010**, *12*, iv5. [CrossRef]

17. Paris, D. Modulation of Angiogenesis by a-Beta Peptide Fragments. Patent US20080031954A1, 7 February 2005.

18. Paris, D.; Gan ey, N.; Banasiak, M.; Laporte, V.; Patel, N.; Mullan, M.; Murphy, S.F.; Yee, G.T.; Bachmeier, C.; Ganey, C.; et al. Impaired orthotopic glioma growth and vascularization in transgenic mouse models of Alzheimer's disease. *J. Neurosci.* **2010**, *30*, 11251–11258. [CrossRef]

19. Inyushin, M.Y.; Sanabria, P.; Rojas, L.; Kucheryavykh, Y.; Kucheryavykh, L. Aβ Peptide Originated from Platelets Promises New Strategy in Anti-Alzheimer's Drug Development. *Biomed. Res. Int.* **2017**, *2017*, 3948360. [CrossRef]

20. Inyushin, M.; Zayas-Santiago, A.; Rojas, L.; Kucheryavykh, Y.; Kucheryavykh, L. Platelet-generated amyloid beta peptides in Alzheimer's disease and glaucoma. *Histol. Histopathol.* **2019**, 18111. [CrossRef]

21. Kucheryavykh, L.Y.; Dávila-Rodríguez, J.; Rivera-Aponte, D.E.; Zueva, L.V.; Washington, A.V.; Sanabria, P.; Inyushin, M.Y. Platelets are responsible for the accumulation of β-amyloid in blood clots inside and around blood vessels in mouse brain after thrombosis. *Brain Res. Bull.* **2017**, *128*, 98–105. [CrossRef]

22. Kucheryavykh, L.Y.; Kucheryavykh, Y.V.; Washington, A.V.; Inyushin, M.Y. Amyloid Beta Peptide Is Released during Thrombosis in the Skin. *Int. J. Mol. Sci.* **2018**, *19*, 1705. [CrossRef]

23. Jurasz, P.; Alonso-Escolano, D.; Radomski, M.W. Platelet–cancer interactions: Mechanisms and pharmacology of tumour cell-induced platelet aggregation. *Br. J. Pharm.* **2004**, *143*, 819–826. [CrossRef] [PubMed]

24. Goubran, H.A.; Burnouf, T.; Radosevic, M.; El-Ekiaby, M. The platelet-cancer loop. *Eur. J. Intern. Med.* **2013**, *24*, 393–400. [CrossRef]

25. Hermanson, M.; Funa, K.; Hartman, M.; Claesson-Welsh, L.; Heldin, C.H.; Westermark, B.; Nistér, M. Platelet-derived growth factor and its receptors in human glioma tissue: Expression of messenger RNA and protein suggests the presence of autocrine and paracrine loops. *Cancer Res.* **1992**, *52*, 3213–3219.

26. Kucheryavykh, L.Y.; Kucheryavykh, Y.V.; Rolón-Reyes, K.; Skatchkov, S.N.; Eaton, M.J.; Cubano, L.A.; Inyushin, M.Y. Visualization of implanted GL261 glioma cells in living mouse brain slices using fluorescent 4-(4-(dimethylamino)-styryl)-*N*-methylpyridinium iodide (ASP+). *Biotechniques* **2012**. [CrossRef] [PubMed]

27. Rolón-Reyes, K.; Kucheryavykh, Y.V.; Cubano, L.A.; Inyushin, M.; Skatchkov, S.N.; Eaton, M.J.; Harrison, J.K.; Kucheryavykh, L.Y. Microglia Activate Migration of Glioma Cells through a Pyk2 Intracellular Pathway. *PLoS ONE* **2015**, *10*, e0131059. [CrossRef] [PubMed]

28. Dubois, L.G.; Campanati, L.; Righy, C.; D'Andrea-Meira, I.; Spohr, T.C.; Porto-Carreiro, I.; Pereira, C.M.; Balça-Silva, J.; Kahn, S.A.; DosSantos, M.F.; et al. Gliomas and the vascular fragility of the blood brain barrier. *Front. Cell Neurosci.* **2014**, *8*, 418. [CrossRef] [PubMed]

29. Lan, J.; Liu, J.; Zhao, Z.; Xue, R.; Zhang, N.; Zhang, P.; Zhao, P.; Zheng, F.; Sun, X. The peripheral blood of Aβ binding RBC as a biomarker for diagnosis of Alzheimer's disease. *Age Ageing* **2015**, *44*, 458–464. [CrossRef]

30. Mohanty, J.G.; Eckley, D.M.; Williamson, J.D.; Launer, L.J.; Rifkind, J.M. Do red blood cell-β-amyloid interactions alter oxygen delivery in Alzheimer's disease? *Adv. Exp. Med. Biol.* **2008**, *614*, 29–35.

31. Kelényi, G. Thioflavin S fluorescent and Congo red anisotropic stainings in the histologic demonstration of amyloid. *Acta Neuropathol.* **1967**, *7*, 336–348. [CrossRef]

32. Biancalana, M.; Koide, S. Molecular mechanism of Thioflavin-T binding to amyloid fibrils. *Biochim. Biophys. Acta* **2010**, *1804*, 1405–1412. [CrossRef]

33. Wu, Y.; Du, S.; Johnson, J.L.; Tung, H.Y.; Landers, C.T.; Liu, Y.; Seman, B.G.; Wheeler, R.T.; Costa-Mattioli, M.; Kheradmand, F.; et al. Microglia and amyloid precursor protein coordinate control of transient Candida cerebritis with memory deficits. *Nat. Commun.* **2019**, *10*, 58. [CrossRef] [PubMed]

34. Brody, A.H.; Strittmatter, S.M. Synaptotoxic Signaling by Amyloid Beta Oligomers in Alzheimer's Disease Through Prion Protein and mGluR5. *Adv. Pharm.* **2018**, *82*, 293–323. [CrossRef]

35. Salazar, S.V.; Cox, T.O.; Lee, S.; Brody, A.H.; Chyung, A.S.; Haas, L.T.; Strittmatter, S.M. Alzheimer's Disease Risk Factor Pyk2 Mediates Amyloid-β-Induced Synaptic Dysfunction and Loss. *J. Neurosci.* **2019**, *39*, 758–772. [CrossRef] [PubMed]

36. Lipinski, C.A.; Tran, N.L.; Menashi, E.; Rohl, C.; Kloss, J.; Bay, R.C.; Berens, M.E.; Loftus, J.C. The tyrosine kinase pyk2 promotes migration and invasion of glioma cells. *Neoplasia* **2005**, *7*, 435–445. [CrossRef]

37. Fan, Q.W.; Weiss, W.A. Targeting the RTK-PI3K-mTOR axis in malignant glioma: Overcoming resistance. *Curr. Top. Microbiol. Immunol.* **2010**, *347*, 279–296. [CrossRef]

38. Zhao, H.F.; Wang, J.; Shao, W.; Wu, C.P.; Chen, Z.P.; To, S.T.; Li, W.P. Recent advances in the use of PI3K inhibitors for glioblastoma multiforme: Current preclinical and clinical development. *Mol. Cancer* **2017**, *16*, 100. [CrossRef] [PubMed]

39. Klippel, A.; Reinhard, C.; Kavanaugh, W.M.; Apell, G.; Escobedo, M.A.; Williams, L.T. Membrane localization of phosphatidylinositol 3-kinase is sufficient to activate multiple signal-transducing kinase pathways. *Mol. Cell. Biol.* **1996**, *16*, 4117–4127. [CrossRef]

40. Gao, X.; Lowry, P.R.; Zhou, X.; Depry, C.; Wei, Z.; Wong, G.W.; Zhang, J. PI3K/Akt signaling requires spatial compartmentalization in plasma membrane microdomains. *Proc. Natl. Acad. Sci. USA* **2011**, *108*, 14509–14514. [CrossRef]

41. Chen, T.J.; Wang, D.C.; Chen, S.S. Amyloid-beta interrupts the PI3K-Akt-mTOR signaling pathway that could be involved in brain-derived neurotrophic factor-induced Arc expression in rat cortical neurons. *J. Neurosci. Res.* **2009**, *87*, 2297–2307. [CrossRef]

42. Mruthinti, S.; Hill, W.D.; Swamy-Mruthinti, S.; Buccafusco, J.J. Relationship between the induction of RAGE cell-surface antigen and the expression of amyloid binding sites. *J. Mol. Neurosci.* **2003**, *20*, 223–232. [CrossRef]

43. Deane, R.; Du Yan, S.; Submamaryan, R.K.; LaRue, B.; Jovanovic, S.; Hogg, E.; Welch, D.; Manness, L.; Lin, C.; Yu, J.; et al. RAGE mediates amyloid-beta peptide transport across the blood-brain barrier and accumulation in brain. *Nat. Med.* **2003**, *9*, 907–913. [CrossRef] [PubMed]

44. Logsdon, C.D.; Fuentes, M.K.; Huang, E.H.; Arumugam, T. RAGE and RAGE ligands in cancer. *Curr. Mol. Med.* **2007**, *7*, 777–789. [CrossRef] [PubMed]

45. Verkhratsky, A.; Olabarria, M.; Noristani, H.N.; Yeh, C.-Y.; Rodriguez, J.J. Astrocytes in Alzheimer's Disease. *Neurotherapeutics* **2010**, *7*, 399–412. [CrossRef] [PubMed]

46. Veeraraghavalu, K.; Zhang, C.; Zhang, X.; Tanzi, R.E.; Sisodia, S.S. Age-dependent non-cell-autonomous deposition of amyloid from synthesis of β-amyloid by cells other than excitatory neurons. *J. Neurosci.* **2014**, *34*, 3668–3673. [CrossRef]

47. Frost, G.R.; Li, Y.M. The role of astrocytes in amyloid production and Alzheimer's disease. *Open Biol.* **2017**, *7*, 170228. [CrossRef] [PubMed]

48. Teich, A.F.; Patel, M.; Arancio, O. A reliable way to detect endogenous murine β-amyloid. *PLoS ONE* **2013**, *8*, e55647. [CrossRef]

49. Okereke, O.I.; Xia, W.; Irizarry, M.C.; Sun, X.; Qiu, W.Q.; Fagan, A.M.; Mehta, P.D.; Hyman, B.T.; Selkoe, D.J.; Grodstein, F. Performance characteristics of plasma amyloid-beta 40 and 42 assays. *J. Alzheimers Dis.* **2009**, *16*, 277–285. [CrossRef]

50. Aluise, C.D.; Sowell, R.A.; Butterfield, D.A. Peptides and proteins in plasma and cerebrospinal fluid as biomarkers for the prediction, diagnosis, and monitoring of therapeutic efficacy of Alzheimer's disease. *Biochim. Biophys. Acta* **2008**, *1782*, 549–558. [CrossRef]

51. Stewart, K.L.; Radford, S.E. Amyloid plaques beyond Aβ: A survey of the diverse modulators of amyloid aggregation. *Biophys. Rev.* **2017**, *9*, 405–419. [CrossRef]

52. Ganesan, A.; Debulpaep, M.; Wilkinson, H.; Van Durme, J.; De Baets, G.; Jonckheere, W.; Ramakers, M.; Ivarsson, Y.; Zimmermann, P.; Van Eldere, J.; et al. Selectivity of aggregation-determining interactions. *J. Mol. Biol.* **2015**, *427*, 236–247. [CrossRef]

53. Bolognesi, B.; Tartaglia, G.G. Physicochemical principles of protein aggregation. *Prog. Mol. Biol. Transl. Sci.* **2013**, *117*, 53–72. [CrossRef] [PubMed]

54. Valenta, L.J.; Michel-Bechet, M.; Mattson, J.C.; Singer, F.R. Microfollicular thyroid carcinoma with amyloid rich stroma, resembling the medullary carcinoma of the thyroid (MCT). *Cancer* **1977**, *39*, 1573–1586. [CrossRef]

55. Khan, I.S.; Loh, K.S.; Petersson, F. Amyloid and hyaline globules in undifferentiated nasopharyngeal carcinoma. *Ann. Diagn. Pathol.* **2019**, *40*, 1–6. [CrossRef] [PubMed]

56. Franklin, C.D.; Martin, M.V.; Clark, A.; Smith, C.J.; Hindle, M.O. An investigation into the origin and nature of 'amyloid' in a calcifying epithelial odontogenic tumour. *J. Oral Pathol.* **1981**, *10*, 417–429. [CrossRef] [PubMed]

57. Delaney, M.A.; Singh, K.; Murphy, C.L.; Solomon, A.; Nel, S.; Boy, S.C. Immunohistochemical and biochemical evidence of ameloblastic origin of amyloid-producing odontogenic tumors in cats. *Vet. Pathol.* **2013**, *50*, 238–242. [CrossRef] [PubMed]

58. Hirayama, K.; Endoh, C.; Kagawa, Y.; Ohmachi, T.; Yamagami, T.; Nomura, K.; Matsuda, K.; Okamoto, M.; Taniyama, H. Amyloid-Producing Odontogenic Tumors of the Facial Skin in Three Cats. *Vet. Pathol.* **2017**, *54*, 218–221. [CrossRef]

59. Silverman, J.F.; Dabbs, D.J.; Norris, H.T.; Pories, W.J.; Legier, J.; Kay, S. Localized primary (AL) amyloid tumor of the breast. Cytologic, histologic, immunocytochemical and ultrastructural observations. *Am. J. Surg. Pathol.* **1986**, *10*, 539–545. [CrossRef] [PubMed]

60. Mori, M.; Kotani, H.; Sawaki, M.; Hattori, M.; Yoshimura, A.; Gondo, N.; Adachi, Y.; Kataoka, A.; Sugino, K.; Horisawa, N.; et al. Amyloid tumor of the breast. *Surg. Case Rep.* **2019**, *5*, 31. [CrossRef]

61. Rosenzweig, M.; Landau, H. Light chain (AL) amyloidosis: Update on diagnosis and management. *J. Hematol. Oncol.* **2011**, *4*, 47. [CrossRef]

62. Zayas-Santiago, A.; Ríos, D.S.; Zueva, L.V.; Inyushin, M.Y. Localization of αA-Crystallin in Rat Retinal Müller Glial Cells and Photoreceptors. *Microsc. Microanal.* **2018**, *24*, 545–552. [CrossRef] [PubMed]

 International Journal of
Molecular Sciences

Article

Hidden Aggregation Hot-Spots on Human Apolipoprotein E: A Structural Study

Paraskevi L. Tsiolaki [†], Aikaterini D. Katsafana [†], Fotis A. Baltoumas, Nikolaos N. Louros and Vassiliki A. Iconomidou *

Section of Cell Biology and Biophysics, Department of Biology, National and Kapodistrian University of Athens, Panepistimiopolis, Athens 15701, Greece; etsiolaki@biol.uoa.gr (P.L.T.); k.katsafana@gmail.com (A.D.K.); fbaltoumas@biol.uoa.gr (F.A.B.); nlouros@biol.uoa.gr (N.N.L.)
* Correspondence: veconom@biol.uoa.gr; Tel.: +30-210-7274871; Fax: +30-210-7274254
† These authors contributed equally to this work.

Received: 10 April 2019; Accepted: 6 May 2019; Published: 8 May 2019

Abstract: Human apolipoprotein E (apoE) is a major component of lipoprotein particles, and under physiological conditions, is involved in plasma cholesterol transport. Human apolipoprotein E found in three isoforms (E2; E3; E4) is a member of a family of apolipoproteins that under pathological conditions are detected in extracellular amyloid depositions in several amyloidoses. Interestingly, the lipid-free apoE form has been shown to be co-localized with the amyloidogenic Aβ peptide in amyloid plaques in Alzheimer's disease, whereas in particular, the apoE4 isoform is a crucial risk factor for late-onset Alzheimer's disease. Evidence at the experimental level proves that apoE self-assembles into amyloid fibrilsin vitro, although the misfolding mechanism has not been clarified yet. Here, we explored the mechanistic insights of apoE misfolding by testing short apoE stretches predicted as amyloidogenic determinants by AMYLPRED, and we computationally investigated the dynamics of apoE and an apoE–Aβ complex. Our in vitro biophysical results prove that apoE peptide–analogues may act as the driving force needed to trigger apoE aggregation and are supported by the computational apoE outcome. Additional computational work concerning the apoE–Aβ complex also designates apoE amyloidogenic regions as important binding sites for oligomeric Aβ; taking an important step forward in the field of Alzheimer's anti-aggregation drug development.

Keywords: apolipoprotein E; amyloid fibrils; Alzheimer's disease; Aβ oligomer

1. Introduction

Human mature apolipoprotein E (apoE) is a 299 amino acid glycoprotein [1,2], taking part in most lipoprotein classes, such as chylomicrons, very low-density lipoproteins (VLDL) and high-density lipoproteins (HDL) [3]. It is a member of an apolipoprotein family, along with apoA-I, apoA-II, apoA-IV, ApoC-I, apoC-II, and apoC-III [4,5]. Each apolipoprotein class has distinct functions and participates actively in the formation of specific lipoprotein scaffolds [6]. Human mature apolipoprotein E is primarily synthesized in the liver, where it is found in higher quantities, but it is also a protein of the brain and other tissues [7]. The functional form of the protein is involved in metabolic pathways that are related to plasma cholesterol and triglyceride transport and distribution among the tissues, by interacting with members of the low-density lipoprotein receptor (LDLR) superfamily [8–11].

The *APOE* gene [12], co-localized with the *APOC1* [12,13] and *APOC2* genes [14–16], has three alleles; *APOE2*, *APOE3* and *APOE4* [17,18]. Each allele exhibits distinct frequencies among the human population, with *APOE3* having the highest (approximately 78%) [19–21]. The expression of these alleles results in three main forms of the protein, namely, apoE2, apoE3, and apoE4. Interestingly, the apoE4 isoform is of great importance, since it is reported to be involved in both hereditary and sporadic types of the Alzheimer's disease (AD) [22–24]. The differences among the three forms are restricted in the positions 112 and 158 of the mature polypeptide chain. More specifically, in apoE2, cysteines are located in both positions, whereas in apoE4 there is an arginine in both positions. In apoE3, on the other hand, there is a cysteine in position 112 and an arginine in position 158 [25].

Apolipoprotein E is found in both lipid-bound and lipid-free forms. Lipid-free species are relatively rare and are possibly the result of transient dissociation events during the lipoprotein creation [26–31]. It has not been yet possible for any lipid-free form of apoE to be crystallized in the monomeric form, due to its tendency to assemble in tetramers or octamers [32]. A nuclear magnetic resonance (NMR) structure, with the addition of several mutations, successfully determined the three-dimensional conformation of an apoE lipid-free monomer [33]. According to the model, supported by the experimental outcome of the NMR structure, apoE has three structural domains: the N-terminal domain (Figure 1a, green), the C-terminal domain (Figure 1a, blue), and the hinge domain (Figure 1a, red). The monomer connectivity includes the association of the N-terminal domain (residues 1–167) [34,35] with the C-terminal domain (residues 206–299) [36] through a short interim hinge domain (residues 168–205) [33]. Part of the N-terminal domain adopts a four-helix bundle conformation, which is proposed to be the domain buried in the interior of the lipid-free particle [33] (Figure 1a, green).

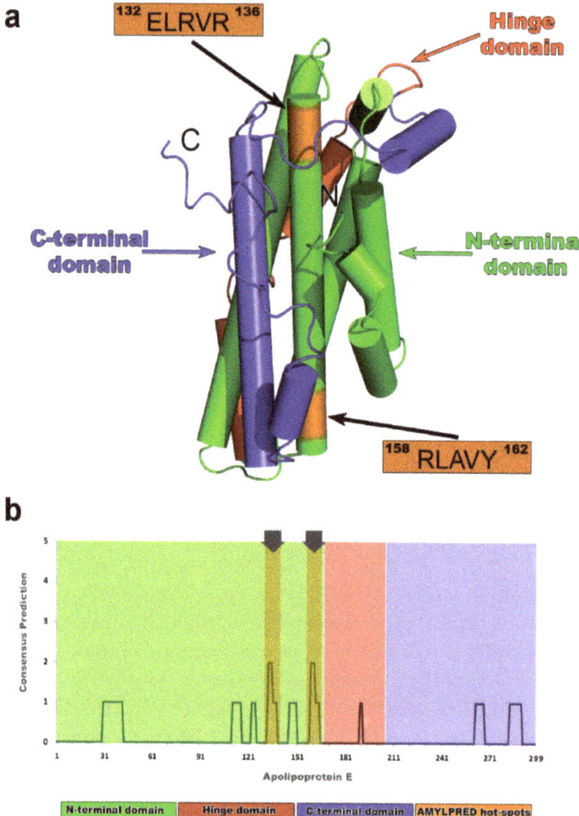

Figure 1. Native nuclear magnetic resonance (NMR) structure of human mature apolipoprotein E (apoE) [33] and apoE amyloidogenic profile by AMYLPRED [37]. (**a**) Different colors show all three structural domains of the apoE3 in solution: the N-terminal domain (green);the C-terminal domain (blue); and the hinge domain (red). Colored regions in orange illustrate "aggregation-prone" segments

^{132}ERLVR^{136} and ^{158}RLAVY^{162}, respectively, both located on the 4th helix of the four-helix bundle. (**b**) Amyloid propensity apoE histogram represents a weak overall amyloidogenicity, since only two segments exceed the consensus AMYLPRED threshold (regions ^{132}ERLVR^{136} and ^{158}RLAVY^{162}). Color scheme follows the rules described in (**a**).

Lipid-free apolipoproteins related to apoE are implicated with several amyloidosis [38] as a result of their proneness to misfold [39]. ApoE self-accumulation properties are still poorly understood, although—as mentioned above—the APOE4 allele is known as a causative risk factor for the neurodegenerative AD [40,41]. ApoE has been characterized as a potential Aβ chaperone in AD, suggesting the strong tendency between these two macromolecules to interact. Interestingly, apoE misfolding was proposed as the first step towards Aβ nucleation and polymerization. In any case, the outstanding appearance of apoE in AD and other neurodegenerative diseases is attributed to the fact that lipid transport in cerebrospinal fluid (CSF) is mediated by HDL particles rich in apoE [42–44].

In the context of the "amyloid stretch hypothesis", which proposes that amyloidogenesis is actually driven by short fragments of misfolded proteins [45], scientists have extensively been studying a variety of short aggregation-prone stretches, with a potential to guide amyloid fibril formation from a soluble globular domain [46–53]. Based on this idea, many algorithms have been developed, in an attempt to extract the information of amyloidogenicity only from primary protein sequences [54]. Among them, AMYLPRED, a consensus prediction algorithm developed in our lab [37], was used to identify regions with amyloidogenic properties in the amino acid sequence of apoE (Figure 1b). The ultimate aim of the present study was to characterize the amyloidogenic properties of apoE3—the most common form in human population. For this purpose, we have used a combination of TEM, X-rays, polarizing microscopy ATR-FTIR spectroscopy, and molecular dynamics simulations to test whether the predicted apoE fragments can influence aggregation of either apoE or the oligomeric Aβ interacting partner. Our biophysical approach indicates that two aggregation-prone apoE hot-spots (Figure 1a, peptides ^{132}ELRVR^{136} and ^{158}RLAVY^{162} shown in orange) have strong self-association properties and destabilize the apoE lipid-free topology. Further, molecular details of the interaction between apoE and oligomeric Aβ, derived by our computational results, also profile the impact of hidden amyloidogenic apoE regions in AD.

2. Results and Discussion

2.1. Computational Identification of apoE Hot-Spots

After a computational scanning, AMYLPRED revealed a weak overall amyloidogenic tendency for apoE, in contrast to other amyloidogenic apolipoproteins studied before [55,56]. The consensus prediction recognized two regions of apoE, namely, ^{133}LRV^{135} and ^{159}LAV^{161}, as peptides with aggregation potency that exceeds the AMYLPRED threshold (Figure 1b). Both peptides werelocated in the same α-helix corresponding to the N-terminal four-helix bundle domain (Figure 1a, orange). According to AMYLPRED, predicted aggregation hot-spots were only found in the helix bundle of apoE that includedthe primary binding epitope for both lipids and Aβ [57], although previous in vitro aggregation assays revealed the C-terminal part as the most amyloidogenic apoE domain [58]. Arginine 112, rendering ApoE4 the least stable apoE isoform [23,59], does not affect the amyloidogenic profile of different apoEs. Analogous hot-spots traced in all apoE forms since the ^{133}LRV^{135} peptide is a commonly predicted segment for all three apoEs, while the ^{159}LAV^{161} was found only in the apoE3 and apoE4 isoforms (Figure S1). The ^{133}LRV^{135} peptide is an important functional region, since it is neighbouring to the LDL receptor binding domain of the molecule [34]. It has been suggested that the C-terminal apoE domain dissociates causing exposure of the four-helix bundle of apoE [33]. This finding is in good agreement with our prediction and verifies the idea that aggregation-prone regions are not buried [37]. We hypothesize that a critical apoE conformational transition can uncover both ^{133}LRV^{135} and ^{159}LAV^{161} aggregation-prone segments, and thus, can initiate apoE misfolding (See MD results below). In this study, predicted regions were extended from both ends, following the idea that five-residue-long peptides are sufficient to independently form amyloid-like fibrils [60], and thus, [132]ELRVR[136] and [158]RLAVY[162] pentapeptide–analogues were experimentally used to pinpoint segments that play crucial role in the self-assembly process of apoE and in the molecular recognition of Aβ.

2.2. Isolated apoE Peptide–Analogues Fulfill All Basic Amyloid Criteria

Designed apoE peptide–analogues [132]ELRVR[136] and [158]RLAVY[162] were thoroughly examined and found to self-assemble, forming fibril-containing gels after an incubation period of one week. As observed by negative staining TEM, both [132]ERLVR[136] and [158]RLAVY[162] fibrillar populations were measured to have similar diameters (Figure 2a,b). The thinnest single fibril of the [132]ELRVR[136] peptide–analogue hadan average diameter of 100 Å, whereas the [158]RLAVY[162] peptide thickness wasapproximately 110 Å. However, the overall arrangement of the fibrils in each gel seems to differ between the two peptides, possibly owing to differences between the peptide–peptide interactions, acting as building blocks of the fibrillar core [61]. Congo red was shown to selectively bind on thin hydrated films derived by both peptides, as seen under bright field illumination. The characteristic yellow/green birefringence wasclearly seen under crossed polars of a polarizing microscope (Figure 2c,d).

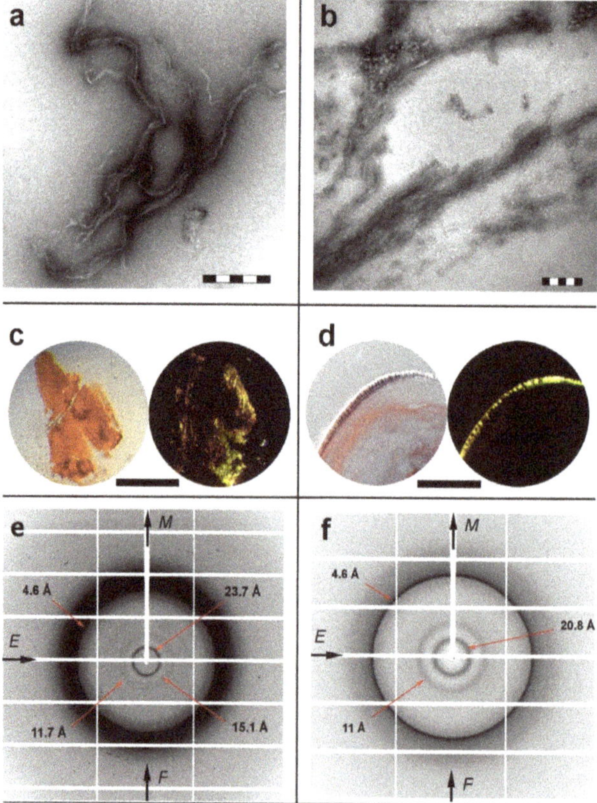

Figure 2. Experimental results of self-aggregation assays for apoE peptide–analogues. (**a**,**b**) Electron micrographs of typical amyloid fibrils, derived by self-assembly of (**a**) [132]ERLVR[136] and (**b**) [158]RLAVY[162] "aggregation-prone" fragments. Scalebars for (**a**) [132]ERLVR[136] and (**b**) [158]RLAVY[162] are 200 nm and

500 nm, respectively. (**c,d**) Photomicrographs of apoE peptide fibrils stained with the amyloid specific Congo red dye ((**c**) [132]ERLVR[136] and (**d**) [158]RLAVY[162]). The apple-green birefringence, characteristic for all amyloid fibrillar materials, is clearly seen (Scale bar 500 μm).(**e,f**) X-ray diffraction patterns from oriented fibers of apoE "aggregation-prone" fragments, (**e**) [132]ERLVR[136] and (**f**) [158]RLAVY[162].

X-ray fiber diffraction and FT-IR experiments have all shown that in their fibrillary form both peptides adopt a well-defined β-sheet conformation. The X-ray patterns indicate that fibrils from both the [132]ELRVR[136] and [158]RLAVY[162] peptide–analogues possess the typical "cross-β" architecture of amyloid fibrils (Figure 2e,f). Concerning the [132]ELRVR[136], a strong -but diffuse- 4.6 Å reflection is seen in the diffraction pattern, in addition to an 11.7 Å structural repeat. The former reflections may be attributed to the periodic distance between consecutive hydrogen-bonded β-strands, which are aligned perpendicular to the fiber axis, and the repetitive distance between packed β-sheets aligned parallel to the fiber axis, respectively. In addition to the typical "cross-β" repetitions, a reflection measured at 23.7 Å could be indicative of the inter-sheet distance (half of the 23.7 Å is approximately 11.7 Å), indicating a long-range order of packed β-sheets in the fiber. Finally, the reflection at 15.1 Å may be attributed to the length of the extended [132]ERLVR[136]peptide. The respective reflections in the diffraction pattern of the [158]RLAVY[162] peptide were measured to be at 4.6 Å, representing the repetitive interchain distance between β-strands and 11 Å, corresponding to the inter-sheet stacking periodicity, both closely resembling typical "cross-β" patterns taken from amyloid fibrils. An additional spacing at 20.8 Å is the evidence for the distance between ordered and packed β-sheets (half of the 20.8 Å is approximately 11 Å). Reflections were also verified utilizing ZipperDB [62] models that overlap with [132]ELRVR[136] and [158]RLAVY[162] peptide–analogues (data not shown). ATR FT-IR was subsequently used to access the secondary structure characteristics of both peptides and to verify the results derived by X-rays. An ATR FT-IR spectrum of a thin-film cast from suspensions of the amyloid-like fibrils of the peptide–analogue[132]ERLVR[136] (Figure 3a) shows prominent bands at 1627 cm[-1]and 1539 cm[-1], in the amide I and II regions, respectively, indicating the presence of β-sheets. A band at 1695 cm[-1]is indicative of anti-parallel β-sheets (Table 1). Similarly, in the spectrum of [158]RLAVY[162] (Figure 3b), the bands at 1631 cm[-1] (amide I) and 1548 cm[-1](amide II) are also attributed to β-sheets, whereas the band at 1689 cm[-1] is attributed to anti-parallel β-sheets (Table 1).

Our experimental analysis reveals that apoE peptide–analogues [132]ELRVR[136] and [158]RLAVY[162] hada strong propensity to independently form β-aggregates, fulfilling the basic amyloid criteria. This finding is compatible with the proposed apoE aggregation pathway suggesting that a minor apoE fraction forms β-strands that stabilize the apoE fibril core [63].

Table 1. Bands observed in the ATR FT-IR spectra obtained from thin films, containing suspensions of fibrils, produced by apoE peptide–analogues, and their tentative assignments.

Wavenumber (cm^{-1})		Assignment
[132]ERLVR[136]	[158]RLAVY[162]	
1134	1137	TFA
1182	1184	TFA
1201	1201	TFA
-	1514	Tyrosine
1539	1548	β-sheet (Amide II)
1627	1631	β-sheet (Amide I)
1666	1666	TFA
1695	1689	Antiparallel β-sheets

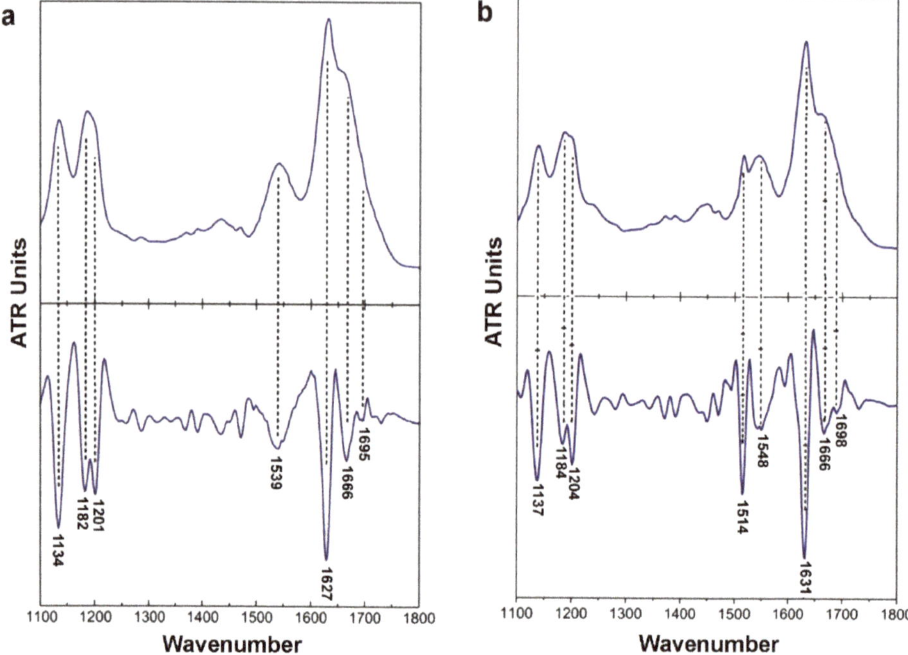

Figure 3. FT-IR spectra (1100–1800 cm^{-1}) derived from suspensions of fibrils, produced from (**a**) ^{132}ERLVR^{136} and (**b**) ^{158}RLAVY^{162}. Each apoE peptide cast on a flat stainless-steel plate and left to air-dry slowly at ambient conditions to form hydrated, thin films. Each film possesses a β-sheet conformation, as it is evident by the presence of strong amide I and II bands.

2.3. Implication of apoE Peptide–Analogues in the Tertiary Structural Stability of apoE

Molecular dynamics simulations were carried out on the most representative NMR conformer of apoE3 [33], putting the spotlight on the implication of the experimentally tested amyloidogenic peptide–analogues ^{132}ELRVR^{136} and ^{158}RLAVY^{162}. Computational tests assessed the structural stability, integrity, and dynamic behavior of apoE over time (300 ns) under physiological pH conditions at 300 K. Structural movements were monitored over the course of the simulations through time-dependent root mean square deviation (RMSD) measurements with respect to the starting configuration, to evaluate apoE overall structural transitions, as well as through per-residue root mean square fluctuation (RMSF) calculations to monitor the mobility of specific regions.

Fibril-forming segments (^{132}ELRVR^{136} and ^{158}RLAVY^{162}) influence the apoE structural features over time, since a noticeable difference found between the starting conformation (Figure 4, 0 ns frames) and the 300 ns conformation (Figure 4, 300 ns frames). The N-terminal domain kept its bundle-structure throughout the simulation, and only a slight conformational tilt was observed in the 3D shape of the molecule (Figure 4, 300 ns frames). Conversely, the C-terminal domain was characterized by large fluctuations (8–10 Å) with respect to the N-terminal domain, possibly due to the higher solvent exposure (Figure S2a, blue curve). Root mean square fluctuation calculations reveal approximately 10 Å deviation between residues Glu270 and His299, corresponding to the C-terminal domain (Figure S3). This result is common for apoE since similar conformational changes allow the four-helix bundle to emerge during lipid binding [33]. It is also believed that C-terminal fluctuations allow new interactions or α-helix to β-sheet conversion, due to partial destabilization of apoE, subsequently resulting in self-assembling. In either case, conformational instability of the C-terminal domain exposes aggregation-prone segments ^{132}ELRVR^{136} and ^{158}RLAVY^{162}, otherwise hidden in the core of apoE. The overall conformational variations of segments ^{132}ELRVR^{136} and ^{158}RLAVY^{162} are visually inspected in Figure S2. ^{158}RLAVY^{162} exhibited higher conformational mobility, meaning that this segment participatedin transient C-terminal conformational changes (Figure S2b, orange triangles). The conformational unraveling of the most aggregation-prone part of apoE (according to AMYLPRED, Figure 1a) explains the intrinsic apoE propensity to form amyloid-like fibrils [63]. Our aggregation assays in combination with computational MD results suggest that the C-terminal domain protects the aggregation-prone part of apoE from misfolding, by covering the aggregation-prone regions ^{132}ELRVR^{136} and ^{158}RLAVY^{162} located at the N-terminal domain. This finding is in agreement with the computational analysis by Das and Gursky [55].

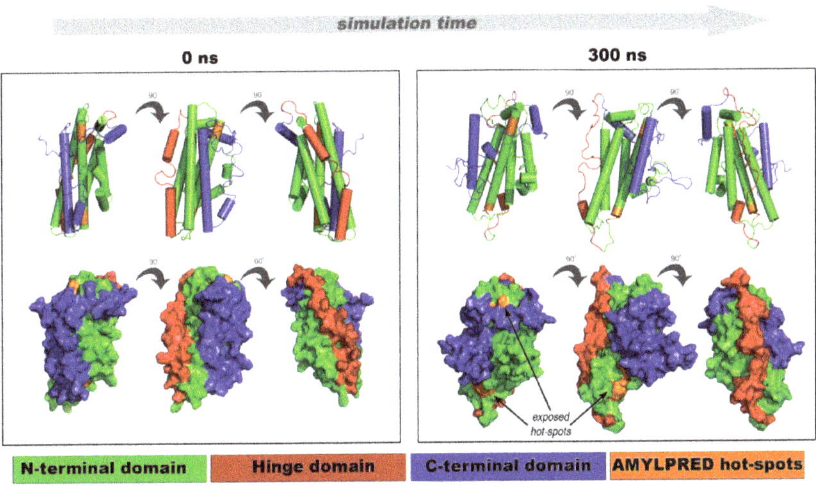

Figure 4. Dynamics simulations of an apoE NMR structure for 300 ns. The N-terminal domain is shown in green, the C-terminal domain is shown in blue, and the hinge domain is shown in red. "Aggregation-prone" hot-spots ^{132}ERLVR^{136} and ^{158}RLAVY^{162} are colored in orange. Structural movements uncover otherwise hidden apoE hot-spots (arrows). Models are represented in 0°, 90°, and 180°.

2.4. An apoE Aggregation Hot-Spot Anchors Oligomeric Aβ

Numerous studies demonstrate that apoE is a component of peripheral deposits and senile plaques of AD patients [64–66]. In vitro experiments have shown that co-incubation of apoE3 and apoE4 with Aβ peptide induces the fibrillation of the peptide [67,68], supporting the idea that apoE specifically interacts with the Aβ. One mechanism by which apoE might be involved in the pathology of AD is by modulating the activity of Aβ and binding in oligomeric Aβ species [69,70]. Molecular docking was employed towards the identification of the Aβ and apoE epitopes in the Aβ–apoE complex. The Aβ aggregation profile was analyzed utilizing AMYLPRED (Figure S4). Two aggregation-prone regions were predicted comprising an N-terminal pentapeptide (KLVFFA) and a longer C-terminal thirteen-residue-long peptide (GAIIGLMVGGVVI) (Figure S4). Previous experimental studies have shown that both regions have self-aggregation properties and have been suggested as crucial amyloidogenic determinants of Aβ [71].

Supervised molecular docking was performed for building the Aβ–apoE complex using the 300 ns apoE conformation as the initial structure for the N-terminal apoE domain (described above) and the 2BEG NMR structure [72] as the 3D structure of Aβ oligomers. The apoE binding epitope was restricted between residues 130 to 165, based on reliable information from experimental and computational studies, pinpointing this apoE domain as the major interacting part of apoE withAβ [57]. Predicted Aβ aggregation-prone regions were used as computational restraints in HADDOCK. The identification of the complex ascertained the interaction between the C-terminal aggregation-prone epitope of Aβ and the amyloidogenic ^{132}ELRVR^{136} peptide, located at the N-terminal apoE domain. This cluster evaluated having the best HADDOCK score, whichcorresponds to the smallest weighted HADDOCK sum (Figure S5, 0 ns).

Having investigated the most favorable Aβ–apoE complex, the next step was to evaluate its dynamics and stability. After 100 ns all-atom MD simulations, a complex dissociation was observed and the structure of the Aβ oligomer changed. Similar results were observed after 200 and 300 ns simulation time (Figure S5). Despite the secondary structure alterations, the interaction interface between Aβ and apoE retained over time. This verifies recursively the spatial position emerged from the molecular docking model (Figure S5, 0 ns). The structure of the apoE N-terminal domain was recorded as the most stable entity, since this domain displayed similar dynamic behavior over time. This behavior is consistent with the simulation results observed for the representative NMR conformer of full-length apoE3, presented above (Please refer to Section 2.3). Except for significant changes in Aβ orientation and stability, the Aβ C-terminal epitope remained constantly attached to the amyloidogenic ^{132}ELRVR^{136} peptide over ns time. The Aβ oligomer's instability wasreasonable, since oligomeric states are significantly unstable compared to the amyloid state of proteins [73]. Given the competitive relationship between lipid-free apoE molecules tending to self-assemble, and Aβ oligomers "willing" to interact with apoE monomers [74,75], we hypothesized that these computational results give new insights into Aβ–apoE's delicate interconnection. This computational outcome provides some details into the intermolecular and intramolecular interactions, associated with the formation of homomeric or heteromeric supramolecular assemblies, which may be the key to target protein misfolding diseases.

3. Materials and Methods

3.1. Identification of Aggregation-Prone Peptides in apoE and Aβ

Human apoE sequence (UniProtKB: P02649/APOE_HUMAN), corresponding to APOE3 allele, and human Amyloid-beta precursor protein (APP) fragment 672–713 (UniProtKB: P05067/A4_HUMAN), namely, Aβ$_{1–42}$, were analyzed with AMYLPRED [37] for identifying fibril-forming aggregation hot-spots. Fibril-forming segments chosen for this study were predicted at least by two predictors (default AMYLPRED threshold). Figure S1 and Figure S4 illustrate the consensus AMYLPRED prediction for all apoE isoforms and Aβ, respectively.

3.2. Peptide Design, Synthesis,and Preparation of Peptide Samples

Based on the amyloidogenic profile of apoE (Figure S1), 2 short pentapeptide–analogues were designed. Since, according to previous studies, sequence stretches in proteins should comprise a minimum of five consecutive residues, and AMYLPRED predictions were extended from both ends. The pentapeptide–analogues ^{132}ELRVR^{136} and ^{158}RLAVY^{162}, corresponding to the 4th helix of the four-helix bundle of apoE (Figure 1, orange), were chemically synthesized in high peptide purity (>98%) by GeneCust© Europe, Luxemburg. Peptide-analogues have free N- and C-terminals. Lyophilized aliquots of both pentapeptides were re-suspended in distilled water (pH 5.5) at concentrations up to 15 mg ml^{-1} and incubated at ambient temperatures for 1–2 weeks. Both pentapeptides were found to produce fibril-containing gels.

3.3. X-ray Diffraction

For each peptide–analogue a droplet (~10µL) of mature fibril suspension was placed between two quartz capillaries covered with wax. Capillaries spaced ~1.5 mm apart and mounted horizontally on a glass substrate, as collinearly as possible, in order to obtain an oriented fiber. The X-ray diffraction pattern from this fiber was collected at a P14 beamline synchrotron (Petra III, EMBL-Hamburg, Germany) operated at a wavelength of 1.23953 Å, with a 6M PILATUS detector. The specimen-to-film distance was set at 225.11 mm and the exposure time was set to 1 s. The X-ray patterns were initially viewed using the program CrysAlisPro [76,77] and subsequently displayed and measured with the aid of the iMosFLM [78] program [78].

3.4. Negative Staining and Transmission Electron Microscopy

For negative staining, droplets (3–5 μL) of the mature fibril suspensions were applied to glow-discharged 400-mesh carbon-coated copper grids for 2 min. The grids were stained with a droplet (5 μL) of 2% (*w/v*) aqueous uranyl acetate for 60 s and the excess staining was removed by blotting with a filter paper. The fibril-containing grids were initially air-dried and subsequently examined with a Morgagni™ 268 transmission electron microscope, operated at 80 kV. Digital acquisitions were performed with an 11-Mpixel side-mounted Morada CCD camera (Soft Imaging System, Muenster, Germany).

3.5. Attenuated Total Reflectance Fourier-Transform Infrared Spectroscopy (ATR FTIR) and Post-Run Computations of the Spectra

A 10-μL droplet of each apoE peptide mature fibril suspension was cast on flat stainless-steel plates, coated with an ultrathin hydrophobic layer (SpectRIM, Tienta Sciences, Inc. Indianapolis, IN, USA) and left to dry slowly at ambient conditions in order to form thin hydrated films. Infrared spectra were obtained from these films at a resolution of 4 cm^{-1}, utilizing an IR microscope (IRScope II by Bruker Optics) equipped with a Ge attenuated total reflectance (ATR) objective lens (20×) and attached to a Fourier-transform infrared (FTIR) spectrometer (Equinox 55, by Bruker Optics). Ten 32-scan spectra were collected from each sample and averaged to improve the sound/noise (S/N) ratio. Both are shown in the absorption mode after correction for the wavelength dependence of the penetration depth (pd~λ). Absorption band maxima were determined from the minima in the second derivative of the corresponding spectra. Derivatives were computed analytically using routines of the Bruker OPUS/OS2 software, including smoothing over a ±13 cm^{-1} range around each data point, performed by the Savitzky–Golay algorithm [79]. Smoothing over narrower ranges resulted in a deterioration of the S/N ratio and did not increase the number of minima that could be determined with confidence.

3.6. Congo Red Staining and Polarized Light Microscopy

Fibril suspensions of the peptide solutions were applied to glass slides and stained with a 10 mM Congo red (Sigma) solution in distilled water (pH 5.5) for ~30 min. Excess staining was removed by several washes with distilled water and left to dry for approximately 10 min. The samples were observed under bright field illumination and between crossed polars, using a Leica MZ_{75} polarizing stereomicroscope, equipped with a JVC GC-X3E camera.

3.7. Molecular Docking and Molecular Dynamics Simulations

For deriving a structural model of the Aβ–apoE complex, the web server version 2.2 of HADDOCK was used [80]. The HADDOCK score was used to rank and evaluate the generated clusters. The scoring was the weighted sum of a linear combination of various energies and buried surface area between molecules constituting the complex. A number of molecular dynamics (MDs) simulations were designed and performed for apoE next, both in its monomeric form and in its complex with Aβ protofibrils. Each protein system was inserted into a cubic solvent box, with a minimum distance of at least 1.5 nm between the box's boundaries and protein coordinates. The solvent was modeled using the TIP3P water model [81] and the systems were ionized using NaCl counter-ions to neutralize unwanted charges and an ambient NaCl ion concentration of 0.15 M, mimicking neutral pH conditions. Each simulation system was subjected to thorough energy minimization, followed by two stages of equilibration simulations with position restraints applied on protein coordinates, namely, a 500 ps simulation in the canonical (NVT) ensemble to equilibrate temperature and a 1 ns simulation in the isothermal-isobaric (NPT) ensemble to equilibrate pressure. An additional 1 ns equilibration simulation was also performed without any restraints. Finally, production simulations were performed in the NPT ensemble for 300 ns.

All simulations were performed using GROMACS v. 2016.3 [82] and the AMBER 99SB-ILDN force field [83]. The LINCS algorithm [84] was applied to model bond constraints, enabling the use of a 2 fs time-step. Short range non-bonded interactions were modeled using a twin-range cutoff at 0.8 nm, while long-range electrostatic interactions were modeled using the Particle Mesh Ewald (PME) method [85], with a Fourier grid spacing at 0.12 nm and a cubic interpolation (PME rank 4). Temperature was maintained at 300 K with separate couplings for the proteins and solvent, using the Berendsen weak coupling algorithm [86] during equilibration and the Nosé–Hoover thermostat [87,88] in the production simulations, with a coupling constant of $\tau_T = 0.1$ ps. Pressure was isotopically controlled at 1.013 bar (1 atm), using the Berendsen weak coupling algorithm [86] during equilibration and the Parrinello–Rahman barostat [89] in the production simulations, with a coupling constant of $\tau_P = 2.0$ ps and a compressibility of 4.5×10^{-5} bar^{-1}. Simulation results were analyzed using various GROMACS utilities, and Visual Molecular Dynamics (VMD) v. 1.9.4 [90]. Pictures were collected with PyMOL [91].

4. Conclusions

The purpose of this study was to investigate the poorly explored amyloidogenic properties of human apolipoprotein E [58,63], a protein closely associated with disorders with worldwide prevalence, such as Alzheimer's disease [23,24]. Wild-type or mutated apolipoproteins, evolutionarily related to apoE, have been found in depositions of amyloid fibrils in vivo in several amyloidosis [92–101]. More specifically, apoCII and apoCIII has been reported to form amyloid fibrils both in vitro [102,103] and recently in vivo, causing rare forms of hereditary systemic amyloidosis [104,105]. These two newly identified fibril proteins expand the list of amyloidogenic apolipoproteins associated with amyloidoses [38] and draw attention to unknown aggregation properties of the apolipoprotein family.

In this work, AMYPRED was used in order to probe hidden amyloidogenic motifs on the apoE polypeptide chain, whereas several biophysical and computational techniques were applied to characterize its properties. The results of our experimental work prove that predicted aggregation-prone apoE peptides self-assemble into amyloid-like fibrillar structures, displaying the main structural and tinctorial features of amyloids [106,107]. Computational tests evaluated the contribution of these peptides into the stability of apoE and explored their affinity to oligomeric Aβ. Molecular dynamic simulations revealed that both predicted apoE peptide–analogues undergo a critical structural transition that under the right in vitro conditions may result in apoE instability. Importantly, the amyloidogenic ^{132}ELRVR^{136} peptide, a commonly predicted segment for all three apoE isoforms, emerged as the most favorable apoE epitope "attracting"the C-terminal epitope of the oligomeric Aβ. Overall, self-aggregation properties of apoE peptides, described here, add considerable further details as they establish a mechanistic explanation of apoE misfolding and involvement with oligomeric Aβ. As an extension to these conclusions, the concept of interacting amyloidogenic regions, by separate partners found in amyloidoses, offers hope of new anti-aggregation treatment directions.

Supplementary Materials: Supplementary materials can be found at http://www.mdpi.com/1422-0067/20/9/2274/s1.

Author Contributions: Conceptualization, P.L.T., V.A.I.; Methodology, P.L.T., F.A.B., N.N.L., V.A.I.; Validation, P.L.T., A.D.K., F.A.B., N.N.L.; Formal analysis, P.L.T., A.D.K.; Investigation, P.L.T., A.D.K., V.A.I.; Resources, V.A.I.; Writing, original draft preparation, P.L.T., A.D.K.; Writing, review and editing, P.L.T., V.A.I.; Visualization, P.L.T., V.A.I.; Supervision, V.A.I.; Funding acquisition, V.A.I.

Funding: The present work was co-funded by the European Union and Greek national funds through the Operational Program "Competitiveness, Entrepreneurship and Innovation", under the call "RESEARCH-CREATE-INNOVATE" (project code: T1EDK-00353).

Acknowledgments: We thank the Institute of Biology, Medicinal Chemistry and Biotechnology at the National Hellenic Research Foundation for access to the X-ray diffraction facility. We acknowledge the help of EvangeliaChrysina with the X-ray diffraction experiments. The help of George Baltatzis and EfstratiosPatsouris and the use of the Morgagni Microscope at the 1st Department of Pathology, Medical School, the National and

Kapodistrian University of Athens is also gratefully acknowledged. We should also like to sincerely thank the two handling editors and the reviewers of this manuscript for their very useful and constructive criticism.

Conflicts of Interest: The authors declare no conflicts of interest.

Abbreviations

AD	Alzheimer's Disease
APP	Amyloid-Beta Precursor Protein
apoE	Apolipoprotein E
CSF	Cerebrospinal Fluid
HDL	High-Density Lipoproteins
LDLR	Low-Density Lipoprotein Receptor
MD	Molecular Dynamics
NMR	Nuclear Magnetic Resonance
RMSD	Root Mean Square Deviation
RMSF	Root Mean Square Fluctuation
TEM	Transmission Electron Microscopy
VLDL	Very Low-Density Lipoproteins

References

1. Rall, S.C.; Weisgraber, K.H.; Mahley, R.W. Human apolipoprotein E. The complete amino acid sequence. *J. Biol. Chem.* **1982**, *257*, 4171–4178.
2. McLean, J.W.; Elshourbagy, N.A.; Chang, D.J.; Mahley, R.W.; Taylor, J.M. Human apolipoprotein E mRNA. cDNA cloning and nucleotide sequencing of a new variant. *J. Biol. Chem.* **1984**, *259*, 6498–6504.
3. Mahley, R. Apolipoprotein E: cholesterol transport protein with expanding role in cell biology. *Science* **1988**, *240*, 622–630. [CrossRef] [PubMed]
4. Luo, C.-C.; Li, W.-H.; Moore, M.N.; Chan, L. Structure and evolution of the apolipoprotein multigene family. *J. Mol. Biol.* **1986**, *187*, 325–340. [CrossRef]
5. Li, W.H.; Tanimura, M.; Luo, C.C.; Datta, S.; Chan, L. The apolipoprotein multigene family: Biosynthesis, structure, structure-function relationships, and evolution. *J. Lipid Res.* **1988**, *29*, 245–271. [PubMed]
6. Mahley, R.W.; Innerarity, T.L.; Rall, S.C.; Weisgraber, K.H. Plasma lipoproteins: apolipoprotein structure and function. *J. Lipid Res.* **1984**, *25*, 1277–1294. [PubMed]
7. Elshourbagy, N.A.; Liao, W.S.; Mahley, R.W.; Taylor, J.M. Apolipoprotein E mRNA is abundant in the brain and adrenals, as well as in the liver, and is present in other peripheral tissues of rats and marmosets. *Proc. Natl. Acad. Sci. USA* **1985**, *82*, 203–207. [CrossRef]
8. Beisiegel, U.; Weber, W.; Ihrke, G.; Herz, J.; Stanley, K.K. The LDL–receptor–related protein, LRP, is an apolipoprotein E-binding protein. *Nat. Cell Biol.* **1989**, *341*, 162–164. [CrossRef]
9. Takahashi, S.; Kawarabayasi, Y.; Nakai, T.; Sakai, J.; Yamamoto, T. Rabbit very low density lipoprotein receptor: a low density lipoprotein receptor-like protein with distinct ligand specificity. *Proc. Natl. Acad. Sci. USA* **1992**, *89*, 9252–9256. [CrossRef]
10. Willnow, T.E.; Goldstein, J.L.; Orth, K.; Brown, M.S.; Herz, J. Low density lipoprotein receptor-related protein and gp330 bind similar ligands, including plasminogen activator-inhibitor complexes and lactoferrin, an inhibitor of chylomicron remnant clearance. *J. Biol. Chem.* **1992**, *267*, 26172–26180. [PubMed]
11. Kim, D.-H.; Iijima, H.; Goto, K.; Sakai, J.; Ishii, H.; Kim, H.-J.; Suzuki, H.; Kondo, H.; Saeki, S.; Yamamoto, T. Human Apolipoprotein E Receptor 2: A Novel Lipoprotein Receptor of the Low Density Lipoprotein Receptor Family Predominantly Expressed in Brain. *J. Biol. Chem.* **1996**, *271*, 8373–8380. [CrossRef]
12. Das, H.K.; McPherson, J.; Bruns, G.A.; Karathanasis, S.K.; Breslow, J.L. Isolation, characterization, and mapping to chromosome 19 of the human apolipoprotein E gene. *J. Biol. Chem.* **1985**, *260*, 6240–6247.
13. Davison, P.; Norton, P.; Wallis, S.; Gill, L.; Cook, M.; Williamson, R.; Humphries, S. There are two gene sequences for human apolipoprotein CI (apo CI) on chromosome 19, one of which is 4 KB from the gene for apo E. *Biochem. Biophys. Commun.* **1986**, *136*, 876–884. [CrossRef]
14. Scott, J.; Knott, T.J.; Shaw, D.J.; Brook, J.D. Localization of genes encoding apolipoproteins CI, CII, and E to the p13?cen region of human chromosome 19. *Hum. Genet.* **1985**, *71*, 144–146. [CrossRef]

15. Humphries, S.E.; Berg, K.; Gill, L.; Cumming, A.M.; Robertson, F.W.; Stalenhoef, A.F.; Williamson, R.; Børresen, A.L. The gene for apolipoprotein C-ll is closely linked to the gene for apolipoprotein E on chromosome 19. *Clin. Genet.* **1984**, *26*, 389–396. [CrossRef]

16. Myklebost, O.; Rogne, S. A physical map of the apolipoprotein gene cluster on human chromosome 19. *Hum. Genet.* **1988**, *78*, 244–247. [CrossRef]

17. Utermann, G.; Langenbeck, U.; Beisiegel, U.; Weber, W. Genetics of the apolipoprotein E-system in man. *Am. J. Hum. Genet.* **1980**, *32*, 339–347. [PubMed]

18. Zannis, V.I.; Breslow, J.L. Human very low density lipoprotein apolipoprotein E isoprotein polymorphism is explained by genetic variation and posttranslational modification. *Biochemistry* **1981**, *20*, 1033–1041. [CrossRef]

19. Hatters, D.M.; Peters-Libeu, C.A.; Weisgraber, K.H. Apolipoprotein E structure: insights into function. *Trends Biochem. Sci.* **2006**, *31*, 445–454. [CrossRef]

20. Poirier, J.; Bertrand, P.; Poirier, J.; Kogan, S.; Gauthier, S.; Poirier, J.; Gauthier, S.; Davignon, J.; Bouthillier, D.; Davignon, J. Apolipoprotein E polymorphism and Alzheimer's disease. *Lancet* **1993**, *342*, 697–699. [CrossRef]

21. Farrer, L.A.; Cupples, L.A.; Haines, J.L.; Hyman, B.; Kukull, W.A.; Mayeux, R.; Myers, R.H.; Pericak-Vance, M.A.; Risch, N.; van Duijn, C.M. Effects of age, sex, and ethnicity on the association between apolipoprotein e genotype and alzheimer disease: A meta-analysis. *JAMA* **1997**, *278*, 1349–1356. [CrossRef] [PubMed]

22. Hauser, P.S.; Ryan, R.O. Impact of apolipoprotein E on Alzheimer's disease. *Curr. Res.* **2013**, *10*, 809–817. [CrossRef]

23. Mahley, R.W.; Weisgraber, K.H.; Huang, Y. Apolipoprotein E4: A causative factor and therapeutic target in neuropathology, including Alzheimer's disease. *Proc. Natl. Acad. Sci. USA* **2006**, *103*, 5644–5651. [CrossRef]

24. Blennow, K.; de Leon, M.J.; Zetterberg, H. Alzheimer's disease. *Lancet* **2006**, *368*, 387–403. [CrossRef]

25. Weisgraber, K.H.; Rall, S.C.; Mahley, R.W. Human E apoprotein heterogeneity. Cysteine-arginine interchanges in the amino acid sequence of the apo-E isoforms. *J. Biol. Chem.* **1981**, *256*, 9077–9083. [PubMed]

26. Innerarity, T.L.; Pitas, R.E.; Mahley, R.W. Binding of arginine-rich (E) apoprotein after recombination with phospholipid vesicles to the low density lipoprotein receptors of fibroblasts. *J. Biol. Chem.* **1979**, *254*, 4186–4190. [PubMed]

27. Peters-Libeu, C.A.; Newhouse, Y.; Hatters, D.M.; Weisgraber, K.H. Model of Biologically Active Apolipoprotein E Bound to Dipalmitoylphosphatidylcholine. *J. Biol. Chem.* **2006**, *281*, 1073–1079. [CrossRef] [PubMed]

28. Hatters, D.M.; Budamagunta, M.S.; Voss, J.C.; Weisgraber, K.H. Modulation of Apolipoprotein E Structure by Domain Interaction: Differences in Lipid-Bound and lipid-Free Forms. *J. Biol. Chem.* **2005**, *280*, 34288–34295. [CrossRef]

29. Narayanaswami, V.; Szeto, S.S.W.; Ryan, R.O. Lipid association-induced N- and C-terminal domain reorganization in human apolipoprotein E3. *J. Biol. Chem.* **2010**. [CrossRef]

30. Narayanaswami, V.; Ryan, R.O. Molecular basis of exchangeable apolipoprotein function. *Biochim. Biophys. Acta* **2000**, *1483*, 15–36. [CrossRef]

31. Peters-Libeu, C.A.; Newhouse, Y.; Hall, S.C.; Witkowska, H.E.; Weisgraber, K.H. Apolipoprotein E*dipalmitoylphosphatidylcholine particles are ellipsoidal in solution. *J. Lipid Res.* **2007**, *48*, 1035–1044. [CrossRef]

32. Patel, A.B.; Khumsupan, P.; Narayanaswami, V. Pyrene Fluorescence Analysis Offers New Insights into the Conformation of the Lipoprotein-Binding Domain of Human Apolipoprotein, E. *Biophys. J.* **2010**, *98*, 23a. [CrossRef]

33. Chen, J.; Li, Q.; Wang, J. Topology of human apolipoprotein E3 uniquely regulates its diverse biological functions. *Proc. Natl. Acad. Sci. USA* **2011**, *108*, 14813–14818. [CrossRef]

34. Wilson, C.; Wardell, M.; Weisgraber, K.; Mahley, R.; Agard, D. Three-dimensional structure of the LDL receptor-binding domain of human apolipoprotein E. *Science* **1991**, *252*, 1817–1822. [CrossRef]

35. Wilson, C.; Mau, T.; Weisgraber, K.H.; Wardell, M.R.; Mahley, R.W.; Agard, D.A. Salt bridge relay triggers defective LDL receptor binding by a mutant apolipoprotein. *Structure* **1994**, *2*, 713–718. [CrossRef]

36. Forstner, M.; Peters-Libeu, C.; Contreras-Forrest, E.; Newhouse, Y.; Knapp, M.; Rupp, B.; Weisgraber, K.H. Carboxyl-Terminal Domain of Human Apolipoprotein E: Expression, Purification, and Crystallization. *ProteinExpr. Purif.* **1999**, *17*, 267–272. [CrossRef] [PubMed]

37. Frousios, K.K.; Iconomidou, V.A.; Karletidi, C.-M.; Hamodrakas, S.J. Amyloidogenic determinants are usually not buried. *BMC Struct. Biol.* **2009**, *9*, 44. [CrossRef] [PubMed]

38. Sipe, J.D.; Benson, M.D.; Buxbaum, J.N.; Ikeda, S.I.; Merlini, G.; Saraiva, M.J.; Westermark, P. Amyloid fibril proteins and amyloidosis: chemical identification and clinical classification International Society of Amyloidosis 2016 Nomenclature Guidelines. *Amyloid* **2016**, *23*, 209–213. [CrossRef]

39. Hatters, D.M.; Howlett, G.J. The structural basis for amyloid formation by plasma apolipoproteins: A review. *Eur. Biophys. J.* **2002**, *31*, 2–8. [CrossRef]

40. Corder, E.; Saunders, A.; Strittmatter, W.; Schmechel, D.; Gaskell, P.; Small, G.; Roses, A.; Haines, J.; Pericak-Vance, M. Gene dose of apolipoprotein E type 4 allele and the risk of Alzheimer's disease in late onset families. *Science* **1993**, *261*, 921–923. [CrossRef]

41. Roses, A.D. Apolipoprotein E genotyping in the differential diagnosis, not prediction, of Alzheimer's disease. *Ann. Neurol.* **1995**, *38*, 6–14. [CrossRef]

42. Strittmatter, W.J.; Saunders, A.M.; Schmechel, D.; Pericak-Vance, M.; Enghild, J.; Salvesen, G.S.; Roses, A.D. Apolipoprotein E: High-avidity binding to beta-amyloid and increased frequency of type 4 allele in late-onset familial Alzheimer disease. *Proc. Natl. Acad. Sci. USA* **1993**, *90*, 1977–1981. [CrossRef]

43. Han, S.H.; Einstein, G.; Weisgraber, K.H.; Strittmatter, W.J.; Saunders, A.M.; Pericak-Vance, M.; Roses, A.D.; Schmechel, D.E. Apolipoprotein E is localized to the cytoplasm of human cortical neurons: A light and electron microscopic study. *J. Neuropathol. Exp. Neurol.* **1994**, *53*, 535–544. [CrossRef] [PubMed]

44. Diedrich, J.F.; Minnigan, H.; Carp, R.I.; Whitaker, I.N.; Race, R.; Frey, W.; Hazse, A.T. Neuropathological changes in scrapIe and AlzheImer's disease are associated with increased expression of apolipoprotein E and cathepsin D in astrocytes. *J. Virol.* **1991**, *65*, 4759–4768.

45. Esteras-Chopo, A.; Serrano, L.; de la Paz, M.L. The amyloid stretch hypothesis: Recruiting proteins toward the dark side. *Proc. Natl. Acad. Sci. USA* **2005**, *102*, 16672–16677. [CrossRef]

46. Iconomidou, V.A.; Pheida, D.; Hamodraka, E.S.; Antony, C.; Hoenger, A.; Hamodrakas, S.J. An amyloidogenic determinant in N-terminal pro-brain natriuretic peptide (NT-proBNP): Implications for cardiac amyloidoses. *Biopolymers* **2012**, *98*, 67–75. [CrossRef]

47. Iconomidou, V.A.; Leontis, A.; Hoenger, A.; Hamodrakas, S.J. Identification of a novel 'aggregation-prone'/ 'amyloidogenic determinant' peptide in the sequence of the highly amyloidogenic human calcitonin. *FEBS Lett.* **2013**, *587*, 569–574. [CrossRef]

48. Louros, N.N.; Iconomidou, V.A.; Tsiolaki, P.L.; Chrysina, E.D.; Baltatzis, G.E.; Patsouris, E.S.; Hamodrakas, S.J. An N-terminal pro-atrial natriuretic peptide (NT-proANP) 'aggregation-prone' segment involved in isolated atrial amyloidosis. *FEBS Lett.* **2014**, *588*, 52–57. [CrossRef]

49. Teng, P.K.; Eisenberg, D. Short protein segments can drive a non-fibrillizing protein into the amyloid state. *Protein Eng. Sel.* **2009**, *22*, 531–536. [CrossRef]

50. Tenidis, K.; Waldner, M.; Bernhagen, J.; Fischle, W.; Bergmann, M.; Weber, M.; Merkle, M.-L.; Voelter, W.; Brunner, H.; Kapurniotu, A. Identification of a penta- and hexapeptide of islet amyloid polypeptide (IAPP) with amyloidogenic and cytotoxic properties11Edited by R. Huber. *J. Mol. Biol.* **2000**, *295*, 1055–1071. [CrossRef]

51. López de la Paz, M.; Serrano, L. Sequence determinants of amyloid fibril formation. *Proc. Natl. Acad. Sci. USA* **2004**, *101*, 87–92. [CrossRef]

52. Tsiolaki, P.L.; Louros, N.N.; Hamodrakas, S.J.; Iconomidou, V.A. Exploring the 'aggregation-prone' core of human Cystatin C: A structural study. *J. Struct. Biol.* **2015**, *191*, 272–280. [CrossRef] [PubMed]

53. Louros, N.; Tsiolaki, P.L.; Zompra, A.A.; Pappa, E.V.; Magafa, V.; Pairas, G.; Cordopatis, P.; Cheimonidou, C.; Trougakos, I.P.; Iconomidou, V.A.; et al. Structural studies and cytotoxicity assays of "aggregation-prone" IAPP 8-16 and its non-amyloidogenic variants suggest its important role in fibrillogenesis and cytotoxicity of human amylin. *Biopolymers* **2015**, *104*, 196–205. [CrossRef] [PubMed]

54. Conchillo-Solé, O.; De Groot, N.S.; Aviles, F.X.; Vendrell, J.; Daura, X.; Ventura, S. AGGRESCAN: A server for the prediction and evaluation of "hot spots" of aggregation in polypeptides. *BMC Bioinform.* **2007**, *8*, 65. [CrossRef]

55. Das, M.; Gursky, O. Amyloid-Forming Properties of Human Apolipoproteins: Sequence Analyses and Structural Insights. *Adv. Exp. Med. Biol.* **2015**, *855*, 175–211. [PubMed]

56. Louros, N.N.; Tsiolaki, P.L.; Griffin, M.D.; Howlett, G.J.; Hamodrakas, S.J.; Iconomidou, V.A. Chameleon 'aggregation-prone' segments of apoA-I: A model of amyloid fibrils formed in apoA-I amyloidosis. *Int. J. Biol. Macromol.* **2015**, *79*, 711–718. [CrossRef] [PubMed]

57. Winkler, K.; Scharnagl, H.; Tisljar, U.; Hoschützky, H.; Friedrich, I.; Hoffmann, M.M.; Hüttinger, M.; Wieland, H.; März, W. Competition of Abeta amyloid peptide and apolipoprotein E for receptor-mediated endocytosis. *J. Lipid Res.* **1999**, *40*, 447–455. [PubMed]

58. Wisniewski, T.; Lalowski, M.; Golabek, A.; Frangione, B.; Vogel, T. Is Alzheimer's disease an apolipoprotein E amyloidosis? *Lancet* **1995**, *345*, 956–958. [CrossRef]

59. Weisgraber, K.H. Apolipoprotein E distribution among human plasma lipoproteins: Role of the cysteine-arginine interchange at residue 112. *J. Lipid Res.* **1990**, *31*, 1503–1511.

60. Azriel, R.; Gazit, E. Analysis of the Minimal Amyloid-forming Fragment of the Islet Amyloid Polypeptide: An Experimental Support for The Key Role of The Phenylalanine Residue in Amyloid Formation. *J. Biol. Chem.* **2001**, *276*, 34156–34161. [CrossRef]

61. Close, W.; Neumann, M.; Schmidt, A.; Hora, M.; Annamalai, K.; Schmidt, M.; Reif, B.; Schmidt, V.; Grigorieff, N.; Fändrich, M. Physical basis of amyloid fibril polymorphism. *Nat. Commun.* **2018**, *9*, 699. [CrossRef] [PubMed]

62. Sawaya, M.R.; Sambashivan, S.; Nelson, R.; Ivanova, M.I.; Sievers, S.A.; Apostol, M.I.; Thompson, M.J.; Balbirnie, M.; Wiltzius, J.J.; McFarlane, H.T.; et al. Atomic structures of amyloid cross-beta spines reveal varied steric zippers. *Nature* **2007**, *447*, 453–457. [CrossRef] [PubMed]

63. Hatters, D.M.; Zhong, N.; Rutenber, E.; Weisgraber, K.H. Amino-terminal Domain Stability Mediates Apolipoprotein E Aggregation into Neurotoxic Fibrils. *J. Mol. Biol.* **2006**, *361*, 932–944. [CrossRef] [PubMed]

64. Namba, Y.; Tomonaga, M.; Kawasaki, H.; Otomo, E.; Ikeda, K. Apolipoprotein E immunoreactivity in cerebral amyloid deposits and neurofibrillary tangles in Alzheimer's disease and kuru plaque amyloid in Creutzfeldt-Jakob disease. *Brain Res.* **1991**, *541*, 163–166. [CrossRef]

65. Wisniewski, T.; Frangione, B. Apolipoprotein E: A pathological chaperone protein in patients with cerebral and systemic amyloid. *Neurosci. Lett.* **1992**, *135*, 235–238. [CrossRef]

66. Choi-Miura, N.H.; Takahashi, Y.; Nakano, Y.; Tobe, T.; Tomita, M. Identification of the Disulfide Bonds in Human Plasma Protein SP-40,40 (Apolipoprotein-J)1. *J. Econ. Èntomol.* **1992**, *112*, 557–561. [CrossRef]

67. Ma, J.; Yee, A.; Brewer, H.B.; Das, S.; Potter, H. Amyloid-associated proteins [alpha]1-antichymotrypsin and apolipoprotein E promote assembly of Alzheimer [beta]-protein into filaments. *Nature* **1994**, *372*, 92–94. [CrossRef]

68. Sanan, D.A.; Weisgraber, K.H.; Russell, S.J.; Mahley, R.W.; Huang, D.; Saunders, A.; Schmechel, D.; Wisniewski, T.; Frangione, B.; Roses, A.D. Apolipoprotein E associates with beta amyloid peptide of Alzheimer's disease to form novel monofibrils. Isoform apoE4 associates more efficiently than apoE3. *J. Clin. Investig.* **1994**, *94*, 860–869. [CrossRef]

69. Wisniewski, T.; Castaño, E.M.; Golabek, A.; Vogel, T.; Frangione, B. Acceleration of Alzheimer's fibril formation by apolipoprotein E in vitro. *Am. J. Pathol.* **1994**, *145*, 1030–1035.

70. Evans, K.C.; Berger, E.P.; Cho, C.G.; Weisgraber, K.H.; Lansbury, P.T. Apolipoprotein E is a kinetic but not a thermodynamic inhibitor of amyloid formation: Implications for the pathogenesis and treatment of Alzheimer disease. *Proc. Natl. Acad. Sci. USA* **1995**, *92*, 763–767. [CrossRef]

71. Paravastu, A.K.; Leapman, R.D.; Yau, W.M.; Tycko, R. Molecular structural basis for polymorphism in Alzheimer's beta-amyloid fibrils. *Proc. Natl. Acad. Sci. USA* **2008**, *105*, 18349–18354. [CrossRef]

72. Luhrs, T.; Ritter, C.; Adrian, M.; Riek-Loher, D.; Bohrmann, B.; Dobeli, H.; Schubert, D.; Riek, R. 3D structure of Alzheimer's amyloid-beta(1-42) fibrils. *Proc. Natl. Acad. Sci. USA* **2005**, *102*, 17342–17347. [CrossRef]

73. Breydo, L.; Uversky, V.N. Structural, morphological, and functional diversity of amyloid oligomers. *FEBS Lett.* **2015**, *589*, 2640–2648. [CrossRef]

74. Pechmann, S.; Tartaglia, G.G.; Vendruscolo, M.; Levy, E.D. Physicochemical principles that regulate the competition between functional and dysfunctional association of proteins. *Proc. Natl. Acad. Sci. USA* **2009**, *106*, 10159–10164. [CrossRef]

75. Castillo, V.; Ventura, S. Amyloidogenic Regions and Interaction Surfaces Overlap in Globular Proteins Related to Conformational Diseases. *PLoS Comput. Biol.* **2009**, *5*, e1000476. [CrossRef]

76. CrysAlisPRO, Agilent Technologies. *Software System*; Agilent Technologies UK Ltd.: Oxford, UK, 2012.

77. CrysAlisPRO Agilent Technologies. *Version 1.171.37.31*; Agilent Technologies UK Ltd.: Oxford, UK, 2014.

78. Leslie, A.G.W.; Powell, H.R. *Processing Diffraction Data with Mosflm*; Springer: Dordrecht, The Netherlands, 2007; pp. 41–51.

79. Savitzky, A.; Golay, M.J.E. Smoothing and Differentiation of Data by Simplified Least Squares Procedures. *Anal. Chem.* **1964**, *36*, 1627–1639. [CrossRef]

80. Van Zundert, G.; Rodrigues, J.; Trellet, M.; Schmitz, C.; Kastritis, P.; Karaca, E.; Melquiond, A.; Van Dijk, M.; De Vries, S.; Bonvin, A.M.; et al. The HADDOCK2.2 Web Server: User-Friendly Integrative Modeling of Biomolecular Complexes. *J. Mol. Biol.* **2016**, *428*, 720–725. [CrossRef]

81. Chandrasekhar, J.; Impey, R.W.; Jorgensen, W.L.; Madura, J.D.; Klein, M.L. Comparison of simple potential functions for simulating liquid water. *J. Chem. Phys.* **1983**, *79*, 926.

82. Abraham, M.J.; Murtola, T.; Schulz, R.; Páll, S.; Smith, J.C.; Hess, B.; Lindahl, E. GROMACS: High performance molecular simulations through multi-level parallelism from laptops to supercomputers. *SoftwareX* **2015**, 19–25. [CrossRef]

83. Piana, S.; Palmo, K.; Maragakis, P.; Klepeis, J.L.; Dror, R.O.; Shaw, D.E.; Lindorff-Larsen, K.; Lindorff-Larsen, K.; Lindorff-Larsen, K. Improved side-chain torsion potentials for the Amber ff99SB protein force field. *Proteins: Struct. Funct. Bioinform.* **2010**, *78*, 1950–1958.

84. Hess, B.; Bekker, H.; Berendsen, H.J.C.; Fraaije, J.G. LINCS: A linear constraint solver for molecular simulations. *J. Comput. Chem.* **1997**, *18*, 1463–1472. [CrossRef]

85. Darden, T.; York, D.; Pedersen, L. Particle mesh Ewald: An N·log(N) method for Ewald sums in large systems. *J. Chem. Phys.* **1993**, *98*, 10089–10092. [CrossRef]

86. Berendsen, H.J.C.; Postma, J.P.M.; DiNola, A.; Haak, J.R.; Van Gunsteren, W.F. Molecular dynamics with coupling to an external bath. *J. Chem. Phys.* **1984**, *81*, 3684. [CrossRef]

87. Nosé, S. A molecular dynamics method for simulations in the canonical ensemble. *Mol. Phys.* **1984**, *52*, 255–268. [CrossRef]

88. Hoover, W.G. Canonical dynamics: Equilibrium phase-space distributions. *Physical A* **1985**, *31*, 1695–1697. [CrossRef]

89. Parrinello, M. Polymorphic transitions in single crystals: A new molecular dynamics method. *J. Appl. Phys.* **1981**, *52*, 7182. [CrossRef]

90. Humphrey, W.; Dalke, A.; Schulten, K. VMD: Visual molecular dynamics. *J. Mol. Gr.* **1996**, *14*, 33–38. [CrossRef]

91. Schrodinger, LLC. *The PyMOL Molecular Graphics System, Version 1.8*; DeLano Scientific: San Carlos, CA, USA, 2015.

92. Wisniewski, T.; Golabek, A.A.; Kida, E.; Wisniewski, K.E.; Frangione, B. Conformational mimicry in Alzheimer's disease. Role of apolipoproteins in amyloidogenesis. *Am. J. Pathol.* **1995**, *147*, 238–244.

93. Nichols, W.C.; Dwulet, F.E.; Liepnieks, J.; Benson, M.D. Variant apolipoprotein AI as a major constituent of a human hereditary amyloid. *Biochem. Biophys. Commun.* **1988**, *156*, 762–768. [CrossRef]

94. Murphy, C.L.; Wang, S.; Weaver, K.; Gertz, M.A.; Weiss, D.T.; Solomon, A. Renal apolipoprotein A-I amyloidosis associated with a novel mutant Leu64Pro. *Am. J. Kidney Dis.* **2004**, *44*, 1103–1109. [CrossRef] [PubMed]

95. Soutar, A.K.; Hawkins, P.N.; Vigushin, D.M.; Tennent, G.A.; Booth, S.E.; Hutton, T.; Nguyen, O.; Totty, N.F.; Feest, T.G.; Hsuan, J.J. Apolipoprotein AI mutation Arg-60 causes autosomal dominant amyloidosis. *Proc. Natl. Acad. Sci. USA* **1992**, *89*, 7389–7393. [CrossRef]

96. Johnson, K.H.; Sletten, K.; Hayden, D.W.; O'Brien, T.D.; Roertgen, K.E.; Westermark, P. Pulmonary vascular amyloidosis in aged dogs. A new form of spontaneously occurring amyloidosis derived from apolipoprotein AI. *Am. J. Pathol.* **1992**, *141*, 1013–1019.

97. Benson, M.D.; Liepnieks, J.J.; Yazaki, M.; Yamashita, T.; Asl, K.H.; Guenther, B.; Kluve-Beckerman, B. A New Human Hereditary Amyloidosis: The Result of a Stop-Codon Mutation in the Apolipoprotein AII Gene. *Genomics* **2001**, *72*, 272–277. [CrossRef] [PubMed]

98. Higuchi, K.; Kitagawa, K.; Naiki, H.; Hanada, K.; Hosokawa, M.; Takeda, T. Polymorphism of apolipoprotein A-II (apoA-II) among inbred strains of mice. Relationship between the molecular type of apoA-II and mouse senile amyloidosis. *Biochem. J.* **1991**, *279*, 427–433. [CrossRef]

99. Bergstrom, J.; Murphy, C.L.; Weiss, D.T.; Solomon, A.; Sletten, K.; Hellman, U.; Westermark, P. Two different types of amyloid deposits[mdash]apolipoprotein A-IV and transthyretin[mdash]in a patient with systemic amyloidosis. *Lab Investig.* **2004**, *84*, 981–988. [CrossRef]

100. Sethi, S.; Theis, J.D.; Shiller, S.M.; Nast, C.C.; Harrison, D.; Rennke, H.G.; Vrana, J.A.; Dogan, A. Medullary amyloidosis associated with apolipoprotein A-IV deposition. *Kidney Int.* **2012**, *81*, 201–206. [CrossRef] [PubMed]

101. Bois, M.C.; Dasari, S.; Mills, J.R.; Theis, J.; Highsmith, W.E.; Vrana, J.A.; Grogan, M.; Dispenzieri, A.; Kurtin, P.J.; Maleszewski, J.J. Apolipoprotein A-IV–Associated Cardiac Amyloidosis. *J. Am. Cardiol.* **2017**, *69*, 2248–2249. [CrossRef]

102. Hatters, D.M.; Macphee, C.E.; Lawrence, L.J.; Sawyer, W.H.; Howlett, G.J. Human Apolipoprotein C-II Forms Twisted Amyloid Ribbons and Closed Loops. *Biochemistry* **2000**, *39*, 8276–8283. [CrossRef] [PubMed]

103. De Messieres, M.; Huang, R.K.; He, Y.; Lee, J.C. Amyloid Triangles, Squares, and Loops of Apolipoprotein C-III. *Biochemistry* **2014**, *53*, 3261–3263. [CrossRef] [PubMed]

104. Nasr, S.H.; Dasari, S.; Hasadsri, L.; Theis, J.D.; Vrana, J.A.; Gertz, M.A.; Muppa, P.; Zimmermann, M.T.; Grogg, K.L.; Dispenzieri, A.; et al. Novel Type of Renal Amyloidosis Derived from Apolipoprotein-CII. *J. Am. Soc. Nephrol.* **2017**, *28*, 439–445. [CrossRef] [PubMed]

105. Valleix, S.; Verona, G.; Jourde-Chiche, N.; Nedelec, B.; Mangione, P.P.; Bridoux, F.; Mangé, A.; Doğan, A.; Goujon, J.-M.; Lhomme, M.; et al. D25V apolipoprotein C-III variant causes dominant hereditary systemic amyloidosis and confers cardiovascular protective lipoprotein profile. *Nat. Commun.* **2016**, *7*, 10353. [CrossRef] [PubMed]

106. Sunde, M.; Blake, C. The Structure of Amyloid Fibrils by Electron Microscopy and X-ray Diffraction. In *Advances in Protein Chemistry*; Elsevier BV: Amsterdam, The Netherlands, 1997; Volume 50, pp. 123–159.

107. Sunde, M.; Serpell, L.C.; Bartlam, M.; Fraser, P.E.; Pepys, M.B.; Blake, C.C.F. Common core structure of amyloid fibrils by synchrotron X-ray diffraction11Edited by F. E. Cohen. *J. Mol. Biol.* **1997**, *273*, 729–739. [CrossRef] [PubMed]

International Journal of
Molecular Sciences

MDPI

Review

A Clinical Approach for the Use of VIP Axis in Inflammatory and Autoimmune Diseases

Carmen Martínez [1,*], Yasmina Juarranz [1], Irene Gutiérrez-Cañas [1], Mar Carrión [1], Selene Pérez-García [1], Raúl Villanueva-Romero [1], David Castro [1], Amalia Lamana [1], Mario Mellado [2], Isidoro González-Álvaro [3] and Rosa P. Gomariz [1]

[1] Departamento de Biología Celular, Facultad de Biología y Facultad de Medicina, Universidad Complutense de Madrid, 28040 Madrid, Spain; yashina@ucm.es (Y.J.); irgutier@ucm.es (I.G.-C.); macarrio@ucm.es (M.C.); selene@ucm.es (S.P.-G.); ravillan@ucm.es (R.V.-R.); dcastr01@ucm.es (D.C.); amaliala@ucm.es (A.L.); gomariz@ucm.es (R.P.G.)

[2] Departamento de Inmunología y Oncología, Centro Nacional de Biotecnología (CNB)/CSIC, 28049 Madrid, Spain; mmellado@cnb.csic.es

[3] Servicio de Reumatología, Instituto de Investigación Médica, Hospital Universitario La Princesa, 28006 Madrid, Spain; isidoro.ga@ser.es

* Correspondence: cmmora@ucm.es; Tel.: +34-913941405

Received: 30 November 2019; Accepted: 18 December 2019; Published: 20 December 2019

Abstract: The neuroendocrine and immune systems are coordinated to maintain the homeostasis of the organism, generating bidirectional communication through shared mediators and receptors. Vasoactive intestinal peptide (VIP) is the paradigm of an endogenous neuropeptide produced by neurons and endocrine and immune cells, involved in the control of both innate and adaptive immune responses. Exogenous administration of VIP exerts therapeutic effects in models of autoimmune/inflammatory diseases mediated by G-protein-coupled receptors (VPAC1 and VPAC2). Currently, there are no curative therapies for inflammatory and autoimmune diseases, and patients present complex diagnostic, therapeutic, and prognostic problems in daily clinical practice due to their heterogeneous nature. This review focuses on the biology of VIP and VIP receptor signaling, as well as its protective effects as an immunomodulatory factor. Recent progress in improving the stability, selectivity, and effectiveness of VIP/receptors analogues and new routes of administration are highlighted, as well as important advances in their use as biomarkers, contributing to their potential application in precision medicine. On the 50th anniversary of VIP's discovery, this review presents a spectrum of potential clinical benefits applied to inflammatory and autoimmune diseases.

Keywords: vasoactive intestinal peptide; VPAC1 receptor; VPAC2 receptor; rheumatic diseases; inflammatory bowel disease; central nervous system diseases; type 1 diabetes; Sjögren's syndrome; biomarkers

1. Introduction

The nervous, endocrine, and immune systems are coordinated to maintain the homeostasis of the organism, generating bidirectional communication through shared mediators and receptors [1,2].

Vasoactive intestinal peptide (VIP) is the paradigm of an endogenous neuropeptide produced during autoimmune responses and processes of systemic and local inflammation. It acts as an immunomodulatory agent to restore homeostasis of the immune system [3]. Synthesized by neurons and endocrine and immune cells, VIP is involved in the control of both innate and adaptive immune responses [4–6]. Exogenous administration of VIP exerts therapeutic effects in models of autoimmune/inflammatory diseases mediated by two G-protein-coupled receptors (VPAC1, VPAC2) [7–12].

Inflammatory and autoimmune diseases include a clinically heterogeneous group of chronic diseases sharing inflammatory mechanisms, as well as a deregulation of the immune system [13,14]. These diseases can affect any organ or system and are often multiorganic. Among these pathologies, we find rheumatic diseases such as rheumatoid arthritis (RA), inflammatory bowel diseases (IBD), and multiple sclerosis (MS).

According to the Autoimmune Diseases Coordinating Committee of the National Institutes of Health in the United States, the prevalence of autoimmune pathologies is estimated at up to 8% of the population. These pathologies are characterized by a complex etiology combining different genetic, epigenetic, and environmental factors, such as tobacco use or history of infections, which result in the alteration of the regulation of the immune system [14–17]. These diseases lead to substantial levels of morbidity, a significant reduction in the quality of life, and premature death [18–21].

Currently, there are no curative therapies for inflammatory and autoimmune diseases, and patients present complex diagnostic, therapeutic, and prognostic problems in daily clinical practice due to their heterogeneous nature. Many of these challenges could be alleviated with appropriate biomarkers, allowing a more efficient use of current therapies, as well as the development of precision medicine.

This review focuses on the biology of VIP and VIP receptor signaling, as well as its protective effects as an immunomodulatory factor. Here, we consider their role in the pathogenesis of autoimmune diseases and inflammatory disorders and address the potential clinical application of the VIP/receptor axis.

2. Biological Characteristics of VIP

2.1. VIP Discovery, Cellular Location, and Structure

Sami Said described, for the first time in 1969, the existence of a peptide vasoactive agent with systemic vasodilator capacity present in the lungs of mammals. In collaboration with Viktor Mutt, Said purified this peptide from pig lungs, but only partially. Challenges in isolating it from the lungs led them to examine the intestine, since both tissues have a common embryonic origin. Thus, using porcine duodenal tissue, they isolated this vasodilator peptide and presented it to the scientific community, calling it the vasoactive intestinal peptide [22].

A few years later, the presence of this peptide was demonstrated in different areas of the central and peripheral nervous system, such as the bodies, axons, and neuronal dendrites [23], as well as in presynaptic endings [24], resulting in the categorization of the VIP as a neuropeptide with neuromodulatory and neurotransmitter functions. This role was confirmed with the characterization of VIP receptors in numerous areas of the central nervous system (CNS) [25].

In the immune system, the first information dates back to 1985, when Felten et al. described VIP-like immunoreactivity in the thymus nerve endings [26]. Since then, VIP-ergic innervation in the spleen, lymph nodes, and mucosal-associated immune system has been demonstrated [27]. It is also important to note that sympathetic nervous system fibers innervate the joints, which explains the role of VIP in rheumatic diseases.

Regarding the cellular source involved in VIP production, the first evidence was reported for cells of myeloid lineage. Expression in mast cells was demonstrated by radioimmunoassay and immunohistochemistry in the rat peritoneum, intestine, and lung [28]. In 1980, O'Dorisio described the presence of VIP in human peripheral blood polymorphonuclear cells, especially neutrophils, but not in mononuclear cells [29]. VIP expression has also been described in human eosinophils [30] and in eosinophils of granulomatous lesions induced by infection with *Schistosomiasis mansoni* [31]. We reported that neither M1 nor M2 human macrophages express transcripts of VIP [32]. Concerning cells of a lymphoid lineage, in the 1990s, our team reported, for the first time, the synthesis and secretion of VIP in murine T and B lymphocytes [33–35]. Since then, information on the important role VIP plays in inflammation and autoimmunity continues to accumulate. Today, VIP is an important player in the circuit formed by the nervous, endocrine, and immune systems. It is also present in rheumatic diseases [36] and is one of the most studied peptides in terms of a physiological role in health and disease, especially in the immune system.

The origin of VIP in the microenvironment of the different pathologies in which its effect has been studied is the nerve endings and cells. In this sense, nerve fibers of the sympathetic nervous system in the joints have been reported in rheumatic diseases. Moreover, a decrease in the number of these nerve endings has been described in osteoarthritis (OA) and RA [37]. Regarding cellular origin, synovial fibroblasts (SF) from OA and RA patients have been found to express and release VIP [38].

VIP belongs to a broad family of neuropeptides and hormones, related both structurally and at sequence level, called the secretin/VIP family. In addition to VIP and secretin, this family includes the adenylate cyclase activating peptide pituitary (PACAP) 27 and PACAP38, helodermin, histidine-methionine peptide (PHM, in humans) or histidine-isoleucine peptide (PHI, in other mammals), the releasing factor of growth hormone (GHFR), glucagon and its related peptides GLP1 and GLP2, and the gastric inhibitor peptide (GIP) [39]. The structural homology observed among the different members of this family is very high, with the following characteristics being common [40]: (I) precursor peptide formed by a signal peptide, from 1 to 3 bioactive peptides and N- and C-terminal peptides; (II) length of the mature peptide comprised of between 25 and 50 aa residues; (III) synthesis and release by nerve, immune, and/or endocrine cells; (IV) patent tendency for the formation of α-helix structures; and (V) presence of a structural motif called N-Cap in the amino terminal region. The helical structure seems to be a key element in the interaction with receptors and signaling and is considered an interesting therapeutic target [41]. These peptides show strong homology in their amino acid sequences on an evolutionary scale, suggesting a common origin from an ancestral gene [42]. VIP is a 3.326 Da molecular weight peptide, with a basic nature and amphipathic character. Its primary structure consists of a single chain with 28 aa whose sequence has been highly conserved throughout evolution [43]. Although the presence of all of these aa is necessary for VIP to perform its biological functions, it has been proven that certain residues are crucial for this performance (His^1, Val^5, Arg^{14}, Lis^{15}, Lis^{21}, Leu^{23}, and Ile^{26}). The secondary structure has a random coil in the N-terminal region and an α-helix structure in the C-terminal region. This structure is similar to that of other family members, especially that of PACAP27, with whom it shares 68% sequence homology [44].

2.2. General Biological Functions

The expression of VIP in the nervous system results in its release in multiple organs by releasing nerve fibers. Thus, VIP is present in the innervation of the heart, kidney, lung, thyroid gland, and gastrointestinal and urogenital tracts. As we have described previously, central and peripheral lymphoid organs, such as the thymus, spleen, and lymph nodes, are also innervated by VIP sympathetic nerve fibers. Moreover, VIP expression in cells of myeloid and lymphoid origin also contributes to its broad distribution, correlating with its functional pleiotropism. Thus, VIP acts as a neurotransmitter, immunoregulator, vasodilator, and stimulator of hormone secretion or secretagogue [6]. VIP contributes to a wide variety of physiological activities related to development, growth, immune response, circadian rhythms, endocrine control, and functions of the digestive, respiratory, cardiovascular, and reproductive systems [39]. Some of VIP's multiple biological activities include increased cardiac output, bronchodilation, smooth muscle relaxation, regulation of secretion processes, and motility in the gastrointestinal tract. In addition, as a secretagogue, VIP promotes the release of prolactin, luteinizing hormone, and growth hormone by the pituitary gland and regulates the release of insulin and glucagon in the pancreas. This peptide also promotes analgesia, hyperthermia, learning, and behavior; it has neurotrophic effects and regulates bone metabolism and embryonic development [45].

3. VIP Receptors, Ligands, and Signaling Pathways

3.1. VIP Receptors

VIP and PACAP were discovered in the 1970s and 1980s, respectively, and cloned in the 1980s and 1990s. The existence of several "VIP receptors" was inferred by pharmacological studies, such as cyclic AMP (cAMP) or radioligand binding assays, long before the actual receptor cloning. In this way, "VIP receptors" were described in normal and tumor cells and tissues [40]. These receptors showed pharmacological differences and were not correctly identified until the description of several ligands, agonists, and antagonists, and the cloning of the receptors.

VIP receptors belong to group B of G-protein-coupled receptors (GPCRs), which include seven transmembrane receptors that represent the most extensive family of signaling proteins. Ligands for class B GPCRs are peptides that bind to the large N-terminal part of the GPCR [46]. There are three receptors recognized by VIP: VPAC1, VPAC2, and PAC1 receptors. VIP binds to VPAC receptors with equal affinity and with much less affinity to PAC1, which is the PACAP-preferred receptor. The structures of several class B receptors have been determined, as well as crystal structures of peptide-bound receptors, which help to understand how the peptides can bind to their receptors and how these receptors undergo conformational changes to allow downstream signaling [47].

3.1.1. VPAC1 Receptor

The rat VPAC1 receptor was cloned in 1991 from a cDNA library, and the human VPAC1 was cloned in 1993 from the HT-29 cell line [48,49]. Only one variant of this receptor has been described thus far, which is expressed in several normal and malignant cells. It has a deletion that results in a receptor with five transmembrane domains lacking the G-protein binding domain. Even so, it can activate protein tyrosine kinase activity, but in a different way than the seven transmembrane domain receptor [50].

The affinity of several peptides for this receptor is as follows: VIP = PACAP > GRF > secretin [40]. In the last decade, advances in the study of molecule structures have allowed the dissection of the physical sites of interaction between VPAC1 and VIP, observing that the side chains of several residues in the VIP sequence are in contact with several others in the receptor sequence, although the whole interaction between the two molecules has yet to be elucidated. Nevertheless, all the models available are in concordance with the mechanism proposed for the ligand–receptor interaction for this family of receptors, the "two domain" model, in which part of the peptide remains inside the N-terminal ectodomain of the receptor, while the N-terminus of the peptide is able to interact with the transmembrane region of the receptor [39].

3.1.2. VPAC2 Receptor

This receptor was also first cloned from rats, with the human and mouse receptors cloned shortly thereafter [51–53]. The first variant of this receptor described in mice tissues showed a deletion in exon 12, which corresponds to the carboxyl-terminal end of the seventh transmembrane domain. This variant lacks its normal function of increasing cAMP [54]. The second variant found in a human malignant T-cell line presented a deletion in exon 11 and had lower affinity for VIP [55]. The most recently described was the same variant as that described for the VPAC1 receptor [50]. The order of affinity for human VPAC2 expressed in different cell lines is VIP = PACAP = helodermin > secretin [40].

3.2. Ligands

During the last four decades, many ligands have been developed, both agonists and antagonists for VIP receptors. Most of them were created by modifying endogenous peptides and displayed different affinities and selectivities, with the first descriptions unable to differentiate between the two receptors [56–58]. Selective agonists for VPAC1 receptor have been generated, such as [K^{15}, R^{16}, L^{27}]VIP(1-7)/GRF(8-27) [59], [Ala^{11,22,28}]VIP [60], [L^{22}]VIP [61], [R_{16}]PACAP(1-23) [62], and LBT-3393 [63]. A selective antagonist is also available: PG97-269 [64]. Regarding VPAC2, cyclic peptides have been demonstrated to be selective agonists, such as Ro25-1553 [65], Ro25-1392 [66], and some other peptides, such as BAY 55-9837 [67] Hexanoyl [A^{19},$K^{27,28}$]VIP, rRBAYL [68], and LBT-3627 [63]. Only two VPAC2 selective antagonists have been described thus far: PG99-465 [69] and VIpep-3 [70]. Some of these approaches aim to identify more metabolically stable peptides [63] in order to ameliorate their in vivo administration. Recently, one in silico study predicted possible structures defining affinities for these receptors, constructing classifiers to predict the bioactivities of novel VPAC ligands and highlighting the importance of electrostatic properties in the interaction of VIP derivatives with the receptors [71]. Moreover, there are some other approaches providing tools for the study of these receptors or for their use as therapeutic tools, such as nanoparticles that enhance the half-life of one VPAC2 selective agonist [72] and specific nanobodies for the VPAC1 receptor that bind at a different site than VIP, thus without interfering with the coupling of the peptide [73].

3.3. Signaling Pathways

The main signaling pathway of VPAC receptors is their coupling to G-proteins, which are heterotrimeric proteins composed of three subunits: α, β, and γ. When stimulated, the α subunit binds to GTP and dissociates from the $\beta\gamma$ dimer. Activated Gα moves through the membrane to its effector, the enzyme adenylate cyclase (AC), which in turn catalyzes cAMP synthesis [40]. This second messenger classically activates protein kinase A (PKA), which phosphorylates and can activate or inactivate different signaling pathways, depending on the cell type. For instance, the typical transcription factor activated by cAMP through PKA is cAMP-response element binding (CREB). PKA can also activate mitogen-activated protein kinases (MAPK) from several subfamilies. Moreover, cAMP in a PKA-independent way can activate exchange proteins directly activated by cAMP (EPAC), which is a G-protein exchange factor (GEF) for Rap small G-protein [74]. The $\beta\gamma$ dimer interacts with several proteins, such as Ras, which in turn activates extracellular regulated kinases (ERK). Subsequently, ERK interacts with and activates phosphoinositide 3-kinase (PI3K) [75,76].

VPAC receptors can also mediate the increase in Ca^{2+} through the activation of either Gi/o or Gq proteins. Although this pathway has shown lower potencies, relative to cAMP, the rise in calcium is of physiological relevance [77]. Less frequently, they can also activate phospholipase C (PLC) through the activation of Gi/o by a mechanism that likely involves the $\beta\gamma$ dimer and that subsequently increases the production of inositol phosphate (IP_3). This rise in IP_3 increases the [Ca^{2+}], which, together with diacylglycerol (DAG), activates PKC, which can also phosphorylate several other kinases [78]. Furthermore, PLD can be activated in a pathway involving the small G protein ARF. This phospholipase hydrolyzes phosphatidylcholine (PC), generating the signaling molecule phosphatidic acid (PA), which functions pleiotropically in several signaling pathways [79].

Receptor activity modifying proteins (RAMPs) are single pass transmembrane proteins that do not bind any known ligand and need to be coupled with any receptor to arrive at the cellular surface. VPAC1 and VPAC2 can bind the three known RAMPs, not modifying their affinity for ligands. Both receptors enhance the cell-surface expression of the three RAMPs, and co-expression of VPAC1 and RAMP2 enhances the response of IP_3 without modifying cAMP signaling [80]. Furthermore, VPAC2 co-expression with RAMP1 increases the basal cAMP and is diminished when VPAC2 is co-expressed with RAMP3. Moreover, co-expression with RAMP1 and 2 enhances coupling to Gi/o/t/z, but does not modify the binding to Gs [81].

Regarding inflammatory pathways, three of the most important transcription factors activated in inflammatory processes are AP-1 (activator protein 1), NFκB (nuclear factor κB), and IRF (interferon regulatory factor). VIP is able to inhibit AP-1 and IRF activation through a PKA dependent mechanism and can also prevent NFκB translocation to the nucleus impeding IKK activation through a cAMP independent mechanism [82–85].

The $\beta\gamma$ dimer has been shown to bind GPCR kinases (GRKs), recruiting them to the membrane. When GPCRs are phosphorylated by GRKs, they bind arrestins, allowing receptor desensitization and internalization and/or signaling from the inside, mainly through activation of different MAPKs [46]. In particular, VPAC1 and VPAC2 exhibit augmented desensitization when co-transfected with GRKs 2, 3, 5, and 6 [86]. Nonetheless, this is not the only way in which GPCRs have effects within the cell. Recently, the presence of intracellular GPCRs has been described, which arrive there through many different processes [87]. VPAC1 has been found to be expressed in the nucleus of several tumor cells, such as human breast cancer [88], human renal carcinoma [89], and human glioblastoma, where there is also a weak expression of nuclear VPAC2 in one of the cell lines studied [90]. More recently, the presence of VPAC1 has been observed to be located on the surface and nuclear membrane of T helper (Th) cells, whereas its expression is limited to the nucleus when these cells are activated [91]. There is no evidence regarding how VPAC2 arrives at the nucleus, and all the options described elsewhere are possible [87]. However, there is work showing that VPAC1 has a nuclear localization signal sequence in the C-terminal domain of the receptor, and in this way is able to exhibit nuclear expression [90]. It has been described that Cys37 in the VPAC1 receptor is essential for the translocation of the receptor to the nucleus and that it must be palmitoylated to be functional [92].

4. A Very Important Peptide in Inflammation and Autoimmunity

4.1. Targeting Balance of Inflammatory Factors

Inflammation is a complex homeostatic process mediated by factors of plasma and cellular origin whereby the effects of harmful stimuli are controlled in the tissues. When the inflammation persists over time, beyond what is necessary, and stops responding to the reparative process, it becomes destructive and chronic.

In chronic inflammation, there is a massive infiltration of cells involved in innate (monocytes–macrophages and dendritic cells) and adaptive immunity (TCD4$^+$ and B cells). A complex network of pro-inflammatory cytokines is established, and the cytokines are secreted primarily by activated macrophages and CD4$^+$ T cells at the site of inflammation.

The macrophage is the main producer of cytokines, and, when activated by different danger signals, it releases several pro-inflammatory products, such as interleukin (IL)-1, tumor necrosis factor (TNF-α), IL-6 and IL-l2, and nitric oxide (NO), followed later by the secretion of anti-inflammatory cytokines such as IL-10 [93]. At the site of inflammation, numerous chemokines are also secreted, exacerbating the inflammatory process by the attraction of more leukocytes. Despite its beneficial effects in the defense of the organism, the sustained production of pro-inflammatory factors can lead to pathological conditions such as septic shock, respiratory distress syndrome, and autoimmune disease [94–96].

Numerous studies, both in animal and human models, show that VIP plays a key role in maintaining homeostasis, by controlling the balance of pro- and anti-inflammatory cytokines by inhibiting the production of pro-inflammatory cytokines and chemokines such as TNF-α, IL-6, IL -12 CXCL8, and CCL2, as well as NO, and stimulating the expression of anti-inflammatory cytokines such as IL-10 [4–6].

In activated macrophages, VIP inhibits the production of TNF-α, IL-12, and NO primarily through VPAC1, expressed constitutively, and, to a lesser degree, through inducible VPAC2. VIP's binding to VPAC1 induces both a cAMP-dependent and a cAMP-independent pathway that regulates cytokine production and NO at the transcriptional level. VIP inhibits the expression of TNF-α, IL-12, and inducible nitric oxide synthase (iNOS) by reducing the binding of the NFκB transcription factor to the promoter and increasing IL-10 by increasing the binding of the CREB factor [97]. Thus, molecular mechanisms and transcription factors involved in the VIP signaling during inflammatory responses include inhibition of interferon (IFN)-γ-induced Jak1/Jak2 phosphorylation and STAT1 activation, inhibition of different MAPK cascades, inhibition of IkB-kinase, and stimulation of CREB factor [3,97] (Figure 1).

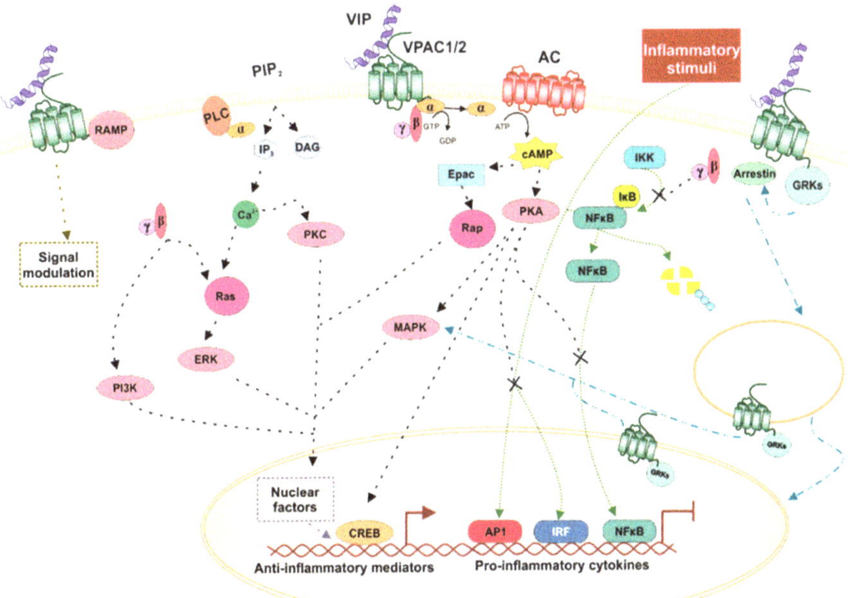

Figure 1. VIP/VPAC receptors' axis signaling pathways. VIP/VPAC receptors' binding activates a cAMP-dependent signaling pathway mediated by the induction of AC (black arrows). Then, cAMP activates PKA, which in turn induces nuclear translocation of CREB (black arrows). Besides, PKA inhibits the activation of pro-inflammatory transcription factors such as AP-1, IRF, or NFκB (black cross). Additionally, cAMP in a PKA-independent way simulates EPAC (black arrows). This second messenger induces anti-inflammatory transcription factors. VPAC receptors can also activate PLC and PI3K (black arrows). Both signaling pathways produce the nuclear translocation of anti-inflammatory transcription factors. These receptors can also interact with accessory RAMPs, modulating the canonical signaling pathways (dark green arrow). Furthermore, VPAC1 is able to translocate to the nucleus by the interaction with GRKs (blue arrows). Inflammatory stimuli activate signaling pathways (light green arrows).

VIP also modulates inflammatory responses through the regulation of different functions of other cells, including mast cells, microglia, dendritic cells, and synovial fibroblasts [98–101]. Moreover, in terms of adaptive immunity, VIP reduces pro-inflammatory Th1 and Th17 responses, as described below.

The importance of endogenous VIP in the regulation of inflammation and autoimmunity has been confirmed in knockout (KO) mouse models showing altered immune responses. At basal conditions, the immune phenotype of the mice studied so far is relatively mild. The role of VIP is mainly highlighted in challenging inflammatory conditions. Thus, VIP-deficient mice develop lung inflammation [102,103]. However, there are discrepancies about the resistance or susceptibility of VIP-KO mice to endotoxemia. Hamidi et al. described an increased susceptibility to death from endotoxemia, while Abad et al. found that VIP-KO mice exhibit resistance to endotoxic shock and decreased pro-inflammatory responses due, in part, to the presence of an intrinsic defect in the responsiveness of inflammatory cells in the chronic absence of VIP, suggesting that these mice may exhibit a defect in the innate arm of the immune system [104,105].

Given the accumulated evidence of VIP anti-inflammatory properties, VIP treatment has been reported to protect against septic shock and various inflammatory and autoimmune diseases, and to act as a survival factor against injury of lung and neuronal cells [45,106–110]. The role of VIP in the inflammatory component of the diseases described in this review is treated in detail in the different pathologies.

4.2. Modulating the Expression of TLRs

Toll-like receptors (TLR) are a large family of type I transmembrane proteins belonging to pattern-recognition receptors, which are specialized in the recognition of extracellular and endosomal pathogen-associated molecular patterns, serving as warning signals for the immune system. Likewise, TLRs are able to specifically bind damage-associated molecular patterns, which are associated with tissue damage, cell stress, and cell death [111–113]. Therefore, TLRs are defined as essential receptors to trigger innate immune response and, subsequently, for the regulation of the adaptive immune response [114]. The expression of these receptors has been detected not only in immune cells, including macrophages, neutrophils, mast cells, dendritic cells, and T and B lymphocytes [115–117], but also in non-immune cells, such as synovial fibroblasts, keratinocytes, pulmonary, and intestinal epithelial cells [118–121]. In humans, these transmembrane receptors are found both on cell membranes (TLR1, 2, 4, 5, and 6) and in endosomes (TLR3, 7, 8, and 9) [111,122].

Upon ligand binding to TLRs, complex signal transduction cascades are triggered, requiring different adapter proteins. Myeloid differentiation primary response protein (MyD88) is involved in signaling by all TLRs, with the exception of TLR3 [123,124]. Both MyD88-dependent and -independent pathways lead to activation of NFκB, IRF3/7, and/or AP-1, which ultimately induce the production of inflammatory mediators, and co-stimulatory molecules [111,113,125]. Thus, an inappropriate or deregulated TLR activation, such as a persistent infection or a failure in their ability to discriminate self from non-self molecules can compromise immunological homeostasis [114,126]. In fact, numerous studies have demonstrated the involvement of TLRs in a wide variety of pathological processes, including both acute and chronic infections, as well as in the induction, progression, or exacerbation of many systemic autoimmune and/or inflammatory conditions [113,127–129]. In this regard, extensive data accumulated from animal models and in vitro human studies have strongly demonstrated the homeostatic effects of VIP on the deregulated expression and signaling of TLR in a context of inflammatory and/or autoimmune disease [130–133].

TLR modulation by VIP was described for the first time in the trinitrobenzene sulfonic acid (TNBS)-induced colitis mouse model, which mimics human Crohn's disease (CD). In that model, VIP reduces the upregulated expression of TLR2 and TLR4, as we describe in Section 5, "Protective Effects of VIP in Inflammatory/Autoimmune Diseases" [134,135]. The inhibitory effect of VIP on TLR2 expression was suggested to be due to its ability to prevent the nuclear translocation of NFκB, which has a binding site in the murine TLR2 gene [85,136]. Moreover, research on primary murine macrophages and the RAW 264.7 cell line showed that VIP exerts its suppressive effects on murine TLR4 expression at the transcriptional level by decreasing the binding of the transcription factor PU.1 via PI3K/Akt1 pathway [137]. In agreement with these findings, VIP was also able to reduce the lipopolysaccharide (LPS)-induced expression of TLR2 and TLR4 in human monocytic THP1 cells and peripheral blood monocytes, as well as to inhibit their differentiation to macrophages [138]. In these cells, VIP inhibited the nuclear translocation of PU.1, which acts as a transcriptional regulator of both TLR2 and TLR4 genes in humans [139,140].

The potent immunomodulatory effect of VIP on TLRs has also been reported in the mice cornea after *Pseudomonas aeruginosa* infection. Data from in vitro studies demonstrated that VIP reduced LPS-stimulated expression of TLR1, TLR4, TLR6, TLR8, and TLR9 in macrophages and Langerhans cells [141].

The effects of VIP on TLRs were also assessed in SF from OA and RA patients. TLR2, TLR4, and TLR3 expression are described to be higher in RA-SF compared with OA, whereas greater levels of TLR7 have been detected in OA-SF [82,142,143]. In vitro data indicated that VIP treatment decreases both LPS- and TNF-induced expression of TLR4 in RA-SF, whereas it has no effect on the elevated constitutive expression of TLR2 and TLR4 [142,144]. Furthermore, VIP also exerts a negative modulation of TLR4 signaling in these cells by the downregulation of important molecules of both the MyD88-dependent and -independent signaling pathways [83]. On the other hand, no effect of VIP on the expression of other TLRs has been detected in OA and RA-SF. However, VIP exerts an inhibitory activity on nuclear translocation of transcription factors activated by TLR3 and TLR7, with the subsequent reduction of antiviral, pro-inflammatory, and joint destruction mediators upregulated by engagement with these receptors [82,142].

All in all, VIP's ability to balance TLR expression and signaling may be of physiological relevance in the specific control of innate and adaptive immune responses.

4.3. Regulating Th Cells

Th cells are specialist cytokine-producing cells that modulate the adaptive immune response. During inflammation or infection, different Th subsets are activated, playing a fundamental role in the type of response and the degree of amplification. These subsets are classified by their cytokine profile and the expression of specific transcription factors (master regulators) that direct their functional activity [145]. Thus, Th subpopulations are organized into two branches, effector Th cells and regulatory T cells (Treg). Th1, Th2, Th17, Th follicular (Tfh), Th9, and Th22 subsets are found within the branch of effector Th cells. In an immune steady-state, the balance between these subsets underwrites the preservation of immune tolerance. When a microbial or viral infection or tissue damage occurs, this balance changes from a tolerant state to an immunogenic/inflammatory state, until the immunogen is eliminated. Then the homeostatic regulatory mechanisms are recovered, and the system returns to its initial state. Inflammatory and/or autoimmune diseases occur when these mechanisms fail [146]. Some of the Th subpopulations play a key role in these pathologies; for example, Th1 and Th17 are the key effector Th cells in RA and Crohn's disease, while a loss in the number or function of the Treg has been described. Not only is the level of presence of each of the subsets important in the development of autoimmune diseases, but so is the plasticity observed between them or even their heterogeneity [147–150]. In this sense, pathogenic Th17 can change its lineage commitment to a Th1 profile, called nonclassical Th1 or ex-Th17. This has been observed in different mouse models of autoimmune diseases or in RA patients [151–154]. Th plasticity is also observed in Treg, which can shift its linage commitment to Th1 or Th17. In turn, nonpathogenic Th17 can acquire a Treg profile [150]. Therapeutically, it is important to know the involvement of each subpopulation in the different pathologies, as well as their possible plasticity or heterogeneity.

VIP is a microenvironment mediator involved in the generation of diversity and plasticity of Th subsets in inflammatory or autoimmune diseases. This claim is supported by numerous experimental studies in both animal models and ex vivo samples of patients [5,155–158]. VIP was able to decrease the cytokine profile and master regulators related to Th1 and Th17 subsets and to increase those of them related to Treg or Th2 in different autoimmunity animal models, such as the collagen-induced (CIA) arthritis mouse model of RA, the TNBS mouse model of Crohn's disease, the nonobese diabetic (NOD) mouse model of autoimmune diabetes, the experimental autoimmune encephalomyelitis (EAE) mouse model of multiple sclerosis, the experimental model of autoimmune myocarditis, and the pristine-induced lupus model of lupus nephritis [7,11,159–165]. In addition, this immunomodulatory role of VIP was observed in two inflammatory animal models, including the models of CNS inflammation or atherosclerosis [108,164]. The same effect was observed in mouse Th cells activated in vitro studies or with Th lymphocytes from patients activated ex vivo, mainly in studies with RA patients, treated with exogenous VIP [153,157,166,167]. VIP not only acts on a specific subset in these pathologies, but is also able to balance the different Th subsets, inducing nonpathogenic phenotypes or modify their plasticity. Studies on different transcription factors, cytokines, cytokine receptors, chemokines, and chemokine receptors in the above mentioned mice models, in vitro or ex vivo, showed that VIP counterbalances the ratio of Th1/Th2, Th17/Treg, Th1/Treg, or Th2/Th9, reducing pathogenicity and increasing tolerance [10,165,168]. Th17 cells are a heterogeneous subset with a nonpathogenic or pathogenic profile, depending on the microenvironment. VIP maintains the nonpathogenic profile of human Th17-polarized cells in vitro from naïve Th cells [169]. Indeed, it lowers the pathogenic Th17 profile in activated/expended memory Th cell ex vivo from early RA patients [153,170]. Taking into consideration the plasticity of Th subsets, this neuropeptide decreases the Th17/1 profile, inducing a negative correlation between Th17 and Th1 in ex vivo cultured cells from early RA patients, but also increases the Th17/Treg profile [153,169,170]. The effect of VIP on the plasticity of Th17 cells is in agreement with its effect on heterogeneity, since the nonpathogenic Th17 phenotype is closely related to Th17/Treg plasticity.

In summary, the generation/differentiation, plasticity, and heterogeneity of Th subsets are crucial events during the development of inflammatory/autoimmune diseases. These processes are susceptible to modulation by different mediators present in the microenvironment of Th cells, an example of which is the VIP neuropeptide that induces a less pathogenic and more tolerogenic response in Th cells.

4.4. Inducing Tolerogenic Dendritic Cells

Conventional or classical dendritic cells (DCs) are critical for initiating the activation and differentiation of T cells during an inflammatory state, mainly due to their co-stimulatory capacity. They can be classified functionally according to their maturation state in immature or mature DCs [146]. Immature DCs, also called lymphoid organ-resident DCs, are phenotypically immature since they show on their surface low amounts of costimulatory receptors. When they migrate, they initiate a maturation process by strongly expressing these receptors. During the maturation process of DCs, they can differentiate into tolerogenic or immunogenic antigen-presenting cells, each distinguished by specific cytokine production and cell-surface receptors. Immunogenic DCs develop an immunogenic/inflammatory state, whereas mature tolerogenic DCs can induce immune tolerance [146,171]. These latter cells are prompted by either anti-inflammatory signals or signals interfering with the function of immunogenic DCs. Their role is to inhibit effector and autoreactive T cells and trigger Treg development. As a consequence, they play a main role in inducing immune tolerance, resolution of ongoing immune responses, and prevention of autoimmunity [172,173].

An increasing body of data indicates that tolerogenic DCs could be promising therapeutic targets in the treatment of autoimmune diseases [172,174]. One of the approaches is to generate ex vivo tolerogenic DCs for DC-based immunotherapy [175,176]. In this sense, VIP-treated DCs retained their tolerogenic ability in vitro and in vivo under different inflammatory situations [45,177,178]. Two strategies have been followed to generate VIP-tolerogenic DCs: VIP treatment during differentiation of DCs derived from bone marrow or monocytes, or using lenti-VIP transduced DCs [6,179]. In either case, the later administration in vivo of these cells produces Ag-specific Treg capable of inducing specific tolerance to naïve recipients. In this way, they cause the attenuation of symptoms of different animal models of autoimmune and/or inflammatory diseases, for example, in CIA arthritis, TNBS-induced colitis, EAE, sepsis, and spontaneous autoimmune peripheral polyneuropathy [177,179–182]. In addition, in vitro studies with VIP have shown that it affects not only the phenotypic and functional maturation of DCs, but also the migration of these cells [183–185].

5. Protective Effects of VIP in Inflammatory/Autoimmune Diseases

5.1. Rheumatic Diseases

5.1.1. VIP in Rheumatoid Arthritis

RA is a systemic inflammatory rheumatic disease of unknown etiology, with a significant autoimmune component, characterized by a persistent synovitis of symmetrical peripheral joints and the presence of auto-antibodies such as rheumatoid factor and anti-citrullinated protein antibodies (ACPA) [186–188]. The natural course of RA is generally associated with progressive destruction of articular cartilage and bone, resulting in a severe functional impairment and serious worsening of the patient's quality of life. However, RA is described as a heterogeneous disease with several subtypes that differ in clinical symptoms, such as age of onset, rate of progression, disease severity, and outcome [187,189,190]. Its complexity is also reflected in the fact that it is considered a multifactorial disease, as genetic background and environmental conditions, including infectious events and dysbiosis in the gut and the lung microbiome, have been indicated as factors involved in triggering the aberrant immune response [186,187].

Although RA pathogenesis is not completely understood, it is widely accepted that local and systemic immune dysregulation, as a result of imbalance in the Th cell subsets, plays an important role in creating a synovial joint microenvironment that favors a hyperactivated phenotype of SF and macrophages. Indeed, both cell types are thought to be central to disease progression by mediating synovial hyperplasia and the release of pro-inflammatory cytokines and tissue damaging enzymes [187,191–194]. In RA, resident synovial cells and both adaptive and innate immune cells establish a positive feed-forward activation loop mediated by pro-inflammatory cytokines such as TNF-α, IL-1β, IL-6, and IL-12, which perpetuate the disease and ultimately lead to joint destruction [195–198].

Experimental evidence accumulated over the last two decades has demonstrated beneficial effects of VIP in all stages of RA development through its anti-inflammatory and immunomodulatory abilities [131,158,199] (Figure 2). In addition, numerous studies have shown the direct antimicrobial activity of VIP against a wide range of bacteria [200], as well as its protective effect in polymicrobial sepsis [201]. Interestingly, VIP is able to counteract the effects of LPS from *Porphyromonas gingivalis* in monocytes, which is an oral bacterium related to increased risk of arthritis associated with periodontal disease [202,203].

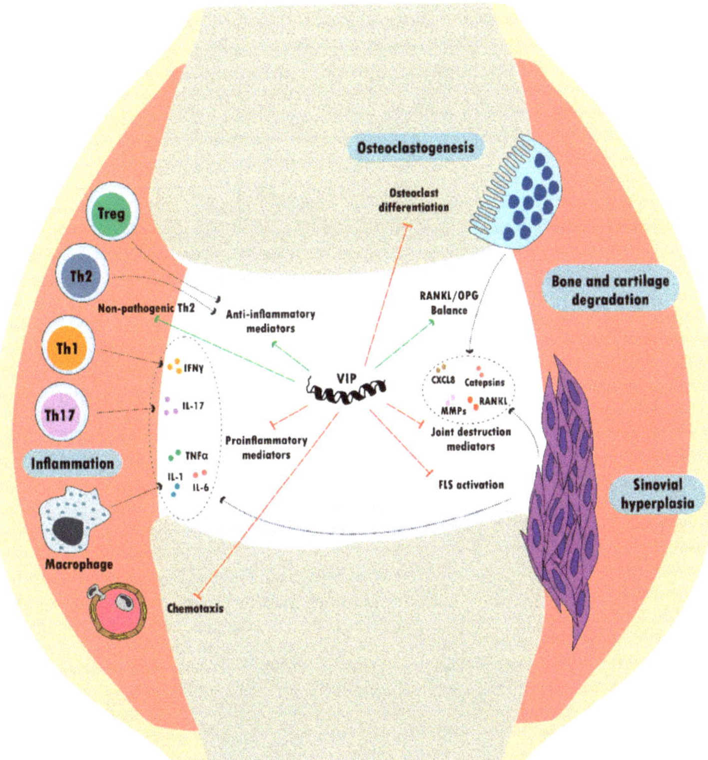

Figure 2. Biological effects of VIP in rheumatoid arthritis. Schematic representation of an RA joint. Green arrows indicate "induction", whereas red arrows indicate "inhibition".

Initial data about the anti-inflammatory properties and therapeutic potential of VIP in the context of RA were obtained in the CIA mouse model [11,199]. Exogenous administration of VIP was able to reduce the incidence and severity of arthritis in mice, inducing a dramatic decrease in cartilage and bone erosion. In that model, VIP has been demonstrated to modulate the subsets of Th lymphocytes by promoting a Th2-type response while expanding CD4$^+$ CD25$^+$ Treg [11,204]. In line with these findings, other studies in the same arthritis mouse model showed protective effects of VIP on bone destruction by modulating the RANK/RANKL/OPG system through the downregulation of the Th17 response and subsequent increase of the Treg/Th17 ratio [84,159,168]. Moreover, an inhibitory action of VIP on osteoclastogenesis has been described in CIA mice, exerting a direct effect on osteoclast progenitor cells purified from bone marrow, as well as through its modulatory action on stromal and osteoblast cells [84,205].

In light of the anti-inflammatory and immunomodulatory effects of VIP in the CIA model detailed above, several studies assessed the role of this neuropeptide in the context of human RA. Accordingly to the murine model, in vitro studies on human synovial fibroblasts, macrophages, peripheral blood lymphocytes, and Th cells from patients with RA confirmed the ability of VIP to regulate components of both innate and adaptive immune responses [158].

In brief, VIP has been shown to significantly attenuate the basal and TNF-α-induced production of pro-inflammatory chemokines and IL-6 in both synovial tissue suspensions and SF from RA patients [98]. Interestingly, such anti-inflammatory effects were later fully reproduced in cultured RA-SF by specific VPAC2 agonists, according to the dominant presence of that receptor described in these cells [38].

Subsequent studies on RA-SF also proved an inhibitory effect of VIP on the expression and signal transduction of some PRRs, which are linked to the pathogenic activation of these synovial cells [193,197] as previously explained. Moreover, VIP is able to downregulate the enhanced expression of the IL-22 specific receptor, preventing the IL-22 stimulatory effects on proliferation and production of matrix metalloproteinase-1 (MMP-1) and S100A8/A9 alarmins involved in RA-SF mediated joint destruction [206]. Likewise, it has been described that VIP counteracts the stimulatory effect of pro-inflammatory mediators, including TLR3 and TLR4 ligands, TNF-α, and IL-17, on the expression of IL-17 receptors and the IL-12 family of cytokines in RA-SF, which, in turn, mediates their cross-talk with Th1/Th17 cells [207]. Along with the anti-inflammatory effects of VIP in RA-SF through its action on TLR, this neuropeptide has been described to decrease the pro-inflammatory peptides corticotropin releasing factor (CRF) and urocortin (UCN)-1, while increasing the expression of the potential anti-inflammatory agents UCN-3 and CRF receptor 2 (CRFR2). Moreover, VIP is able to inhibit CREB activation, cyclooxygenase 2 expression, and prostaglandin 2 (PGE2) secretion in RA-SF [208].

In line with these findings, the potent anti-inflammatory role for VIP on cellular components of the immune system in the context of RA has also been validated by in vitro studies [158]. Upregulated levels of pro-inflammatory mediators, including TNF-α, IL-6, and CXCL8 and CCL2 chemokines, in polyclonally stimulated peripheral blood lymphocytes from RA patients were reduced after treatment with VIP [167]. Furthermore, regarding its effects on macrophages, VIP was able to impair the acquisition of the pro-inflammatory polarization profile described for macrophages in RA synovium, favoring instead an anti-inflammatory phenotype [32]. Additionally, the involvement of VIP in the modulation of Th subsets has been extensively studied, as previously detailed.

Apart from the effects of VIP treatment in animal models and in cultured cells from RA patients, recent studies have focused on evaluating the potential value of endogenous VIP as a biomarker in RA, as we discuss later.

5.1.2. VIP in Osteoarthritis

OA is a chronic rheumatic disease and is considered the most prevalent in developed countries and the main cause of incapacity in the elderly population. It is a complex multifactorial disease and is the clinical endpoint of heterogeneous disorders with common clinical, pathological, and radiological characteristics, resulting in the alteration of one or more joints [209–213]. Although it is usually an age-related disease, OA is also associated with other multiple risk factors that culminate in joint dysfunction, including genetic predisposition, epigenetic factors, gender, obesity, exercise, work related injury, and trauma [209,214–216].

OA is characterized by cell stress and extracellular-matrix (ECM) degradation, resulting in an imbalance in joint-tissue metabolism, which culminates in a progressive loss of synovial joint function, with pain and disability. While cartilage degradation is the main event, the view of OA as solely a pathology of cartilage has changed in recent years. This pathology affects the whole joint, resulting in the remodeling of adjacent subchondral bone, osteophyte formation, and synovial inflammation [213,217–226]. Although OA has an important mechanical component, it is currently also considered as a low-grade inflammatory disease. The biological imbalance and the mechanical stress lead to a pathological situation, with altered chondrocyte behavior, which results in the release of inflammatory mediators and ECM-degrading enzymes [19–22]. All of these factors, along with the inhibition of cartilage biosynthesis, increase the fragility and loss of cartilage integrity [23]. Although synovitis is usually localized and may be asymptomatic in OA [24], synovial activation causes the release of inflammatory mediators and proteases that accelerate the progression of the disease [2,25,26]. Moreover, the subchondral bone is also affected and is involved in the progression of OA through the release of catabolic mediators that promote an altered metabolism in chondrocytes [14,27].

The majority of available therapies for OA focus on relieving symptoms rather than slowing the progression of the disease. Therefore, it is important to find new therapeutic targets for the development of new drugs to treat the disease [227,228].

While the association of VIP with RA has been widely studied, its role in OA is not well established, although it is the second rheumatic pathology in which more advances have been obtained in the study of the VIP function [229] (Figure 3). Less is known about the role of VIP in other disorders such as systemic lupus erythematosus or spondyloarthritis (SpA). The effects of VIP reported in rheumatic diseases could be mediated in part by its action on the SF, as has been described in several in vitro studies [82,98,142,207,230]. OA-SF expresses and releases VIP, with a greater expression than RA-SF [38]. However, its expression is decreased in the synovial fluid and cartilage of OA patients compared to healthy controls, which could contribute to the pathology [231,232]. Regarding VIP receptors, both VPACs are detected in OA-SF with a greater expression of VPAC1. Pro-inflammatory mediators released to the joint microenvironment during the disease, such as TNF-α, decrease the expression of VIP, and modulate the VPAC1/VPAC2 ratio, therefore approaching its profile to that of RA-SF [38].

VIP is able to counteract the action of pro-inflammatory mediators, alleviating the inflammation and the pain in OA. VIP reduces the serum levels of TNF-α and IL-2 and increases serum IL-4 in a rat model of knee OA. In this model, VIP also inhibits proliferation of OA-SF and decreases the production of TNF-α, IL-2, MMP-13, and ADAMTS-5 (a disintegrin and metalloproteinase with thrombospondin motifs-5), at the same time that it induces the expression of type II collagen and osteoprotegerin, by inhibition of NFκB signaling [232,233]. In addition, VIP modulates the corticotropin-releasing factor family of neuropeptides, also increasing the expression of the potential anti-inflammatory mediators UCN-2 and -3, as well as CRFR2 in OA-SF. Moreover, VIP increases cAMP and induces CREB activation in OA-SF [208], which would support its anti-inflammatory role through the inhibition of other signaling pathways, involving JNK-MAPK, NFκB, or c-Jun, inhibiting the production of pro-inflammatory mediators and promoting the expression of anti-inflammatory cytokines [85,234–236].

On the other hand, some studies reported that the accumulation of VIP in joints can also contribute to the pathogenesis of OA. Thus, VIP treatment in rat OA knees promotes synovial hyperemia, as well as sensitization of joint afferent fibers via AC/cAMP/PKA, also increasing firing rate and decreasing mechanical threshold during movement. Therefore, VIP might promote mechanosensitivity and pain in rat OA models [37,225,232,237,238]. Moreover, Rahman et al. reported that VIP stimulates PGE2 production in human articular chondrocytes, human osteoblast-like cells, and human SF, as well as cAMP production in human osteoblast-like cells, suggesting a pro-inflammatory role for this peptide [239]. Another study also related increased VIP levels in the synovial fluid to the presence of synovitis in OA patients [240], suggesting that both downregulation and upregulation of VIP could contribute to the OA pathology [232].

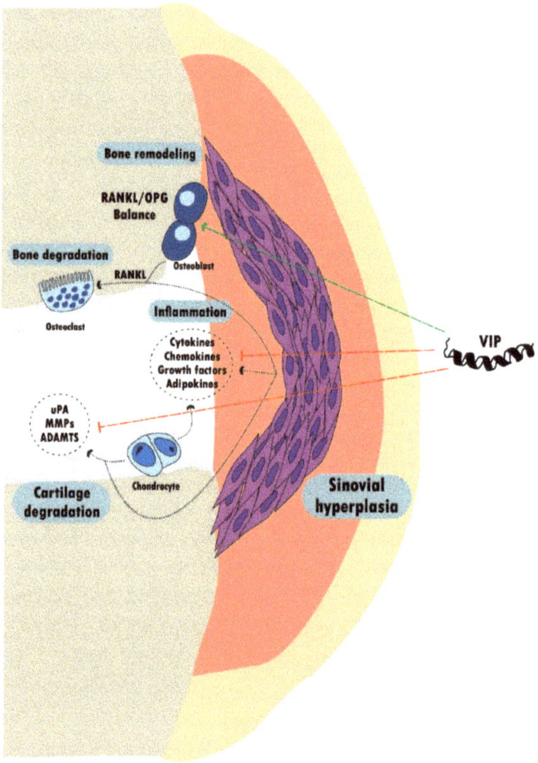

Figure 3. VIP effects in OA. Schematic overview of VIP effects in OA. Green arrows indicate "induction" whereas red arrows indicate "inhibition".

In addition to the inflammatory process, ECM degradation and cartilage loss is a key factor in the OA pathology. In this regard, VIP might prevent cartilage damage, since VIP modulates the profile of ECM-degrading enzymes released to the joint microenvironment by SF from OA patients. Thus, VIP decreases the expression and activity of the proteinase urokinase-type plasminogen activator (uPA), as well as the production of its receptor (uPAR), after stimulation with the pro-inflammatory cytokine IL-1β or the degradative mediators 45kDa fibronectin-fragments (Fn-fs). On the other hand, VIP induces the production of the plasminogen activator inhibitor-1 (PAI-1) under basal conditions in these cells. Furthermore, VIP reduces the production of MMP-9 in IL-1β- or Fn-fs-stimulated OA-SF, as well as MMP-13, the main proteinase involved in the degradation of type II collagen, after stimulation with Fn-fs [220]. In addition, VIP decreases the production of ADAMTS in OA-SF, including the aggrecanases ADAMTS-4 and -5, key proteinases in the degradation of aggrecan from the cartilage ECM, after IL-1β or Fn-fs stimulation, as well as the cartilage oligomeric matrix protein (COMP)-degrading ADAMTS, ADAMTS-7 after both stimuli, and -12 after Fn-fs treatment. In this sense, VIP also reduces COMP degradation from cartilage explants cultured with IL-1β- or Fn-fs-stimulated OA-SF, as well as the aggrecanase activity and glycosaminoglycans (GAGs) release only after Fn-fs stimulation. Moreover, VIP inhibits the activation of Runx2 transcription factor and Wnt/β-catenin signaling involved in ECM remodeling and proteinase expression, after the stimulation of these cells with both stimuli [241].

Few studies have focused on the presence of VIP and its receptors in chondrocytes [37]. As previously described in SF, articular cartilage from OA patients also has lower VIP levels compared to controls. Moreover, VIP expression in synovial fluid is positively correlated to its optical density in articular cartilage [231]. Juhász et al. described the expression of VPAC1, VPAC2, and PAC1 in chicken chondrogenic cells [242,243].

Concerning subchondral bone, VIP receptors have been described on osteoclasts and osteoblasts of several species, including human, mouse, and rat [37,244]. VIP inhibits osteoclast-mediated bone resorption and induces the production of IL-6 from osteoblasts, regulates the expression of osteoclastogenic factors like RANKL and OPG in osteoblasts, and seems to be involved in osteoblastogenesis [37,242,245,246]. In addition, VIP promotes osteoblast activity and proliferation and stimulates bone remodeling [244,247]. Furthermore, Xiao et al. showed higher VIP levels in the femoral bone from OA postmenopausal women compared to those with osteoporosis, where VIP was also positively associated with pain [248].

Recent studies also associate VIP levels to the progression of rheumatic diseases. In this regard, VIP levels in synovial fluid and cartilage of OA patients are negatively associated with progressive joint damage, being a potential indicator of disease severity [231]. In addition, VIP could be postulated as a potential therapeutic target in OA, since it is involved in the activation of several anabolic signaling pathways in the synovial joint [245].

5.2. Inflammatory Bowel Disease

Inflammatory bowel disease is the prevailing gut autoimmune disorder and comprises Crohn's disease and ulcerative colitis (UC). As with other autoimmune diseases, the origin is multifactorial and comprises genetic, environmental, and host-related factors that affect the development of bowel inflammation [249].

UC lesions are located within the colon, while CD is a relapsing remitting granulomatous disease, which can affect any segment of the digestive tract, producing transmural lesions. Although IBD pathogenesis is unclear, an atypical immune response to intestinal microbial products and/or food allergens represents an important causal factor. Moreover, interactions between the enteric nervous system (ENS) and the immune system play an important role in its pathophysiology [250,251]. These communications include the secretion of neuropeptides, which conduct signals bidirectionally between enteric neurons and immune effectors [252]. VIP and its receptors are expressed in the gastrointestinal tract to perform its anti-inflammatory/immunomodulatory action. The source of endogenous VIP in the gut could be of nervous origin, or from lymphoid cells. Concerning receptors, as we previously described, VPAC receptors are expressed in monocytes, macrophages, and T and B cells, as well as in myeloid cells, such as mast and polymorphonuclear cells.

To date, different models of chemically induced IBD have been characterized, showing several clinical, histological, and immune-response characteristic of UC and CD: the Dextran Sodium Sulfate (DSS), the oxazolone-induced colitis, TNBS, and Dinitrobenzene sulfonic acid (DNBS). The murine model has benefits, as well as limitations, in some characteristics of their clinical, immunological, and histopathological relevance to IBD. Administration of 3–10% DSS in the drinking water is one of the most common chemical methods used to induce colitis in rodents [253]. The oxazolone-induced colitis represents a model of Th2-driven inflammation. In this model, colitis is induced by intracolonic instillation of the haptenating agent oxazolone dissolved in ethanol after a skin pre-sensitization step [254]. However, there is limited information about the time course and cytokine profile of the immune response involved in this model of colitis.

Other models of colitis include the hapten-induced DNBS or TNBS that are administered by rectal instillation diluted in ethanol. The haptenization of host proteins induced infiltration of neutrophils, macrophages, and Th1 lymphocytes in the injured mucosa. In comparison to DNBS, TNBS is considered to be a hazardous chemical due to its highly oxidative properties. Since it was developed more recently, research using the DNBS-induced model is less common [255].

Gut inflammation and a differential expression profile of cytokines are key properties of their immune response. The TNBS colitis model develops with elevated Th1–Th17 response (increased IL-12 and IL-17), while DSS colitis switches from a Th1–Th17-mediated acute inflammation (increased TNF-α, IL6, and IL-17) to a central Th2-mediated inflammatory response (increment in IL-4 and IL-10 and associated reduction in TNF-α, IL6, and IL-17) [256]. This dissimilar cytokine profile has been used to establish an equivalence with human IBD. Thus, TNBS colitis mimics CD, while chronic DSS-colitis mimics UC [257].

The first report on the role of VIP as a therapeutic agent in IBD was published in 2003, in the TNBS colitis model [7]. VIP treatment reduced the clinical and histopathologic severity of TNBS-induced colitis, abolishing body-weight loss, diarrhea, and macroscopic and microscopic gut inflammation. The VIP effect is mediated by both innate and acquired immune responses. Regarding innate immunity, administration of VIP in the TNBS model decreased myeloperoxidase activity in colon extracts, a specific marker of neutrophils, and reduced the expression of receptors involved in neutrophil recruitment, such as CXCR1 and CXCR2 [7,156,258]. CD4[+] T helper cells are major initiators of IBD. CD4[+] T cells are enriched in the gut of patients with CD and UC and blockade or reduction of CD4[+] T is effective in treating patients with IBD [259].

Th1 and Th17 subsets are important players in the development of CD [260]. In the TNBS model, we reported that VIP reduced IFN-γ and TNF-α enhancing IL-4 and IL-10 levels in colon and cell cultures from splenocytes and lamina propria immune cells, thus promoting Th2 vs Th1 responses. VIP diminished IL-17, IL-21, and IL-17R mRNA expression in the colon, supporting an inhibitory action over the Th17 cell subpopulation. Interestingly, VIP increased the Foxp3 and transforming growth factor (TGF)-β mRNA expression in CD4[+] cells from mesenteric lymph nodes, as well as the IL-10 expression in the colon upregulating Treg responses [7,156,258]. Finally, VIP also reduced the TNBS-induced numbers of TCD4 lymphocytes, whereas it induced an increase in the number of B-lymphocytes (CD19[+]) in mesenteric lymph nodes.

To date, different mechanisms involved in the therapeutic effect of VIP have been described. One of the first actions described was the VIP modulation of TLR. Among the environmental factors, the modification of gut microbiota or dysbiosis has been reported as a key element in the development of IBD [261].

Additionally, the receptors of the innate immune system, TLRs, affect many aspects of IBD etiology, including immune responses and microbiota. Differential expression of TLRs in IBD patients in comparison with healthy donors has been characterized. Modification of TLR expression or signaling has been reported, not only in experimental models of IBD in mice, but also in human IBD. Most TLR signaling pathways participate in the development of IBD and are sometimes beneficial and other times harmful [262]. Nevertheless, much of the evidence has indicated that the TLR2 and TLR4 signaling pathway has a negative role in IBD. It was reported that the inhibition of TLR2–TLR6/1 activity ameliorated DSS-induced colitis. In healthy patients, TLR4 is expressed at a low level in intestinal epithelial cells; however, its expression was upregulated in the intestinal epithelia of patients with active UC, suggesting that TLR4 could be involved in UC disease development [262].

In the TNBS-induced colitis model, VIP treatment exerts a time-course inhibition of TLR2 and TLR4 expression in colon epithelial and mononuclear cells. Moreover, VIP acts at a systemic level in lymph nodes. Mesenteric lymph nodes are the draining nodes of the intestinal tract that regulate the traffic of lymphoid cells. VIP inhibits the TNBS-induced TLR2 and TLR4 overexpression in macrophages, dendritic cells and the lymphocyte subpopulations, T CD4[+], T CD8[+], and B CD19[+] [134,135]. The peptide also enhances the expression of Foxp3 and TGF-β, which are both involved in regulatory T-cell function. Overall, we reported that, after specific stimulation of TLR2 and TLR4, VIP exerts homeostatic function, balancing innate and adaptive immune responses in the murine model of CD, both locally in the colon and at the periphery in lymphoid nodes [131,156,263,264].

Another study using the TNBS-induced colitis model reported that treatment with VIP did not modify the clinical and histological parameters [265]. However, another recent study confirmed our results. Because the nanocarrier sterically stabilized micelles (SSM) protect peptides from enzymatic degradation, ameliorating their bioavailability and half-life, Jayawardena et al. developed sterically stabilized micelles of VIP (VIP-SSM). They characterized the healthy role of VIP and VIP-SSM in the DSS-induced colitis model. At clinical and histological levels, VIP and nanoparticles of VIP treatment decreased the pro-inflammatory cytokine profile in the colon, reducing tight junction and ion-transporter protein expression associated with severe DSS colitis [266].

It is also important to note that VIP has shown beneficial effects in other models of colitis, such as colitis induced by *Citrobacter rodentium* [267] and the oxazolone-induced colitis [268].

Results are variable in knockout mice of the VIP/VPAC receptor axis, [269]. The VPAC2 receptor KO mice showed worse progression of DSS-induced colitis, whereas VPAC1 knockout DSS-induced colitis in VPAC1-KO mice was resistant to colitis [270]. Concerning VIP knockout mice, the results using the chemical-induced colitis models are contradictory. Thus, DNBS and DSS-induced colitis were more severe in VIP-KO than wild-type mice. VIP treatment recovered the phenotype, protecting VIP-KO mice against DSS colitis. Moreover, VIP is beneficial for the development and maintenance of a colonic epithelial barrier structure under physiological conditions and promoting epithelial repair and homeostasis during colitis [271]. Abad et al. reported in the TNBS model that mice lacking VIP developed reduced colitis [272]. These discordant results using the KO model of the VIP axis could be explained by the presence of differential microbiota, by alterations in the development of the chemical-induced model of colitis or by the existence of compensatory mechanisms in VIP-KO, by PACAP, a related peptide, or by another mediator.

Despite the scarce dissimilar results about the effect of VIP in the TNBS model of CD and those obtained with the KO mice, the conclusive results are robust and are summarized in Figure 4.

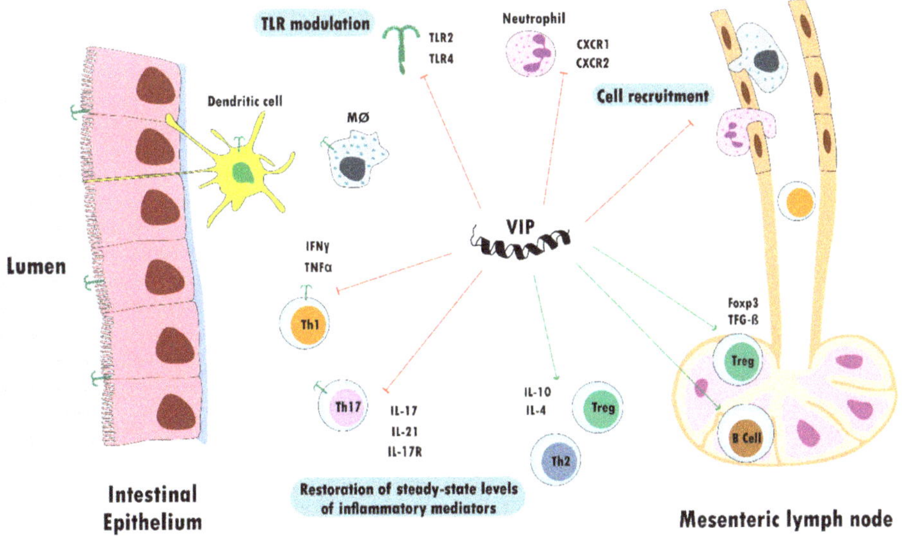

Figure 4. Biological effects of VIP in the TNBS-induced murine model of Crohn's disease. Main effects of VIP on disease's development are represented schematically. Green arrows indicate "induction", whereas red arrows indicate "inhibition".

Data on the role of VIP in humans are scarce and are relative to the presence of the peptide in health conditions and in IBD patients. Contradictory results about alterations in gut VIP innervation in IBD patients have been reported. Several studies described enhanced VIP expression in the intestine in IBD patients [273]. Moreover, an increase in VIP immunoreactivity both in nerve fibers and neurons were characterized in CD patients [264]. Conversely, other studies have characterized a reduction in the abundance of VIP-immunoreactive nerve fibers in the lamina propria and submucosa in both CD and UC patients. Remarkably, the variation in the decrease was significantly related to the severity of the disease [274]. In general, it is well recognized that a broad loss of mucosal neuropeptidic innervation may be related to areas of high inflammation; thus, the contradictory results could be explained by the experimental conditions.

In a recent elegant study, Sun et al. described the beneficial role of VIP in UC patients. In agreement with data reported in several rheumatic diseases [245,275,276], they found that serum VIP levels are lower in UC patients than in healthy controls. The study provided evidence showing that VIP serum levels are lower in IgE$^+$ UC patients than that in IgE$^-$ UC patients [268].

The same authors found that in the regulatory B cells (Bregs) from peripheral blood of UC patients, immune suppressive function is impaired, probably due to lower serum VIP levels and lower IL-10 expression in Bregs. This expression increases with the presence of VIP, which stabilizes IL-10 expression in Bregs. In brief, they demonstrated that VIP administration restored Breg function, inhibited pro-inflammatory cytokine production, prevented the allergen-specific T-cell response, and reestablished colon tissue structure in experimental colitis. All of these results suggest that VIP is a potential therapeutic agent for UC patients with atypical immune responses to food allergens [268].

To date, no therapy is yet available for the treatment of IBD and combined therapy seems to be the best approach [261]. Despite the contradictory data from KO models, the VIP axis represents a promising candidate for use in combined therapy due to its multistep action on the immune response.

5.3. Central Nervous System Diseases

The involvement of inflammatory processes and the adaptive immune system in the pathophysiology of neurodegenerative diseases is supported by evidence from a variety of studies [277–280]. Thus, inflammatory responses may be involved in both regenerative and degenerative processes, e.g., in multiple sclerosis and Parkinson's disease (PD). During neuroinflammation, the activation of the glial cells of the brain, mainly microglia and astrocytes, release several factors, many of which are pro-inflammatory, neurotoxic, and damaging to nerve cells.

In vitro and in vivo investigations have described potent neuroprotective features for VIP promoting neural cell proliferation, survival, axon regeneration, and production of neurotrophic factors, as well as inhibition of inflammation [108,281,282]. All of this indicates that the VIP/receptors axis could be a novel therapeutic target in multiple sclerosis and Parkinson's disease.

Multiple sclerosis is a chronic inflammatory autoimmune and neurodegenerative pathology of the CNS that leads to demyelination. Experimental autoimmune encephalomyelitis is the most common animal model for MS sharing many clinical and pathophysiological features resulting in the generation of autoreactive T-cells, eventually culminating in myelin destruction. Like that observed in other inflammatory diseases, treatment with VIP reduced the clinical and pathological scores in EAE with a blockade of symptoms lasting 60 days. These effects were associated with decreased spinal cord levels of pro-inflammatory cytokines (TNF-α, IL-6, IL-1β, IL-18, and IL-12), iNOS, chemokines (CCL5, CCL3, CXCL1, CCL2, and CXCL10), and CC chemokine receptors (CCR-1, CCR-2, and CCR-5) and increased levels of the anti-inflammatory cytokines IL-10, IL-1Ra, and TGF-β [180].

These investigations revealed that VIP treatment decreases the presence of encephalitogenic Th1 cells in the periphery and the CNS. As a consequence, VIP reduces the appearance of inflammatory infiltrates in the CNS, the loss of oligodendrocytes, and the subsequent demyelination and axonal damage typical of EAE [283].

Several studies point to Tregs as key agents in MS and EAE controlling self-reactive cells and inducing the decrease in inflammation [284–287]. Administration of VIP to EAE mice induces the expansion of Treg cells that express CD4$^+$ CD25$^+$ Foxp3 and produces IL-10/TGF-β in the periphery and the CNS [288].

In accordance with these results, it has been described that mice with a genetic deletion of the VPAC2 gene exhibit an exacerbation of EAE induced by MOG35-55 compared to wild-type mice, presenting an increased pro-inflammatory cytokine profile (TNF-α, IL-6, IFN γ, and IL-17) and reduced production of anti-inflammatory cytokines (IL-10, TGF-β, and IL-4) in the CNS and lymph nodes. In addition, the proliferative index and the in vivo suppressor activity of CD4$^+$ CD25$^+$ FoxP3$^+$ Tregs are markedly reduced in KO VPAC2 mice with EAE [289]. These results point toward an important protective role for the VPAC2 receptor against autoimmunity and as an anti-inflammatory mediator [289].

Unexpectedly, and in contrast, Abad et al. found that VIP-KO mice are highly resistant to EAE. This finding was confirmed by histopathology and clinical evaluation. Supporting this phenotype, the levels of multiple pro-inflammatory cytokines in the spinal cord were strikingly reduced in the KO. The authors found that immune cells were trapped in the meninges of the brain and spinal cords and failed to invade the CNS parenchyma, suggesting a defect in immune-cell migration [290]. Clinical disease in these mice was blocked at a step downstream from immunization. Similarly, EAE clinical symptoms are significantly ameliorated in VPAC1-KO mice. The results demonstrate stronger Th1 and Th17 responses, which are known to induce the pathogenesis of EAE, but reduced Th2 responses in these mice. As the phenotype of VPAC1 is opposite that of VPAC2-KO mice, it has been suggested that, in addition to Th polarization, other events are differentially mediated by VPAC1 and VPAC2, and this may depend on different factors, such as their level of expression, the state of cellular activation, the interaction with other mediators present in the microenvironment, such as inflammatory factors, and finally, the disease phase [291].

Studies in patients with MS have reported alterations in components of the VIP/receptors signaling system. Andersen et al. found a reduced VIP immunoreactivity in the cerebrospinal fluid of patients diagnosed with MS [292]. Similarly, Baranowska-Bik et al. observed a tendency toward reduced levels of VIP in the cerebrospinal fluid of multiple sclerosis patients, although this difference was not statistically significant [293]. Interestingly, CD4$^+$ cells derived from the peripheral blood of patients with MS show a differential expression of the VPAC1 and VPAC2 receptors, compared to those from healthy controls, depending on the activation status of the cells. Without stimulation, similar patterns of VIP receptor expression are detected in CD4$^+$ cells of subjects with MS and healthy controls with a visible expression of VPAC1 and minimum levels of VPAC2. However, the markedly decreased expression of VPAC1 observed after stimulation and activation of CD4$^+$ T cells is compensated for by a higher expression of VPAC2 in healthy individuals. Nevertheless, activated CD4$^+$ T cells from SM patients exhibit an altered expression of VPAC2 as a result of altered gene regulation in the promoter region of the VIP receptor. As a consequence, CD4$^+$ T cells were less sensitive to VIP and biased the system predominantly in a Th1 direction [294].

Finally, the role of the VIP/VPAC axis has also been investigated in Parkinson's disease. In this progressive degenerative movement disorder, the key roles played by the subsets of CD4$^+$ cells, especially the Treg whose number or activity is reduced in this pathology, have recently been highlighted. This, together with microglial activation, leads to changes in the microenvironment of the affected brain with oxidative stress, inflammation, and defective protein folding [295–297].

Using murine models of Parkinson's disease, administration of VPAC2 agonists has been shown to increase Treg activity without altering cell numbers, reduce microglial inflammatory responses, increase survival of dopaminergic neurons, and improve striatal densities [63,296].

5.4. Other Autoimmune Disorders

Type 1 diabetes and Sjögren's syndrome (SS) represent other autoimmune diseases in which the beneficial effects of VIP have been shown. Type I diabetes is an autoimmune disease mediated by T cells associated with the overexpression of inflammatory mediators and the alteration of different subsets of T cells that attack the insulin-producing cells in the pancreas.

Several animal models that develop spontaneous type 1 diabetes have been described, such as NOD mice that exhibit T-cell-mediated insulitis linked to the genes of the major histocompatibility complex. In the NOD mouse model, VIP prevents the increase in the proportion of Th1 to Th17 cells, changes the Tregs/Th17 ratio that leads to tolerance, and reverses the proportion of subsets of Th1/Th2 cells associated with autoimmune pathology. These effects add to the decrease in pro-inflammatory mediators, resulting in a reduction in the destruction of β cells in the pancreas [10].

Studies in KO mice have confirmed the importance of the VIP/VPAC axis in the functionality of the endocrine pancreas. Thus, Martin et al. found that VIP-KO mice exhibit elevated plasma glucose, insulin, and leptin levels [298]. In VPAC2-KO mice, glucose-induced insulin secretion is decreased, with no change in glucose tolerance and mice deficient in VPAC1 show small dysmorphic islets of Langerhans and exhibit impaired neonatal growth that leads to intestinal obstruction and hypoglycemia [299]. Moreover, selective overexpression of the human VIP gene increases glucose-induced insulin secretion in pancreatic β-cells and ameliorates glucose intolerance of 70% depancreatized mice [300].

SS is an autoimmune disease characterized by the infiltration of T lymphocytes at the level of the salivary and lacrimal glands that causes their destruction and the appearance of symptoms related to dry mucous membranes. Recently, the effects of VIP on the immune response and secretory function of the submandibular glands have been investigated by using the NOD model of SS, which develops secretory dysfunction and early loss of glandular homeostatic mechanisms, with mild infiltration in the glands.

Li et al. showed that VIP treatment was able to reduce immune lesions in the exocrine glands and improve the secretory function of these glands by negatively regulating the expression of IL-17A in the exocrine glands. It also improved the secretory function of the exocrine glands by increasing the expression of AQP5, a protein that participates in the transport of water through the glandular epithelium [301].

In the course of salivary function impairment in the NOD mouse model, a progressive decrease in VIP expression in the submandibular glands is observed compared to normal mice. The loss of endogenous VIP is associated with a loss of acinar cells through apoptotic mechanisms that could be further induced by TNF-α and reversed by VIP through a PKA-mediated pathway. The clearance of apoptotic acinar cells by macrophages is impaired by NOD macrophages contributing to the loss of gland homeostasis [302].

Lodde et al. constructed the vector recombinant serotype 2 adeno-associated virus, encoding the human VIP transgene (rAAV2hVIP), to explore its usefulness in SS management. Instillation of rAAV2hVIP in the submandibular glands of NOD mice leads to higher salivary flow rates and increased expression of VIP in the glands and serum, as well as to a reduction of cytokines IL-2, IL-10, and IL-12 (p70) and TNF-α in SMG extracts, and of serum CCL5, compared to the control vector. This work indicates that VIP may be a promising agent for the treatment of the salivary component of SS [9].

Finally, data from humans reveal that monocytes from SS patients show increased expression of VPAC2, which is absent in the monocytes of normal subjects without changes in the expression of VPAC1. This altered expression correlates with an impaired phagocytosis of apoptotic epithelial cells, with reduced engulfment capacity and failure to express an immunosuppressant cytokine profile that is not restored by VIP. This differential expression of VPAC2 associated with phagocytic dysfunction suggests its potential as a functional biomarker in SS [303].

6. VIP as a Therapeutic Agent: Limitations and Perspectives

The treatment of inflammatory and autoimmune diseases is a challenge. Advances in knowledge of the underlying pathophysiological mechanisms, as well as the discovery of biological therapies against potent mediators of inflammation, have revolutionized the way these diseases are treated. Newly developed molecules aim to diminish the impact of these diseases on the quality of life of patients, although, to date, there is no curative treatment.

Since the discovery of VIP half a century ago, more knowledge about its biology, its signaling mechanisms, and its powerful anti-inflammatory effects, as well as its immunoregulatory capacity, have made it a potential therapeutic agent for diverse diseases. Among these disease are asthma [304], pulmonary hypertension [305], sarcoidosis [306], neurological diseases such as Alzheimer's and

Parkinson's [307,308], inflammatory bowel diseases such as Crohn's [7,264], autoimmune diabetes [10], and cancer [309].

Marketed under the name Aviptadil, VIP has been used in the clinic successfully in the treatment of pulmonary hypertension and sarcoidosis. However, the potential of this peptide at the therapeutic level in clinical practice is still far from its theoretical potential. This is due to its high sensitivity to degradation by proteases, spontaneous hydrolysis, and catalytic antibodies [310,311]. A second limitation for the use of VIP in humans is due to cross-interactions, given their ability to bind to different GPCRs, their functional pleiotropism, and their ubiquity. In addition, systemic administration of VIP with binding to multiple cell targets with high affinity could cause unwanted adverse effects [312,313].

Therefore, strategies have been developed to overcome these difficulties. Thus, distribution systems directed against specific targets that also protect the peptide against its degradation are desirable options. Recent advances in this field include the following: the use of metal nanoparticles, which seem to increase the therapeutic potential of VIP both in terms of target and distribution [314]; the use of modified liposomes with lipopeptides conjugated with VIP, which have demonstrated a selective recognition of VPACs and a more effective antitumor activity in a recent study with human osteosarcoma lines [315], as well as nanomicelles, tested in breast cancer [316].

Interestingly, a single subcutaneous injection of a low dose of camptothecin sterically stabilized micelles conjugated with VIP (CPT-SSM-VIP) administered to mice with collagen-induced arthritis was able to abrogate joint inflammation, with no apparent systemic toxicity and with similar efficacy and safety compared to methotrexate, used clinically for RA treatment [317]. The efficacy of this therapeutic approach has been confirmed in the murine model of colitis induced by DSS. Similarly, the anti-inflammatory and antidiarrheal effects of VIP can be achieved effectively when administered as a nanomedicine.

In relation to the problem posed by VIP's binding to different receptors, stable analogues of the VPAC1 and VPAC2 receptors have been developed recently, based on the technology of the peptidases-resistant foldamers (Longevity Biotech), LBT3627 and LBT3393. LBT-3627 is a VPAC2 selective neuroprotective agent that has been successfully investigated in the preclinical phase in a Parkinson's model (Olson et al. 2015 [63]. The results obtained highlight the therapeutic immunomodulatory potential of this agonist to restore Treg activity, attenuate neuroinflammation, and intercept dopaminergic neurodegeneration in PD, as we mentioned above [296].

Finally, gene therapy with VIP, using lentiviral vectors, has yielded good results in the CIA model [318], and VIP adenoviral vectors have also been developed [9]. However, these approaches continue to lack cellular and tissue specificity. Thus, another possibility under study is cell therapy with dendritic cells transduced with a VIP lentiviral vector (LentiVIP-CDs), whose therapeutic effects in sepsis and EAE models have been very positive with a single local administration [179].

Looking toward the future, despite advances in therapeutic options, there is still a need to continue researching the design and transfer to the clinic of stable VIP analogues and specific VPAC1- and VPAC2-receptor drugs, directed against specific objectives, as well as biomarker field approaches to intervene earlier in the course of the disease.

7. Potential of VIP Axis as a Biomarker for Personalized Treatment in Rheumatic Diseases

In addition to its potential use as a therapeutic agent, the VIP axis could be used in a second potential translational strategy as a prognostic biomarker. Different studies have described an altered expression in the VIP/VPAC axis in autoimmune diseases and in the modulation of the inflammatory immune response in rheumatic diseases [82,98,153,167]. In juvenile idiopathic arthritis, serum VIP levels are decreased in patients who manifest more disease activity characterized by cardiac autonomic neuropathy associated with parasympathetic dysfunction [319]. In RA, the expression of VPAC1 is decreased in peripheral blood mononuclear cells (PBMCs) [318], and a lower expression of VPAC1 mRNA is also observed in patients with early arthritis (EA) [320]. However, the expression of VPAC2 is increased in SF and PBMCs show an increased expression of VPAC2 mRNA [38,320]. A deregulated expression of VPAC2 has also been described in monocytes isolated from patients with SS and in activated CD4$^+$ T cells from patients with multiple sclerosis [294].

The abnormal expression of the VIP axis in autoimmune diseases directed the investigations toward the study of its association with the clinical course of some of these diseases and their possible prognostic value. In RA, early diagnosis and the establishment of immediate and effective therapy are essential to prevent greater disease severity [321,322]. Although different parameters have been proposed as prognostic markers for RA (such as rheumatoid factor ACPAs, erythrocyte sedimentation rate, and C-reactive protein), they are only capable of classifying 65% of patients [323–326]. In this sense, patients with RA with high or moderate activity after two years of follow-up had lower levels of VIP at baseline [275]. In multivariate analyses, it was observed that ACPA-negative patients had an odds ratio (a statistic that quantifies the strength of the association between two events) of 6.1, having high activity at two years of follow-up if their initial serum VIP levels were low. This allows the classification of a group of patients with a greater need for treatment within the ACPA seronegative RA patients. Another factor that makes VIP a potential prognostic marker worthy of further study is the fact that several single nucleotide polymorphisms (SNPs) in the VIP gene are associated with differences in serum VIP levels in patients with EA. The combination of three SNPs (minor alleles in rs688136 and absence of minor alleles in rs35643203 and rs12201140) in the VIP gene allows the identification of patients with less-severe disease, and thus possibly good candidates for less-intensive therapy [327].

Regarding the VIP receptors, it has been observed that the expression of VPAC1 and VPAC2 could reflect the clinical status in patients with EA with a significantly lower expression of VPAC1 when patients have systemic inflammatory activation characterized by high serum levels of IL6 and higher levels of Disease Activity Score 28 (DAS28). DAS28 is an index of the disease activity developed and validated by the EULAR (European League Against Rheumatism) [320]. In addition, the VPAC2 expression prevailed over VPAC1 in cells polarized toward Th17 of EA patients [170]. VPAC2 can also mediate anti-inflammatory effects when the expression of VPAC1 is low [38].

Serum VIP levels also showed a prognostic value in spondyloarthritis, a family of rheumatic diseases that share clinical and radiological manifestations where the most prevalent group is ankylosing spondylitis. These patients are HLA-B27 positive, and their inflammation usually occurs with enthesitis and bone formation that can lead to ankylosis. Patients with SpA presented a wide heterogeneity in terms of clinical manifestations, and there are no good biomarkers that predict progression. In early SpA, patients with lower VIP levels showed more disability and factors related to increased inflammation (bone edema on MRI scan, anemia, enthesitis, and cutaneous psoriasis) [276]. Finally, it has been described that VIP levels may have a protective role in the progression of OA [231].

In summary, VIP is an excellent aspirant to be used in clinical practice as a prognostic biomarker that would complement existing markers, such as ACPAs. Concerning receptors, they emerge as good candidates for activity biomarkers, and current studies would expand their potential as a severity biomarker.

Figure 5 summarizes the current advances in the role of the VIP/receptors axis as biomarkers in rheumatic diseases.

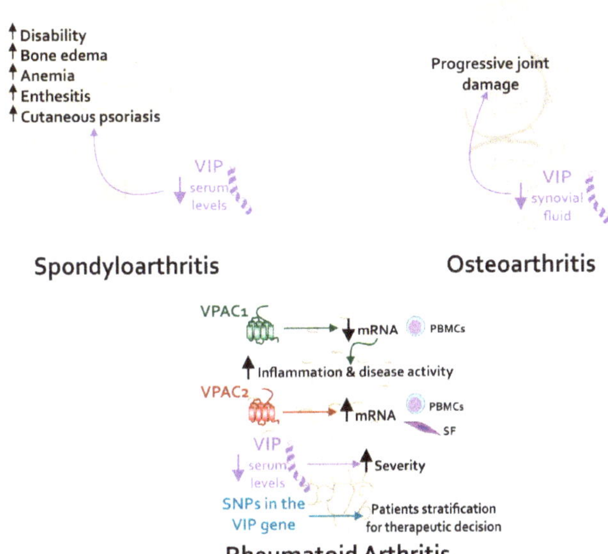

Figure 5. VIP and its receptors as biomarkers in rheumatic diseases. The scheme shows the current advances on the role of the VIP axis and VPAC receptors as biomarkers in spondyloarthritis (SpA), osteoarthritis (OA), and rheumatoid arthritis (RA). ↑Higher. ↓Lower. Purple arrows: association of VIP levels; green arrow: association of VPAC1 expression; red arrow: VPAC2 expression; blue arrow: clinical utility of SNPs (single nucleotide polymorphisms) in VIP gene.

8. Conclusions

On the 50th anniversary of VIP's discovery, this review updates our knowledge about the regulatory functions of the VIP/receptors axis in the immune system and presents a spectrum of potential clinical benefits applied to inflammatory and autoimmune diseases. This article gathers the findings and advances achieved in this field, thanks to the work of numerous researchers, from both basic and translational research areas.

Recent progress in improving the stability, selectivity, and effectiveness of VIP/receptors analogues and new routes of administration are highlighted, as well as important advances in their use as biomarkers, contributing to their potential application in precision medicine.

Despite the achievements, it is necessary to continue researching the design of analogue drugs that are stable, safe, and directed against specific objectives and in the validation of the VIP/receptors axis as biomarkers such that their application in clinical practice becomes a reality for our Very Important Patients.

Author Contributions: C.M., Y.J., I.G.-C., M.C., S.P.-G., A.L., M.M., I.G.-Á., and R.P.G. wrote the manuscript; D.C. and R.V.-R. performed the figures; C.M. coordinated the manuscript; S.P.-G., R.V.-R., and D.C. coordinated the bibliography and the format of the manuscript. All authors have read and agreed to the published version of the manuscript.

Funding: This work was supported by the Fondo de Investigación Sanitaria, Instituto de Salud Carlos III (Grants N°: PI14/ 00477, PI17/00027, RD16/0012/0008, RD16/0012/0011) and by the Ministerio de Economía y Competitividad (RTC-2015-3562-1), co-financed by Fondo Europeo de Desarrollo Regional (FEDER).

Acknowledgments: We are grateful to all patients and the collaborating clinicians for their participation in this study. We are also grateful to Sarah Young for her contribution to the editing of the English manuscript.

Conflicts of Interest: The authors declare that the research was conducted in the absence of any commercial or financial relationships that could be construed as a potential conflict of interest.

Abbreviations

AC	Adenylate cyclase
ACPA	Anti-citrullinated protein antibodies
ADAMTS	A disintegrin and metalloproteinase with thrombospondin motifs
Breg	Regulatory B cells
cAMP	Adenylate cyclase activity
CD	Crohn's disease
CIA	Collagen-induced arthritis
CNS	Central nervous system
CREB	cAMP response element-binding
CRF	Corticotropin-releasing factor
DC	Dendritic cells
DNBS	Dinitrobenzene sulfonic acid
DSS	Dextran Sodium Sulfate
EAE	Experimental autoimmune encephalomyelitis
ECM	Extracellular matrix
ERK	Extracellular regulated kinases
Fn-fs	Fibronectin-fragments
GHRF	Growth hormone–releasing hormone
GPCRs	G-protein-coupled receptors
GRK	GPCR kinase
IBD	Inflammatory bowel disease
IFN	Interferon
IL-	Interleukin
iNOS	Inducible nitric oxide synthase
IRF	Interferon regulatory factor
KO	Knockout
LPS	Lipopolysaccharide
MAPK	Mitogen-activated protein kinase
MMP	Matrix metalloproteinase
MS	Multiple Sclerosis
NFκB	Nuclear factor κB
NO	Nitric oxide
NOD	Non-obese diabetic
OA	Osteoarthritis
PACAP	Pituitary adenylate cyclase-activating polypeptide
PBMC	Peripheral blood mononuclear cell
PG	Prostaglandin
PKA	Protein kinase A
RA	Rheumatoid arthritis
RAMP	Receptor activity modifying protein
SF	Synovial fibroblasts
SNP	Single-nucleotide polymorphism
SpA	Spondyloarthritis
SS	Sjögren's syndrome
Th	T helper
TLR	Toll-like receptors
TNBS	Trinitrobenzenesulfonic acid
TNF	Tumor necrosis factor
Treg	Regulatory T cells
UC	Ulcerative Colitis
UCN	Urocortin
uPA	Urokinase-type plasminogen activator
uPAR	Urokinase-type plasminogen activator receptor
VIP	Vasoactive intestinal peptide

References

1. Weigent, D.A.; Carr, D.J.; Blalock, J.E. Bidirectional communication between the neuroendocrine and immune systems. Common hormones and hormone receptors. *Ann. N. Y. Acad. Sci.* **1990**, *579*, 17–27. [CrossRef] [PubMed]
2. Goetzl, E.J.; Sreedharan, S.P. Mediators of communication and adaptation in the neuroendocrine and immune systems. *FASEB J.* **1992**, *6*, 2646–2652. [CrossRef] [PubMed]
3. Souza-Moreira, L.; Campos-Salinas, J.; Caro, M.; Gonzalez-Rey, E. Neuropeptides as pleiotropic modulators of the immune response. *Neuroendocrinology* **2011**, *94*, 89–100. [CrossRef] [PubMed]
4. Gomariz, R.P.; Martinez, C.; Abad, C.; Leceta, J.; Delgado, M. Immunology of VIP: A review and therapeutical perspectives. *Curr. Pharm. Des.* **2001**, *7*, 89–111. [CrossRef]
5. Gomariz, R.P.; Juarranz, Y.; Abad, C.; Arranz, A.; Leceta, J.; Martinez, C. VIP-PACAP system in immunity: New insights for multitarget therapy. *Ann. N. Y. Acad. Sci.* **2006**, *1070*, 51–74. [CrossRef]
6. Delgado, M.; Ganea, D. Vasoactive intestinal peptide: A neuropeptide with pleiotropic immune functions. *Amino Acids* **2013**, *45*, 25–39. [CrossRef]
7. Abad, C.; Martinez, C.; Juarranz, M.G.; Arranz, A.; Leceta, J.; Delgado, M.; Gomariz, R.P. Therapeutic effects of vasoactive intestinal peptide in the trinitrobenzene sulfonic acid mice model of Crohn's disease. *Gastroenterology* **2003**, *124*, 961–971. [CrossRef]
8. Li, H.; Mei, Y.; Wang, Y.; Xu, L. Vasoactive intestinal polypeptide suppressed experimental autoimmune encephalomyelitis by inhibiting T helper 1 responses. *J. Clin. Immunol.* **2006**, *26*, 430–437. [CrossRef]
9. Lodde, B.M.; Mineshiba, F.; Wang, J.; Cotrim, A.P.; Afione, S.; Tak, P.P.; Baum, B.J. Effect of human vasoactive intestinal peptide gene transfer in a murine model of Sjogren's syndrome. *Ann. Rheum. Dis.* **2006**, *65*, 195–200. [CrossRef]
10. Jimeno, R.; Gomariz, R.P.; Gutierrez-Canas, I.; Martinez, C.; Juarranz, Y.; Leceta, J. New insights into the role of VIP on the ratio of T-cell subsets during the development of autoimmune diabetes. *Immunol. Cell Biol.* **2010**, *88*, 734–745. [CrossRef]
11. Delgado, M.; Abad, C.; Martinez, C.; Leceta, J.; Gomariz, R.P. Vasoactive intestinal peptide prevents experimental arthritis by downregulating both autoimmune and inflammatory components of the disease. *Nat. Med.* **2001**, *7*, 563–568. [CrossRef] [PubMed]
12. Gomariz, R.P.; Juarranz, Y.; Carrión, M.; Pérez-García, S.; Villanueva-Romero, R.; González-Álvaro, I.; Gutiérrez-Cañas, I.; Lamana, A.; Martínez, C. An Overview of VPAC Receptors in Rheumatoid Arthritis: Biological Role and Clinical Significance. *Front. Endocrinol.* **2019**, *10*, 729. [CrossRef] [PubMed]
13. Schett, G.; Elewaut, D.; McInnes, I.B.; Dayer, J.M.; Neurath, M.F. How cytokine networks fuel inflammation: Toward a cytokine-based disease taxonomy. *Nat. Med.* **2013**, *19*, 822–824. [CrossRef] [PubMed]
14. Rahman, P.; Inman, R.D.; El-Gabalawy, H.; Krause, D.O. Pathophysiology and pathogenesis of immune-mediated inflammatory diseases: Commonalities and differences. *J. Rheumatol. Suppl.* **2010**, *85*, 11–26. [CrossRef] [PubMed]
15. David, T.; Ling, S.F.; Barton, A. Genetics of immune-mediated inflammatory diseases. *Clin. Exp. Immunol.* **2018**, *193*, 3–12. [CrossRef] [PubMed]
16. Perricone, C.; Versini, M.; Ben-Ami, D.; Gertel, S.; Watad, A.; Segel, M.J.; Ceccarelli, F.; Conti, F.; Cantarini, L.; Bogdanos, D.P.; et al. Smoke and autoimmunity: The fire behind the disease. *Autoimmun. Rev.* **2016**, *15*, 354–374. [CrossRef]
17. Surace, A.E.A.; Hedrich, C.M. The Role of Epigenetics in Autoimmune/Inflammatory Disease. *Front. Immunol.* **2019**, *10*, 1525. [CrossRef]
18. El-Gabalawy, H.; Guenther, L.C.; Bernstein, C.N. Epidemiology of immune-mediated inflammatory diseases: Incidence, prevalence, natural history, and comorbidities. *J. Rheumatol. Suppl.* **2010**, *85*, 2–10. [CrossRef]
19. Cutolo, M.; Kitas, G.D.; van Riel, P.L. Burden of disease in treated rheumatoid arthritis patients: Going beyond the joint. *Semin. Arthritis Rheum.* **2014**, *43*, 479–488. [CrossRef]
20. Burisch, J.; Jess, T.; Martinato, M.; Lakatos, P.L.; EpiCom, E. The burden of inflammatory bowel disease in Europe. *J. Crohns Colitis* **2013**, *7*, 322–337. [CrossRef]
21. Marrie, R.A. Comorbidity in multiple sclerosis: Implications for patient care. *Nat. Rev. Neurol.* **2017**, *13*, 375–382. [CrossRef] [PubMed]

22. Said, S.I.; Mutt, V. Polypeptide with broad biological activity: Isolation from small intestine. *Science* **1970**, *169*, 1217–1218. [CrossRef] [PubMed]

23. Said, S.I.; Rosenberg, R.N. Vasoactive intestinal polypeptide: Abundant immunoreactivity in neural cell lines and normal nervous tissue. *Science* **1976**, *192*, 907–908. [CrossRef] [PubMed]

24. Johansson, O.; Lundberg, J.M. Ultrastructural localization of VIP-like immunoreactivity in large dense-core vesicles of 'cholinergic-type' nerve terminals in cat exocrine glands. *Neuroscience* **1981**, *6*, 847–862. [CrossRef]

25. Staun-Olsen, P.; Ottesen, B.; Bartels, P.D.; Nielsen, M.H.; Gammeltoft, S.; Fahrenkrug, J. Receptors for vasoactive intestinal polypeptide on isolated synaptosomes from rat cerebral cortex. Heterogeneity of binding and desensitization of receptors. *J. Neurochem.* **1982**, *39*, 1242–1251. [CrossRef]

26. Felten, D.L.; Felten, S.Y.; Carlson, S.L.; Olschowka, J.A.; Livnat, S. Noradrenergic and peptidergic innervation of lymphoid tissue. *J. Immunol.* **1985**, *135*, 755s–765s.

27. Bellinger, D.L.; Lorton, D.; Brouxhon, S.; Felten, S.; Felten, D.L. The significance of vasoactive intestinal polypeptide (VIP) in immunomodulation. *Adv. Neuroimmunol.* **1996**, *6*, 5–27. [CrossRef]

28. Cutz, E.; Chan, W.; Track, N.S.; Goth, A.; Said, S.I. Release of vasoactive intestinal polypeptide in mast cells by histamine liberators. *Nature* **1978**, *275*, 661–662. [CrossRef]

29. O'Dorisio, M.S.; O'Dorisio, T.M.; Cataland, S.; Balcerzak, S.P. Vasoactive intestinal polypeptide as a biochemical marker for polymorphonuclear leukocytes. *J. Lab. Clin. Med.* **1980**, *96*, 666–672. [CrossRef]

30. Aliakbari, J.; Sreedharan, S.P.; Turck, C.W.; Goetzl, E.J. Selective localization of vasoactive intestinal peptide and substance P in human eosinophils. *Biochem. Biophys. Res. Commun.* **1987**, *148*, 1440–1445. [CrossRef]

31. Weinstock, J.V.; Blum, A.M. Detection of vasoactive intestinal peptide and localization of its mRNA within granulomas of murine schistosomiasis. *Cell Immunol.* **1990**, *125*, 291–300. [CrossRef]

32. Carrión, M.; Pérez-García, S.; Martínez, C.; Juarranz, Y.; Estrada-Capetillo, L.; Puig-Kroger, A.; Gomariz, R.P.; Gutiérrez-Cañas, I. VIP impairs acquisition of the macrophage proinflammatory polarization profile. *J. Leukoc. Biol.* **2016**, *100*, 1385–1393. [CrossRef] [PubMed]

33. Gomariz, R.P.; Lorenzo, M.J.; Cacicedo, L.; Vicente, A.; Zapata, A.G. Demonstration of immunoreactive vasoactive intestinal peptide (IR-VIP) and somatostatin (IR-SOM) in rat thymus. *Brain Behav. Immun.* **1990**, *4*, 151–161. [CrossRef]

34. Gomariz, R.P.; Delgado, M.; Naranjo, J.R.; Mellstrom, B.; Tormo, A.; Mata, F.; Leceta, J. VIP gene expression in rat thymus and spleen. *Brain Behav. Immun.* **1993**, *7*, 271–278. [CrossRef] [PubMed]

35. Gomariz, R.P.; Garrido, E.; Leceta, J.; Martinez, C.; Abalo, R.; Delgado, M. Gene expression of VIP receptor in rat lymphocytes. *Biochem. Biophys. Res. Commun.* **1994**, *203*, 1599–1604. [CrossRef]

36. Straub, R.H.; Bijlsma, J.W.; Masi, A.; Cutolo, M. Role of neuroendocrine and neuroimmune mechanisms in chronic inflammatory rheumatic diseases—The 10-year update. *Semin. Arthritis Rheum.* **2013**, *43*, 392–404. [CrossRef]

37. Grassel, S.; Muschter, D. Peripheral Nerve Fibers and Their Neurotransmitters in Osteoarthritis Pathology. *Int. J. Mol. Sci.* **2017**, *18*, 931. [CrossRef]

38. Juarranz, Y.; Gutiérrez-Cañas, I.; Santiago, B.; Carrión, M.; Pablos, J.L.; Gomariz, R.P. Differential expression of vasoactive intestinal peptide and its functional receptors in human osteoarthritic and rheumatoid synovial fibroblasts. *Arthritis Rheum.* **2008**, *58*, 1086–1095. [CrossRef]

39. Couvineau, A.; Laburthe, M. VPAC receptors: Structure, molecular pharmacology and interaction with accessory proteins. *Br. J. Pharmacol.* **2012**, *166*, 42–50. [CrossRef]

40. Dickson, L.; Finlayson, K. VPAC and PAC receptors: From ligands to function. *Pharmacol. Ther.* **2009**, *121*, 294–316. [CrossRef]

41. Neumann, J.M.; Couvineau, A.; Murail, S.; Lacapere, J.J.; Jamin, N.; Laburthe, M. Class-B GPCR activation: Is ligand helix-capping the key? *Trends Biochem. Sci.* **2008**, *33*, 314–319. [CrossRef] [PubMed]

42. Sherwood, N.M.; Krueckl, S.L.; McRory, J.E. The origin and function of the pituitary adenylate cyclase-activating polypeptide (PACAP)/glucagon superfamily. *Endocr. Rev.* **2000**, *21*, 619–670. [PubMed]

43. Nicole, P.; Rouyer-Fessard, C.; Couvineau, A.; Drouot, C.; Fulcrand, P.; Martinez, J.; Laburthe, M. Alanine scanning of VIP. Structure-function relationship for binding to human recombinant VPAC1 receptor. *Ann. N. Y. Acad. Sci.* **2000**, *921*, 352–356. [CrossRef] [PubMed]

44. Wray, V.; Kakoschke, C.; Nokihara, K.; Naruse, S. Solution structure of pituitary adenylate cyclase activating polypeptide by nuclear magnetic resonance spectroscopy. *Biochemistry* **1993**, *32*, 5832–5841. [CrossRef]

45. Ganea, D.; Hooper, K.M.; Kong, W. The neuropeptide vasoactive intestinal peptide: Direct effects on immune cells and involvement in inflammatory and autoimmune diseases. *Acta Physiol.* **2015**, *213*, 442–452. [CrossRef]

46. Gurevich, V.V.; Gurevich, E.V. GPCR Signaling Regulation: The Role of GRKs and Arrestins. *Front. Pharmacol.* **2019**, *10*, 125. [CrossRef]

47. Liao, C.; de Molliens, M.P.; Schneebeli, S.T.; Brewer, M.; Song, G.; Chatenet, D.; Braas, K.M.; May, V.; Li, J. Targeting the PAC1 Receptor for Neurological and Metabolic Disorders. *Curr. Top. Med. Chem.* **2019**, *19*, 1399–1417. [CrossRef]

48. Ishihara, T.; Shigemoto, R.; Mori, K.; Takahashi, K.; Nagata, S. Functional expression and tissue distribution of a novel receptor for vasoactive intestinal polypeptide. *Neuron* **1992**, *8*, 811–819. [CrossRef]

49. Sreedharan, S.P.; Patel, D.R.; Huang, J.X.; Goetzl, E.J. Cloning and functional expression of a human neuroendocrine vasoactive intestinal peptide receptor. *Biochem. Biophys. Res. Commun.* **1993**, *193*, 546–553. [CrossRef]

50. Bokaei, P.B.; Ma, X.Z.; Byczynski, B.; Keller, J.; Sakac, D.; Fahim, S.; Branch, D.R. Identification and characterization of five-transmembrane isoforms of human vasoactive intestinal peptide and pituitary adenylate cyclase-activating polypeptide receptors. *Genomics* **2006**, *88*, 791–800. [CrossRef]

51. Lutz, E.M.; Sheward, W.J.; West, K.M.; Morrow, J.A.; Fink, G.; Harmar, A.J. The VIP2 receptor: Molecular characterisation of a cDNA encoding a novel receptor for vasoactive intestinal peptide. *FEBS Lett.* **1993**, *334*, 3–8. [CrossRef]

52. Svoboda, M.; Tastenoy, M.; Van Rampelbergh, J.; Goossens, J.F.; De Neef, P.; Waelbroeck, M.; Robberecht, P. Molecular cloning and functional characterization of a human VIP receptor from SUP-T1 lymphoblasts. *Biochem. Biophys. Res. Commun.* **1994**, *205*, 1617–1624. [CrossRef] [PubMed]

53. Inagaki, N.; Yoshida, H.; Mizuta, M.; Mizuno, N.; Fujii, Y.; Gonoi, T.; Miyazaki, J.; Seino, S. Cloning and functional characterization of a third pituitary adenylate cyclase-activating polypeptide receptor subtype expressed in insulin-secreting cells. *Proc. Natl. Acad. Sci. USA* **1994**, *91*, 2679–2683. [CrossRef] [PubMed]

54. Grinninger, C.; Wang, W.; Oskoui, K.B.; Voice, J.K.; Goetzl, E.J. A natural variant type II G protein-coupled receptor for vasoactive intestinal peptide with altered function. *J. Biol. Chem.* **2004**, *279*, 40259–40262. [CrossRef] [PubMed]

55. Miller, A.L.; Verma, D.; Grinninger, C.; Huang, M.C.; Goetzl, E.J. Functional splice variants of the type II G protein-coupled receptor (VPAC2) for vasoactive intestinal peptide in mouse and human lymphocytes. *Ann. N. Y. Acad. Sci.* **2006**, *1070*, 422–426. [CrossRef] [PubMed]

56. Robberecht, P.; Coy, D.H.; De Neef, P.; Camus, J.C.; Cauvin, A.; Waelbroeck, M.; Christophe, J. [D-Phe4] peptide histidine-isoleucinamide ([D-Phe4]PHI), a highly selective vasoactive-intestinal-peptide (VIP) agonist, discriminates VIP-preferring from secretin-preferring receptors in rat pancreatic membranes. *Eur. J. Biochem.* **1987**, *165*, 243–249. [CrossRef]

57. Rouyer-Fessard, C.; Couvineau, A.; Voisin, T.; Laburthe, M. Ac-Tyr1hGRF discriminates between VIP receptors from rat liver and intestinal epithelium. *Life Sci.* **1989**, *45*, 829–833. [CrossRef]

58. Fishbein, V.A.; Coy, D.H.; Hocart, S.J.; Jiang, N.Y.; Mrozinski, J.E., Jr.; Mantey, S.A.; Jensen, R.T. A chimeric VIP-PACAP analogue but not VIP pseudopeptides function as VIP receptor antagonists. *Peptides* **1994**, *15*, 95–100. [CrossRef]

59. Gourlet, P.; Vandermeers, A.; Vertongen, P.; Rathe, J.; De Neef, P.; Cnudde, J.; Waelbroeck, M.; Robberecht, P. Development of high affinity selective VIP1 receptor agonists. *Peptides* **1997**, *18*, 1539–1545. [CrossRef]

60. Nicole, P.; Lins, L.; Rouyer-Fessard, C.; Drouot, C.; Fulcrand, P.; Thomas, A.; Couvineau, A.; Martinez, J.; Brasseur, R.; Laburthe, M. Identification of key residues for interaction of vasoactive intestinal peptide with human VPAC1 and VPAC2 receptors and development of a highly selective VPAC1 receptor agonist. Alanine scanning and molecular modeling of the peptide. *J. Biol. Chem.* **2000**, *275*, 24003–24012. [CrossRef]

61. Gourlet, P.; Vandermeers-Piret, M.C.; Rathe, J.; De Neef, P.; Cnudde, J.; Robberecht, P.; Waelbroeck, M. Vasoactive intestinal peptide modification at position 22 allows discrimination between receptor subtypes. *Eur. J. Pharmacol.* **1998**, *348*, 95–99. [CrossRef]

62. Van Rampelbergh, J.; Juarranz, M.G.; Perret, J.; Bondue, A.; Solano, R.M.; Delporte, C.; De Neef, P.; Robberecht, P.; Waelbroeck, M. Characterization of a novel VPAC(1) selective agonist and identification of the receptor domains implicated in the carboxyl-terminal peptide recognition. *Br. J. Pharmacol.* **2000**, *130*, 819–826. [CrossRef] [PubMed]

63. Olson, K.E.; Kosloski-Bilek, L.M.; Anderson, K.M.; Diggs, B.J.; Clark, B.E.; Gledhill, J.M., Jr.; Shandler, S.J.; Mosley, R.L.; Gendelman, H.E. Selective VIP Receptor Agonists Facilitate Immune Transformation for Dopaminergic Neuroprotection in MPTP-Intoxicated Mice. *J. Neurosci.* **2015**, *35*, 16463–16478. [CrossRef] [PubMed]

64. Gourlet, P.; De Neef, P.; Cnudde, J.; Waelbroeck, M.; Robberecht, P. In vitro properties of a high affinity selective antagonist of the VIP1 receptor. *Peptides* **1997**, *18*, 1555–1560. [CrossRef]

65. O'Donnell, M.; Garippa, R.J.; Rinaldi, N.; Selig, W.M.; Simko, B.; Renzetti, L.; Tannu, S.A.; Wasserman, M.A.; Welton, A.; Bolin, D.R. Ro 25-1553: A novel, long-acting vasoactive intestinal peptide agonist. Part I: In vitro and in vivo bronchodilator studies. *J. Pharmacol. Exp. Ther.* **1994**, *270*, 1282–1288.

66. Xia, M.; Sreedharan, S.P.; Bolin, D.R.; Gaufo, G.O.; Goetzl, E.J. Novel cyclic peptide agonist of high potency and selectivity for the type II vasoactive intestinal peptide receptor. *J. Pharmacol. Exp. Ther.* **1997**, *281*, 629–633.

67. Tsutsumi, M.; Claus, T.H.; Liang, Y.; Li, Y.; Yang, L.; Zhu, J.; Dela Cruz, F.; Peng, X.; Chen, H.; Yung, S.L.; et al. A potent and highly selective VPAC2 agonist enhances glucose-induced insulin release and glucose disposal: A potential therapy for type 2 diabetes. *Diabetes* **2002**, *51*, 1453–1460. [CrossRef]

68. Ma, Y.; Ma, M.; Dai, Y.; Hong, A. Expression, identification and biological effects of a novel VPAC2-specific agonist with high stability and bioactivity. *Acta Biochim. Biophys. Sin.* **2010**, *42*, 21–29. [CrossRef]

69. Moreno, D.; Gourlet, P.; De Neef, P.; Cnudde, J.; Waelbroeck, M.; Robberecht, P. Development of selective agonists and antagonists for the human vasoactive intestinal polypeptide VPAC(2) receptor. *Peptides* **2000**, *21*, 1543–1549. [CrossRef]

70. Sakamoto, K.; Koyama, R.; Kamada, Y.; Miwa, M.; Tani, A. Discovery of artificial VIPR2-antagonist peptides possessing receptor- and ligand-selectivity. *Biochem. Biophys. Res. Commun.* **2018**, *503*, 1973–1979. [CrossRef]

71. Li, J.; Wang, X.; Liu, H.; Li, H. In silico classification and prediction of VIP derivatives as VPAC1/ VPAC2 receptor agonists/antagonists. *Comb. Chem. High Throughput Screen.* **2015**, *18*, 33–41. [CrossRef] [PubMed]

72. Zhao, S.J.; Wang, D.H.; Li, Y.W.; Han, L.; Xiao, X.; Ma, M.; Wan, D.C.; Hong, A.; Ma, Y. A novel selective VPAC2 agonist peptide-conjugated chitosan modified selenium nanoparticles with enhanced anti-type 2 diabetes synergy effects. *Int. J. Nanomed.* **2017**, *12*, 2143–2160. [CrossRef] [PubMed]

73. Peyrassol, X.; Laeremans, T.; Lahura, V.; Debulpaep, M.; El Hassan, H.; Steyaert, J.; Parmentier, M.; Langer, I. Development by Genetic Immunization of Monovalent Antibodies Against Human Vasoactive Intestinal Peptide Receptor 1 (VPAC1), New Innovative, and Versatile Tools to Study VPAC1 Receptor Function. *Front. Endocrinol.* **2018**, *9*, 153. [CrossRef] [PubMed]

74. Gutiérrez-Cañas, I.; Juarranz, M.G.; Collado, B.; Rodríguez-Henche, N.; Chiloeches, A.; Prieto, J.C.; Carmena, M.J. Vasoactive intestinal peptide induces neuroendocrine differentiation in the LNCaP prostate cancer cell line through PKA, ERK, and PI3K. *Prostate* **2005**, *63*, 44–55. [CrossRef]

75. Crespo, P.; Xu, N.; Simonds, W.F.; Gutkind, J.S. Ras-dependent activation of MAP kinase pathway mediated by G-protein beta gamma subunits. *Nature* **1994**, *369*, 418–420. [CrossRef]

76. Maier, U.; Babich, A.; Nurnberg, B. Roles of non-catalytic subunits in gbetagamma-induced activation of class I phosphoinositide 3-kinase isoforms beta and gamma. *J. Biol. Chem.* **1999**, *274*, 29311–29317. [CrossRef]

77. Dickson, L.; Aramori, I.; McCulloch, J.; Sharkey, J.; Finlayson, K. A systematic comparison of intracellular cyclic AMP and calcium signalling highlights complexities in human VPAC/PAC receptor pharmacology. *Neuropharmacology* **2006**, *51*, 1086–1098. [CrossRef]

78. MacKenzie, C.J.; Lutz, E.M.; Johnson, M.S.; Robertson, D.N.; Holland, P.J.; Mitchell, R. Mechanisms of phospholipase C activation by the vasoactive intestinal polypeptide/pituitary adenylate cyclase-activating polypeptide type 2 receptor. *Endocrinology* **2001**, *142*, 1209–1217. [CrossRef]

79. McCulloch, D.A.; Lutz, E.M.; Johnson, M.S.; MacKenzie, C.J.; Mitchell, R. Differential activation of phospholipase D by VPAC and PAC1 receptors. *Ann. N. Y. Acad. Sci.* **2000**, *921*, 175–185. [CrossRef]

80. Christopoulos, A.; Christopoulos, G.; Morfis, M.; Udawela, M.; Laburthe, M.; Couvineau, A.; Kuwasako, K.; Tilakaratne, N.; Sexton, P.M. Novel receptor partners and function of receptor activity-modifying proteins. *J. Biol. Chem.* **2003**, *278*, 3293–3297. [CrossRef]

81. Wootten, D.; Lindmark, H.; Kadmiel, M.; Willcockson, H.; Caron, K.M.; Barwell, J.; Drmota, T.; Poyner, D.R. Receptor activity modifying proteins (RAMPs) interact with the VPAC2 receptor and CRF1 receptors and modulate their function. *Br. J. Pharmacol.* **2013**, *168*, 822–834. [CrossRef]

82. Carrión, M.; Juarranz, Y.; Pérez-García, S.; Jimeno, R.; Pablos, J.L.; Gomariz, R.P.; Gutiérrez-Cañas, I. RNA sensors in human osteoarthritis and rheumatoid arthritis synovial fibroblasts: Immune regulation by vasoactive intestinal peptide. *Arthritis Rheum.* **2011**, *63*, 1626–1636. [CrossRef] [PubMed]

83. Arranz, A.; Gutiérrez-Cañas, I.; Carrión, M.; Juarranz, Y.; Pablos, J.L.; Martínez, C.; Gomariz, R.P. VIP reverses the expression profiling of TLR4-stimulated signaling pathway in rheumatoid arthritis synovial fibroblasts. *Mol. Immunol.* **2008**, *45*, 3065–3073. [CrossRef] [PubMed]

84. Juarranz, Y.; Abad, C.; Martínez, C.; Arranz, A.; Gutiérrez-Cañas, I.; Rosignoli, F.; Gomariz, R.P.; Leceta, J. Protective effect of vasoactive intestinal peptide on bone destruction in the collagen-induced arthritis model of rheumatoid arthritis. *Arthritis Res. Ther.* **2005**, *7*, R1034–R1045. [CrossRef] [PubMed]

85. Delgado, M.; Munoz-Elias, E.J.; Kan, Y.; Gozes, I.; Fridkin, M.; Brenneman, D.E.; Gomariz, R.P.; Ganea, D. Vasoactive intestinal peptide and pituitary adenylate cyclase-activating polypeptide inhibit tumor necrosis factor alpha transcriptional activation by regulating nuclear factor-kB and cAMP response element-binding protein/c-Jun. *J. Biol. Chem.* **1998**, *273*, 31427–31436. [CrossRef]

86. Shetzline, M.A.; Walker, J.K.; Valenzano, K.J.; Premont, R.T. Vasoactive intestinal polypeptide type-1 receptor regulation. Desensitization, phosphorylation, and sequestration. *J. Biol. Chem.* **2002**, *277*, 25519–25526. [CrossRef]

87. Jong, Y.I.; Harmon, S.K.; O'Malley, K.L. GPCR signalling from within the cell. *Br. J. Pharmacol.* **2018**, *175*, 4026–4035. [CrossRef]

88. Valdehita, A.; Bajo, A.M.; Fernandez-Martinez, A.B.; Arenas, M.I.; Vacas, E.; Valenzuela, P.; Ruiz-Villaespesa, A.; Prieto, J.C.; Carmena, M.J. Nuclear localization of vasoactive intestinal peptide (VIP) receptors in human breast cancer. *Peptides* **2010**, *31*, 2035–2045. [CrossRef]

89. Vacas, E.; Arenas, M.I.; Munoz-Moreno, L.; Bajo, A.M.; Sanchez-Chapado, M.; Prieto, J.C.; Carmena, M.J. Antitumoral effects of vasoactive intestinal peptide in human renal cell carcinoma xenografts in athymic nude mice. *Cancer Lett.* **2013**, *336*, 196–203. [CrossRef]

90. Barbarin, A.; Seite, P.; Godet, J.; Bensalma, S.; Muller, J.M.; Chadeneau, C. Atypical nuclear localization of VIP receptors in glioma cell lines and patients. *Biochem. Biophys. Res. Commun.* **2014**, *454*, 524–530. [CrossRef]

91. Villanueva-Romero, R.; Gutierrez-Canas, I.; Carrion, M.; Gonzalez-Alvaro, I.; Rodriguez-Frade, J.M.; Mellado, M.; Martinez, C.; Gomariz, R.P.; Juarranz, Y. Activation of Th lymphocytes alters pattern expression and cellular location of VIP receptors in healthy donors and early arthritis patients. *Sci. Rep.* **2019**, *9*, 7383. [CrossRef] [PubMed]

92. Yu, R.; Liu, H.; Peng, X.; Cui, Y.; Song, S.; Wang, L.; Zhang, H.; Hong, A.; Zhou, T. The palmitoylation of the N-terminal extracellular Cys37 mediates the nuclear translocation of VPAC1 contributing to its anti-apoptotic activity. *Oncotarget* **2017**, *8*, 42728. [CrossRef] [PubMed]

93. Di Benedetto, P.; Ruscitti, P.; Vadasz, Z.; Toubi, E.; Giacomelli, R. Macrophages with regulatory functions, a possible new therapeutic perspective in autoimmune diseases. *Autoimmun. Rev.* **2019**, *18*, 102369. [CrossRef] [PubMed]

94. Van Amersfoort, E.S.; Van Berkel, T.J.; Kuiper, J. Receptors, mediators, and mechanisms involved in bacterial sepsis and septic shock. *Clin. Microbiol. Rev.* **2003**, *16*, 379–414. [CrossRef] [PubMed]

95. Meduri, G.U. Clinical review: A paradigm shift: The bidirectional effect of inflammation on bacterial growth. Clinical implications for patients with acute respiratory distress syndrome. *Crit. Care* **2002**, *6*, 24–29. [CrossRef]

96. Navegantes, K.C.; de Souza Gomes, R.; Pereira, P.A.T.; Czaikoski, P.G.; Azevedo, C.H.M.; Monteiro, M.C. Immune modulation of some autoimmune diseases: The critical role of macrophages and neutrophils in the innate and adaptive immunity. *J. Transl. Med.* **2017**, *15*, 36. [CrossRef]

97. Leceta, J.; Gomariz, R.P.; Martinez, C.; Abad, C.; Ganea, D.; Delgado, M. Receptors and transcriptional factors involved in the anti-inflammatory activity of VIP and PACAP. *Ann. N. Y. Acad. Sci.* **2000**, *921*, 92–102. [CrossRef]

98. Juarranz, M.G.; Santiago, B.; Torroba, M.; Gutiérrez-Cañas, I.; Palao, G.; Galindo, M.; Abad, C.; Martínez, C.; Leceta, J.; Pablos, J.L.; et al. Vasoactive intestinal peptide modulates proinflammatory mediator synthesis in osteoarthritic and rheumatoid synovial cells. *Rheumatology* **2004**, *43*, 416–422. [CrossRef]

99. Tuncel, N.; Tore, F.; Sahinturk, V.; Ak, D.; Tuncel, M. Vasoactive intestinal peptide inhibits degranulation and changes granular content of mast cells: A potential therapeutic strategy in controlling septic shock. *Peptides* **2000**, *21*, 81–89. [CrossRef]

100. Delgado, M.; Ganea, D. Anti-inflammatory neuropeptides: A new class of endogenous immunoregulatory agents. *Brain Behav. Immun.* **2008**, *22*, 1146–1151. [CrossRef]

101. Delgado, M.; Jonakait, G.M.; Ganea, D. Vasoactive intestinal peptide and pituitary adenylate cyclase-activating polypeptide inhibit chemokine production in activated microglia. *Glia* **2002**, *39*, 148–161. [CrossRef] [PubMed]

102. Szema, A.M.; Hamidi, S.A.; Lyubsky, S.; Dickman, K.G.; Mathew, S.; Abdel-Razek, T.; Chen, J.J.; Waschek, J.A.; Said, S.I. Mice lacking the VIP gene show airway hyperresponsiveness and airway inflammation, partially reversible by VIP. *Am. J. Physiol. Lung Cell. Mol. Physiol.* **2006**, *291*, L880–L886. [CrossRef] [PubMed]

103. Said, S.I.; Hamidi, S.A.; Dickman, K.G.; Szema, A.M.; Lyubsky, S.; Lin, R.Z.; Jiang, Y.P.; Chen, J.J.; Waschek, J.A.; Kort, S. Moderate pulmonary arterial hypertension in male mice lacking the vasoactive intestinal peptide gene. *Circulation* **2007**, *115*, 1260–1268. [CrossRef] [PubMed]

104. Abad, C.; Tan, Y.V.; Cheung-Lau, G.; Nobuta, H.; Waschek, J.A. VIP deficient mice exhibit resistance to lipopolysaccharide induced endotoxemia with an intrinsic defect in proinflammatory cellular responses. *PLoS ONE* **2012**, *7*, e36922. [CrossRef]

105. Hamidi, S.A.; Szema, A.M.; Lyubsky, S.; Dickman, K.G.; Degene, A.; Mathew, S.M.; Waschek, J.A.; Said, S.I. Clues to VIP function from knockout mice. *Ann. N. Y. Acad. Sci.* **2006**, *1070*, 5–9. [CrossRef]

106. Said, S.I. Vasoactive intestinal peptide and nitric oxide: Divergent roles in relation to tissue injury. *Ann. N. Y. Acad. Sci.* **1996**, *805*, 379–387; discussion 387-8. [CrossRef]

107. Said, S.I. Molecules that protect: The defense of neurons and other cells. *J. Clin. Investig.* **1996**, *97*, 2163–2164. [CrossRef]

108. Waschek, J.A. VIP and PACAP: Neuropeptide modulators of CNS inflammation, injury, and repair. *Br. J. Pharmacol.* **2013**, *169*, 512–523. [CrossRef]

109. Delgado, M.; Martinez, C.; Pozo, D.; Calvo, J.R.; Leceta, J.; Ganea, D.; Gomariz, R.P. Vasoactive intestinal peptide (VIP) and pituitary adenylate cyclase-activation polypeptide (PACAP) protect mice from lethal endotoxemia through the inhibition of TNF-alpha and IL-6. *J. Immunol.* **1999**, *162*, 1200–1205.

110. Said, S.I.; Dickman, K.; Dey, R.D.; Bandyopadhyay, A.; De Stefanis, P.; Raza, S.; Pakbaz, H.; Berisha, H.I. Glutamate toxicity in the lung and neuronal cells: Prevention or attenuation by VIP and PACAP. *Ann. N. Y. Acad. Sci.* **1998**, *865*, 226–237. [CrossRef]

111. Akira, S.; Uematsu, S.; Takeuchi, O. Pathogen recognition and innate immunity. *Cell* **2006**, *124*, 783–801. [CrossRef] [PubMed]

112. Medzhitov, R. Recognition of microorganisms and activation of the immune response. *Nature* **2007**, *449*, 819–826. [CrossRef] [PubMed]

113. Kawai, T.; Akira, S. The role of pattern-recognition receptors in innate immunity: Update on Toll-like receptors. *Nat. Immunol.* **2010**, *11*, 373–384. [CrossRef] [PubMed]

114. Takeuchi, O.; Akira, S. Pattern recognition receptors and inflammation. *Cell* **2010**, *140*, 805–820. [CrossRef] [PubMed]

115. Takeda, K.; Akira, S. Toll-like receptors. *Curr. Protoc. Immunol.* **2015**, *109*, 14.12.1–14.12.10. [CrossRef] [PubMed]

116. Hornung, V.; Rothenfusser, S.; Britsch, S.; Krug, A.; Jahrsdorfer, B.; Giese, T.; Endres, S.; Hartmann, G. Quantitative expression of toll-like receptor 1–10 mRNA in cellular subsets of human peripheral blood mononuclear cells and sensitivity to CpG oligodeoxynucleotides. *J. Immunol.* **2002**, *168*, 4531–4537. [CrossRef] [PubMed]

117. Kawai, T.; Akira, S. TLR signaling. *Cell Death Differ.* **2006**, *13*, 816–825. [CrossRef]

118. Ospelt, C.; Gay, S. TLRs and chronic inflammation. *Int. J. Biochem. Cell Biol.* **2010**, *42*, 495–505. [CrossRef]

119. Lebre, M.C.; van der Aar, A.M.; van Baarsen, L.; van Capel, T.M.; Schuitemaker, J.H.; Kapsenberg, M.L.; de Jong, E.C. Human keratinocytes express functional Toll-like receptor 3, 4, 5, and 9. *J. Investig. Dermatol.* **2007**, *127*, 331–341. [CrossRef]

120. Platz, J.; Beisswenger, C.; Dalpke, A.; Koczulla, R.; Pinkenburg, O.; Vogelmeier, C.; Bals, R. Microbial DNA induces a host defense reaction of human respiratory epithelial cells. *J. Immunol.* **2004**, *173*, 1219–1223. [CrossRef]

121. Gewirtz, A.T. Intestinal epithelial toll-like receptors: To protect. And serve? *Curr. Pharm. Des.* **2003**, *9*, 1–5. [CrossRef] [PubMed]

122. Blasius, A.L.; Beutler, B. Intracellular toll-like receptors. *Immunity* **2010**, *32*, 305–315. [CrossRef] [PubMed]

123. Kawasaki, T.; Kawai, T. Toll-like receptor signaling pathways. *Front. Immunol.* **2014**, *5*, 461. [CrossRef] [PubMed]

124. Kawai, T.; Akira, S. Toll-like receptor and RIG-I-like receptor signaling. *Ann. N. Y. Acad. Sci.* **2008**, *1143*, 1–20. [CrossRef]

125. Brasier, A.R. *The NF-κB Signaling Network: Insights from Systems Approaches*; American Society for Microbiology: Washington, DC, USA, 2008; pp. 119–135.

126. Goh, F.G.; Midwood, K.S. Intrinsic danger: Activation of Toll-like receptors in rheumatoid arthritis. *Rheumatology* **2012**, *51*, 7–23. [CrossRef]

127. Fischer, M.; Ehlers, M. Toll-like receptors in autoimmunity. *Ann. N. Y. Acad. Sci.* **2008**, *1143*, 21–34. [CrossRef]

128. Kawasaki, T.; Kawai, T.; Akira, S. Recognition of nucleic acids by pattern-recognition receptors and its relevance in autoimmunity. *Immunol. Rev.* **2011**, *243*, 61–73. [CrossRef]

129. Farrugia, M.; Baron, B. The Role of Toll-Like Receptors in Autoimmune Diseases through Failure of the Self-Recognition Mechanism. *Int. J. Inflam.* **2017**, *2017*, 1–12. [CrossRef]

130. Gomariz, R.P.; Arranz, A.; Juarranz, Y.; Gutiérrez-Cañas, I.; García-Gómez, M.; Leceta, J.; Martínez, C. Regulation of TLR expression, a new perspective for the role of VIP in immunity. *Peptides* **2007**, *28*, 1825–1832. [CrossRef]

131. Gomariz, R.P.; Gutierrez-Canas, I.; Arranz, A.; Carrion, M.; Juarranz, Y.; Leceta, J.; Martinez, C. Peptides targeting Toll-like receptor signalling pathways for novel immune therapeutics. *Curr. Pharm. Des.* **2010**, *16*, 1063–1080. [CrossRef]

132. Smalley, S.G.; Barrow, P.A.; Foster, N. Immunomodulation of innate immune responses by vasoactive intestinal peptide (VIP): Its therapeutic potential in inflammatory disease. *Clin. Exp. Immunol.* **2009**, *157*, 225–234. [CrossRef] [PubMed]

133. Ibrahim, H.; Barrow, P.; Foster, N. Transcriptional modulation by VIP: A rational target against inflammatory disease. *Clin. Epigenetics* **2011**, *2*, 213–222. [CrossRef] [PubMed]

134. Gomariz, R.P.; Arranz, A.; Abad, C.; Torroba, M.; Martínez, C.; Rosignoli, F.; García-Gómez, M.; Leceta, J.; Juarranz, Y. Time-course expression of Toll-like receptors 2 and 4 in inflammatory bowel disease and homeostatic effect of VIP. *J. Leukoc. Biol.* **2005**, *78*, 491–502. [CrossRef] [PubMed]

135. Arranz, A.; Abad, C.; Juarranz, Y.; Torroba, M.; Rosignoli, F.; Leceta, J.; Gomariz, R.P.; Martínez, C. Effect of VIP on TLR2 and TLR4 expression in lymph node immune cells during TNBS-induced colitis. *Ann. N. Y. Acad. Sci.* **2006**, *1070*, 129–134. [CrossRef]

136. Musikacharoen, T.; Matsuguchi, T.; Kikuchi, T.; Yoshikai, Y. NF-kappa B and STAT5 play important roles in the regulation of mouse Toll-like receptor 2 gene expression. *J. Immunol.* **2001**, *166*, 4516–4524. [CrossRef]

137. Arranz, A.; Androulidaki, A.; Zacharioudaki, V.; Martínez, C.; Margioris, A.N.; Gomariz, R.P.; Tsatsanis, C. Vasoactive intestinal peptide suppresses toll-like receptor 4 expression in macrophages via Akt1 reducing their responsiveness to lipopolysaccharide. *Mol. Immunol.* **2008**, *45*, 2970–2980. [CrossRef]

138. Foster, N.; Lea, S.R.; Preshaw, P.M.; Taylor, J.J. Pivotal advance: Vasoactive intestinal peptide inhibits up-regulation of human monocyte TLR2 and TLR4 by LPS and differentiation of monocytes to macrophages. *J. Leukoc. Biol.* **2007**, *81*, 893–903. [CrossRef]

139. Haehnel, V.; Schwarzfischer, L.; Fenton, M.J.; Rehli, M. Transcriptional regulation of the human toll-like receptor 2 gene in monocytes and macrophages. *J. Immunol.* **2002**, *168*, 5629–5637. [CrossRef]

140. Rehli, M.; Poltorak, A.; Schwarzfischer, L.; Krause, S.W.; Andreesen, R.; Beutler, B. PU. 1 and interferon consensus sequence-binding protein regulate the myeloid expression of the human Toll-like receptor 4 gene. *J. Biol. Chem.* **2000**, *275*, 9773–9781. [CrossRef]

141. Jiang, X.; McClellan, S.A.; Barrett, R.P.; Zhang, Y.; Hazlett, L.D. Vasoactive intestinal peptide downregulates proinflammatory TLRs while upregulating anti-inflammatory TLRs in the infected cornea. *J. Immunol.* **2012**, *189*, 269–278. [CrossRef]

142. Gutiérrez-Cañas, I.; Juarranz, Y.; Santiago, B.; Arranz, A.; Martínez, C.; Galindo, M.; Paya, M.; Gomariz, R.P.; Pablos, J.L. VIP down-regulates TLR4 expression and TLR4-mediated chemokine production in human rheumatoid synovial fibroblasts. *Rheumatology* **2006**, *45*, 527–532. [CrossRef] [PubMed]

143. Ospelt, C.; Neidhart, M.; Gay, R.E.; Gay, S. Synovial activation in rheumatoid arthritis. *Front. Biosci.* **2004**, *9*, 2323–2334. [CrossRef] [PubMed]

144. Juarranz, Y.; Gutiérrez-Cañas, I.; Arranz, A.; Martínez, C.; Abad, C.; Leceta, J.; Pablos, J.L.; Gomariz, R.P. VIP decreases TLR4 expression induced by LPS and TNF-alpha treatment in human synovial fibroblasts. *Ann. N. Y. Acad. Sci.* **2006**, *1070*, 359–364. [CrossRef] [PubMed]

145. Zhu, J. T Helper Cell Differentiation, Heterogeneity, and Plasticity. *Cold Spring Harb. Perspect. Biol.* **2018**, *10*, a030338. [CrossRef] [PubMed]

146. Horwitz, D.A.; Fahmy, T.M.; Piccirillo, C.A.; La Cava, A. Rebalancing Immune Homeostasis to Treat Autoimmune Diseases. *Trends Immunol.* **2019**, *40*, 888–908. [CrossRef]

147. Zhou, L.; Chong, M.M.; Littman, D.R. Plasticity of CD4$^+$ T cell lineage differentiation. *Immunity* **2009**, *30*, 646–655. [CrossRef]

148. Stadhouders, R.; Lubberts, E.; Hendriks, R.W. A cellular and molecular view of T helper 17 cell plasticity in autoimmunity. *J. Autoimmun.* **2018**, *87*, 1–15. [CrossRef]

149. Van Hamburg, J.P.; Tas, S.W. Molecular mechanisms underpinning T helper 17 cell heterogeneity and functions in rheumatoid arthritis. *J. Autoimmun.* **2018**, *87*, 69–81. [CrossRef]

150. Sallusto, F. Heterogeneity of Human CD4$^+$ T Cells Against Microbes. *Annu. Rev. Immunol.* **2016**, *34*, 317–334. [CrossRef]

151. Bending, D.; De la Pena, H.; Veldhoen, M.; Phillips, J.M.; Uyttenhove, C.; Stockinger, B.; Cooke, A. Highly purified Th17 cells from BDC2.5NOD mice convert into Th1-like cells in NOD/SCID recipient mice. *J. Clin. Investig.* **2009**, *119*, 565–572. [CrossRef]

152. Ghoreschi, K.; Laurence, A.; Yang, X.P.; Hirahara, K.; O'Shea, J.J. T helper 17 cell heterogeneity and pathogenicity in autoimmune disease. *Trends Immunol.* **2011**, *32*, 395–401. [CrossRef]

153. Jimeno, R.; Gomariz, R.P.; Garin, M.; Gutierrez-Canas, I.; Gonzalez-Alvaro, I.; Carrion, M.; Galindo, M.; Leceta, J.; Juarranz, Y. The pathogenic Th profile of human activated memory Th cells in early rheumatoid arthritis can be modulated by VIP. *J. Mol. Med.* **2015**, *93*, 457–467. [CrossRef] [PubMed]

154. Kotake, S.; Yago, T.; Kobashigawa, T.; Nanke, Y. The Plasticity of Th17 Cells in the Pathogenesis of Rheumatoid Arthritis. *J. Clin. Med.* **2017**, *6*, 67. [CrossRef] [PubMed]

155. Delgado, M. VIP: A very important peptide in T helper differentiation. *Trends Immunol.* **2003**, *24*, 221–224. [CrossRef]

156. Arranz, A.; Abad, C.; Juarranz, Y.; Leceta, J.; Martinez, C.; Gomariz, R.P. Vasoactive intestinal peptide as a healing mediator in Crohn's disease. *Neuroimmunomodulation* **2008**, *15*, 46–53. [CrossRef]

157. Jimeno, R.; Leceta, J.; Martinez, C.; Gutierrez-Canas, I.; Perez-Garcia, S.; Carrion, M.; Gomariz, R.P.; Juarranz, Y. Effect of VIP on the balance between cytokines and master regulators of activated helper T cells. *Immunol. Cell Biol.* **2012**, *90*, 178–186. [CrossRef]

158. Villanueva-Romero, R.; Gutierrez-Canas, I.; Carrion, M.; Perez-Garcia, S.; Seoane, I.V.; Martinez, C.; Gomariz, R.P.; Juarranz, Y. The Anti-Inflammatory Mediator, Vasoactive Intestinal Peptide, Modulates the Differentiation and Function of Th Subsets in Rheumatoid Arthritis. *J. Immunol. Res.* **2018**, *2018*, 6043710. [CrossRef]

159. Leceta, J.; Gomariz, R.P.; Martinez, C.; Carrion, M.; Arranz, A.; Juarranz, Y. Vasoactive intestinal peptide regulates Th17 function in autoimmune inflammation. *Neuroimmunomodulation* **2007**, *14*, 134–138. [CrossRef]

160. Rosignoli, F.; Torroba, M.; Juarranz, Y.; Garcia-Gomez, M.; Martinez, C.; Gomariz, R.P.; Perez-Leiros, C.; Leceta, J. VIP and tolerance induction in autoimmunity. *Ann. N. Y. Acad. Sci.* **2006**, *1070*, 525–530. [CrossRef]

161. Roca, V.; Calafat, M.; Larocca, L.; Ramhorst, R.; Farina, M.; Franchi, A.M.; Leiros, C.P. Potential immunomodulatory role of VIP in the implantation sites of prediabetic nonobese diabetic mice. *Reproduction* **2009**, *138*, 733–742. [CrossRef]

162. Gonzalez-Rey, E.; Fernandez-Martin, A.; Chorny, A.; Martin, J.; Pozo, D.; Ganea, D.; Delgado, M. Therapeutic effect of vasoactive intestinal peptide on experimental autoimmune encephalomyelitis: Down-regulation of inflammatory and autoimmune responses. *Am. J. Pathol.* **2006**, *168*, 1179–1188. [CrossRef] [PubMed]

163. Abad, C.; Waschek, J.A. Immunomodulatory roles of VIP and PACAP in models of multiple sclerosis. *Curr. Pharm. Des.* **2011**, *17*, 1025–1035. [CrossRef] [PubMed]

164. Benitez, R.; Delgado-Maroto, V.; Caro, M.; Forte-Lago, I.; Duran-Prado, M.; O'Valle, F.; Lichtman, A.H.; Gonzalez-Rey, E.; Delgado, M. Vasoactive Intestinal Peptide Ameliorates Acute Myocarditis and Atherosclerosis by Regulating Inflammatory and Autoimmune Responses. *J. Immunol.* **2018**, *200*, 3697–3710. [CrossRef] [PubMed]

165. Fu, D.; Senouthai, S.; Wang, J.; You, Y. Vasoactive intestinal peptide ameliorates renal injury in a pristane-induced lupus mouse model by modulating Th17/Treg balance. *BMC Nephrol.* **2019**, *20*, 350. [CrossRef] [PubMed]

166. Delgado, M.; Leceta, J.; Gomariz, R.P.; Ganea, D. Vasoactive intestinal peptide and pituitary adenylate cyclase-activating polypeptide stimulate the induction of Th2 responses by up-regulating B7.2 expression. *J. Immunol.* **1999**, *163*, 3629–3635.

167. Gutiérrez-Cañas, I.; Juarranz, Y.; Santiago, B.; Martínez, C.; Gomariz, R.P.; Pablos, J.L.; Leceta, J. Immunoregulatory properties of vasoactive intestinal peptide in human T cell subsets: Implications for rheumatoid arthritis. *Brain Behav. Immun.* **2008**, *22*, 312–317. [CrossRef] [PubMed]

168. Deng, S.; Xi, Y.; Wang, H.; Hao, J.; Niu, X.; Li, W.; Tao, Y.; Chen, G. Regulatory effect of vasoactive intestinal peptide on the balance of Treg and Th17 in collagen-induced arthritis. *Cell Immunol.* **2010**, *265*, 105–110. [CrossRef] [PubMed]

169. Jimeno, R.; Leceta, J.; Martinez, C.; Gutierrez-Canas, I.; Carrion, M.; Perez-Garcia, S.; Garin, M.; Mellado, M.; Gomariz, R.P.; Juarranz, Y. Vasoactive intestinal peptide maintains the nonpathogenic profile of human th17-polarized cells. *J. Mol. Neurosci.* **2014**, *54*, 512–525. [CrossRef]

170. Jimeno, R.; Leceta, J.; Garin, M.; Ortiz, A.M.; Mellado, M.; Rodriguez-Frade, J.M.; Martínez, C.; Pérez-García, S.; Gomariz, R.P.; Juarranz, Y. Th17 polarization of memory Th cells in early arthritis: The vasoactive intestinal peptide effect. *J. Leukoc. Biol.* **2015**, *98*, 257–269. [CrossRef]

171. Hubo, M.; Trinschek, B.; Kryczanowsky, F.; Tuettenberg, A.; Steinbrink, K.; Jonuleit, H. Costimulatory molecules on immunogenic versus tolerogenic human dendritic cells. *Front. Immunol.* **2013**, *4*, 82. [CrossRef]

172. Ritprajak, P.; Kaewraemruaen, C.; Hirankarn, N. Current Paradigms of Tolerogenic Dendritic Cells and Clinical Implications for Systemic Lupus Erythematosus. *Cells* **2019**, *8*, 1291. [CrossRef] [PubMed]

173. Rutella, S.; Danese, S.; Leone, G. Tolerogenic dendritic cells: Cytokine modulation comes of age. *Blood* **2006**, *108*, 1435–1440. [CrossRef] [PubMed]

174. Berkun, Y.; Verbovetski, I.; Ben-Ami, A.; Paran, D.; Caspi, D.; Krispin, A.; Trahtemberg, U.; Gill, O.; Naparstek, Y.; Mevorach, D. Altered dendritic cells with tolerizing phenotype in patients with systemic lupus erythematosus. *Eur. J. Immunol.* **2008**, *38*, 2896–2904. [CrossRef] [PubMed]

175. Hackstein, H.; Thomson, A.W. Dendritic cells: Emerging pharmacological targets of immunosuppressive drugs. *Nat. Rev. Immunol.* **2004**, *4*, 24–34. [CrossRef] [PubMed]

176. Svajger, U.; Rozman, P. Induction of Tolerogenic Dendritic Cells by Endogenous Biomolecules: An Update. *Front. Immunol.* **2018**, *9*, 2482. [CrossRef] [PubMed]

177. Chorny, A.; Gonzalez-Rey, E.; Fernandez-Martin, A.; Pozo, D.; Ganea, D.; Delgado, M. Vasoactive intestinal peptide induces regulatory dendritic cells with therapeutic effects on autoimmune disorders. *Proc. Natl. Acad. Sci. USA* **2005**, *102*, 13562–13567. [CrossRef] [PubMed]

178. Gonzalez-Rey, E.; Chorny, A.; Fernandez-Martin, A.; Ganea, D.; Delgado, M. Vasoactive intestinal peptide generates human tolerogenic dendritic cells that induce CD4 and CD8 regulatory T cells. *Blood* **2006**, *107*, 3632–3638. [CrossRef] [PubMed]

179. Toscano, M.G.; Delgado, M.; Kong, W.; Martin, F.; Skarica, M.; Ganea, D. Dendritic cells transduced with lentiviral vectors expressing VIP differentiate into VIP-secreting tolerogenic-like DCs. *Mol. Ther.* **2010**, *18*, 1035–1045. [CrossRef]

180. Gonzalez-Rey, E.; Delgado, M. Therapeutic treatment of experimental colitis with regulatory dendritic cells generated with vasoactive intestinal peptide. *Gastroenterology* **2006**, *131*, 1799–1811. [CrossRef]

181. Wu, H.; Shen, J.; Liu, L.; Lu, X.; Xue, J. Vasoactive intestinal peptide-induced tolerogenic dendritic cells attenuated arthritis in experimental collagen-induced arthritic mice. *Int. J. Rheum. Dis.* **2019**, *22*, 1255–1262. [CrossRef]

182. Yalvac, M.E.; Arnold, W.D.; Hussain, S.A.; Braganza, C.; Shontz, K.M.; Clark, K.R.; Walker, C.M.; Ubogu, E.E.; Mendell, J.R.; Sahenk, Z. VIP-expressing dendritic cells protect against spontaneous autoimmune peripheral polyneuropathy. *Mol. Ther.* **2014**, *22*, 1353–1363. [CrossRef] [PubMed]

183. Lu, J.; Zheng, M.H.; Yan, J.; Chen, Y.P.; Pan, J.P. Effects of vasoactive intestinal peptide on phenotypic and functional maturation of dendritic cells. *Int. Immunopharmacol.* **2008**, *8*, 1449–1454. [CrossRef] [PubMed]

184. Weng, Y.; Sun, J.; Wu, Q.; Pan, J. Regulatory effects of vasoactive intestinal peptide on the migration of mature dendritic cells. *J. Neuroimmunol.* **2007**, *182*, 48–54. [CrossRef] [PubMed]

185. Delgado, M.; Reduta, A.; Sharma, V.; Ganea, D. VIP/PACAP oppositely affects immature and mature dendritic cell expression of CD80/CD86 and the stimulatory activity for CD4$^+$ T cells. *J. Leukoc. Biol.* **2004**, *75*, 1122–1130. [CrossRef] [PubMed]

186. Smolen, J.S.; Aletaha, D.; Barton, A.; Burmester, G.R.; Emery, P.; Firestein, G.S.; Kavanaugh, A.; McInnes, I.B.; Solomon, D.H.; Strand, V.; et al. Rheumatoid arthritis. *Nat. Rev. Dis. Primers* **2018**, *4*, 18001. [CrossRef]

187. Firestein, G.S.; McInnes, I.B. Immunopathogenesis of Rheumatoid Arthritis. *Immunity* **2017**, *46*, 183–196. [CrossRef]

188. Scott, D.L.; Wolfe, F.; Huizinga, T.W. Rheumatoid arthritis. *Lancet* **2010**, *376*, 1094–1108. [CrossRef]

189. Chen, Z.; Bozec, A.; Ramming, A.; Schett, G. Anti-inflammatory and immune-regulatory cytokines in rheumatoid arthritis. *Nat. Rev. Rheumatol.* **2019**, *15*, 9–17. [CrossRef]

190. Smolen, J.S.; Aletaha, D.; McInnes, I.B. Rheumatoid arthritis. *Lancet* **2016**, *388*, 2023–2038. [CrossRef]

191. Orr, C.; Vieira-Sousa, E.; Boyle, D.L.; Buch, M.H.; Buckley, C.D.; Canete, J.D.; Catrina, A.I.; Choy, E.H.S.; Emery, P.; Fearon, U.; et al. Synovial tissue research: A state-of-the-art review. *Nat. Rev. Rheumatol.* **2017**, *13*, 630. [CrossRef]

192. Dakin, S.G.; Coles, M.; Sherlock, J.P.; Powrie, F.; Carr, A.J.; Buckley, C.D. Pathogenic stromal cells as therapeutic targets in joint inflammation. *Nat. Rev. Rheumatol.* **2018**, *14*, 714–726. [CrossRef] [PubMed]

193. Ospelt, C. Synovial fibroblasts in 2017. *RMD Open* **2017**, *3*, e000471. [CrossRef] [PubMed]

194. Neumann, E.; Lefevre, S.; Zimmermann, B.; Gay, S.; Muller-Ladner, U. Rheumatoid arthritis progression mediated by activated synovial fibroblasts. *Trends Mol. Med.* **2010**, *16*, 458–468. [CrossRef] [PubMed]

195. McGettrick, H.M.; Butler, L.M.; Buckley, C.D.; Rainger, G.E.; Nash, G.B. Tissue stroma as a regulator of leukocyte recruitment in inflammation. *J. Leukoc. Biol.* **2012**, *91*, 385–400. [CrossRef] [PubMed]

196. Bartok, B.; Firestein, G.S. Fibroblast-like synoviocytes: Key effector cells in rheumatoid arthritis. *Immunol. Rev.* **2010**, *233*, 233–255. [CrossRef] [PubMed]

197. Yoshitomi, H. Regulation of Immune Responses and Chronic Inflammation by Fibroblast-Like Synoviocytes. *Front. Immunol.* **2019**, *10*, 1395. [CrossRef]

198. Mulherin, D.; Fitzgerald, O.; Bresnihan, B. Synovial tissue macrophage populations and articular damage in rheumatoid arthritis. *Arthritis Rheum.* **1996**, *39*, 115–124. [CrossRef]

199. Firestein, G.S. VIP: A very important protein in arthritis. *Nat. Med.* **2001**, *7*, 537–538. [CrossRef]

200. El Karim, I.A.; Linden, G.J.; Orr, D.F.; Lundy, F.T. Antimicrobial activity of neuropeptides against a range of micro-organisms from skin, oral, respiratory and gastrointestinal tract sites. *J. Neuroimmunol.* **2008**, *200*, 11–16. [CrossRef]

201. Campos-Salinas, J.; Cavazzuti, A.; O'Valle, F.; Forte-Lago, I.; Caro, M.; Beverley, S.M.; Delgado, M.; Gonzalez-Rey, E. Therapeutic efficacy of stable analogues of vasoactive intestinal peptide against pathogens. *J. Biol. Chem.* **2014**, *289*, 14583–14599. [CrossRef]

202. Hajishengallis, G. Periodontitis: From microbial immune subversion to systemic inflammation. *Nat. Rev. Immunol.* **2015**, *15*, 30–44. [CrossRef] [PubMed]

203. Kharlamova, N.; Jiang, X.; Sherina, N.; Potempa, B.; Israelsson, L.; Quirke, A.M.; Eriksson, K.; Yucel-Lindberg, T.; Venables, P.J.; Potempa, J.; et al. Antibodies to Porphyromonas gingivalis Indicate Interaction Between Oral Infection, Smoking, and Risk Genes in Rheumatoid Arthritis Etiology. *Arthritis Rheumatol.* **2016**, *68*, 604–613. [CrossRef] [PubMed]

204. Chen, G.; Hao, J.; Xi, Y.; Wang, W.; Wang, Z.; Li, N.; Li, W. The therapeutic effect of vasoactive intestinal peptide on experimental arthritis is associated with CD4$^+$CD25$^+$ T regulatory cells. *Scand. J. Immunol.* **2008**, *68*, 572–578. [CrossRef] [PubMed]

205. Muschter, D.; Schafer, N.; Stangl, H.; Straub, R.H.; Grassel, S. Sympathetic Neurotransmitters Modulate Osteoclastogenesis and Osteoclast Activity in the Context of Collagen-Induced Arthritis. *PLoS ONE* **2015**, *10*, e0139726. [CrossRef]

206. Carrión, M.; Juarranz, Y.; Seoane, I.V.; Martínez, C.; González-Álvaro, I.; Pablos, J.L.; Gutiérrez-Cañas, I.; Gomariz, R.P. VIP modulates IL-22R1 expression and prevents the contribution of rheumatoid synovial fibroblasts to IL-22-mediated joint destruction. *J. Mol. Neurosci.* **2014**, *52*, 10–17. [CrossRef]

207. Carrión, M.; Pérez-García, S.; Jimeno, R.; Juarranz, Y.; González-Álvaro, I.; Pablos, J.L.; Gutiérrez-Cañas, I.; Gomariz, R.P. Inflammatory mediators alter interleukin-17 receptor, interleukin-12 and -23 expression in human osteoarthritic and rheumatoid arthritis synovial fibroblasts: Immunomodulation by vasoactive intestinal Peptide. *Neuroimmunomodulation* **2013**, *20*, 274–284. [CrossRef]

208. Pérez-García, S.; Juarranz, Y.; Carrión, M.; Gutiérrez-Cañas, I.; Margioris, A.; Pablos, J.L.; Tsatsanis, C.; Gomariz, R.P. Mapping the CRF-urocortins system in human osteoarthritic and rheumatoid synovial fibroblasts: Effect of vasoactive intestinal peptide. *J. Cell. Physiol.* **2011**, *226*, 3261–3269. [CrossRef]

209. Monfort, J. *Artrosis: Fisiopatología, Diagnóstico y Tratamiento*; Médica Panamericana: Madrid, Spain, 2010.

210. Sokolove, J.; Lepus, C.M. Role of inflammation in the pathogenesis of osteoarthritis: Latest findings and interpretations. *Ther. Adv. Musculoskelet. Dis.* **2013**, *5*, 77–94. [CrossRef]

211. Mimpen, J.Y.; Snelling, S.J.B. Chondroprotective Factors in Osteoarthritis: A Joint Affair. *Curr. Rheumatol. Rep.* **2019**, *21*, 41. [CrossRef]

212. Brooks, P. Inflammation as an important feature of osteoarthritis. *Bull. World Health Organ.* **2003**, *81*, 689–690.

213. Goldring, M.B.; Goldring, S.R. Osteoarthritis. *J. Cell. Physiol.* **2007**, *213*, 626–634. [CrossRef] [PubMed]

214. Bijlsma, J.W.; Berenbaum, F.; Lafeber, F.P. Osteoarthritis: An update with relevance for clinical practice. *Lancet* **2011**, *377*, 2115–2126. [CrossRef]

215. Kraus, V.B.; Blanco, F.J.; Englund, M.; Karsdal, M.A.; Lohmander, L.S. Call for standardized definitions of osteoarthritis and risk stratification for clinical trials and clinical use. *Osteoarthr. Cartil.* **2015**, *23*, 1233–1241. [CrossRef] [PubMed]

216. Raman, S.; FitzGerald, U.; Murphy, J.M. Interplay of Inflammatory Mediators with Epigenetics and Cartilage Modifications in Osteoarthritis. *Front. Bioeng. Biotechnol.* **2018**, *6*, 22. [CrossRef] [PubMed]

217. Goldring, M.B.; Marcu, K.B. Cartilage homeostasis in health and rheumatic diseases. *Arthritis Res. Ther.* **2009**, *11*, 224. [CrossRef] [PubMed]

218. Benito, M.J.; Veale, D.J.; FitzGerald, O.; van den Berg, W.B.; Bresnihan, B. Synovial tissue inflammation in early and late osteoarthritis. *Ann. Rheum. Dis.* **2005**, *64*, 1263–1267. [CrossRef]

219. Batlle-Gualda, E.; Benito-Ruiz, P.; Blanco, F.J.; Martín, E. *Manual SER de la Artrosis*; IM&C: Madrid, Spain, 2002.

220. Pérez-García, S.; Carrión, M.; Jimeno, R.; Ortiz, A.M.; González-Álvaro, I.; Fernández, J.; Gomariz, R.P.; Juarranz, Y. Urokinase plasminogen activator system in synovial fibroblasts from osteoarthritis patients: Modulation by inflammatory mediators and neuropeptides. *J. Mol. Neurosci.* **2014**, *52*, 18–27. [CrossRef]

221. Pérez-García, S.; Gutiérrez-Cañas, I.; Seoane, I.V.; Fernández, J.; Mellado, M.; Leceta, J.; Tío, L.; Villanueva-Romero, R.; Juarranz, Y.; Gomariz, R.P. Healthy and Osteoarthritic Synovial Fibroblasts Produce a Disintegrin and Metalloproteinase with Thrombospondin Motifs 4, 5, 7, and 12: Induction by IL-1beta and Fibronectin and Contribution to Cartilage Damage. *Am. J. Pathol.* **2016**, *186*, 2449–2461. [CrossRef]

222. Martel-Pelletier, J.; Pelletier, J.P. Is osteoarthritis a disease involving only cartilage or other articular tissues? *Eklem Hastalik Cerrahisi* **2010**, *21*, 2–14.

223. Sharma, L. Osteoarthritis year in review 2015: Clinical. *Osteoarthr. Cartil.* **2016**, *24*, 36–48. [CrossRef]

224. Goldring, M.B.; Otero, M. Inflammation in osteoarthritis. *Curr. Opin. Rheumatol.* **2011**, *23*, 471–478. [CrossRef] [PubMed]

225. Hunter, D.J.; McDougall, J.J.; Keefe, F.J. The symptoms of osteoarthritis and the genesis of pain. *Rheum. Dis. Clin. N. Am.* **2008**, *34*, 623–643. [CrossRef] [PubMed]

226. Abramson, S.B.; Attur, M. Developments in the scientific understanding of osteoarthritis. *Arthritis Res. Ther.* **2009**, *11*, 227. [CrossRef] [PubMed]

227. Martel-Pelletier, J.; Wildi, L.M.; Pelletier, J.P. Future therapeutics for osteoarthritis. *Bone* **2012**, *51*, 297–311. [CrossRef]

228. Ghouri, A.; Conaghan, P.G. Update on novel pharmacological therapies for osteoarthritis. *Ther. Adv. Musculoskelet. Dis.* **2019**, *11*, 1759720X19864492. [CrossRef]

229. Sutton, S.; Clutterbuck, A.; Harris, P.; Gent, T.; Freeman, S.; Foster, N.; Barrett-Jolley, R.; Mobasheri, A. The contribution of the synovium, synovial derived inflammatory cytokines and neuropeptides to the pathogenesis of osteoarthritis. *Vet. J.* **2009**, *179*, 10–24. [CrossRef]

230. Carrión, M.; Juarranz, Y.; Martínez, C.; González-Álvaro, I.; Pablos, J.L.; Gutiérrez-Cañas, I.; Gomariz, R.P. IL-22/IL-22R1 axis and S100A8/A9 alarmins in human osteoarthritic and rheumatoid arthritis synovial fibroblasts. *Rheumatology* **2013**, *52*, 2177–2186. [CrossRef]

231. Jiang, W.; Gao, S.G.; Chen, X.G.; Xu, X.C.; Xu, M.; Luo, W.; Tu, M.; Zhang, F.J.; Zeng, C.; Lei, G.H. Expression of synovial fluid and articular cartilage VIP in human osteoarthritic knee: A new indicator of disease severity? *Clin. Biochem.* **2012**, *45*, 1607–1612. [CrossRef]

232. Jiang, W.; Wang, H.; Li, Y.S.; Luo, W. Role of vasoactive intestinal peptide in osteoarthritis. *J. Biomed. Sci.* **2016**, *23*, 63. [CrossRef]

233. Liang, Y.; Chen, S.; Yang, Y.; Lan, C.; Zhang, G.; Ji, Z.; Lin, H. Vasoactive intestinal peptide alleviates osteoarthritis effectively via inhibiting NF-kappaB signaling pathway. *J. Biomed. Sci.* **2018**, *25*, 25. [CrossRef]

234. Delgado, M.; Munoz-Elias, E.J.; Gomariz, R.P.; Ganea, D. Vasoactive intestinal peptide and pituitary adenylate cyclase-activating polypeptide enhance IL-10 production by murine macrophages: In vitro and in vivo studies. *J. Immunol.* **1999**, *162*, 1707–1716. [PubMed]

235. Kambayashi, T.; Jacob, C.O.; Zhou, D.; Mazurek, N.; Fong, M.; Strassmann, G. Cyclic nucleotide phosphodiesterase type IV participates in the regulation of IL-10 and in the subsequent inhibition of TNF-alpha and IL-6 release by endotoxin-stimulated macrophages. *J. Immunol.* **1995**, *155*, 4909–4916. [PubMed]

236. Strassmann, G.; Patil-Koota, V.; Finkelman, F.; Fong, M.; Kambayashi, T. Evidence for the involvement of interleukin 10 in the differential deactivation of murine peritoneal macrophages by prostaglandin E2. *J. Exp. Med.* **1994**, *180*, 2365–2370. [CrossRef] [PubMed]

237. Hernanz, A.; Medina, S.; de Miguel, E.; Martin-Mola, E. Effect of calcitonin gene-related peptide, neuropeptide Y, substance P, and vasoactive intestinal peptide on interleukin-1beta, interleukin-6 and tumor necrosis factor-alpha production by peripheral whole blood cells from rheumatoid arthritis and osteoarthritis patients. *Regul. Pept.* **2003**, *115*, 19–24. [PubMed]

238. Schuelert, N.; McDougall, J.J. Electrophysiological evidence that the vasoactive intestinal peptide receptor antagonist VIP6-28 reduces nociception in an animal model of osteoarthritis. *Osteoarthr. Cartil.* **2006**, *14*, 1155–1162. [CrossRef]

239. Rahman, S.; Dobson, P.R.; Bunning, R.A.; Russell, R.G.; Brown, B.L. The regulation of connective tissue metabolism by vasoactive intestinal polypeptide. *Regul. Pept.* **1992**, *37*, 111–121. [CrossRef]

240. Tío, L.; Orellana, C.; Pérez-García, S.; Piqueras, L.; Escudero, P.; Juarranz, Y.; García-Giralt, N.; Montañés, F.; Farran, A.; Benito, P.; et al. Effect of chondroitin sulphate on synovitis of knee osteoarthritic patients. *Med. Clin.* **2017**, *149*, 9–16. [CrossRef]

241. Pérez-García, S.; Carrión, M.; Gutiérrez-Cañas, I.; González-Álvaro, I.; Gomariz, R.P.; Juarranz, Y. VIP and CRF reduce ADAMTS expression and function in osteoarthritis synovial fibroblasts. *J. Cell. Mol. Med.* **2016**, *20*, 678–687. [CrossRef]

242. Juhász, T.; Helgadottir, S.L.; Tamás, A.; Reglődi, D.; Zákány, R. PACAP and VIP signaling in chondrogenesis and osteogenesis. *Peptides* **2015**, *66*, 51–57. [CrossRef]

243. Juhasz, T.; Matta, C.; Katona, E.; Somogyi, C.; Takacs, R.; Gergely, P.; Csernoch, L.; Panyi, G.; Toth, G.; Reglodi, D.; et al. Pituitary adenylate cyclase activating polypeptide (PACAP) signalling exerts chondrogenesis promoting and protecting effects: Implication of calcineurin as a downstream target. *PLoS ONE* **2014**, *9*, e91541. [CrossRef]

244. Lerner, U.H.; Persson, E. Osteotropic effects by the neuropeptides calcitonin gene-related peptide, substance P and vasoactive intestinal peptide. *J. Musculoskelet Neuronal Interact* **2008**, *8*, 154–165. [PubMed]

245. Grassel, S.; Muschter, D. Do Neuroendocrine Peptides and Their Receptors Qualify as Novel Therapeutic Targets in Osteoarthritis? *Int. J. Mol. Sci.* **2018**, *19*, 367. [CrossRef]

246. Persson, E.; Lerner, U.H. The neuropeptide VIP regulates the expression of osteoclastogenic factors in osteoblasts. *J. Cell. Biochem.* **2011**, *112*, 3732–3741. [CrossRef] [PubMed]

247. Shih, C.; Bernard, G.W. Neurogenic substance P stimulates osteogenesis in vitro. *Peptides* **1997**, *18*, 323–326. [CrossRef]

248. Xiao, J.; Yu, W.; Wang, X.; Wang, B.; Chen, J.; Liu, Y.; Li, Z. Correlation between neuropeptide distribution, cancellous bone microstructure and joint pain in postmenopausal women with osteoarthritis and osteoporosis. *Neuropeptides* **2016**, *56*, 97–104. [CrossRef] [PubMed]

249. Kaplan, G.G. The global burden of IBD: From 2015 to 2025. *Nat. Rev. Gastroenterol. Hepatol.* **2015**, *12*, 720–727. [CrossRef] [PubMed]

250. Goyal, R.K.; Hirano, I. The enteric nervous system. *N. Engl. J. Med.* **1996**, *334*, 1106–1115. [CrossRef] [PubMed]

251. Gross, K.J.; Pothoulakis, C. Role of neuropeptides in inflammatory bowel disease. *Inflamm. Bowel Dis.* **2007**, *13*, 918–932. [CrossRef]

252. Margolis, K.G.; Gershon, M.D. Neuropeptides and inflammatory bowel disease. *Curr. Opin. Gastroenterol.* **2009**, *25*, 503–511. [CrossRef]

253. Perse, M.; Cerar, A. Dextran sodium sulphate colitis mouse model: Traps and tricks. *J. Biomed Biotechnol.* **2012**, *2012*, 718617. [CrossRef]

254. Wirtz, S.; Neufert, C.; Weigmann, B.; Neurath, M.F. Chemically induced mouse models of intestinal inflammation. *Nat. Protoc.* **2007**, *2*, 541–546. [CrossRef] [PubMed]

255. Morampudi, V.; Bhinder, G.; Wu, X.; Dai, C.; Sham, H.P.; Vallance, B.A.; Jacobson, K. DNBS/TNBS colitis models: Providing insights into inflammatory bowel disease and effects of dietary fat. *J. Vis. Exp.* **2014**, *84*, e51297. [CrossRef] [PubMed]

256. Alex, P.; Zachos, N.C.; Nguyen, T.; Gonzales, L.; Chen, T.E.; Conklin, L.S.; Centola, M.; Li, X. Distinct cytokine patterns identified from multiplex profiles of murine DSS and TNBS-induced colitis. *Inflamm. Bowel Dis.* **2009**, *15*, 341–352. [CrossRef] [PubMed]

257. Brand, S. Crohn's disease: Th1, Th17 or both? The change of a paradigm: New immunological and genetic insights implicate Th17 cells in the pathogenesis of Crohn's disease. *Gut* **2009**, *58*, 1152–1167. [CrossRef] [PubMed]

258. Abad, C.; Juarranz, Y.; Martinez, C.; Arranz, A.; Rosignoli, F.; Garcia-Gomez, M.; Leceta, J.; Gomariz, R.P. cDNA array analysis of cytokines, chemokines, and receptors involved in the development of TNBS-induced colitis: Homeostatic role of VIP. *Inflamm. Bowel Dis.* **2005**, *11*, 674–684. [CrossRef]

259. Imam, T.; Park, S.; Kaplan, M.H.; Olson, M.R. Effector T Helper Cell Subsets in Inflammatory Bowel Diseases. *Front. Immunol.* **2018**, *9*, 1212. [CrossRef]

260. Fitzpatrick, L.R. Inhibition of IL-17 as a pharmacological approach for IBD. *Int. Rev. Immunol.* **2013**, *32*, 544–555. [CrossRef]

261. Ananthakrishnan, A.N.; Bernstein, C.N.; Iliopoulos, D.; Macpherson, A.; Neurath, M.F.; Ali, R.A.R.; Vavricka, S.R.; Fiocchi, C. Environmental triggers in IBD: A review of progress and evidence. *Nat. Rev. Gastroenterol. Hepatol.* **2018**, *15*, 39–49. [CrossRef]

262. Lu, Y.; Li, X.; Liu, S.; Zhang, Y.; Zhang, D. Toll-like Receptors and Inflammatory Bowel Disease. *Front. Immunol.* **2018**, *9*, 72. [CrossRef]

263. Arranz, A.; Juarranz, Y.; Leceta, J.; Gomariz, R.P.; Martínez, C. VIP balances innate and adaptive immune responses induced by specific stimulation of TLR2 and TLR4. *Peptides* **2008**, *29*, 948–956. [CrossRef]

264. Abad, C.; Gomariz, R.; Waschek, J.; Leceta, J.; Martinez, C.; Juarranz, Y.; Arranz, A. VIP in inflammatory bowel disease: State of the art. *Endocr. Metab. Immune Disords Drug Targets* **2012**, *12*, 316–322. [CrossRef] [PubMed]

265. Newman, R.; Cuan, N.; Hampartzoumian, T.; Connor, S.J.; Lloyd, A.R.; Grimm, M.C. Vasoactive intestinal peptide impairs leucocyte migration but fails to modify experimental murine colitis. *Clin. Exp. Immunol.* **2005**, *139*, 411–420. [CrossRef] [PubMed]

266. Jayawardena, D.; Anbazhagan, A.N.; Guzman, G.; Dudeja, P.K.; Onyuksel, H. Vasoactive Intestinal Peptide Nanomedicine for the Management of Inflammatory Bowel Disease. *Mol. Pharm.* **2017**, *14*, 3698–3708. [CrossRef] [PubMed]

267. Conlin, V.S.; Wu, X.; Nguyen, C.; Dai, C.; Vallance, B.A.; Buchan, A.M.; Boyer, L.; Jacobson, K. Vasoactive intestinal peptide ameliorates intestinal barrier disruption associated with Citrobacter rodentium-induced colitis. *Am. J. Physiol. Gastrointest. Liver Physiol.* **2009**, *297*, G735–G750. [CrossRef]

268. Sun, X.; Guo, C.; Zhao, F.; Zhu, J.; Xu, Y.; Liu, Z.Q.; Yang, G.; Zhang, Y.Y.; Gu, X.; Xiao, L.; et al. Vasoactive intestinal peptide stabilizes intestinal immune homeostasis through maintaining interleukin-10 expression in regulatory B cells. *Theranostics* **2019**, *9*, 2800–2811. [CrossRef]

269. Abad, C.; Tan, Y.V. Immunomodulatory Roles of PACAP and VIP: Lessons from Knockout Mice. *J. Mol. Neurosci.* **2018**, *66*, 102–113. [CrossRef]

270. Yadav, M.; Huang, M.C.; Goetzl, E.J. VPAC1 (vasoactive intestinal peptide (VIP) receptor type 1) G protein-coupled receptor mediation of VIP enhancement of murine experimental colitis. *Cell Immunol.* **2011**, *267*, 124–132. [CrossRef]

271. Wu, X.; Conlin, V.S.; Morampudi, V.; Ryz, N.R.; Nasser, Y.; Bhinder, G.; Bergstrom, K.S.; Yu, H.B.; Waterhouse, C.C.; Buchan, A.M.; et al. Vasoactive intestinal polypeptide promotes intestinal barrier homeostasis and protection against colitis in mice. *PLoS ONE* **2015**, *10*, e0125225. [CrossRef]

272. Abad, C.; Cheung-Lau, G.; Coute-Monvoisin, A.C.; Waschek, J.A. Vasoactive intestinal peptide-deficient mice exhibit reduced pathology in trinitrobenzene sulfonic acid-induced colitis. *Neuroimmunomodulation* **2015**, *22*, 203–212. [CrossRef]

273. Schulte-Bockholt, A.; Fink, J.G.; Meier, D.A.; Otterson, M.F.; Telford, G.L.; Hopp, K.; Koch, T.R. Expression of mRNA for vasoactive intestinal peptide in normal human colon and during inflammation. *Mol. Cell Biochem.* **1995**, *142*, 1–7. [CrossRef]

274. Kubota, Y.; Petras, R.E.; Ottaway, C.A.; Tubbs, R.R.; Farmer, R.G.; Fiocchi, C. Colonic vasoactive intestinal peptide nerves in inflammatory bowel disease. *Gastroenterology* **1992**, *102*, 1242–1251. [CrossRef]

275. Martinez, C.; Ortiz, A.M.; Juarranz, Y.; Lamana, A.; Seoane, I.V.; Leceta, J.; Garcia-Vicuna, R.; Gomariz, R.P.; Gonzalez-Alvaro, I. Serum levels of vasoactive intestinal peptide as a prognostic marker in early arthritis. *PLoS ONE* **2014**, *9*, e85248. [CrossRef] [PubMed]

276. Seoane, I.V.; Tomero, E.; Martinez, C.; Garcia-Vicuna, R.; Juarranz, Y.; Lamana, A.; Ocon, E.; Ortiz, A.M.; Gomez-Leon, N.; Gonzalez-Alvaro, I.; et al. Vasoactive Intestinal Peptide in Early Spondyloarthritis: Low Serum Levels as a Potential Biomarker for Disease Severity. *J. Mol. Neurosci.* **2015**, *56*, 577–584. [CrossRef] [PubMed]

277. Caggiu, E.; Arru, G.; Hosseini, S.; Niegowska, M.; Sechi, G.; Zarbo, I.R.; Sechi, L.A. Inflammation, Infectious Triggers, and Parkinson's Disease. *Front. Neurol.* **2019**, *10*, 122. [CrossRef]

278. Garretti, F.; Agalliu, D.; Lindestam Arlehamn, C.S.; Sette, A.; Sulzer, D. Autoimmunity in Parkinson's Disease: The Role of alpha-Synuclein-Specific T Cells. *Front. Immunol.* **2019**, *10*, 303. [CrossRef]

279. Martinez, B.; Peplow, P.V. Neuroprotection by immunomodulatory agents in animal models of Parkinson's disease. *Neural Regen. Res.* **2018**, *13*, 1493–1506.

280. Ransohoff, R.M. How neuroinflammation contributes to neurodegeneration. *Science* **2016**, *353*, 777–783. [CrossRef]

281. Deng, G.; Jin, L. The effects of vasoactive intestinal peptide in neurodegenerative disorders. *Neurol Res.* **2017**, *39*, 65–72. [CrossRef]

282. Tan, Y.V.; Waschek, J.A. Targeting VIP and PACAP receptor signalling: New therapeutic strategies in multiple sclerosis. *ASN Neuro* **2011**, *3*, e00065. [CrossRef]

283. Fernandez-Martin, A.; Gonzalez-Rey, E.; Chorny, A.; Martin, J.; Pozo, D.; Ganea, D.; Delgado, M. VIP prevents experimental multiple sclerosis by downregulating both inflammatory and autoimmune components of the disease. *Ann. N. Y. Acad. Sci.* **2006**, *1070*, 276–281. [CrossRef]

284. Reddy, J.; Illes, Z.; Zhang, X.; Encinas, J.; Pyrdol, J.; Nicholson, L.; Sobel, R.A.; Wucherpfennig, K.W.; Kuchroo, V.K. Myelin proteolipid protein-specific CD4$^+$CD25$^+$ regulatory cells mediate genetic resistance to experimental autoimmune encephalomyelitis. *Proc. Natl. Acad. Sci. USA* **2004**, *101*, 15434–15439. [CrossRef] [PubMed]

285. Kohm, A.P.; Carpentier, P.A.; Anger, H.A.; Miller, S.D. Cutting edge: CD4$^+$CD25$^+$ regulatory T cells suppress antigen-specific autoreactive immune responses and central nervous system inflammation during active experimental autoimmune encephalomyelitis. *J. Immunol.* **2002**, *169*, 4712–4716. [CrossRef] [PubMed]

286. Yu, P.; Gregg, R.K.; Bell, J.J.; Ellis, J.S.; Divekar, R.; Lee, H.H.; Jain, R.; Waldner, H.; Hardaway, J.C.; Collins, M.; et al. Specific T regulatory cells display broad suppressive functions against experimental allergic encephalomyelitis upon activation with cognate antigen. *J. Immunol.* **2005**, *174*, 6772–6780. [CrossRef] [PubMed]

287. Steinman, L. Immunology of relapse and remission in multiple sclerosis. *Annu. Rev. Immunol.* **2014**, *32*, 257–281. [CrossRef] [PubMed]

288. Fernandez-Martin, A.; Gonzalez-Rey, E.; Chorny, A.; Ganea, D.; Delgado, M. Vasoactive intestinal peptide induces regulatory T cells during experimental autoimmune encephalomyelitis. *Eur. J. Immunol.* **2006**, *36*, 318–326. [CrossRef]

289. Tan, Y.V.; Abad, C.; Wang, Y.; Lopez, R.; Waschek, J. VPAC2 (vasoactive intestinal peptide receptor type 2) receptor deficient mice develop exacerbated experimental autoimmune encephalomyelitis with increased Th1/Th17 and reduced Th2/Treg responses. *Brain Behav. Immun.* **2015**, *44*, 167–175. [CrossRef]

290. Abad, C.; Tan, Y.V.; Lopez, R.; Nobuta, H.; Dong, H.; Phan, P.; Feng, J.M.; Campagnoni, A.T.; Waschek, J.A. Vasoactive intestinal peptide loss leads to impaired CNS parenchymal T-cell infiltration and resistance to experimental autoimmune encephalomyelitis. *Proc. Natl. Acad. Sci. USA* **2010**, *107*, 19555–19560. [CrossRef]

291. Abad, C.; Jayaram, B.; Becquet, L.; Wang, Y.; O'Dorisio, M.S.; Waschek, J.A.; Tan, Y.V. VPAC1 receptor (Vipr1)-deficient mice exhibit ameliorated experimental autoimmune encephalomyelitis, with specific deficits in the effector stage. *J. Neuroinflamm.* **2016**, *13*, 169. [CrossRef]

292. Andersen, O.; Fahrenkrug, J.; Wikkelso, C.; Johansson, B.B. VIP in cerebrospinal fluid of patients with multiple sclerosis. *Peptides* **1984**, *5*, 435–437. [CrossRef]

293. Baranowska-Bik, A.; Kochanowski, J.; Uchman, D.; Wolinska-Witort, E.; Kalisz, M.; Martynska, L.; Baranowska, B.; Bik, W. Vasoactive intestinal peptide (VIP) and pituitary adenylate cyclase activating polypeptide (PACAP) in humans with multiple sclerosis. *J. Neuroimmunol.* **2013**, *263*, 159–161. [CrossRef]

294. Sun, W.; Hong, J.; Zang, Y.C.; Liu, X.; Zhang, J.Z. Altered expression of vasoactive intestinal peptide receptors in T lymphocytes and aberrant Th1 immunity in multiple sclerosis. *Int. Immunol.* **2006**, *18*, 1691–1700. [CrossRef] [PubMed]

295. Appel, S.H.; Beers, D.R.; Henkel, J.S. T cell-microglial dialogue in Parkinson's disease and amyotrophic lateral sclerosis: Are we listening? *Trends Immunol.* **2010**, *31*, 7–17. [CrossRef] [PubMed]

296. Mosley, R.L.; Lu, Y.; Olson, K.E.; Machhi, J.; Yan, W.; Namminga, K.L.; Smith, J.R.; Shandler, S.J.; Gendelman, H.E. A Synthetic Agonist to Vasoactive Intestinal Peptide Receptor-2 Induces Regulatory T Cell Neuroprotective Activities in Models of Parkinson's Disease. *Front. Cell Neurosci.* **2019**, *13*, 421. [CrossRef] [PubMed]

297. Gendelman, H.E.; Mosley, R.L. A Perspective on Roles Played by Innate and Adaptive Immunity in the Pathobiology of Neurodegenerative Disorders. *J. Neuroimmune Pharmacol.* **2015**, *10*, 645–650. [CrossRef] [PubMed]

298. Martin, B.; Shin, Y.K.; White, C.M.; Ji, S.; Kim, W.; Carlson, O.D.; Napora, J.K.; Chadwick, W.; Chapter, M.; Waschek, J.A.; et al. Vasoactive intestinal peptide-null mice demonstrate enhanced sweet taste preference, dysglycemia, and reduced taste bud leptin receptor expression. *Diabetes* **2010**, *59*, 1143–1152. [CrossRef] [PubMed]

299. Fabricius, D.; Karacay, B.; Shutt, D.; Leverich, W.; Schafer, B.; Takle, E.; Thedens, D.; Khanna, G.; Raikwar, S.; Yang, B.; et al. Characterization of intestinal and pancreatic dysfunction in VPAC1-null mutant mouse. *Pancreas* **2011**, *40*, 861–871. [CrossRef]

300. Kato, I.; Suzuki, Y.; Akabane, A.; Yonekura, H.; Tanaka, O.; Kondo, H.; Takasawa, S.; Yoshimoto, T.; Okamoto, H. Transgenic mice overexpressing human vasoactive intestinal peptide (VIP) gene in pancreatic beta cells. Evidence for improved glucose tolerance and enhanced insulin secretion by VIP and PHM-27 in vivo. *J. Biol. Chem.* **1994**, *269*, 21223–21228.

301. Li, C.; Zhu, F.; Wu, B.; Wang, Y. Vasoactive Intestinal Peptide Protects Salivary Glands against Structural Injury and Secretory Dysfunction via IL-17A and AQP5 Regulation in a Model of Sjogren Syndrome. *Neuroimmunomodulation* **2017**, *24*, 300–309. [CrossRef]

302. Hauk, V.; Calafat, M.; Larocca, L.; Fraccaroli, L.; Grasso, E.; Ramhorst, R.; Leiros, C.P. Vasoactive intestinal peptide/vasoactive intestinal peptide receptor relative expression in salivary glands as one endogenous modulator of acinar cell apoptosis in a murine model of Sjogren's syndrome. *Clin. Exp. Immunol.* **2011**, *166*, 309–316. [CrossRef]

303. Hauk, V.; Fraccaroli, L.; Grasso, E.; Eimon, A.; Ramhorst, R.; Hubscher, O.; Perez Leiros, C. Monocytes from Sjogren's syndrome patients display increased vasoactive intestinal peptide receptor 2 expression and impaired apoptotic cell phagocytosis. *Clin. Exp. Immunol.* **2014**, *177*, 662–670. [CrossRef]

304. Groneberg, D.A.; Springer, J.; Fischer, A. Vasoactive intestinal polypeptide as mediator of asthma. *Pulm. Pharmacol. Ther.* **2001**, *14*, 391–401. [CrossRef] [PubMed]

305. Petkov, V.; Mosgoeller, W.; Ziesche, R.; Raderer, M.; Stiebellehner, L.; Vonbank, K.; Funk, G.C.; Hamilton, G.; Novotny, C.; Burian, B.; et al. Vasoactive intestinal peptide as a new drug for treatment of primary pulmonary hypertension. *J. Clin. Investig.* **2003**, *111*, 1339–1346. [CrossRef] [PubMed]

306. Prasse, A.; Zissel, G.; Lutzen, N.; Schupp, J.; Schmiedlin, R.; Gonzalez-Rey, E.; Rensing-Ehl, A.; Bacher, G.; Cavalli, V.; Bevec, D.; et al. Inhaled vasoactive intestinal peptide exerts immunoregulatory effects in sarcoidosis. *Am. J. Respir. Crit. Care Med.* **2010**, *182*, 540–548. [CrossRef] [PubMed]

307. Gozes, I.; Bardea, A.; Reshef, A.; Zamostiano, R.; Zhukovsky, S.; Rubinraut, S.; Fridkin, M.; Brenneman, D.E. Neuroprotective strategy for Alzheimer disease: Intranasal administration of a fatty neuropeptide. *Proc. Natl. Acad. Sci. USA* **1996**, *93*, 427–432. [CrossRef]

308. Korkmaz, O.T.; Tuncel, N.; Tuncel, M.; Oncu, E.M.; Sahinturk, V.; Celik, M. Vasoactive intestinal peptide (VIP) treatment of Parkinsonian rats increases thalamic gamma-aminobutyric acid (GABA) levels and alters the release of nerve growth factor (NGF) by mast cells. *J. Mol. Neurosci.* **2010**, *41*, 278–287. [CrossRef]

309. Vacas, E.; Bajo, A.M.; Schally, A.V.; Sanchez-Chapado, M.; Prieto, J.C.; Carmena, M.J. Vasoactive intestinal peptide induces oxidative stress and suppresses metastatic potential in human clear cell renal cell carcinoma. *Mol. Cell Endocrinol.* **2013**, *365*, 212–222. [CrossRef]

310. Chu, T.G.; Orlowski, M. Soluble metalloendopeptidase from rat brain: Action on enkephalin-containing peptides and other bioactive peptides. *Endocrinology* **1985**, *116*, 1418–1425. [CrossRef]

311. Vessillier, S.; Adams, G.; Montero-Melendez, T.; Jones, R.; Seed, M.; Perretti, M.; Chernajovsky, Y. Molecular engineering of short half-life small peptides (VIP, alphaMSH and gamma(3)MSH) fused to latency-associated peptide results in improved anti-inflammatory therapeutics. *Ann. Rheum. Dis.* **2012**, *71*, 143–149. [CrossRef]

312. Bloom, S.R.; Polak, J.M.; Pearse, A.G. Vasoactive intestinal peptide and watery-diarrhoea syndrome. *Lancet* **1973**, *2*, 14–16. [CrossRef]

313. Henning, R.J.; Sawmiller, D.R. Vasoactive intestinal peptide: Cardiovascular effects. *Cardiovasc. Res.* **2001**, *49*, 27–37. [CrossRef]

314. Fernandez-Montesinos, R.; Castillo, P.M.; Klippstein, R.; Gonzalez-Rey, E.; Mejias, J.A.; Zaderenko, A.P.; Pozo, D. Chemical synthesis and characterization of silver-protected vasoactive intestinal peptide nanoparticles. *Nanomedicine* **2009**, *4*, 919–930. [CrossRef] [PubMed]

315. Masaka, T.; Li, Y.; Kawatobi, S.; Koide, Y.; Takami, A.; Yano, K.; Imai, R.; Yagi, N.; Suzuki, H.; Hikawa, H.; et al. Liposome modified with VIP-lipopeptide as a new drug delivery system. *Yakugaku Zasshi* **2014**, *134*, 987–995. [CrossRef] [PubMed]

316. Onyuksel, H.; Mohanty, P.S.; Rubinstein, I. VIP-grafted sterically stabilized phospholipid nanomicellar 17-allylamino-17-demethoxy geldanamycin: A novel targeted nanomedicine for breast cancer. *Int. J. Pharm.* **2009**, *365*, 157–161. [CrossRef] [PubMed]

317. Koo, O.M.; Rubinstein, I.; Onyuksel, H. Actively targeted low-dose camptothecin as a safe, long-acting, disease-modifying nanomedicine for rheumatoid arthritis. *Pharm. Res.* **2011**, *28*, 776–787. [CrossRef] [PubMed]

318. Delgado, M.; Toscano, M.G.; Benabdellah, K.; Cobo, M.; O'Valle, F.; Gonzalez-Rey, E.; Martin, F. In vivo delivery of lentiviral vectors expressing vasoactive intestinal peptide complementary DNA as gene therapy for collagen-induced arthritis. *Arthritis Rheum.* **2008**, *58*, 1026–1037. [CrossRef] [PubMed]

319. El-Sayed, Z.A.; Mostafa, G.A.; Aly, G.S.; El-Shahed, G.S.; El-Aziz, M.M.; El-Emam, S.M. Cardiovascular autonomic function assessed by autonomic function tests and serum autonomic neuropeptides in Egyptian children and adolescents with rheumatic diseases. *Rheumatology* **2009**, *48*, 843–848. [CrossRef] [PubMed]

320. Seoane, I.V.; Ortiz, A.M.; Piris, L.; Lamana, A.; Juarranz, Y.; Garcia-Vicuna, R.; Gonzalez-Alvaro, I.; Gomariz, R.P.; Martinez, C. Clinical Relevance of VPAC1 Receptor Expression in Early Arthritis: Association with IL-6 and Disease Activity. *PLoS ONE* **2016**, *11*, e0149141. [CrossRef]

321. Choy, E.; Taylor, P.; McAuliffe, S.; Roberts, K.; Sargeant, I. Variation in the use of biologics in the management of rheumatoid arthritis across the UK. *Curr. Med. Res. Opin.* **2012**, *28*, 1733–1741. [CrossRef]

322. Felson, D.T.; Smolen, J.S.; Wells, G.; Zhang, B.; van Tuyl, L.H.; Funovits, J.; Aletaha, D.; Allaart, C.F.; Bathon, J.; Bombardieri, S.; et al. American College of Rheumatology/European League Against Rheumatism provisional definition of remission in rheumatoid arthritis for clinical trials. *Arthritis Rheum.* **2011**, *63*, 573–586. [CrossRef]

323. Van Leeuwen, M.A.; van Rijswijk, M.H.; van der Heijde, D.M.; Te Meerman, G.J.; van Riel, P.L.; Houtman, P.M.; van De Putte, L.B.; Limburg, P.C. The acute-phase response in relation to radiographic progression in early rheumatoid arthritis: A prospective study during the first three years of the disease. *Br. J. Rheumatol.* **1993**, *32*, 9–13. [CrossRef]

324. Aletaha, D.; Neogi, T.; Silman, A.J.; Funovits, J.; Felson, D.T.; Bingham, C.O., III; Birnbaum, N.S.; Burmester, G.R.; Bykerk, V.P.; Cohen, M.D.; et al. 2010 Rheumatoid arthritis classification criteria: An American College of Rheumatology/European League Against Rheumatism collaborative initiative. *Arthritis Rheum.* **2010**, *62*, 2569–2581. [CrossRef] [PubMed]

325. Castrejon, I.; Ortiz, A.M.; Garcia-Vicuna, R.; Lopez-Bote, J.P.; Humbria, A.; Carmona, L.; Gonzalez-Alvaro, I. Are the C-reactive protein values and erythrocyte sedimentation rate equivalent when estimating the 28-joint disease activity score in rheumatoid arthritis? *Clin. Exp. Rheumatol.* **2008**, *26*, 769–775. [PubMed]

326. Centola, M.; Cavet, G.; Shen, Y.; Ramanujan, S.; Knowlton, N.; Swan, K.A.; Turner, M.; Sutton, C.; Smith, D.R.; Haney, D.J.; et al. Development of a multi-biomarker disease activity test for rheumatoid arthritis. *PLoS ONE* **2013**, *8*, e60635. [CrossRef] [PubMed]
327. Seoane, I.V.; Martinez, C.; Garcia-Vicuna, R.; Ortiz, A.M.; Juarranz, Y.; Talayero, V.C.; Gonzalez-Alvaro, I.; Gomariz, R.P.; Lamana, A. Vasoactive intestinal peptide gene polymorphisms, associated with its serum levels, predict treatment requirements in early rheumatoid arthritis. *Sci. Rep.* **2018**, *8*, 2035. [CrossRef]

International Journal of
Molecular Sciences

MDPI

Review

Prolactin-Releasing Peptide: Physiological and Pharmacological Properties

Veronika Pražienková [1], Andrea Popelová [1], Jaroslav Kuneš [1,2] and Lenka Maletínská [1,*]

[1] Biochemistry and Molecular Biology, Institute of Organic Chemistry and Biochemistry of the Czech Academy of Sciences 16610 Prague, Czech Republic; prazienkova@uochb.cas.cz (V.P.); andrea.popelova@uochb.cas.cz (A.P.); kunes@biomed.cas.cz (J.K.)

[2] Experimental Hypertension, Institute of Physiology of the Czech Academy of Sciences, 14200 Prague, Czech Republic

[*] Correspondence: maletin@uochb.cas.cz; Tel.: +420-220-183-567

Received: 2 October 2019; Accepted: 23 October 2019; Published: 24 October 2019

Abstract: Prolactin-releasing peptide (PrRP) belongs to the large RF-amide neuropeptide family with a conserved Arg-Phe-amide motif at the C-terminus. PrRP plays a main role in the regulation of food intake and energy expenditure. This review focuses not only on the physiological functions of PrRP, but also on its pharmacological properties and the actions of its G-protein coupled receptor, GPR10. Special attention is paid to structure-activity relationship studies on PrRP and its analogs as well as to their effect on different physiological functions, mainly their anorexigenic and neuroprotective features and the regulation of the cardiovascular system, pain, and stress. Additionally, the therapeutic potential of this peptide and its analogs is explored.

Keywords: prolactin-releasing peptide; GPR10; RF-amide peptides; food intake regulation; energy expenditure; neuroprotection; signaling

---

1. Introduction

There is no doubt that the function of prolactin-releasing peptide (PrRP) in organisms is quite important as its structure is well conserved within different animal species. PrRP is reported to regulate food intake and energy metabolism, but it could have several other specific functions, such as the regulation of cardiac output, stress response, reproduction, the release of endocrine factors, and recently neuroprotective features. The site of the main action of PrRP is the brain, where its release is regulated by a number of stimuli, including those coming from the periphery.

PrRP binds with high affinity to the GPR10 receptor and also has lesser activity towards the neuropeptide FF (NPFF) receptor type 2 (NPFF-R2). In addition, cooperation with other food intake regulating neuropeptides, especially leptin, cholecystokinin (CCK), or neuropeptide Y (NPY), is very important for the effects of PrRP.

In structure-activity relationship (SAR) studies, novel PrRP analogs with attached fatty acids and changes in the amino acid chain were synthetized to overcome the blood-brain barrier and to improve the stability and bioavailability from the periphery, thus representing interesting targets for therapeutic use.

2. Discovery and Structure of PrRP

PrRP was first isolated in 1998 by Hinuma and colleagues from an extract of bovine hypothalamus and was described as a ligand for the orphan seven-transmembrane-domain receptor (7TM) GPR10 (also known as hGR3 or rat ortholog UHR-1) using reverse pharmacology ([1,2] and reviewed in [3]). The cloned full-length cDNA of the *PrRP* gene is 435 bp in length and encodes an 87 amino acid long precursor [4]. The *PrRP* rat gene contains three exons and two introns and spans a region of approximately 2.4 kb [5].

The average precursor length is 105 amino acids with two cleavage sites [6]. From the protein precursors, at least two isoforms of different lengths, PrRP20 and PrRP31 (Table 1), are produced. Shorter PrRP20 shares identical C-termini with the longer form of PrRP31. The fish ortholog of PrRP20, C-RFa, was isolated and described by Fujimoto et al. from the brain of *Carassius auratus langsdorfii* in the same year that PrRP was discovered [7]. The cloned cDNA of the *C-RFa* gene is 997 bp in length and encodes a precursor of 108 amino acids [4]. Subsequently, PrRP was identified in amphibians in *Xenopus laevis* in both isoforms [8]. In birds, specifically in *Gallus gallus*, PrRP has a similar sequence as in fishes and amphibians and is also expressed in the brain [9]. Moreover, Wang et al. measured the expression of both PrRP and C-RFa in chickens, as well as in *Xenopus* and zebrafish, suggesting that those peptides are encoded by two separate genes and may play similar yet distinctive roles in nonmammalian vertebrate species [4].

PrRPs in vertebrates share very conserved homology and there is evidence that PrRP evolved from a common ancestry precursor in nonmammalian and mammalian species [10]. The precursor is composed of a hydrophobic N-terminal sequence, paired basic amino acids for the recognition of endopeptidases, and a very conserved C-terminal sequence, where the amino acid glycine is a donor for the amide group. The bovine/human C-terminal octapeptide is Gly-Ile-Arg-Pro-Val-Gly-Arg-Phe-NH$_2$; in fish C-RFa, isoleucine and valine are swapped (Table 1) [6].

The name of PrRP was suggested on the basis of its prolactin-releasing activity in a rat pituitary adenoma-derived cell line and in pituitary cells obtained from lactating rats [1]. Additionally, another study reported that the injection of PrRP stimulated plasma prolactin levels in female rats in proestrus, estrus, and metestrus, and increased doses of PrRP were necessary to increase plasma prolactin in male rats [11]. Nevertheless, this prolactin-releasing function was later questioned because it did not have typical features for hypophysiotropic hormones [12,13]. Currently, PrRP is considered likely to be an anorexigenic (i.e., food-intake-lowering) neuropeptide, which mainly plays a role in the regulation of food intake and energy expenditure [12,14–16], but also regulates stress [17,18], sleep [19,20], and the cardiovascular system [21–23]. In addition, its potential neuroprotective properties have been described [24–26].

Table 1. Sequences of prolactin-releasing peptide (PrRP): PrRP20 and PrRP31 in different animal species [1,4,7].

Peptide	Species	Sequence
PrRP20	carp	S P E I D P F W Y V G R G V R P I G R F - NH$_2$
	chicken	S P E I D P F W Y V G R G V R P I G R F - NH$_2$
	rat	T P D I N P A W Y T G R G I R P V G R F - NH$_2$
	bovine	T P D I N P A W Y A G R G I R P V G R F - NH$_2$
	human	T P D I N P A W Y A S R G I R P V G R F - NH$_2$
PrRP31	carp	G T T V E H D L H I V H N V D N R S P E I D P F W Y V G R G V R P I G R F - NH$_2$
	chicken	S R P F K H Q I D N R S P E I D P F W Y V G R G V R P I G R F - NH$_2$
	rat	S R A H Q H S M E T R T P D I N P A W Y T G R G I R P V G R F - NH$_2$
	bovine	S R A H R H S M E I R T P D I N P A W Y A G R G I R P V G R F - NH$_2$
	human	S R T H R H S M E I R T P D I N P A W Y A S R G I R P V G R F - NH$_2$

Grey color marks same amino acids.

3. GPR10 Discovery and Gene Location

Using polymerase chain reaction (PCR), Marchese et al. discovered genes encoding novel G-protein coupled receptors (GPCRs), including the human gene for *GPR10* [27]. GPR10 shares high amino acid identity with NPY receptor 1 (NPY-1R) and orphan receptor induced by glucocorticoids (GIR) [27]. The overall amino acid identity is 31% and 46% in the transmembrane domains for NPY-1R and 30% and 46% in those for GIR. This GPR10 receptor was later confirmed to be identical to orphan hGR3 reported as a receptor for PrRP by Hinuma et al. [1]. Human GPR10 shares high homology (89%) with rat ortholog UHR-1 [2]. The human 1107 bp long gene for *GPR10* is located on chromosome 10 q25.3–q26.1 and a related sequence on chromosome 13 q14.3–q21.1, encoding a 370 amino acid long protein [27].

In nonmammalian vertebrates, fish and chicken PrRP receptor genes are located on chromosome 17 and chromosome 5, respectively [27,28]. GPR10 is well conserved in mammals with more than 90% identity, however in chickens, it is only 54% identical compared with the mammalian counterpart, probably because of phylogenetic differences. The most conserved sequence is on the C-terminus of the receptor, particularly the last six amino acid peptides that could interact with a ligand [29,30]. Both isoforms PrRP20 and PrRP31 bind with high affinity to the GPR10 receptor and rat UHR-1 [31].

Later, it was discovered that PrRP has an affinity for NPFF-R2 [32]. Different studies confirmed the molecular and functional identity of the HLWAR77 receptor, which is a common target for NPFF and neuropeptide AF (NPAF), with NPFF-R2 [33]. Human NPFF-R2 shares 89% amino acid identity with its rat ortholog, high homology with NPY receptors [34], and 37% homology with the orexin-A receptor [33].

4. Distribution of PrRP and its Receptor GPR10

4.1. Distribution of PrRP

The highest expression of *PrRP* mRNA was measured in the brainstem in the nucleus of the solitary tract (NTS) and a moderate level was detected in the dorsomedial hypothalamic nucleus (DMN), ventrolateral reticular nucleus of the thalamus (VRT) (Figure 1), and in the periphery in the intestine both in rats and humans when analyzed with reverse transcription-PCR [31,35,36]. Immunoreactive cell bodies were found mainly in the DMN, ventromedial hypothalamic nucleus (VMN), NTS, and ventrolateral medulla oblongata (ME), and nerve projections were present in the paraventricular hypothalamic nucleus (PVN), supraoptic nucleus (SON), DMN, lateral hypothalamic area (LHA), thalamic nucleus, amygdala, and area postrema (AP) (Figure 1) [31]. Immunoreactive fibers were also detected in high concentrations in the posterior pituitary [37,38]. Using enzyme immunoassay for PrRP distribution, immunoreactive PrRP was widely present in the hypothalamus, midbrain and posterior pituitary, and ME [37]. In mammals, rats, and humans, peripheral tissue *PrRP* mRNA was found mainly in the adrenal gland, lung, pancreas, liver, kidney, reproductive organs, and gut [35,37,39,40]. Concentration of PrRP in rat plasma was very low (0,13 fmol/mL) [37]. In chicken tissue, *C-RFa* mRNA was detected in the kidney, lung, reproductive organs, heart, intestine, liver, and pituitary [4]. In the amphibious fish, mudskipper, *PrRP* mRNA expression was observed in the brain, liver, gut, and ovary, with lower levels detected in the skin and kidney [41].

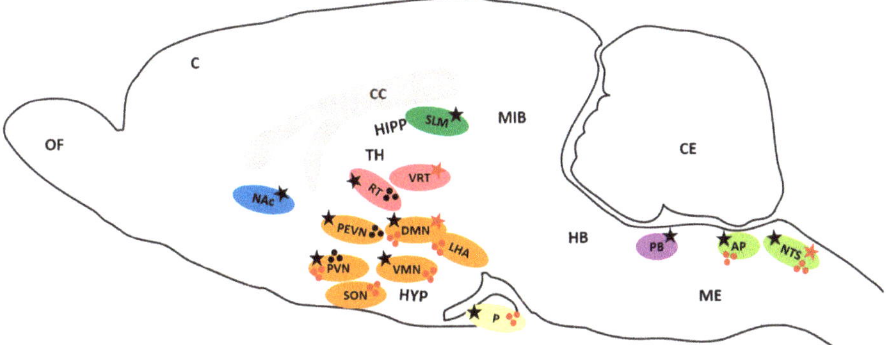

Figure 1. Distribution of PrRP and GPR10. Ellipses represent distinct brain areas (blue—nucleus accumbens, grey—corpus callosum, green—hippocampus, red—thalamus, orange—hypothalamus, yellow—pituitary, violet—parabrachial nucleus, light green—medulla oblongata). Stars mark the expression of mRNA (red star—*PrRP*, black star—*GPR10*). Spots represent the distribution of PrRP (red), GPR10 (black) cell bodies and fibers. AP: area postrema, C: cerebral cortex, CC: corpus callosum, CE: cerebellum, DMN: dorsomedial hypothalamic nucleus, HB: hindbrain, HIPP: hippocampus, HYP: hypothalamus, LHA: lateral hypothalamic area, ME: medulla oblongata, MIB: midbrain, NAc: nucleus accumbens, NTS: nucleus of the solitary tract, OF: olfactory bulb, P: pituitary, PB: parabrachial nucleus, PEVN: periventricular hypothalamic nucleus, PVN: paraventricular hypothalamic nucleus, RT: reticular nucleus of the thalamus, SON: supraoptic nucleus, SLM: stratum lacunosum-moleculare, TH: thalamus, VMN: ventromedial hypothalamic nucleus, VRT: ventrolateral reticular nucleus of the thalamus.

4.2. Distribution of GPR10

The highest expression of *GPR10* mRNA was detected in several parts of the rat brain, mainly in the reticular nucleus of the thalamus (RT), PVN, periventricular hypothalamic nucleus (PEVN) and DMN, AP, and NTS. A moderate level of expression of the receptor was also detected in the anterior pituitary and VMN (Figure 1) [31,42]. Radiolabeled ^{125}I-PrRP31 bound in a specific pattern to the reticular thalamic nucleus and PEVN [31]. GPR10 was also found in the parabrachial nucleus (PB) or nucleus accumbens (NAc), which are areas that are involved in pain processing [31], and in low levels in the hippocampus (stratum lacunosum-moleculare; SLM), which involves areas that are involved in memory [2,26]. In the periphery, *GPR10* mRNA was found in the rat adrenal medulla [35,43,44]. Through the detection of mRNA and in situ hybridization or immunohistochemical studies, PrRP and its receptor were found in discrete areas within the brain and periphery. Indeed, PrRP nerve fibers are in close proximity to areas where GPR10 is present, but PrRP still has to be transported to other sites to be released. This fact may also support the hypothesis that PrRP binding and signaling are not restricted to the GPR10 receptor.

5. PrRP Intracellular Signaling Pathways

To explore signal transduction pathways and the potential agonist or antagonist properties of PrRP action at GPR10, several studies have been published. Hinuma et al. first reported that PrRP promoted arachidonic acid metabolite release in Chinese hamster ovary (CHO) cells expressing GPR10 [1]. PrRP was able to dose-dependently stimulate calcium release in cells that were transfected with GPR10 in a calcium mobilization assay (Figure 2) [31].

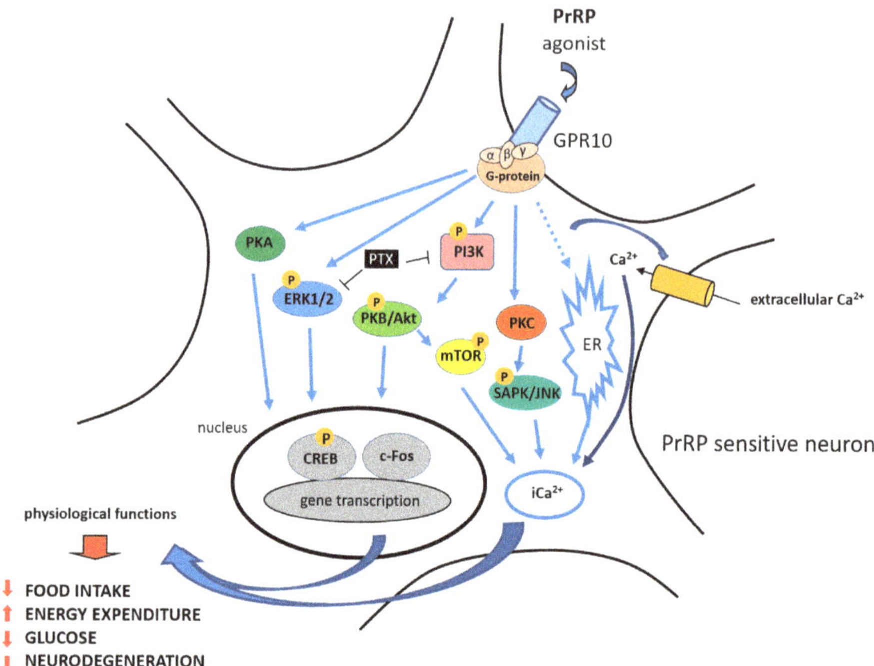

Figure 2. PrRP physiological functions and signaling—summary. PrRP and its agonist exerts its effect through GPR10. Blue arrow represents activation of the signaling pathway, T-bar represents blocking of the signaling pathway. PrRP stimulated calcium release (Ca^{2+}) in calcium mobilization assay and rapidly activated extracellular signal-regulated protein kinase (ERK – blue). It also activated c-Jun N-terminal protein kinase (JNK—light blue) and phosphorylated cAMP response element-binding protein (CREB—grey). Pertussis toxin (PTX—black) blocked the ERK and Akt activation induced by PrRP. PrRP activated the PI3K B/Akt-mammalian target of rapamycin (PI3K-Akt-mTOR) pathways in leiomyoma cells (PI3K—pink, PKB/Akt—green,.mTOR—yellow). PrRP significantly stimulated both the PKA (dark green) and PKC (orange) pathways.

PrRP rapidly activated extracellular signal-regulated protein kinase (ERK) from the mitogen-activated protein kinase (MAPK) family in GH3 rat pituitary tumor cells and in primary rat anterior pituitary cultures (Figure 2) [45]. Moreover, pertussis toxin (PTX), which inactivates Gi/Go proteins, completely blocked the ERK activation induced by PrRP, suggesting that at least part of the coupling of GPR10 is through Gi/Go proteins [45]. Kimura et al. also demonstrated that PrRP activated c-Jun N-terminal protein kinase (JNK) in a protein kinase C (PKC)-dependent manner in GH3 rat pituitary tumor cells [45].

PrRP20 was then reported not to alter basal levels of intracellular cyclic AMP in human embryonic kidney HEK293 cells that were transfected with GPR10, suggesting that in this system, GPR10 does not couple through Gs protein, which would activate adenylyl cyclase to increase the cyclic AMP concentration [46]. In addition, PrRP20 did not decrease forskolin-stimulated cyclic AMP levels, indicating that GPR10 does not couple via Gi, which would inhibit adenylyl cyclase and decrease cyclic AMP levels [46]. Therefore, the possible involvement of GPR10 signaling through the Gq pathway was proposed.

Engstrom et al. tested the ability of PrRP20 or PrRP31 to stimulate [^{35}S]GTPγS binding to membranes of CHO cells expressing GPR10; more than 80% of the binding of PrRP was prevented by PTX [32]. Taken together, these data suggest that a large part of the GPR10 coupling occurs via Gi/Go proteins, however this depends on the cellular system in which the receptor is expressed [32,45,46]. In the study from Engstrom et al., intracellular calcium assays also confirmed the full agonist properties of both PrRP20 and PrRP31 at GPR10 [46].

PrRP rapidly and transiently stimulated the activation of protein kinase B (Akt) in GH3 cells, and a phosphoinositide 3-kinase-protein kinase (PI3K) inhibitor blocked the PrRP-induced activation of Akt (Figure 2) (reviewed in [47]). Additionally, PTX completely blocked the Akt activation induced by PrRP, suggesting the involvement of Gi/Go proteins [48]. PrRP31 significantly induced an increase in the activity of ERKs and JNK, but not p38 MAPK in the rat PC12 pheochromocytoma cell line [49]. Moreover, PrRP stimulated dopamine release and catecholamine secretion and increased tyrosine hydroxylase levels via the protein kinase A (PKA) and PKC pathway in PC12 cells [49,50]. PrRP has also been shown to stimulate adenylyl cyclase in the PC12 cell line and promote the proliferation of cultured cells [51]. The stimulation of the chicken PrRP receptor expressed in CHO cells by PrRP also leads to the activation of the intracellular PKA signaling pathway [4,52].

PrRP activated the PI3K B/Akt-mammalian target of rapamycin (PI3K-Akt-mTOR) pathways and cell proliferation in primary leiomyoma cells, where GPR10 is aberrantly expressed [53]. Maixnerova et al. showed that both PrRP20 and PrRP31 activated ERK and cAMP response element-binding protein (CREB) signaling and induced prolactin release in the rat pituitary cell line RC-4B/C with equal potency (Figure 2) [54]. Additionally, modified analogs of PrRP20 and PrRP31, either with changes in the amino acids at the C-terminus or with lipidization, strongly induced the phosphorylation of the ERK pathway in CHO cells expressing GPR10 [55].

6. Structure-Activity Relationship Studies

Two isoforms of PrRP with either 20 or 31 amino acids sharing identical C-termini showed comparable in vitro and in vivo activity [1]. Several SAR studies with PrRP analogs were performed [31,56–58]. No study about selective antagonists of PrRP has been published yet, but in 2010, Otsuka Pharmaceuticals patented nonpeptide heterocyclic antagonists derived from tetrahydropyridol [4,3-d]pyrimidinone developed for stress-related diseases (reviewed in [59]).

First, Roland et al. demonstrated that N-terminal deletions from PrRP20 slightly decreased the affinity of the PrRP analogs for GPR10 [31]. The shortest analog that was still able to bind to GPR10 was C-terminal heptapeptide PrRP(25–31). However, this fragment displayed a two order of magnitude decrease in binding affinity compared to that of PrRP20 and PrRP31, which exhibited affinity in the nanomolar range. The replacement of the C-terminal amide group with an acid resulted in a complete loss of binding affinity [1,31]. Moreover, an alanine scan through PrRP(25–31) showed that the arginine at positions 26 and 30 is crucial for binding to the receptor, and their change results in a loss of affinity [31]. D'Ursi et al. described a conformational analysis of PrRP20 using circular dichroism (CD) and nuclear magnetic resonance (NMR) spectroscopies and molecular modeling calculations. The C-terminal region consisted of amphipathic helices with hydrophobic nonpolar side chains of Ala^{21}, Ile^{25}, Val^{28}, and Phe^{31} and hydrophilic side chains of Arg^{23}, Arg^{26}, and Arg^{30} [60].

PrRP could be shortened without a loss of in vitro activity to the tridecapeptide PrRP(19–31), H-Trp^{19}-Tyr^{20}-Ala^{21}-Ser^{22}-Arg^{23}-Gly^{24}-Ile^{25}-Arg^{26}-Pro^{27}-Val^{28}-Gly^{29}-Arg^{30}-Phe^{31}-NH_2, which has the minimal length for retaining binding affinity and agonist properties [56]. The binding affinity was significantly decreased by further truncation of the peptide; therefore, the active site is located within the C-terminal region. This large SAR study focused on the replacement of amino acids at positions 21 to 31, with a main focus on the phenylalanine at position 31. Nineteen different amino acids were used, but only a bulky side chain His(Bzl), Trp, Cys(Bzl), Glu(Obzl), norleucine (Nle) or a halogenated aromatic ring (Phe(4-Cl)) led to similar or improved binding affinity and good agonist activity [56]. Replacement of Arg^{23} by Pro significantly decreased the affinity. The results confirmed that the functionally important residues are located within the C-terminal segment with the essential and irretrievable arginine 30 and the high importance of phenylalanine 31.

Based on a previous study by Boyle et al., Maletínská et al. [58] designed PrRP20 analogs with modifications of Phe[31] by amino acids with different aromatic rings. Phe[31] was replaced by (3,4-dichlor)phenylalanine (PheCl$_2$[31]), (4-nitro)phenylalanine (PheNO$_2$[31]), pentafluoro-phenylalanine (PheF$_5$[31]), napthylalanine (1-Nal[31], 2-Nal[31]), or Tyr[31]. In addition, the amino acids cyclohexylalanine (Cha[31]) and phenylglycine (Phg[31]) were included [58]. This study showed that all analogs except [Cha[31]]PrRP20 and [Phg[31]]PrRP20 preserved high binding affinity to rat RC-4B/C pituitary cells and increased the phosphorylation of ERK and CREB in this cell line.

DeLuca et al. performed a structural study based on NMR and CD spectroscopy, where they determined the α-helical conformation in trifluoroethanol of the C-terminal sequence of PrRP20 [57]. Shorter PrRP20 analogs, PrRP(4–20), PrRP13 (PrRP(8–20)), and heptapeptide PrRP(14–20), decreased the stability of the helical segment and their biological activity was reduced. Therefore, this stable C-terminal α-helical structure facilitates ligand recognition by the receptor and enables its activation [57].

The lipidization of peptides (i.e., the attachment of fatty acids to peptides through an ester or amide bond) is a useful strategy for designing new peptide drugs. This modification may enhance potency, selectivity, and therapeutic efficacy because it can increase stability and prolong the half-life in an organism. Moreover, it could enable delivery across the blood-brain barrier (reviewed in [61]). This lipidized peptide is liraglutide, an analog of glucagon-like peptide 1 (GLP-1) that is palmitoylated at position 26 via a γ-glutamyl linker [62], with a strongly prolonged half-life [63]. Therefore, the lipidization of neuropeptides that is involved in food intake regulation might be a new way for the development of drugs for the treatment of obesity (reviewed in [64]).

Maletínská et al. designed novel lipidized PrRP analogs with fatty acids of different lengths attached to the N-terminus [55]. All PrRP20 and PrRP31 analogs lipidized with octanoic, decanoic, dodecanoic, myristic, palmitic, and stearic acid had agonist characteristics and preserved high binding affinity to GPR10 compared to native PrRP20 or PrRP31 [55].

Lipidized PrRP31 analogs with noncoded amino acids 1-Nal, PheCl$_2$, PheNO$_2$, PheF$_5$, or Tyr at position 31 and myristoylated or palmitoylated on the N-terminus revealed high binding potency to GPR10. The original methionine at position 8 was replaced by the more stable Nle to avoid oxidation of Met without any loss of binding and signaling activity [65].

Analogs of PrRP31 where palmitic acid was attached through the γ-glutamyl linker or the short chain of polyethylene glycol at Lys[11] or analog with two palmitic acids at Lys[11] and at the N-terminus were tested both in in vitro and in vivo studies. Binding and signaling experiments showed preserved binding affinity to GPR10, although the analog with two palmitic acids was less potent. The attachment of the single palmitic acid could be performed on different positions of the chain without the loss of binding affinity [66].

Recently, a new study by Pflimlin et al. was published in which novel long-lasting PrRP analogs with staples incorporating multiple ethylene glycol-fatty acids (MEG-FAs) were synthetized [67]. Crucial arginines at positions 23 and 30 were replaced with homoarginine (hArg), beta-homoarginine (β-hArg), and N-methylarginine (Nme-Arg). All modifications at Arg[30] significantly affected the potency. In Arg[23], only substitution by Nme-Arg, but not by hArg or β-hArg, decreased the affinity. All synthetized analogs contained dicysteine mutations, the best tolerated of which occurred at positions 6–13, 15–22, and 18–25. As lead compounds, they chose the PrRP analog 18-S4, an analog with Cys[6], Cys[13], Nle[11], and hArg[23] and stapled at cysteines by staple featuring four ethylene glycol units attached to octadecanedioic acid via a lysine linker incorporating a carboxylated moiety. They generated analogs with in vitro selective agonist activity towards GPR10 [67]. The structure of all of the mentioned PrRP analogs is described in Figure 3.

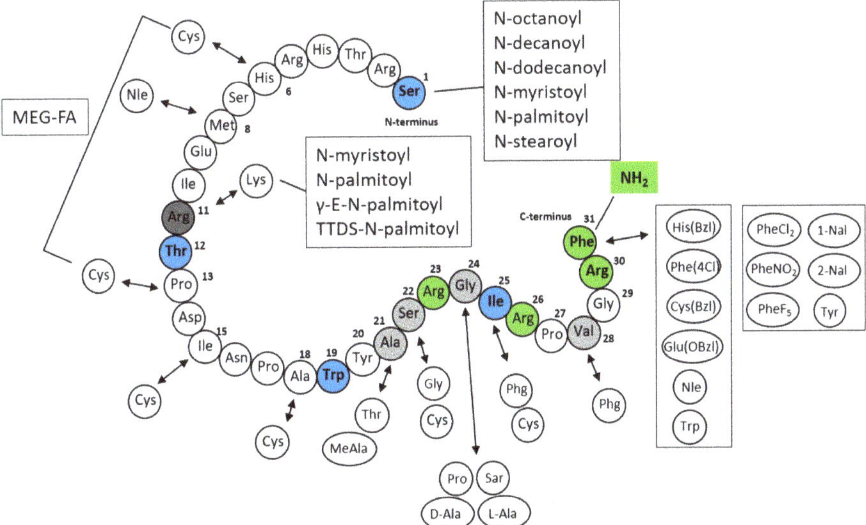

Figure 3. Structure of PrRP20 and PrRP31 and its analogs [31,55,56,58,65,67]. Blue amino acid Ser[1] marks the beginning of PrRP31 from the N-terminus. Blue amino acid Thr[12] marks the beginning of shorter isoform PrRP20 (also known as PrRP(12–31)). Blue amino acid Trp[19] marks tridecapeptide PrRP(19–31) and blue Ile[25] marks the shortest fragment heptapeptide PrRP(25–31). Green amide group NH2 and green Arg[23], Arg[26], Arg[30], and Phe[31] amino acids are essential for the functionality of the peptide. Light grey amino acids mark changes of amino acids that preserved good functional activity. Dark grey Arg[11] could be substituted by Lys[11] and its secondary amino group, fatty acids, were attached through different linkers. γ-E: γ-glutamic acid, MEG-FA: multiple ethylene glycol-fatty acid (four ethylene glycol units attached to octadecanedioic acid via lysine linker incorporating carboxylated moiety), PheCl2: (3,4-dichlor)phenylalanine, PheNO2: (4-nitro)phenylalanine, PheF5: pentafluoro-phenylalanine, 1-Nal, 2-Nal: napthylalanine, Phg: phenylglycine, TTDS: short chain of polyethylene glycol (1,13-diamino-4,7,10-trioxadecan-succinamic acid).

7. PrRP in the Regulation of Food Intake and Energy Expenditure

7.1. PrRP Decreases Food Intake and Regulates Energy Homeostasis

First, it was shown that PrRP caused the release of prolactin from cultured pituitary cells [1], but later, other studies showed the main role of PrRP to be in food intake regulation ([12,14,68] and this is reviewed in [64]).

Lawrence et al. suggested an alternative role for PrRP as a regulator of energy homeostasis and food intake [14]. Intracerebroventricular (ICV) injection of PrRP caused a reduction in food intake in fasted and free-fed rats [14]. Moreover, the subsequent decrease in body weight was not only due to the reduction in food intake, which implies an effect on energy expenditure. They supported the findings by measuring *PrRP* mRNA, which was highly expressed in the hypothalamus, NTS, and ventrolateral ME, and *GPR10* mRNA in the RT, PEVN and DMN, and NTS, all areas of which are implicated in the regulation of food intake (reviewed in [69–71]).

PrRP also mediated some of the central satiating actions of the gut peptide hormone CCK [12]. The measurement of the induction of c-Fos protein showed that PrRP neurons were strongly activated by the intraperitoneal injection of CCK, and central PrRP administration activated areas of the brain that are common for both PrRP and CCK [12]. Ellacott et al. suggested that the anorexigenic action of PrRP is regulated by the adiposity signal leptin [72]. ICV administration of PrRP and leptin resulted in reduced food intake in rats and an increase in body temperature compared with each peptide alone.

Additionally, using in situ hybridization, *PrRP* mRNA levels were reduced in fasting and obese Zucker rats, indicating that *PrRP* expression is regulated by leptin [72].

Repeated ICV injection of PrRP strongly reduced food intake and body weight in rats without causing any adverse behavior on locomotor or sensor motor activity [73]. PrRP exerted an effect on energy homeostasis in the short to medium term and increased energy expenditure [74].

Through the generation of *GPR10* knockout (KO) mice with targeted deletion of the *GPR10* gene, GPR10 was confirmed to be a major receptor for PrRP in the hypothalamus because this deletion completely prevented PrRP binding to hypothalamic cell membranes [75]. *GPR10* KO mice become hyperphagic and mildly obese at older ages and develop decreased glucose tolerance with elevated levels of insulin and leptin [75]. Male and female *GPR10* KO mice had increased body weight as a consequence of increased fat mass compared to their wild-type (WT) controls [76]. The total levels of plasma leptin and cholesterol were increased, and a decrease in energy expenditure was observed in *GPR10* KO mice [76]. In fasted or satiated *GPR10* KO mice, ICV administration of PrRP did not reduce food intake in contrast to their WT controls. The administration of CCK did not result in the inhibition of food intake in *GPR10* KO mice, suggesting that PrRP is involved in central satiating actions of CCK [77]. *PrRP* KO mice had higher blood glucose levels and corticosterone levels and became obese with higher amounts of adipose or liver tissue than control WT animals [78]. Under stress conditions, *PrRP* KO mice showed increased levels of plasma corticosterone compared to WT mice, which might indicate that PrRP regulates glucose metabolism through corticosterone secretion and/or catecholamine synthesis [78].

PrRP was also shown to mediate its anorexigenic effect through corticotropin-releasing hormone (CRH) receptors, but not through melanocortin receptors [68]. ICV administration of PrRP elevated adrenocorticotropin (ACTH) levels in plasma, and c-Fos protein was increased in the nuclei of CRH-positive cells in the PVN [79,80]. PrRP-positive neurons have synapse-like contact with CRH cell bodies in the PVN [79]. Furthermore, the injection of PrRP directly into the PVN caused an increase in plasma ACTH [81]. Using hypothalamic explant incubations, researchers showed that PrRP increased hypothalamic CRH release, which is one of the principal ACTH secretagogues, and the subsequent secretion of ACTH. Therefore, an additional potential role of PrRP in the function of the hypothalamic-pituitary-adrenal (HPA) axis and in the cardiovascular system was suggested [23,81].

7.2. Ortholog C-RFa in Food Intake Regulation

Similar to the anorexigenic action of PrRP in mammals, ICV injection of ortholog C-RFa also inhibited food intake in goldfish [82]. However, a completely opposite result was observed in chicks, where ICV injection of rat PrRP31 significantly increased food intake, and the orexigenic effect of NPY was enhanced with the coadministration of PrRP [83]. ICV injection of ortholog C-RFa did not affect food intake in chickens [84].

7.3. PrRP Analogs in the Regulation of Food Intake and Energy Expenditure

The C-terminal 20 amino acids of PrRP (PrRP20) are crucial for preserving the full food-intake-lowering effect. ICV administration of PrRP20 analogs with $PheCl_2{}^{31}$, $PheNO_2{}^{31}$, $PheF_5{}^{31}$, 1-Nal^{31}, 2-Nal^{31}, or Tyr^{31} resulted in decreased food intake in fasted mice [58]. In particular, $[PheNO_2{}^{31}]PrRP20$, $[1\text{-Nal}^{31}]PrRP20$, $[2\text{-Nal}^{31}]PrRP20$, and $[Tyr^{31}]PrRP20$ showed the most significant and long-lasting anorexigenic effect after ICV administration in fasted lean mice. This study showed that a bulky aromatic ring, not necessarily phenylalanine at the C-terminus, enabled full anorexigenic activity [58].

PrRP acts centrally; therefore, the potential of PrRP to decrease food intake after peripheral administration depends on reaching the receptors in the brain and enabling the central effect. Of the analogs with different length fatty acids attached at the N-terminus of PrRP, only myristoylated PrRP20 (myr-PrRP20), palmitoylated (palm-PrRP31), and stearoylated PrRP31 significantly lowered food intake in fasted or freely fed lean mice after subcutaneous (SC) administration [55]. Therefore, those analogs were suggested to probably cross the blood-brain barrier because they caused the central effect after peripheral administration. Analogs containing shorter fatty acids had no effect on food intake. Moreover, analogs palm-PrRP31 and myr-PrRP20, but not natural PrRP20 and PrRP31 or octanoylated PrRP31, showed longer stability in rat plasma and significantly increased c-Fos immunoreactivity in hypothalamic and brainstem nuclei that are involved in food intake regulation, such as PVN, ARC, and NTS.

A significant increase in c-Fos was observed in the PVN, ARC, NTS, and DMN after SC administration of palm-PrRP31. Moreover, palm-PrRP31 administration significantly increased c-Fos in the LHA hypocretin neurons and PVN oxytocin neurons [85].

Palmitoylated or myristoylated PrRP31 analogs with C-terminal changes reduced acute food intake after SC administration in fasted lean mice [65] (reviewed in [64]). Of all the lipidized PrRP analogs, [PheCl$_2$^{31}]PrRP31 palmitoylated or myristoylated at the N-terminus showed the strongest and long-lasting anorexigenic effect in fasted mice [65]. In free-fed Wistar rats, palm-PrRP31 strongly reduced food intake when injected peripherally. Peripheral injection of palm-PrRP31 induced the increase of c-Fos protein in the PVN, NTS, and ARC, which are specific brain regions that are involved in food intake regulation [86].

In diet-induced obese (DIO) mice, a 2-week-long SC administration of palm-PrRP31 and myr-PrRP20 significantly lowered food intake, decreased body weight, improved metabolic parameters such as plasma insulin and leptin, and attenuated lipogenesis compared to lean controls [55].

Repeated administration of PrRP analogs palmitoylated through different linkers to Lys^{11} but not analog with two palmitic acids reduced body and liver weights and the levels of plasma insulin, leptin, triglycerides, cholesterol, and free fatty acid in DIO mice. Moreover, the expression of *uncoupling protein 1* (*UCP-1*) was increased in brown adipose tissue (BAT), suggesting an increase in energy expenditure [66]. A single dose of PrRP31 palmitoylated at Lys^{11} through a γ-glutamyl linker (palm^{11}-PrRP31) again caused neuronal activation and decreased food intake, suggesting its central effect after peripheral administration [66]. This lipidized analog palm^{11}-PrRP31 increased the neural activity, represented by increased FosB immunostaining, only in the DMN and in VMN among the analyzed brain nuclei involved in food intake regulation [87].

The chronic effect of palm-PrRP31 was studied in DIO Sprague-Dawley rats and leptin receptor-deficient Zucker diabetic fatty (ZDF) rats, where palm-PrRP31 was intraperitoneally administered for two weeks. Palm-PrRP31 lowered food intake and body weight, improved glucose tolerance, and tended to decrease leptin levels and adipose tissue in DIO rats [88]. In contrast, the administration of palm-PrRP31 lowered food intake, but it did not significantly affect body weight or glucose tolerance in ZDF rats.

Repeated administration of the lipidized PrRP analog palm^{11}-PrRP31 improved glucose tolerance in Koletsky-spontaneously hypertensive obese (SHROB) rats, which have mutations in their leptin receptor and, therefore, impaired leptin signaling [89]. These findings suggest that the effect of palm^{11}-PrRP31 on glucose metabolism is independent of leptin signaling and body weight lowering. Treatment with palm^{11}-PrRP31 also decreased body weight in control spontaneously hypertensive rats (SHRs), but not in SHROB rats. It seems that the palm^{11}-PrRP anorexigenic effect depends on the proper leptin signaling. Moreover, in SHROB rats, palm^{11}-PrRP31 ameliorated the insulin/glucagon ratio and increased *insulin receptor substrate 1* and 2 expression in fat and insulin signaling in the hypothalamus, while it had no effect on blood pressure [89]. An increase in all parameters mentioned pointed to a beneficial effect of palm^{11}-PrRP on the diabetic state. Additionally, in SHRs and normotensive Wistar Kyoto (WKY) rats on a high-fat diet, treatment with palm^{11}-PrRP31 lowered body weight and improved biochemical and biometric parameters. Palm^{11}-PrRP31 also improved glucose tolerance in WKY rats [90].

Novel long-lasting PrRP analogs with cysteine mutations and staples with attached octadecanedioic acid enhanced plasma stability and half-life in mice. In a 12-day SC administration, the 18-S4 analog significantly reduced body weight in DIO mice [67].

Taken together, PrRP and palmitoylated PrRP analogs are anorexigenic peptides that strongly reduce food intake by reducing appetite and impact energy expenditure under the control of leptin. Proper leptin signaling is necessary for the anorexigenic effect of PrRP and its analogs. Palmitoylated PrRP analogs activate c-Fos in specific neuron populations that are connected to the regulation of food intake. Moreover, lipidization prolonged the half-life of PrRP analogs and enabled central action, leading to a strong food-intake-lowering effect after peripheral administration in mice and rats [55,64,66].

8. Neuroprotective Properties of PrRP

Obesity and type 2 diabetes mellitus were recently identified as risk factors for the development of neurological disorders, such as Alzheimer's disease (AD). Thus, anorexigenic and/or antidiabetic substances began to be examined as compounds with potential neuroprotective properties. This potential is supported by the finding that receptors of anorexigenic peptides, such as GPR10 or the GLP-1 receptor, are expressed in the hippocampus, which is the first brain region affected during AD.

Extracellular senile plaques formed by aggregated β-amyloid protein (Aβ) and intracellular neurofibrillary tangles formed by hyperphosphorylated tau protein are two hallmarks of AD [91,92]. However, other pathological features are observed in AD patients, such as decreased synaptic plasticity and neurogenesis or increased neuroinflammation [93].

The neuroprotective properties of the lipidized PrRP analogs palm-PrRP31 and palm[11]-PrRP31 were examined in vitro as well as in vivo in several rodent models of neurodegeneration. The results were reviewed in depth by Maletínská et al. [94].

The effect of human PrRP31 and its lipidized analog palm[11]-PrRP31 on tau hyperphosphorylation was examined in vitro using a model of hypothermia in the neuroblastoma cell line SH-SY5Y and on rat primary cortical neurons. Hypothermic conditions resulted in increased tau hyperphosphorylation at several epitopes, including pThr212 and pSer396/pSer404, in both cellular models. In SH-SY5Y, incubation with palm[11]-PrRP31, as well as with PrRP31, attenuated tau hyperphosphorylation at pThr212. In primary cortical neurons, palm[11]-PrRP31 decreased tau hyperphosphorylation at both pThr212 and pSer396/pSer404. On the other hand, human PrRP did not affect phosphorylation at pThr212 or at pSer396/Ser404 in primary cortical neurons [95].

The effect of PrRP on tau hyperphosphorylation was extensively studied in vivo using different mouse models. Mice with obesity induced by monosodium glutamate (MSG mice) [96,97] develop increased tau hyperphosphorylation due to central insulin resistance manifested by decreased activation of the insulin signaling cascade. Palm-PrRP31 ameliorated the activation of the insulin signaling cascade and subsequently decreased tau phosphorylation at several epitopes, such as pThr231 and pSer396 [26]. A similar effect on tau hyperphosphorylation was observed in the THY-Tau22 mouse model, where the intervention with palm[11]-PrRP31 also improved short-term spatial memory in the Y-maze test and increased synaptic plasticity compared to the vehicle-treated group [25]. The modulation of synaptic transduction was also examined in a study by Lin et al. [30], where they showed that GPR10 modulates the scaffolding and trafficking of the glutamate-gated cation channel α-amino-3-hydroxy-5-methylisoxazole-4-propionic acid receptor to the postsynaptic membrane, which is necessary to mediate fast excitatory transmission in the brain.

APP/PS1 mice, which are double transgenic mice expressing mutated amyloid precursor protein (APP) (Swedish mutation, K595N/M596L) and mutated presenilin (PS1) (deltaE9 PS1) exon deletion, are one of the most frequently used models to study Aβ pathology [98]. Treatment with the lipidized analog palm[11]-PrRP31 decreased the amount of senile Aβ plaques in APP/PS1 mice. Moreover, palm[11]-PrRP31 lowered the markers of neuroinflammation that are colocalized with Aβ plaques—ionized calcium-binding adapter molecule 1 (Iba1), which is a marker of activated microglial cells, and glial fibrillary acidic protein (GFAP), which is a marker of reactive astrocytes. Potential neuroprotective properties are further manifested by increased levels of doublecortin, a marker of neurogenesis, in hippocampi [24].

In conclusion, palmitoylated analogs of PrRP31 seem to be potential tools to treat neurological disorders. However, the mechanism of action remains unclear and must be further studied.

9. Other Physiological Functions of PrRP

PrRP and *GPR10* are expressed in many brain regions that control different physiological functions. It seems that PrRP plays an important role in the stress response (reviewed in [99]). PrRP-producing neurons in the ME were activated in response to some stressful stimuli, such as foot shock stress [100]. Moreover, *PrRP* KO mice were found to react differently to restraint stress than their WT littermates; *PrRP* KO mice have increased blood glucose and corticosterone levels [78]. This study was supported by the finding that neurons producing noradrenaline, which is known as a stress mediator in the CNS, are colocalized with PrRP neurons in the NTS and ventral and lateral reticular nuclei in the ME, and coadministration of PrRP and noradrenaline synergistically increased the release of pituitary ACTH [18]. In NTS, PrRP immunopositive neurons are located in close proximity to GLP-1 immunopositive neurons and signaling, though GLP-1R modulates the activity of PrRP neurons [101]. Both neuronal populations are activated after exposure to stressors and seem to contribute to the central control of stress. The PrRP neural populations from ME were projected to the PVN in the hypothalamus, where CRH and oxytocin, both of which are modulators of the stress response, are produced [79]. Consistent with this, ICV administration of PrRP increased the level of corticosterone and oxytocin in the blood. In addition, the administration of PrRP antibodies abolishes stress-induced activation of PVN and attenuated oxytocin release to the blood [102]. The coadministration of PrRP and astressin, a CRH receptor antagonist, blocked ACTH release; thus, the CRH receptor is important for PrRP action [68]. The physiological role of PrRP is well reviewed by Lin [29], Dodd et al. [3], and Quillet et al. [103].

The effect of PrRP on CRH release could be responsible for the increased heart rate and blood pressure that was observed after central PrRP administration [23]; thus, PrRP could be involved in the regulation of the cardiovascular system (reviewed by [22]). It seems that the effect of PrRP on the increase in blood pressure is not mediated by GPR10 since PrRP was able to increase blood pressure in Otsuka Long-Evans Tokushima Fatty (OLETF) rats that have mutated *GPR10* [104].

A high density of GPR10-producing neurons is observed in the PB, which is responsible for the regulation of nociception. These neurons also produced enkephalins, which are pain suppressors that bind to opioid receptors, which suggests the control of enkephalin production by PrRP [105]. The role of PrRP in nociception is supported by the finding that *GPR10* KO mice have a higher nociceptive threshold and increased stress-induced analgesia. Thus, PrRP could act as a potential antagonist of the opioid system [106].

It was also demonstrated that PrRP may affect the function of chromaffin cells because PrRP and its receptor are highly expressed in the adrenal medulla [39,44]. Moreover, PrRP-immunopositive cells were found in the rat adrenal gland [107]. On the basis of these results, it was suggested that PrRP may play an important role in modulating catecholamine secretion [49].

Due to its distribution, PrRP could also be involved in sexual and reproductive function or in sleep and the control of circadian rhythms (in the ME) [19,20]. PrRP is expressed in brain areas that are implicated in reproduction (in the DMN, ME) and also in periphery in rat testis and epididymis [39,59]. Feng et al. suggested that PrRP could be involved in the regulation of the female rat estrous cycle [108]. Brain *PrRP* mRNA level was higher in the proestrus and estrus in female rats. Moreover, they found colocalization of GPR10 immunoreactive neurons and gonadotropin-releasing hormone in the hypothalamic medial preoptic area. The study by Maruyama et al. showed that ICV administration of PrRP increases plasma oxytocin in rats and they suggested the role of PrRP as a neuromodulator of oxytocin neurons in the brain [109]. There is also some evidence that PrRP is involved in lactation and that PrRP levels are regulated by hormonal changes [100].

10. Conclusions

PrRP, with its conservative RF-amide sequence on the C-terminus, is a potent anorexigenic neuropeptide, decreasing food intake and enhancing energy metabolism. Moreover, it regulates other physiological functions, such as the cardiovascular system, stress, and reproduction, and has neuroprotective properties. These functions are mainly mediated through the receptor GPR10.

The use of specific model systems, particularly *PrRP/GPR10* KO animals, can contribute to an understanding of the molecular mechanisms of PrRP action, thereby contributing to a faster use of PrRP analogs for potential therapy. From our several recent studies, it is clear that lipidized PrRP analogs could have therapeutic potential. Further progress in the development of selective PrRP analogs may contribute to their use not only in the treatment of obesity, but also in the treatment of other metabolic or neurodegenerative diseases.

Author Contributions: V.P. collected the bibliography, wrote the manuscript, and prepared the figures; A.P. contributed to the writing; J.K. contributed to the figure design; J.K. and L.M. conceived the topic and revised the review.

Funding: This research was funded by the Grant Agency of the Czech Republic (grant number 18-10591S) and the Academy of Sciences of the Czech Republic (RVO: 61388963).

Conflicts of Interest: The authors declare no conflict of interest. The funders had no role in the design of the study; in the collection, analyses, or interpretation of data; in the writing of the manuscript, or in the decision to publish the results.

Abbreviations

Aβ	β-amyloid protein
ACTH	Adrenocorticotropin
AD	Alzheimer's disease
Akt	Protein kinase B
AP	Area postrema
APP	Amyloid precursor protein
BAT	Brown adipose tissue
β-hArg	beta-homoarginine
C	Cerebral cortex
CC	Corpus callosum
CE	Cerebellum
CCK	Cholecystokinin
CD	Circular dichroism
CHO	Chinese hamster ovary
Cha	Cyclohexylalanine
CREB	cAMP response element-binding protein
CRH	Corticotropin-releasing hormone
Cys(Bzl)	Boc-S-benzyl-L-cysteine
DIO	Diet-induced obese
DMN	Dorsomedial hypothalamic nucleus
ERK	Extracellular signal-regulated protein kinase
GFAP	Glial fibrillary acidic protein
GIR	Receptor induced by glucocorticoids
Glu(Obzl)	Boc-L-glutamic acid 1-benzyl ester
GLP-1	Glucagon-like peptide 1
GPCR	G-protein coupled receptor
GPR10	Receptor for prolactin-releasing peptide (known as hGR3 or rat ortholog UHR-1)
hArg	Homoarginine
HB	Hindbrain
HIPP	Hippocampus

His(Bzl)	Nα-Fmoc-N(im)-benzyl-ʟ-histidine
HPA	Hypothalamic-pituitary-adrenal
HYP	Hypothalamus
Iba1	Ionized calcium-binding adapter molecule 1
ICV	Intracerebroventricular
JNK	C-Jun N-terminal protein kinase
KO	Knockout
LHA	Lateral hypothalamic area
MAPK	Mitogen-activated protein kinase
ME	Medulla oblongata
MEG-FA	Multiple ethylene glycol-fatty acid
MIB	Midbrain
MSG	Monosodium glutamate
myr	Myristic acid
NAc	Nucleus accumbens
1-Nal	1-napthylalanine
2-Nal	2-napthylalanine
Nle	Norleucine
Nme-Arg	N-methylarginine
NMR	Nuclear magnetic resonance
NPAF	Neuropeptide AF
NPFF	Neuropeptide FF
NPFF-R2	Receptor 2 for neuropeptide FF (known as GPR74 or HLWAR77)
NPY	Neuropeptide Y
NPY-1R	Neuropeptide Y receptor 1
NTS	Nucleus of the solitary tract
OF	Olfactory bulb
OLETF	Otsuka Long-Evans Tokushima Fatty
palm	Palmitic acid
P	Pituitary
PB	Parabrachial nucleus
PCR	Polymerase chain reaction
PEVN	Periventricular hypothalamic nucleus
PheCl$_2$	(3,4-dichlor)phenylalanine
PheF$_5$	pentafluoro-phenylalanine
PheNO$_2$	(4-nitro)phenylalanine
Phg	Phenylglycine
PI3K	Phosphoinositide 3-kinase-protein kinase
PI3K-Akt-mTOR	Phosphoinositide 3-kinase-protein kinase B/Akt-mammalian target of rapamycin
PKA	Protein kinase A
PKC	Protein kinase C
PrRP	Prolactin-releasing peptide
PVN	Paraventricular hypothalamic nucleus
PS1	Presenilin 1
PTX	Pertussis toxin
RT	Reticular nucleus of the thalamus
SAR	Structure-activity relationship
SC	Subcutaneous
SHR	Spontaneously hypertensive rats

SHROB	Spontaneously hypertensive obese rats
SLM	Stratum lacunosum-moleculare
SON	Supraoptic nucleus
TH	Thalamus
TTDS	1,13-diamino-4,7,10-trioxadecan-succinamic acid
UCP-1	Uncoupling-protein 1
UHR-1	Rat ortholog of GPR10
VMN	Ventromedial hypothalamic nucleus
VRT	Ventrolateral reticular nucleus of the thalamus
WKY	Wistar Kyoto
WT	Wild-type
ZDF	Zucker diabetic fatty
7TM	Seven-transmembrane-domain receptor

References

1. Hinuma, S.; Habata, Y.; Fujii, R.; Kawamata, Y.; Hosoya, M.; Fukusumi, S.; Kitada, C.; Masuo, Y.; Asano, T.; Matsumoto, H.; et al. A prolactin-releasing peptide in the brain. *Nature* **1998**, *393*, 272–276. [CrossRef] [PubMed]
2. Welch, S.K.; O'Hara, B.F.; Kilduff, T.S.; Heller, H.C. Sequence and tissue distribution of a candidate G-coupled receptor cloned from rat hypothalamus. *Biochem. Biophys. Res. Commun.* **1995**, *209*, 606–613. [CrossRef] [PubMed]
3. Dodd, G.T.; Luckman, S.M. Physiological Roles of GPR10 and PrRP Signaling. *Front. Endocrinol.* **2013**, *4*, 20–28. [CrossRef] [PubMed]
4. Wang, Y.; Wang, C.Y.; Wu, Y.; Huang, G.; Li, J.; Leung, F.C. Identification of the receptors for prolactin-releasing peptide (PrRP) and Carassius RFamide peptide (C-RFa) in chickens. *Endocrinology* **2012**, *153*, 1861–1874. [CrossRef]
5. Yamada, M.; Ozawa, A.; Ishii, S.; Shibusawa, N.; Hashida, T.; Ishizuka, T.; Hosoya, T.; Monden, T.; Satoh, T.; Mori, M. Isolation and characterization of the rat prolactin-releasing peptide gene: Multiple TATA boxes in the promoter region. *Biochem. Biophys. Res. Commun.* **2001**, *281*, 53–56. [CrossRef]
6. Southey, B.R.; Rodriguez-Zas, S.L.; Sweedler, J.V. Prediction of neuropeptide prohormone cleavages with application to RFamides. *Peptides* **2006**, *27*, 1087–1098. [CrossRef]
7. Fujimoto, M.; Takeshita, K.; Wang, X.; Takabatake, I.; Fujisawa, Y.; Teranishi, H.; Ohtani, M.; Muneoka, Y.; Ohta, S. Isolation and characterization of a novel bioactive peptide, Carassius RFamide (C-RFa), from the brain of the Japanese crucian carp. *Biochem. Biophys. Res. Commun.* **1998**, *242*, 436–440. [CrossRef]
8. Sakamoto, T.; Oda, A.; Yamamoto, K.; Kaneko, M.; Kikuyama, S.; Nishikawa, A.; Takahashi, A.; Kawauchi, H.; Tsutsui, K.; Fujimoto, M. Molecular cloning and functional characterization of a prolactin-releasing peptide homolog from Xenopus laevis. *Peptides* **2006**, *27*, 3347–3351. [CrossRef]
9. Tachibana, T.; Moriyama, S.; Takahashi, A.; Tsukada, A.; Oda, A.; Takeuchi, S.; Sakamoto, T. Isolation and characterisation of prolactin-releasing peptide in chicks and its effect on prolactin release and feeding behaviour. *J. Neuroendocrinol.* **2011**, *23*, 74–81. [CrossRef]
10. Lagerstrom, M.C.; Fredriksson, R.; Bjarnadottir, T.K.; Schioth, H.B. The ancestry of the prolactin-releasing hormone precursor. *Ann. N. Y. Acad. Sci.* **2005**, *1040*, 368–370. [CrossRef]
11. Matsumoto, H.; Noguchi, J.; Horikoshi, Y.; Kawamata, Y.; Kitada, C.; Hinuma, S.; Onda, H.; Nishimura, O.; Fujino, M. Stimulation of prolactin release by prolactin-releasing peptide in rats. *Biochem. Biophys. Res. Commun.* **1999**, *259*, 321–324. [CrossRef] [PubMed]
12. Lawrence, C.B.; Ellacott, K.L.; Luckman, S.M. PRL-releasing peptide reduces food intake and may mediate satiety signaling. *Endocrinology* **2002**, *143*, 360–367. [CrossRef] [PubMed]
13. Samson, W.K.; Resch, Z.T.; Murphy, T.C.; Chang, J.K. Gender-biased activity of the novel prolactin releasing peptides: Comparison with thyrotropin releasing hormone reveals only pharmacologic effects. *Endocrine* **1998**, *9*, 289–291. [CrossRef]
14. Lawrence, C.B.; Celsi, F.; Brennand, J.; Luckman, S.M. Alternative role for prolactin-releasing peptide in the regulation of food intake. *Nat. Neurosci.* **2000**, *3*, 645–646. [CrossRef] [PubMed]

15. Seal, L.J.; Small, C.J.; Dhillo, W.S.; Stanley, S.A.; Abbott, C.R.; Ghatei, M.A.; Bloom, S.R. PRL-releasing peptide inhibits food intake in male rats via the dorsomedial hypothalamic nucleus and not the paraventricular hypothalamic nucleus. *Endocrinology* **2001**, *142*, 4236–4243. [CrossRef] [PubMed]

16. Takayanagi, Y.; Matsumoto, H.; Nakata, M.; Mera, T.; Fukusumi, S.; Hinuma, S.; Ueta, Y.; Yada, T.; Leng, G.; Onaka, T. Endogenous prolactin-releasing peptide regulates food intake in rodents. *J. Clin. Investig.* **2008**, *118*, 4014–4024. [CrossRef]

17. Onaka, T.; Takayanagi, Y.; Leng, G. Metabolic and stress-related roles of prolactin-releasing peptide. *Trends Endocrinol. Metab.* **2010**, *21*, 287–293. [CrossRef]

18. Maruyama, M.; Matsumoto, H.; Fujiwara, K.; Noguchi, J.; Kitada, C.; Fujino, M.; Inoue, K. Prolactin-releasing peptide as a novel stress mediator in the central nervous system. *Endocrinology* **2001**, *142*, 2032–2038. [CrossRef]

19. Zhang, S.Q.; Inoue, S.; Kimura, M. Sleep-promoting activity of prolactin-releasing peptide (PrRP) in the rat. *Neuroreport* **2001**, *12*, 3173–3176. [CrossRef]

20. Zhang, S.Q.; Kimura, M.; Inoue, S. Effects of prolactin-releasing peptide (PrRP) on sleep regulation in rats. *Psychiatry Clin. Neurosci.* **2000**, *54*, 262–264. [CrossRef]

21. Samson, W.K.; Resch, Z.T.; Murphy, T.C. A novel action of the newly described prolactin-releasing peptides: Cardiovascular regulation. *Brain Res.* **2000**, *858*, 19–25. [CrossRef]

22. Mikulaskova, B.; Maletinska, L.; Zicha, J.; Kunes, J. The role of food intake regulating peptides in cardiovascular regulation. *Mol. Cell. Endocrinol.* **2016**, *436*, 78–92. [CrossRef] [PubMed]

23. Yamada, T.; Mochiduki, A.; Sugimoto, Y.; Suzuki, Y.; Itoi, K.; Inoue, K. Prolactin-releasing peptide regulates the cardiovascular system via corticotrophin-releasing hormone. *J. Neuroendocrinol.* **2009**, *21*, 586–593. [CrossRef] [PubMed]

24. Holubova, M.; Hruba, L.; Popelova, A.; Bencze, M.; Prazienkova, V.; Gengler, S.; Kratochvilova, H.; Haluzik, M.; Zelezna, B.; Kunes, J.; et al. Liraglutide and a lipidized analog of prolactin-releasing peptide show neuroprotective effects in a mouse model of beta-amyloid pathology. *Neuropharmacology* **2019**, *144*, 377–387. [CrossRef]

25. Popelova, A.; Prazienkova, V.; Neprasova, B.; Kasperova, B.J.; Hruba, L.; Holubova, M.; Zemenova, J.; Blum, D.; Zelezna, B.; Galas, M.C.; et al. Novel Lipidized Analog of Prolactin-Releasing Peptide Improves Memory Impairment and Attenuates Hyperphosphorylation of Tau Protein in a Mouse Model of Tauopathy. *J. Alzheimers Dis.* **2018**, *62*, 1725–1736. [CrossRef]

26. Spolcova, A.; Mikulaskova, B.; Holubova, M.; Nagelova, V.; Pirnik, Z.; Zemenova, J.; Haluzik, M.; Zelezna, B.; Galas, M.C.; Maletinska, L. Anorexigenic lipopeptides ameliorate central insulin signaling and attenuate tau phosphorylation in hippocampi of mice with monosodium glutamate-induced obesity. *J. Alzheimers Dis.* **2015**, *45*, 823–835. [CrossRef]

27. Marchese, A.; Heiber, M.; Nguyen, T.; Heng, H.H.; Saldivia, V.R.; Cheng, R.; Murphy, P.M.; Tsui, L.C.; Shi, X.; Gregor, P.; et al. Cloning and chromosomal mapping of three novel genes, GPR9, GPR10, and GPR14, encoding receptors related to interleukin 8, neuropeptide Y, and somatostatin receptors. *Genomics* **1995**, *29*, 335–344. [CrossRef]

28. Lagerstrom, M.C.; Fredriksson, R.; Bjarnadottir, T.K.; Fridmanis, D.; Holmquist, T.; Andersson, J.; Yan, Y.L.; Raudsepp, T.; Zoorob, R.; Kukkonen, J.P.; et al. Origin of the prolactin-releasing hormone (PRLH) receptors: Evidence of coevolution between PRLH and a redundant neuropeptide Y receptor during vertebrate evolution. *Genomics* **2005**, *85*, 688–703. [CrossRef]

29. Lin, S.H. Prolactin-releasing peptide. In *Results and Problems in Cell Differentiation*; Springer: Berlin/Heidelberg, Germany, 2008; Volume 46, pp. 57–88. [CrossRef]

30. Lin, S.H.; Arai, A.C.; Wang, Z.; Nothacker, H.P.; Civelli, O. The carboxyl terminus of the prolactin-releasing peptide receptor interacts with PDZ domain proteins involved in alpha-amino-3-hydroxy-5-methylisoxazole-4-propionic acid receptor clustering. *Mol. Pharmacol.* **2001**, *60*, 916–923. [CrossRef]

31. Roland, B.L.; Sutton, S.W.; Wilson, S.J.; Luo, L.; Pyati, J.; Huvar, R.; Erlander, M.G.; Lovenberg, T.W. Anatomical distribution of prolactin-releasing peptide and its receptor suggests additional functions in the central nervous system and periphery. *Endocrinology* **1999**, *140*, 5736–5745. [CrossRef]

32. Engstrom, M.; Brandt, A.; Wurster, S.; Savola, J.M.; Panula, P. Prolactin releasing peptide has high affinity and efficacy at neuropeptide FF2 receptors. *J. Pharmacol. Exp. Ther.* **2003**, *305*, 825–832. [CrossRef] [PubMed]

33. Elshourbagy, N.A.; Ames, R.S.; Fitzgerald, L.R.; Foley, J.J.; Chambers, J.K.; Szekeres, P.G.; Evans, N.A.; Schmidt, D.B.; Buckley, P.T.; Dytko, G.M.; et al. Receptor for the pain modulatory neuropeptides FF and AF is an orphan G protein-coupled receptor. *J. Biol. Chem.* **2000**, *275*, 25965–25971. [CrossRef] [PubMed]

34. Bonini, J.A.; Jones, K.A.; Adham, N.; Forray, C.; Artymyshyn, R.; Durkin, M.M.; Smith, K.E.; Tamm, J.A.; Boteju, L.W.; Lakhlani, P.P.; et al. Identification and characterization of two G protein-coupled receptors for neuropeptide FF. *J. Biol. Chem.* **2000**, *275*, 39324–39331. [CrossRef] [PubMed]

35. Fujii, R.; Fukusumi, S.; Hosoya, M.; Kawamata, Y.; Habata, Y.; Hinuma, S.; Sekiguchi, M.; Kitada, C.; Kurokawa, T.; Nishimura, O.; et al. Tissue distribution of prolactin-releasing peptide (PrRP) and its receptor. *Regul. Pept.* **1999**, *83*, 1–10. [CrossRef]

36. Morales, T.; Hinuma, S.; Sawchenko, P.E. Prolactin-releasing peptide is expressed in afferents to the endocrine hypothalamus, but not in neurosecretory neurones. *J. Neuroendocrinol.* **2000**, *12*, 131–140. [CrossRef]

37. Matsumoto, H.; Murakami, Y.; Horikoshi, Y.; Noguchi, J.; Habata, Y.; Kitada, C.; Hinuma, S.; Onda, H.; Fujino, M. Distribution and characterization of immunoreactive prolactin-releasing peptide (PrRP) in rat tissue and plasma. *Biochem. Biophys. Res. Commun.* **1999**, *257*, 264–268. [CrossRef]

38. Maruyama, M.; Matsumoto, H.; Fujiwara, K.; Kitada, C.; Hinuma, S.; Onda, H.; Fujino, M.; Inoue, K. Immunocytochemical localization of prolactin-releasing peptide in the rat brain. *Endocrinology* **1999**, *140*, 2326–2333. [CrossRef]

39. Nieminen, M.L.; Brandt, A.; Pietila, P.; Panula, P. Expression of mammalian RF-amide peptides neuropeptide FF (NPFF), prolactin-releasing peptide (PrRP) and the PrRP receptor in the peripheral tissues of the rat. *Peptides* **2000**, *21*, 1695–1701. [CrossRef]

40. Reis, F.M.; Vigano, P.; Arnaboldi, E.; Spritzer, P.M.; Petraglia, F.; Di Blasio, A.M. Expression of prolactin-releasing peptide and its receptor in the human decidua. *Mol. Hum. Reprod.* **2002**, *8*, 356–362. [CrossRef]

41. Sakamoto, T.; Amano, M.; Hyodo, S.; Moriyama, S.; Takahashi, A.; Kawauchi, H.; Ando, M. Expression of prolactin-releasing peptide and prolactin in the euryhaline mudskippers (Periophthalmus modestus): Prolactin-releasing peptide as a primary regulator of prolactin. *J. Mol. Endocrinol.* **2005**, *34*, 825–834. [CrossRef]

42. Ibata, Y.; Iijima, N.; Kataoka, Y.; Kakihara, K.; Tanaka, M.; Hosoya, M.; Hinuma, S. Morphological survey of prolactin-releasing peptide and its receptor with special reference to their functional roles in the brain. *Neurosci. Res.* **2000**, *38*, 223–230. [CrossRef]

43. Nieminen, M.L.; Nystedt, J.; Panula, P. Expression of neuropeptide FF, prolactin-releasing peptide, and the receptor UHR1/GPR10 genes during embryogenesis in the rat. *Dev. Dyn.* **2003**, *226*, 561–569. [CrossRef] [PubMed]

44. Takahashi, K.; Totsune, K.; Murakami, O.; Sone, M.; Noshiro, T.; Hayashi, Y.; Sasano, H.; Shibahara, S. Expression of prolactin-releasing peptide and its receptor in the human adrenal glands and tumor tissues of adrenocortical tumors, pheochromocytomas and neuroblastomas. *Peptides* **2002**, *23*, 1135–1140. [CrossRef]

45. Kimura, A.; Ohmichi, M.; Tasaka, K.; Kanda, Y.; Ikegami, H.; Hayakawa, J.; Hisamoto, K.; Morishige, K.; Hinuma, S.; Kurachi, H.; et al. Prolactin-releasing peptide activation of the prolactin promoter is differentially mediated by extracellular signal-regulated protein kinase and c-Jun N-terminal protein kinase. *J. Biol. Chem.* **2000**, *275*, 3667–3674. [CrossRef] [PubMed]

46. Langmead, C.J.; Szekeres, P.G.; Chambers, J.K.; Ratcliffe, S.J.; Jones, D.N.; Hirst, W.D.; Price, G.W.; Herdon, H.J. Characterization of the binding of [(125)I]-human prolactin releasing peptide (PrRP) to GPR10, a novel G protein coupled receptor. *Br. J. Pharmacol.* **2000**, *131*, 683–688. [CrossRef] [PubMed]

47. Duval, D.L.; Gutierrez-Hartmann, A. PRL-releasing peptide stimulation of PRL gene transcription–enter AKT. *Endocrinology* **2002**, *143*, 11–12. [CrossRef] [PubMed]

48. Hayakawa, J.; Ohmichi, M.; Tasaka, K.; Kanda, Y.; Adachi, K.; Nishio, Y.; Hisamoto, K.; Mabuchi, S.; Hinuma, S.; Murata, Y. Regulation of the PRL promoter by Akt through cAMP response element binding protein. *Endocrinology* **2002**, *143*, 13–22. [CrossRef]

49. Nanmoku, T.; Takekoshi, K.; Fukuda, T.; Ishii, K.; Isobe, K.; Kawakami, Y. Stimulation of catecholamine biosynthesis via the PKC pathway by prolactin-releasing peptide in PC12 rat pheochromocytoma cells. *J. Endocrinol.* **2005**, *186*, 233–239. [CrossRef]

50. Nanmoku, T.; Takekoshi, K.; Isobe, K.; Kawakami, Y.; Nakai, T.; Okuda, Y. Prolactin-releasing peptide stimulates catecholamine release but not proliferation in rat pheochromocytoma PC12 cells. *Neurosci. Lett.* **2003**, *350*, 33–36. [CrossRef]

51. Samson, W.K.; Taylor, M.M. Prolactin releasing peptide (PrRP): An endogenous regulator of cell growth. *Peptides* **2006**, *27*, 1099–1103. [CrossRef]

52. Tachibana, T.; Sakamoto, T. Functions of two distinct "prolactin-releasing peptides" evolved from a common ancestral gene. *Front. Endocrinol.* **2014**, *5*, 170. [CrossRef] [PubMed]

53. Varghese, B.V.; Koohestani, F.; McWilliams, M.; Colvin, A.; Gunewardena, S.; Kinsey, W.H.; Nowak, R.A.; Nothnick, W.B.; Chennathukuzhi, V.M. Loss of the repressor REST in uterine fibroids promotes aberrant G protein-coupled receptor 10 expression and activates mammalian target of rapamycin pathway. *Proc. Natl. Acad. Sci. USA* **2013**, *110*, 2187–2192. [CrossRef] [PubMed]

54. Maixnerova, J.; Spolcova, A.; Pychova, M.; Blechova, M.; Elbert, T.; Rezacova, M.; Zelezna, B.; Maletinska, L. Characterization of prolactin-releasing peptide: Binding, signaling and hormone secretion in rodent pituitary cell lines endogenously expressing its receptor. *Peptides* **2011**, *32*, 811–817. [CrossRef] [PubMed]

55. Maletinska, L.; Nagelova, V.; Ticha, A.; Zemenova, J.; Pirnik, Z.; Holubova, M.; Spolcova, A.; Mikulaskova, B.; Blechova, M.; Sykora, D.; et al. Novel lipidized analogs of prolactin-releasing peptide have prolonged half-lives and exert anti-obesity effects after peripheral administration. *Int. J. Obes. (Lond.)* **2015**, *39*, 986–993. [CrossRef] [PubMed]

56. Boyle, R.G.; Downham, R.; Ganguly, T.; Humphries, J.; Smith, J.; Travers, S. Structure-activity studies on prolactin-releasing peptide (PrRP). Analogues of PrRP-(19-31)-peptide. *J. Pept. Sci. Off. Publ. Eur. Pept. Soc.* **2005**, *11*, 161–165. [CrossRef] [PubMed]

57. Deluca, S.H.; Rathmann, D.; Beck-Sickinger, A.G.; Meiler, J. The activity of prolactin releasing peptide correlates with its helicity. *Biopolymers* **2013**, *99*, 314–325. [CrossRef] [PubMed]

58. Maletinska, L.; Spolcova, A.; Maixnerova, J.; Blechova, M.; Zelezna, B. Biological properties of prolactin-releasing peptide analogs with a modified aromatic ring of a C-terminal phenylalanine amide. *Peptides* **2011**, *32*, 1887–1892. [CrossRef] [PubMed]

59. Quillet, R.; Ayachi, S.; Bihel, F.; Elhabazi, K.; Ilien, B.; Simonin, F. RF-amide neuropeptides and their receptors in Mammals: Pharmacological properties, drug development and main physiological functions. *Pharmacol. Ther.* **2016**, *160*, 84–132. [CrossRef] [PubMed]

60. D'Ursi, A.M.; Albrizio, S.; Di Fenza, A.; Crescenzi, O.; Carotenuto, A.; Picone, D.; Novellino, E.; Rovero, P. Structural studies on Hgr3 orphan receptor ligand prolactin-releasing peptide. *J. Med. Chem.* **2002**, *45*, 5483–5491. [CrossRef]

61. Brasnjevic, I.; Steinbusch, H.W.; Schmitz, C.; Martinez-Martinez, P.; European NanoBioPharmaceutics Research, I. Delivery of peptide and protein drugs over the blood-brain barrier. *Prog. Neurobiol.* **2009**, *87*, 212–251. [CrossRef]

62. Knudsen, L.B.; Nielsen, P.F.; Huusfeldt, P.O.; Johansen, N.L.; Madsen, K.; Pedersen, F.Z.; Thogersen, H.; Wilken, M.; Agerso, H. Potent derivatives of glucagon-like peptide-1 with pharmacokinetic properties suitable for once daily administration. *J. Med. Chem.* **2000**, *43*, 1664–1669. [CrossRef]

63. Gault, V.A.; Kerr, B.D.; Harriott, P.; Flatt, P.R. Administration of an acylated GLP-1 and GIP preparation provides added beneficial glucose-lowering and insulinotropic actions over single incretins in mice with Type 2 diabetes and obesity. *Clin. Sci. (Lond.)* **2011**, *121*, 107–117. [CrossRef]

64. Kunes, J.; Prazienkova, V.; Popelova, A.; Mikulaskova, B.; Zemenova, J.; Maletinska, L. Prolactin-releasing peptide: A new tool for obesity treatment. *J. Endocrinol.* **2016**, *230*, R51–R58. [CrossRef]

65. Prazienkova, V.; Ticha, A.; Blechova, M.; Spolcova, A.; Zelezna, B.; Maletinska, L. Pharmacological characterization of lipidized analogs of prolactin-releasing peptide with a modified C- terminal aromatic ring. *J. Physiol. Pharmacol. Off. J. Pol. Physiol. Soc.* **2016**, *67*, 121–128.

66. Prazienkova, V.; Holubova, M.; Pelantova, H.; Buganova, M.; Pirnik, Z.; Mikulaskova, B.; Popelova, A.; Blechova, M.; Haluzik, M.; Zelezna, B.; et al. Impact of novel palmitoylated prolactin-releasing peptide analogs on metabolic changes in mice with diet-induced obesity. *PLoS ONE* **2017**, *12*, e0183449. [CrossRef]

67. Pflimlin, E.; Lear, S.; Lee, C.; Yu, S.; Zou, H.; To, A.; Joseph, S.; Nguyen-Tran, V.; Tremblay, M.S.; Shen, W. Design of a Long-Acting and Selective MEG-Fatty Acid Stapled Prolactin-Releasing Peptide Analog. *ACS Med. Chem. Lett.* **2019**, *10*, 1166–1172. [CrossRef]

68. Lawrence, C.B.; Liu, Y.L.; Stock, M.J.; Luckman, S.M. Anorectic actions of prolactin-releasing peptide are mediated by corticotropin-releasing hormone receptors. *Am. J. Physiol. Regul. Integr. Comp. Physiol.* **2004**, *286*, R101–R107. [CrossRef]

69. Arora, S.; Anubhuti. Role of neuropeptides in appetite regulation and obesity—A review. *Neuropeptides* **2006**, *40*, 375–401. [CrossRef]

70. Sohn, J.W. Network of hypothalamic neurons that control appetite. *BMB Rep.* **2015**, *48*, 229–233. [CrossRef]

71. Schwartz, M.W.; Woods, S.C.; Porte, D., Jr.; Seeley, R.J.; Baskin, D.G. Central nervous system control of food intake. *Nature* **2000**, *404*, 661–671. [CrossRef]

72. Ellacott, K.L.; Lawrence, C.B.; Rothwell, N.J.; Luckman, S.M. PRL-releasing peptide interacts with leptin to reduce food intake and body weight. *Endocrinology* **2002**, *143*, 368–374. [CrossRef] [PubMed]

73. Vergoni, A.V.; Watanobe, H.; Guidetti, G.; Savino, G.; Bertolini, A.; Schioth, H.B. Effect of repeated administration of prolactin releasing peptide on feeding behavior in rats. *Brain Res.* **2002**, *955*, 207–213. [CrossRef]

74. Ellacott, K.L.; Lawrence, C.B.; Pritchard, L.E.; Luckman, S.M. Repeated administration of the anorectic factor prolactin-releasing peptide leads to tolerance to its effects on energy homeostasis. *Am. J. Physiol. Regul. Integr. Comp. Physiol.* **2003**, *285*, R1005–R1010. [CrossRef]

75. Gu, W.; Geddes, B.J.; Zhang, C.; Foley, K.P.; Stricker-Krongrad, A. The prolactin-releasing peptide receptor (GPR10) regulates body weight homeostasis in mice. *J. Mol. Neurosci.* **2004**, *22*, 93–103. [CrossRef]

76. Bjursell, M.; Lenneras, M.; Goransson, M.; Elmgren, A.; Bohlooly, Y.M. GPR10 deficiency in mice results in altered energy expenditure and obesity. *Biochem. Biophys. Res. Commun.* **2007**, *363*, 633–638. [CrossRef]

77. Bechtold, D.A.; Luckman, S.M. Prolactin-releasing Peptide mediates cholecystokinin-induced satiety in mice. *Endocrinology* **2006**, *147*, 4723–4729. [CrossRef]

78. Mochiduki, A.; Takeda, T.; Kaga, S.; Inoue, K. Stress response of prolactin-releasing peptide knockout mice as to glucocorticoid secretion. *J. Neuroendocrinol.* **2010**, *22*, 576–584. [CrossRef]

79. Matsumoto, H.; Maruyama, M.; Noguchi, J.; Horikoshi, Y.; Fujiwara, K.; Kitada, C.; Hinuma, S.; Onda, H.; Nishimura, O.; Inoue, K.; et al. Stimulation of corticotropin-releasing hormone-mediated adrenocorticotropin secretion by central administration of prolactin-releasing peptide in rats. *Neurosci. Lett.* **2000**, *285*, 234–238. [CrossRef]

80. Mera, T.; Fujihara, H.; Kawasaki, M.; Hashimoto, H.; Saito, T.; Shibata, M.; Saito, J.; Oka, T.; Tsuji, S.; Onaka, T.; et al. Prolactin-releasing peptide is a potent mediator of stress responses in the brain through the hypothalamic paraventricular nucleus. *Neuroscience* **2006**, *141*, 1069–1086. [CrossRef]

81. Seal, L.J.; Small, C.J.; Dhillo, W.S.; Kennedy, A.R.; Ghatei, M.A.; Bloom, S.R. Prolactin-releasing peptide releases corticotropin-releasing hormone and increases plasma adrenocorticotropin via the paraventricular nucleus of the hypothalamus. *Neuroendocrinology* **2002**, *76*, 70–78. [CrossRef]

82. Kelly, S.P.; Peter, R.E. Prolactin-releasing peptide, food intake, and hydromineral balance in goldfish. *Am. J. Physiol. Regul. Integr. Comp. Physiol.* **2006**, *291*, R1474–R1481. [CrossRef]

83. Tachibana, T.; Saito, S.; Tomonaga, S.; Takagi, T.; Saito, E.S.; Nakanishi, T.; Koutoku, T.; Tsukada, A.; Ohkubo, T.; Boswell, T.; et al. Effect of central administration of prolactin-releasing peptide on feeding in chicks. *Physiol. Behav.* **2004**, *80*, 713–719. [CrossRef]

84. Tachibana, T.; Tsukada, A.; Fujimoto, M.; Takahashi, H.; Ohkubo, T.; Boswell, T.; Furuse, M. Comparison of mammalian prolactin-releasing peptide and Carassius RFamide for feeding behavior and prolactin secretion in chicks. *Gen. Comp. Endocrinol.* **2005**, *144*, 264–269. [CrossRef]

85. Pirnik, Z.; Zelezna, B.; Kiss, A.; Maletinska, L. Peripheral administration of palmitoylated prolactin-releasing peptide induces Fos expression in hypothalamic neurons involved in energy homeostasis in NMRI male mice. *Brain Res.* **2015**, *1625*, 151–158. [CrossRef]

86. Mikulaskova, B.; Zemenova, J.; Pirnik, Z.; Prazienkova, V.; Bednarova, L.; Zelezna, B.; Maletinska, L.; Kunes, J. Effect of palmitoylated prolactin-releasing peptide on food intake and neural activation after different routes of peripheral administration in rats. *Peptides* **2016**, *75*, 109–117. [CrossRef]

87. Pirnik, Z.; Kolesarova, M.; Zelezna, B.; Maletinska, L. Repeated peripheral administration of lipidized prolactin-releasing peptide analog induces c-fos and FosB expression in neurons of dorsomedial hypothalamic nucleus in male C57 mice. *Neurochem. Int.* **2018**, *116*, 77–84. [CrossRef]

88. Holubova, M.; Zemenova, J.; Mikulaskova, B.; Panajotova, V.; Stohr, J.; Haluzik, M.; Kunes, J.; Zelezna, B.; Maletinska, L. Palmitoylated PrRP analog decreases body weight in rats with DIO but not in ZDF rats. *J. Endocrinol.* **2016**. [CrossRef]

89. Mikulaskova, B.; Holubova, M.; Prazienkova, V.; Zemenova, J.; Hruba, L.; Haluzik, M.; Zelezna, B.; Kunes, J.; Maletinska, L. Lipidized prolactin-releasing peptide improved glucose tolerance in metabolic syndrome: Koletsky and spontaneously hypertensive rat study. *Nutr. Diabetes* **2018**, *8*, 5. [CrossRef]

90. Cermakova, M.; Pelantova, H.; Neprasova, B.; Sediva, B.; Maletinska, L.; Kunes, J.; Tomasova, P.; Zelezna, B.; Kuzma, M. Metabolomic Study of Obesity and Its Treatment with Palmitoylated Prolactin-Releasing Peptide Analog in Spontaneously Hypertensive and Normotensive Rats. *J. Proteome Res.* **2019**, *18*, 1735–1750. [CrossRef]

91. Hassing, L.B.; Dahl, A.K.; Thorvaldsson, V.; Berg, S.; Gatz, M.; Pedersen, N.L.; Johansson, B. Overweight in midlife and risk of dementia: A 40-year follow-up study. *Int. J. Obes. (Lond.)* **2009**, *33*, 893–898. [CrossRef]

92. Kadohara, K.; Sato, I.; Kawakami, K. Diabetes mellitus and risk of early-onset Alzheimer's disease: A population-based case-control study. *Eur. J. Neurol.* **2017**, *24*, 944–949. [CrossRef]

93. Serrano-Pozo, A.; Frosch, M.P.; Masliah, E.; Hyman, B.T. Neuropathological alterations in Alzheimer disease. *Cold Spring Harb. Perspect. Med.* **2011**, *1*, a006189. [CrossRef]

94. Maletinska, L.; Popelova, A.; Zelezna, B.; Bencze, M.; Kunes, J. The impact of anorexigenic peptides in experimental models of Alzheimer's disease pathology. *J. Endocrinol.* **2019**, *240*, R47–R72. [CrossRef]

95. Prazienkova, V.; Schirmer, C.; Holubova, M.; Zelezna, B.; Kunes, J.; Galas, M.C.; Maletinska, L. Lipidized Prolactin-Releasing Peptide Agonist Attenuates Hypothermia-Induced Tau Hyperphosphorylation in Neurons. *J. Alzheimers Dis.* **2019**, *67*, 1187–1200. [CrossRef]

96. Hirata, A.E.; Andrade, I.S.; Vaskevicius, P.; Dolnikoff, M.S. Monosodium glutamate (MSG)-obese rats develop glucose intolerance and insulin resistance to peripheral glucose uptake. *Braz. J. Med Biol. Res. Rev. Bras. Pesqui. Med. Biol. Soc. Bras. Biofisica* **1997**, *30*, 671–674. [CrossRef]

97. Matyskova, R.; Maletinska, L.; Maixnerova, J.; Pirnik, Z.; Kiss, A.; Zelezna, B. Comparison of the obesity phenotypes related to monosodium glutamate effect on arcuate nucleus and/or the high fat diet feeding in C57BL/6 and NMRI mice. *Physiol. Res.* **2008**, *57*, 727–734.

98. Jankowsky, J.L.; Slunt, H.H.; Ratovitski, T.; Jenkins, N.A.; Copeland, N.G.; Borchelt, D.R. Co-expression of multiple transgenes in mouse CNS: A comparison of strategies. *Biomol. Eng.* **2001**, *17*, 157–165. [CrossRef]

99. Maniscalco, J.W.; Rinaman, L. Interoceptive modulation of neuroendocrine, emotional, and hypophagic responses to stress. *Physiol. Behav.* **2017**, *176*, 195–206. [CrossRef]

100. Morales, T.; Sawchenko, P.E. Brainstem prolactin-releasing peptide neurons are sensitive to stress and lactation. *Neuroscience* **2003**, *121*, 771–778. [CrossRef]

101. Card, J.P.; Johnson, A.L.; Llewellyn-Smith, I.J.; Zheng, H.; Anand, R.; Brierley, D.I.; Trapp, S.; Rinaman, L. GLP-1 neurons form a local synaptic circuit within the rodent nucleus of the solitary tract. *J. Comp. Neurol.* **2018**, *526*, 2149–2164. [CrossRef]

102. Zhu, L.L.; Onaka, T. Facilitative role of prolactin-releasing peptide neurons in oxytocin cell activation after conditioned-fear stimuli. *Neuroscience* **2003**, *118*, 1045–1053. [CrossRef]

103. Quillet, R.; Simonin, F. The neuropeptide FF: A signal for M1 to M2 type switching in macrophages from the adipose tissue. *Med. Sci. (Paris)* **2018**, *34*, 27–29. [CrossRef] [PubMed]

104. Ma, L.; MacTavish, D.; Simonin, F.; Bourguignon, J.J.; Watanabe, T.; Jhamandas, J.H. Prolactin-releasing peptide effects in the rat brain are mediated through the Neuropeptide FF receptor. *Eur. J. Neurosci.* **2009**, *30*, 1585–1593. [CrossRef]

105. Lin, S.H.; Leslie, F.M.; Civelli, O. Neurochemical properties of the prolactin releasing peptide (PrRP) receptor expressing neurons: Evidence for a role of PrRP as a regulator of stress and nociception. *Brain Res.* **2002**, *952*, 15–30. [CrossRef]

106. Laurent, P.; Becker, J.A.; Valverde, O.; Ledent, C.; de Kerchove d'Exaerde, A.; Schiffmann, S.N.; Maldonado, R.; Vassart, G.; Parmentier, M. The prolactin-releasing peptide antagonizes the opioid system through its receptor GPR10. *Nat. Neurosci.* **2005**, *8*, 1735–1741. [CrossRef]

107. Fujiwara, K.; Matsumoto, H.; Yada, T.; Inoue, K. Identification of the prolactin-releasing peptide-producing cell in the rat adrenal gland. *Regul. Pept.* **2005**, *126*, 97–102. [CrossRef]

108. Feng, Y.; Zhao, H.; An, X.F.; Ma, S.L.; Chen, B.Y. Expression of brain prolactin releasing peptide (PrRP) changes in the estrous cycle of female rats. *Neurosci. Lett.* **2007**, *419*, 38–42. [CrossRef]
109. Maruyama, M.; Matsumoto, H.; Fujiwara, K.; Noguchi, J.; Kitada, C.; Hinuma, S.; Onda, H.; Nishimura, O.; Fujino, M.; Higuchi, T.; et al. Central administration of prolactin-releasing peptide stimulates oxytocin release in rats. *Neurosci. Lett.* **1999**, *276*, 193–196. [CrossRef]

 International Journal of *Molecular Sciences*

Review

Chemotactic Ligands that Activate G-Protein-Coupled Formylpeptide Receptors

Stacey A Krepel and Ji Ming Wang *

Cancer and Inflammation Program, Center for Cancer Research, National Cancer Institute at Frederick, Frederick, MD 21702, USA
* Correspondence: wangji@mail.nih.gov; Tel.: +1-301-846-6979

Received: 19 June 2019; Accepted: 5 July 2019; Published: 12 July 2019

Abstract: Leukocyte infiltration is a hallmark of inflammatory responses. This process depends on the bacterial and host tissue-derived chemotactic factors interacting with G-protein-coupled seven-transmembrane receptors (GPCRs) expressed on the cell surface. Formylpeptide receptors (FPRs in human and Fprs in mice) belong to the family of chemoattractant GPCRs that are critical mediators of myeloid cell trafficking in microbial infection, inflammation, immune responses and cancer progression. Both murine Fprs and human FPRs participate in many patho-physiological processes due to their expression on a variety of cell types in addition to myeloid cells. FPR contribution to numerous pathologies is in part due to its capacity to interact with a plethora of structurally diverse chemotactic ligands. One of the murine Fpr members, Fpr2, and its endogenous agonist peptide, Cathelicidin-related antimicrobial peptide (CRAMP), control normal mouse colon epithelial growth, repair and protection against inflammation-associated tumorigenesis. Recent developments in FPR (Fpr) and ligand studies have greatly expanded the scope of these receptors and ligands in host homeostasis and disease conditions, therefore helping to establish these molecules as potential targets for therapeutic intervention.

Keywords: formyl peptide receptors; ligands; diseases

1. Properties of FPRs

Formylpeptide receptors (FPR, Fpr for expression in mice) are G-protein-coupled receptors and were incidentally the first GPCRs to be identified in neutrophils [1]. Though initially cloned from neutrophils, FPRs have since been identified in macrophages, endothelial cells, intestinal epithelial cells, fibroblasts, and others [2–5]. Humans possess three different forms of FPRs: FPR1, FPR2, and FPR3. FPR1 was the first named of the receptors, and it was initially discovered as the receptor for the formylated bacterial product formyl-methionine-leucyl-phenylalanine (fMLF), the name of which gave rise to the naming of the receptor in question [6]. FPR1 is most highly expressed in cells in the bone marrow and immune system, though it has some expression notable in cells of the lungs, brain, and gastrointestinal (GI) tract, among others [7,8]. FPR2 was the second discovered of these receptors, but it tends to be the more ubiquitously expressed of the two. The expression is primarily in cells of the bone marrow, immune system, GI tract, female organ tissues, and endocrine glands, though there are also some lower levels of expression in cells of the brain, liver, gallbladder, and pancreas [8,9]. There is little known about the biological significance of FPR3, and very little research has been done to elucidate its role. This receptor is mainly expressed in monocytes and dendritic cells but not neutrophils, and it resides in intracellular vesicles rather than on the cell surface like its counterpart receptors. FPR3, contrary to the other FPR variants, also has only one known endogenous peptide agonist [10,11].

The primary roles of FPRs involve cell chemotaxis in response to agonists, and new research has shown that they even contribute to direct phagocytosis of bacteria by neutrophils [12,13]. Activation of such receptors is also important for wound healing and gut development [14,15]. However, while FPRs were initially thought to only be responsible for neutrophil chemotaxis, FPR1 and FPR2 have both been shown to play pivotal roles in the progression of multiple diseases. For example, FPR2 may promote the malignancy of colon cancer, while FPR1 has similarly been tied to the progression of glioblastoma [16,17]. Conversely, FPR1 has demonstrated tumor suppressor functions in gastric cancer [18]. With such dual roles in cancer progression, it is clear that further mechanistic studies would contribute greatly to our understanding of FPRs, thus potentially leading to new therapeutics. While aberrant expression or activation of FPRs can be detrimental, somewhat contradictory findings demonstrate that constitutively active FPR was indispensable in the defense against the formation of biofilms by *Candida albicans*, as well as aggressive infiltration by *Vibrio harveyi* [19]. FPR-mediated cell activation does not solely include chemotactic and pro-inflammatory responses to pathogens, as activation also plays a very important role in the protection against other pathologies. Gobbetti et al. [20] demonstrated that Fpr2 confers protection against sepsis-mediated damage in mice. Furthermore, FPRs have been reported to mediate anxiety-related disorders and the resultant altered behaviors [21]. Lastly, a novel discovery for FPRs suggests that they act as mechanoreceptors in arteries, making them critical for proper arterial plasticity [22]. Given such diverse functions of these receptors, it should come as no surprise that FPRs respond to a plethora of ligands with diverse classifications.

While most FPR ligands induce cell chemotaxis, calcium flux, and even phagocytosis, they stimulate many other cell functions as well [23]. For instance, some ligands elicit inflammatory processes for the clearance of infection, recruitment of immune cells, etc. while other ligands activate pro-resolving, anti-inflammatory pathways. There are few chemoattractant GPCRs capable of transmitting both pro- and anti-inflammatory signals. This duality in FPR2 is initially determined by the nature of the ligands. Bacterial and mitochondrial formylated peptides are among those that classically activate a pro-inflammatory cell response to clear invaders and tissue damage, while Annexin A1 (Anx A1) and Lipoxin A4 (LXA$_4$) are some of the better-known anti-inflammatory FPR2 ligands [24]. Cooray et al. [25] demonstrated that the switch between FPR2-mediated pro- and anti-inflammatory cell responses is due to conformational changes of the receptor upon ligand binding: binding of anti-inflammatory ligands such as Anx A1 caused FPRs to form homodimers, which led to the release of inflammation-resolving cytokines like IL-10. Conversely, inflammatory ligands such as serum-amyloid alpha (SAA) did not cause receptor homodimerization.

Though some of the diverse FPR (Fpr) ligands are small-molecules or non-peptides, the majority are small peptides that are either synthetic or natural with origins ranging from host and multicellular organisms to viruses and bacteria. These peptides have been extensively studied and patterns of recognized elements have begun to emerge. As is demonstrated below, the presence of formylated methionine in the peptide is generally an activator of FPR1, while FPR2 is less dependent upon this particular residue [26]. Expanding upon this, Bufe et al. [27] concluded that FPR recognition of bacterial peptides requires either a formylated methionine at the N-terminus or an amidated methionine at the C-terminus of a peptide, though they believe that as a general principle, the secondary structure rather than the primary sequence is important for recognition of the highly diverse ligands by FPR. One class of ligands which shall be shortly discussed is comprised of phenol-soluble modulins, and examination of these has led to the conclusion that FPR1 favors short, flexible structures while FPR2 has binding preference for longer peptides which are amphipathic in nature and may contain alpha helices [28]. The class of ligands mentioned above represents a small proportion of the known FPR agonists today, and this review will focus on formylated, microbe-derived, and mitochondrial peptides, as well as host and non-microbial, non-host peptides. Host-derived non-peptides, as well as synthetic or small-molecule ligands will also be discussed.

2. Formylated Peptides

The prototypic FPR ligand is fMLF, which was the first classified FPR agonist and also represents the shortest sequence to elicit a potent receptor response. While Fpr2 is generally considered the more promiscuous FPR, fMLF preferentially activates FPR1 with high affinity [29]. It has been shown that non-formylated bacterial peptides are much less potent than their formylated counterparts; a suggested reason for this is due to the use of formyl-methionine as a Gram-negative start codon, therefore marking a protein as pathogenic from the perspective of the host immune system [30]. There are many derivatives of fMLF which elicit FPR responses, many of which preferentially activate FPR2 rather than FPR1. Such derivatives include the peptide sequences fMLFII, fMLFIK, fMLFK, and fMLFW, among others [31]. Liu et al. [32] summarized many other formylated peptides that elicit responses through both FPR1 and FPR2, including f-MIFL, f-MIVIL, f-MIGWII, and f-MFEDAVAWF. Dozens of similar peptides have been isolated from equally numerous bacterial genera including *Streptococcus*, *Haemophilus*, *Salmonella*, *Hydrogenobacter*, *Listeria*, *Neisseria*, *Staphylococcus*, and others [27]. PSMα is another formylated peptide that has shown great efficacy in FPR activation. Thus far, the phenol-soluble modulins that have demonstrated a capacity for FPR activation include β2, α1, and α2, all of which are virulence factors isolated from *Staphylococcus aureus*. All three peptides activate FPR2, though the structural basis for activation remains unknown [33].

It is interesting to note that mammalian cell mitochondria, well-known for being bacterial in origin, also contain peptides that elicit FPR-mediated responses. However, while bacterial formylated peptides are considered pathogen-associated molecular patterns (PAMPs), the mitochondrial peptides are generally associated with cellular damage and are thus considered damage-associated molecular patterns (DAMPs) that elicit an inflammatory response [34]. These peptides, some of which include MMYALF, MFADRW, and Nle-LF-Nle-YK, have been tied to constriction of airways in the lungs, as well as neutrophil accumulation and other proinflammatory responses [35,36]. Likewise, lung diseases—including acute respiratory distress syndrome—have been found to have a higher presence of formylated mitochondrial peptides in bronchoalveolar lavage fluid, suggesting that lung inflammation is tightly tied to Fpr1 activation via these DAMPs [37]. Mitocryptide-2 (MCT-2) is another mitochondrial peptide. It is related to Cytochrome B of the electron transport chain and activates FPR2. Interestingly, while the N-terminal formylated methionine is entirely necessary for FPR2 activation by MCT-2, the carboxy-terminal residues are also important for receptor activation. Additionally, sequence analysis has revealed that the presence of the residues Thr7 and Ser8 can activate FPR2, though the peptides were most potent with Ile7 and Asn8 residues [38].

Another important mitochondrial peptide is the nicotinamide adenine dinucleotide (NADH) reductase subunit 1 (ND-1), which elicits a strong inflammatory response via FPR1 [39]. Bufe and Zufall [40] have successfully created a model that accurately predicts known mitochondrial peptide agonists for FPRs. Therefore, there is a strong likelihood for the discovery of many new mitochondrial, host-derived FPR agonist peptides. Another DAMP that elicits FPR-mediated responses is Mitochondrial Transcription Factor A (Tfam). While this necrosis-associated peptide has been previously shown to activate FPRs, activation does not appear to play a crucial role in the inflammatory responses of monocytic microglia in the brain; this may be due to lower FPR expression in this cell type [41,42]. The formylated peptide ligands for FPRs (Fprs) are listed in Table 1, and the mitochondrial peptides are listed in Table 2.

Table 1. Formylated bacterial peptide agonists for formylpeptide receptors (FPRs) (Fprs).

Agonist	Classification	Receptor	Citation
MLF		FPR1	[29]
MIFL		FPR1, FPR2	
MIVIL		FPR1, FPR2	
MIGWI		FPR1, FPR2	[32]
MIVTLF		FPR1, FPR2	
MIGWII		FPR1, FPR2	
MFEDAVAWF		FPR1, FPR2	
MLFK (-II, -IK, -K, -W)		FPR2	[31]
Streptococcus f-MGFFIS		FPR1, FPR2	
bacillus f-MKNFKG		FPR 2	
Salmonella f-MAMKKL		FPR 1	
Haemophilus f-MVMKFK		FPR1, FPR2	
Psychrosomonas f-MLFYFS		FPR1, FPR2	
Desulotomaculum f-MLFYLA		FPR1, FPR2	[27]
Borrelia f-MLKKVY		FPR1, FPR2	
Vibrio F-MVKIIF		FPR1, FPR2	
Clostridium f-MKKNLV		FPR 2	
Streptomyces f-MVPISI		FPR1, FPR2	
Hydrogenobacter MKKFLL		FPR 2	
Listeria MKKIML		FPR1, FPR2	
Desulfovibrio MKFCTA		FPR1, FPR2	
Neisseria MKTSIR		FPR1, FPR2	
Staphylococcus MFIYYCK		FPR 1	
f-NLPNTL		FPR 1	[43]
PSMα peptides β2, α1, and α2	Staphylococcus aureus	FPR 2	[33]

Table 2. Mitochondrial peptide ligands for FPRs (Fprs).

Agonist	Classification	Receptor	Citation
MMYALF	Formylated	FPR 2	
MLKIV	formylated	FPR 2	
MYFINILTL	Formylated	FPR 2	[35]
MFADRW	Formylated	FPR 2	
Nle-LF-Nle-YK		FPR 2	
Mitocryptide-2 [MCT-2]	Cytochrome B	FPR 2	[38]
ND1	NADH reductase subunit 1	FPR1	[39]
TFam	Mitochondrial Transcription factor	FPR1	[41]

3. Microbe-Derived Peptides

Table 3 lists the microbe-derived FPR (Fprs) ligands. While formylated peptides first drew the attention of the scientific community to FPRs (Fprs), there are many other bacterial/viral peptides that are not necessarily formylated but which nevertheless elicit receptor responses. Although the majority of formylated microbial peptides preferentially activate FPR1, the preferred receptor for non-formylated peptides is FPR2 [26]. A large percentage of these non-formylated microbe-derived peptides are viral, and many of them are derived from the Human Immunodeficiency Virus (HIV) envelope proteins, including gp41 T20/DP178, gp41 T21/DP107, gp120 V3 loop, gp41 N36, gp120 F, and gp41 MAT-1 [44–46]. Despite the potential importance of FPRs in HIV research, very little work has been done to further explore this connection. However, Li et al. [47] demonstrated that persistent FPR activation desensitized host CCR5 and CXCR4 co-receptors to HIV proteins, thus reducing viral entry and subsequent replication. Still other viruses, including Hepatitis C Virus, HKU-1 coronavirus, and Herpes Simplex Virus, produce chemotactic ligands C5a, N-formyl HKU-1 coronavirus peptide, and gG-2p20, respectively, for FPR1 or FPR2 activation [48–50]. There is, however, some argument as to the efficacy of the Herpes Simplex viral peptide as an FPR agonist, as the overlapping sequence gG-2p19 was unable to definitively demonstrate that FPR activation played a significant role in the NK response to this virus [51].

Table 3. Microbe-derived peptide ligands for FPRs (Fprs).

Agonist	Classification	Receptor	Citation
Hp2-20	Helicobacter	FPR2	[53]
T20/DP178	HIV gp41	FPR1	[45]
T21/DP107	HIV gp41	FPR1, FPR2	[54]
V3 peptide	HIV Gp 120	FPR2	[55]
N36 peptide	HIV Gp 41	FPR2	[56]
F peptide	HIV Gp 120	FPR2	[57]
MAT-1	HIV gp41	FPR2	[46]
gG-2p20	Herpes Simplex Virus	FPR1	[50]
N-formyl HKU-1 coronavirus peptide	HKU-1 virus	FPR1	[49]
Listeria peptides		FPR1	[26]
Enterococcus faecium proteins		FPR2	[52]
PrP$_{106\text{-}126}$		FPR2	[58]
C5a HCV peptide	Hep C Virus	FPR2	[48]
OC43 coronavirus protein	OC43 Coronovirus	Unknown	[49]
229E Coronavirus protein	229E Coronavirus	Unknown	[49]
NL36 Coronavirus protein	NL36 Coronavirus	Unknown	[49]
Spike Protein	Ebola virus	Unknown	[49]
Influenza A Virus	Annexin A1 surface protein	FPR2	[59]

Mills [49] used the sequence homology of T20/DP178 to further determine that the OC43 Coronavirus, 229E Coronavirus, NL36 Coronavirus, and even the Ebola Spike Protein were all peptides with aromatic-rich domains that elicited FPR-dependent cell activation. Interestingly, when examined from the context of the FPRs rather than the ligands, it was found that domain variability in the receptors determined ligand binding and subsequent cellular responses. This led to the conclusion that the variability of receptors among individuals might predispose or protect against certain viral infections, the susceptibility of which may be determined by receptor activation. In terms of non-viral and non-formylated microbe-derived peptides, there are few FPR agonists. Certain peptides from different strains of *Enterococcus faecium* have demonstrated FPR activation properties, though the ligand activity is not entirely predictable based on structure. Interestingly, *E. faecium* strains that are resistant to vancomycin contain potent FPR2 agonists, suggesting a potential role for FPR2 in antibiotic-resistant infections [52].

4. Host-Derived FPR Ligands

There are many different host-derived ligands that elicit strong FPR responses, though they have different biological implications. Misfolded proteins implicated in a variety of pathologies are one class of host-derived proteins eliciting FPR activation. Amyloid β-42 (Aβ-42), a peptide fragment well-documented in Alzheimer's Disease (AD), interacts with FPR2. Additionally, part of FPR2's role includes interaction with the Macrophage Receptor with Collagenous Structure (MARCO) scavenger receptor, which is responsible for reducing inflammation and alleviating inflammation-associated symptoms [60]. Interestingly, the symptoms that are viewed as a hallmark of AD are due to this same ligand binding an entirely different group of receptors which do not internalize them, therefore leading to an increased inflammatory pathology [61]. The prion peptide fragment, $PrP_{106-126}$ also interacts with FPR2 on astrocytes and microglia, and the internalization of this peptide is detrimental to the host and contributes to disease progression [62]. Another neuropeptide that activates FPR2—though other studies claim it also activates FPR1—is the pituitary adenylate cyclase-activating polypeptide 27 (PACAP27), which has been shown to induce migration and Ca^{2+} mobilization, as well as upregulation of CD11b in neutrophils [63,64]. Another FPR agonist, the Vasoactive Intestinal Peptide (VIP), activates monocytes via FPR2 and may initiate an inflammation-resolving process [64,65].

Other host-derived FPR ligands are CKβ8-1 and the SHAAGtide sequence, as well as various uPAR domains from the Urokinase-Type 1 Plasminogen Activator Receptor (uPAR). CKβ8-1 is also known as the CCL23 chemokine, and it acts as an agonist for FPR2, along with its truncated N-terminal peptide called the SHAAGtide sequence that activates a chemokine GPCR [66,67]. It has been demonstrated that several uPAR peptides elicit FPR responses, including $uPAR_{88-92}$, $uPAR_{84-95}$, D2D3, and the SRSRYp sequence [68–70]. These sequences may foster the transition between fibroblasts to myofibroblasts, therefore increasing the pathology of fibrosis via FPR2 activation [68]. Another host-derived peptide, F2L, is derived from the N-terminus of the heme-binding protein, HEPB1. As the sole agonist specific for FPR3, F2L activates macrophages and possibly dendritic cells (DCs) as well [71]. However, other studies have demonstrated that FPR2 in neutrophils also exhibits a moderate affinity for F2L, though in this scenario F2L appears to have an inhibitory rather than a stimulatory effect [72,73]. A newer FPR ligand is Family with Sequence Similarity 3 (Member D), or FAM3D. This chemokine-like peptide is most highly expressed in the GI tract, though it is also expressed in cells of the immune system [8,74]. FAM3D has demonstrated a high affinity for both FPR1 and FPR2 and has been implicated in playing an important role in both inflammation and GI homeostasis via FPR activation [75]. Additional studies have found that FAM3D may also be involved in the beneficial role of glucagon secretion in Type 2 diabetes, as well as the detrimental development of abdominal aortic aneurysms [76,77].

In addition to endogenous peptides associated with cell-surface proteins and functional units, there are many ligands that are secreted by cells in response to tissue damage. Annexin-1 (AnxA1), also called Lipocortin-1, is an anti-inflammatory protein which is upregulated as a result of the stress responses of multiple host systems [78]. Some studies demonstrate AnxA1 to be an FPR1 agonist, while others show it as an FPR2 agonist; hence, it likely activates both. One study demonstrated its role in the attenuation of rheumatoid arthritis symptoms by decreasing fibroblast-like synoviocyte proliferation via FPR2 [79]. However, another study showed that AnxA1 initiated autocrine signaling in breast cancer via FPR1 and led to an increase in tumor growth and metastasis [80]. Additionally, the absence of AnxA1 has been tied to increased disease severity in both rheumatoid arthritis and obstructive pulmonary disease, leading to the hypothesis that treatment with exogenous AnxA1 may help reduce symptoms associated with different inflammatory pathologies [81,82]. In addition to AnxA1, multiple derivatives of the parent protein, including Ac_{1-25}, Ac_{2-26}, and Ac_{9-25}, activate FPR2 [83–85]. These peptides have protective effects in ischemia-induced lung injury and atherosclerosis [84,86]. As with its pro-survival property, AnxA1 acts as a double-edged sword: secretion of either AnxA1 or Ac_{2-26} by tumor-associated fibroblasts induces the acquisition of stem-like features in prostate cancer cells, thus leading to a worse prognosis [87].

Serum-amyloid alpha (SAA) is an endogenous FPR2 agonist secreted by liver or macrophages in response to inflammatory stress and, more notably, tissue damage. In endothelial cells, SAA enhances the expression and activity of Tissue Factor—a protein necessary for clotting and wound repair—while additionally inhibiting the activity of Tissue Factor Pathway Inhibitor. Both functions were demonstrated to be the result of FPR2 activation [88]. SAA, via FPR2 activation, additionally increases the production of the wound-healing chemokine, CCL2, by vascular endothelial cells [89]. More recent studies have demonstrated the role of an SAA-FPR2 axis in neovascularization in the cornea as well [90,91].

Human LL-37 is an antimicrobial peptide that induces Cxcl13 and Tnfsf13b transcription, as well as B cell activation and proliferation via FPR2. It also contributes to the maintenance of B-cell germinal centers in Peyer's Patches of the gut [92]. LL-37 also promotes the growth of both colorectal and ovarian cancer cells [93,94]. The murine homologue of LL-37, Cathelicidin-related antimicrobial peptide (CRAMP), is similarly an Fpr2 agonist and has been shown to promote atherosclerosis and DC maturation [95,96]. CRAMP also plays a pivotal role in maintaining the homeostasis of the colon mucosa and microbiota balance, demonstrating its potential as a therapeutic molecule [97]. The list of host-derived FPR (Fpr) ligands is shown in Table 4.

Table 4. Host-derived FPR (Fpr) ligands and their classification.

Agonist	Classification	Receptor	Citation
CKβ 8-1 [human CCL23]	CCL23 chemokine	FPR2	[66]
SHAAGtide	CKβ 8-1	FPR2	
Humanin		FPR2	[98]
F2L	Heme binding protein	FPR2, FPR3	[71–73]
SAA		FPR2	[90]
Annexin 1/ Lipocortin 1		FPR2	[78]
Ac$_{1-25}$	Annexin 1	FPR2	[83]
Ac$_{2-26}$	Annexin 1	FPR2	[84]
Ac$_{9-25}$	Annexin 1	FPR2	[85]
Antiflammin-2 (AF-2)		FPR2	[99]
Aβ-42	Amyloid peptide	FPR2	[60]
D2D3	Urokinase receptor	FPR2	[70]
SRSRYp	Urokinase receptor	FPR2	
LL-37	Antimicrobial peptide	FPR2	[92]
PrP$_{106-126}$	Prion protein	FPR2	[62]
PACAP27	Pituitary adenylate cyclase	FPR2, FPR1	[63,64]
uPAR$_{88-92}$	Urokinase receptor	FPR2	[68]
uPAR$_{84-95}$	Urokinase receptor	FPR2, FPR3	[69]
Cathepsin G		FPR1	[100]
FAM3D	Chemokine-like	FPR1, FPR2	[75]
FAM19A4	Chemokine-like	FPR1	[13]
ATLXA4 acetylsalicyclic acid-triggered	Arachidonic acid derivative	FPR2	[101]
Resolvin D1 (RvD1) and aspirin-triggered RvD1	Specialized pro-resolving lipid mediator	FPR2	[102]
Vasoactive Intestinal protein		FPR2	[65]
Lipoxin-A4/Apsirin-triggered lipoxins	Fatty Acid	FPR2	[103]

5. Synthetic Peptides and Non-Peptide Small Molecules

By far the most extensive category of FPR ligands includes the synthetic and small-molecule ligands, of which there are over 40 currently known, as shown in Table 5. W-peptides are among the better-known synthetic peptides acting as FPR agonists, and they include the sequences WKYMVm-NH$_2$, WKYMVM-NH$_2$, as well as many derivatives. Bufe et al. [27] demonstrated that both FPR1 and FPR2 mediate Ca^{2+} mobilization responses in leukocytes to more than 20 different combinations and derivatives of the W-peptide. A breakdown of the data suggests that certain residues in the peptide sequence are more important than others: C3 tyrosine, C4 methionine, and C6 D-methionine are all required for ligand activity, as is the carboxy-terminal NH$_2$. Peptides may additionally be shortened on the N-terminus by two amino acids or elongated by three amino acids before FPR activation capacity is severely diminished. While the applications of WKYMV-sequences have been little explored, one recent study showed that the activation of FPR2 by WKYMVm may enhance the homing of endothelial cells, thus improving tissue healing, especially in ischemic neovasculogenesis in injured limbs [3]. Interestingly, another study showed that WKYMVm was capable of desensitizing HIV coreceptors CXCR4 and CCR5, therefore decreasing the entry of HIV-1 into macrophages and CD4+ T cells [47].

M-peptides are another subclass of synthetic/peptide library isolates and include MMK-1 and MMWLL. MMK-1, an FPR2 agonist, is by far the more commonly used peptide, and studies have shown that it may be useful as an anti-anxiety drug, as well as a drug to counteract hair loss from chemotherapy [98,104]. However, there has been some concern about its use in certain drug regimens, as it may amplify the response of monocytes to SIO$_2$-coated nanoparticles, making it an important player in calculating the proper dosing when using such nanoparticles [105]. MMWLL is another M-peptide specific for FPR1. It is not classically a formylated peptide, though the addition of a formylated methionine induces a more potent FPR1 response than even the classic prototypic fMLF [106]. A much newer class of synthetic peptides are the FPR1-agonistic AApeptides, based on the general structure of N-acylated-N-aminoethyl amino acid residues. There are three different AApeptide subgroups called the α-peptides, α-AApeptides, and γ-AApeptide, all of which have different R-groups at the designated α or γ position. Most of these derivatives of AApeptides induce Ca^{2+} mobilization in rat basophil leukemia (RBL) cells transfected with human FPR1, though the γ-AApeptide Compound 7 at 10 μM elicited a more potent cell response than fMLF at the same concentration, making it a reasonably high-affinity ligand for FPR1 at this concentration, though not at lower concentrations [107]. See Table 5 for all synthetic peptides.

Table 5. Synthetic peptide ligands for FPRs (Fprs).

Agonist	Classification	Receptor	Citation
AApeptidesα-peptides, α-AApeptides, and γ-AApeptide	N-acylated-N-aminoethyl amino acid residue	FPR1	[107]
CGEN-855A	TIPMFVPESTSKLQKFTSWFM-amide	FPR2	[108]
P2Y2Pal$_{IC2}$ pepducin	P2Y$_2$R pepducin	FPR2	[109]
F2Pal$_{10}$	FPR2 pepducin	FPR2	[110]
GMMWAI	Gly-Met-Met-Trp-Ala-Ile-CONH$_2$	FPR1	[111]
Lau-[[S]-Aoc]-[Lys-Bnphe]$_6$-NH2	Lipidated peptomimetic peptide	FPR2	[112]
L-37pA	ApoA-I mimetic peptide	FPR2	[113]
MMHWFM	Met-Met-His-Trp-Phe-Met-CONH$_2$	FPR1	[111]
MMHWAM	Met-Met-His-Trp-Ala-Met-CONH$_2$	FPR2	[111]
MMK-1	Peptide library	FPR2	[98]
MMWLL	Peptide library	FPR1	[106]
WKYMVm(-NH2)	Peptide library	FPR1, FPR2	[27]
WKYMVM	Peptide library	FPR1, FPR2	
F2Pal$_{16}$	FPR2 Pepducin	FPR2	[110]

Some newer, non-peptide synthetic compounds being studied are various derivatives of ureidopropanamide molecules. They have demonstrated a capacity for protection against LPS-induced microglial death via an FPR2-dependent pathway, and pre-treatment with concentrations as low as 1 μM showed the protective effects [24]. Thus, these ligands show promising potential as treatment for diseases associated with inflammation in the Central Nervous System (CNS). Another pro-resolving synthetic ligand is the quinazolinone derivative Quin-C1. It is an Fpr2 agonist and has been shown to be effective at reducing inflammatory cytokines and clearing neutrophils and lymphocytes in murine models of lung injury [114]. Schepetkin et al. [115] demonstrated that three different synthetic molecules are agonists for both FPR1 and FPR2. Two of these are bombesin-related BB_1/BB_2 antagonists called PD168368 and PD176252, and they induce Ca^{2+} flux as well as neutrophil degranulation with EC50 values in the nanomolar range, thus making them potent FPR agonists. The third agonist is the Cholecystokinin-1 receptor agonist A-71623, which exhibits FPR1 and FPR2 agonism, though with a much higher EC50. In structural studies, these ligands were cross-reactive with FPR1/2 and possessed both Trp and N-phenylurea moieties. This led to the hypothesis that the combination of moieties greatly increases the chance that an agonist will activate both receptors.

Kirpotina et al. [116] screened over 6000 compounds and isolated nearly 30 different FPR1 and/or FPR2 agonists, all of which are derivatives of acetohydrazide, 2-(N-piperazinyl)acetamide, N'-phenylurea, and benzimidazole. The acetohydrazide derivatives (compounds AG-07/7, AG-09/92, AG-09/96, AG-09/101, and AG-09/102) and N-phenylurea derivatives (AG-09/3, AG-09/4, AG-09/73 through AG-09/77, and AG-09/82) are all FPR2-specific, though the acetohydrazide compounds tend to have lower efficacies on average. Also, the benzimidazole derivatives are either FPR1-specific (AG-09/1, AG-09/2, AG-09/13, AG-09/18, AG-09/19, and AG-09/21) or are agonists for both FPR1 and FPR2 (AG-09/16, AG-09/17, AG-09/20, and AG-09/22 through AG-09/24); no FPR2-specific benzimidazole derivatives have yet been identified. Pyridazines are another class of non-peptide, synthetic molecules which can have many different derivatives. Currently, only two compounds have been identified as potent mixed FPR1 and FPR2 agonists with an EC50 of around 2 μM each. These two compounds are referred to as compounds 8b and 8c with the Pyridazin-3(2H)-one structure. Additionally, they have R substitutions of SCH3 and OCH3, as well as R1 substitutions of I and SCH3, respectively. Both R and R1 substitutions are on substituted benzene rings [117]. (See Table 6 for the list of synthetic/small molecule non-peptide agonists).

Table 6. Small/molecule compounds functioning as FPR (Fpr) agonists.

Agonist	Classification	Receptor	Citation
Quin-C1	Quinazolinone derivative	FPR2	[114]
N-substituted benzimidazole 11		FPR2	[118]
Compound 8a	Pyridazin-3(2H)-one derivative 1	FPR1, FPR2	[117]
Compound 8b	Pyridazin-3(2H)-one derivative 2	FPR1, FPR2	
Chiral pyridazines	6-methyl-2,4-disubstituted pyridazine-3(2H)-ones	FPR1 [weak], FPR2	[119]
PD176252	Bombesin-related BB_1/BB_2 antagonist	FPR1, FPR2	
PD168368	Bombesin-related BB_1/BB_2 antagonist	FPR1, FPR2	[115]
A-71623	Cholescystokinin-1 receptor agonist	FPR1, FPR2	
1753-103	Synthetic library	FPR1	[120]
Compound (S)-17	Ureidopropanamide derivative (S)-3-(4-cyanophenyl)-N-((1-(3-chloro4-fluorophenyl]cyclopropyl]methyl)-2-(3-(4-fluorophenyl)ureido)propaniamide ((S)-17)	FPR2	[24]

Table 6. *Cont.*

Agonist	Classification	Receptor	Citation
1754-49	Synthetic library	FPR2	[120]
(S)-9a	3-(1H-Indol-3-yl)-2-(3-(4-nitrophenyl)ureido) propenamide derivative	FPR2	[121]
IDR1	KSRIVPAIPVSLL-NH2	FPR1	[122]
IDR-1002	VQRWLIVWRIRK-NH2	FPR2-uncertain	[123]
AG-09/1, AG-09/2, AG-09/13, AG-09/18, AG-09/19, and AG-09/21	Benzimidazole derivatives	FPR1	
AG-09/16, AG-09/17, AG-09/20, and AG-09/22 through AG-09/24	Benzimidazole derivatives	FPR1, FPR2	[116]
AG-09/3, AG-09/4, AG-09/73 through AG-09/77, and AG-09/82	N-phenylurea derivatives	FPR2	
AG-09/7, AG-09/92, AG-09/92, AG-09/96, AG-09/101, and AG-09/102	Acetohydrazide derivatives	FPR2 [low efficacy]	
Compound17a/17b/14a	Synthetic library	FPR1, FPR2	[124,125]
Compound43	Synthetic library	FPR1, FPR2	[79]

6. Ligands from Non-Human Sources

As the search for disease treatments continues, many investigators have turned to developing compounds isolated from various plants and animals for potential therapeutic uses. Some of these new compounds have been shown to activate human or murine FPRs. The first of these is a series of compounds isolated from the centipede *Scolopendra subspinipes mutilans*, which has classically been used in Oriental medicine and is now being studied for therapeutic potential [126]. New studies show that compounds Scolopendrasin III and V both cause human neutrophil migration via FPR1, while Scolopendrasin IX seems to work through FPR2 to promote neutrophil chemotaxis. The two former compounds have not yet been studied for effectiveness against particular pathologies, though Compound IX has traditionally been an effective treatment for rheumatoid arthritis in Oriental medicine. New evidence confirms this activity, citing the activation of FPR2 as the mechanism [9,127]. Temporins are another class of FPR agonists and consist of antimicrobial peptides isolated from the *Rana Temoraria* frogs. Temporin A and Rana-6 are two such peptides, and both activate FPR2 to promotion leukocyte migration. There are also two distinct synthetic peptides, I4S10-C and I4G10-C, that are modeled after temporins and activate FPR2 [128].

Rubimetide is a peptide (Met-Arg-Trp) isolated from the digest of Rubisco in spinach. While it has been studied for some time, it has just recently been classified as an FPR2 agonist and has further demonstrated an ability to produce anxiolytic-like effects, thus alleviating some symptoms associated with anxiety [98]. The same investigators also isolated soymetide from the α' subunit of β-conglycinin from soybeans and then demonstrated its activity as an FPR1 agonist [129]. Furthermore, the antimicrobial peptides Piscidin-1 and -3, which were isolated from fish, have been shown to induce myeloid cell chemotaxis via both FPR1 and FPR2. As a testament to the harms of water pollution by metals, this study also demonstrated that conjugation of Cu^{2+} with either of these compounds reduces the chemotactic activity of mammalian neutrophils [130] (See Table 7 for the list of peptides from other non-human sources).

Table 7. Non-human, non-microbe-derived FPR (Fpr) ligands.

Agonist	Classification	Receptor	Citation
Scholopendrasin IX	Centipede Peptides	FPR2	[131]
Scholopendrasin III and V		FPR1	[127]
Rubimetide	Spinach	FPR2	[98]
Piscidin 1 and 3	From fish	FPR2	[130]
Temporin A	From frogs	FPR2	[128]
Rana-6		FPR2	

7. Concluding Remarks

FPRs are a class of seven-transmembrane, G-protein-coupled receptors (GPCRs) that interact with a remarkably diverse range of ligands. As demonstrated, these ligands may originate from pathogens, the host, the synthetic peptide or compound library, or even non-host multicellular organisms. With such diverse agonist binding capacity, it is not surprising that FPRs may be either detrimental or beneficial in different pathophysiological conditions. Though the majority of these agonists have been known for more than a decade, newer studies are finding novel roles for these ligands in treatments for conditions ranging from anxiety and mental health disorders to arthritis and wounds. The field of FPR agonist studies has demonstrated the potential of these molecules to have therapeutic mechanisms useful for medicine. In addition to the vast number of agonists summarized here, there are also extensive lists of antagonistic ligands that may also provide protective mechanisms in various diseases [23,44]. Thus, further exploration of FPRs and ligands as therapeutic targets would be highly beneficial to diseases including cancer, sceptic shock, arthritis, and many other inflammatory pathologies.

Funding: This project has been funded in part by Federal funds from the National Cancer Institute (NCI), National Institutes of Health (NIH), under Contract No. HSN261200800001E, and is also supported in part by the Intramural Research Program of NCI, NIH.

Acknowledgments: The authors thank Cheri Rhoderick for secretarial assistance.

Conflicts of Interest: The authors declare no conflict of interest.

References

1. Dorward, D.; Lucas, C.; Chapman, G.; Haslett, C.; Dhaliwal, K.; Rossi, A. The role of formylated peptides and formyl peptide receptor 1 in governing neutrophil function during acute inflammation. *Am. J. Pathol.* **2015**, *185*, 1172–1184. [CrossRef] [PubMed]
2. Le, Y.; Murphy, P.; Wang, J. Formyl-peptide receptors revisited. *Trends Immunol.* **2002**, *23*, 541–548. [CrossRef]
3. Heo, S.; Kwon, Y.; Jang, I.; Jeong, G.; Yoon, J.; Kim, C.; Kwon, S.; Bae, Y.; Kim, J. WKYMVm-induced activation of formyl peptide receptor 2 stimulates ischemic neovasculogenesis by promoting homing of endothelial colony-forming cells. *Stem Cells* **2014**, *32*, 779–790. [CrossRef] [PubMed]
4. Kim, M.; Min, S.; Park, Y.; Kim, J.; Ryu, S.; Bae, Y. Expression and functional role of formyl peptide receptor in human bone marrow-derived mesenchymal stem cells. *FEBS Lett.* **2007**, *581*, 1917–1922. [CrossRef] [PubMed]
5. Van Compernolle, S.; Clark, K.; Rummel, K.; Todd, S. Expression and function of formyl peptide receptors on human fibroblast cells. *J. Immunol.* **2003**, *171*, 2050–2056. [CrossRef] [PubMed]
6. Boulay, F.; Tardif, M.; Brouchon, L.; Viganis, P. Synthesis and use of a novel N-formyl peptide derivative to isolate a human N-formyl peptide receptor cDNA *Biochem. Biophys. Res. Commun.* **1990**, *168*, 1103–1109. [CrossRef]
7. V18.1.proteinatlas.org. FPR1 Tissue Expression. Available online: https://www.proteinatlas.org/ENSG00000171051-FPR1/tissue (accessed on 19 May 2019).

8. Uhlen, M.; Fagerberg, L.; Hallstrom, B.; Lindskog, C.; Oksvold, P.; Mardinoglu, A.; Sivertsson, A.; Kampf, C.; Sjostedt, E.; Asplund, A.; et al. Proteomics. Tissue-based map of the human proteome. *Science* **2015**, *347*, 1260419. [CrossRef] [PubMed]

9. V18.1.proteinatlas.org. FPR2 Tissue Expression. Available online: https://www.proteinatlas.org/ENSG00000171049-FPR2/tissue (accessed on 19 May 2019).

10. Rabiet, M.; Macari, L.; Dahlgren, C.; Boulay, F. N-formyl peptide receptor 3 (FPR3) departs from the homologous FPR2/ALD receptor with regard to the major processes governing chemoattractant receptor regulation, expression at the cell surface, and phosphorylation. *J. Biol. Chem.* **2011**, *286*, 26718–26731. [CrossRef] [PubMed]

11. Kim, S.; Jim, J.; Jo, S.; Lee, H.; Lee, S.; Shim, J.; Seo, S.; Yun, J.; Bae, Y. Functional expression of formyl peptide receptor family in human NK cells. *J. Immunol.* **2009**, *183*, 5511–5517. [CrossRef] [PubMed]

12. Wen, X.; Xu, X.; Sun, W.; Chen, K.; Pan, W.; Wang, J.; Bolland, S.; Jin, T. G-protein-coupled formyl peptide receptors play a dual role in neutrophil chemotaxis and bacterial phagocytosis. *Mol. Biol. Cell* **2019**, *30*, 346–356. [CrossRef]

13. Wang, W.; Li, T.; Wang, X.; Yuan, W.; Cheng, Y.; Zhang, H.; Xu, E.; Zhang, Y.; Shi, S.; Ma, D.; et al. FAM19A4 is a novel cytokine ligand of formyl peptide receptor 1 (FPR1) and is able to promote the migration and phagocytosis of macrophages. *Cell Mol. Immunol.* **2015**, *12*, 615–624. [CrossRef]

14. Liu, M.; Chen, K.; Yoshimura, T.; Liu, Y.; Gong, W.; Le, Y.; Gao, J.; Zhao, J.; Wang, J.; Wang, A. Formylpeptide receptors mediate rapid neutrophil mobilization to accelerate wound healing. *PLoS ONE* **2014**, *9*, e90613. [CrossRef]

15. Chen, K.; Liu, M.; Liu, Y.; Yoshimura, T.; Shen, W.; Le, Y.; Durum, S.; Gong, W.; Wang, C.; Gao, J.; et al. Formylpeptide receptor-2 contributes to colonic epithelial homeostasis, inflammation, and tumorigenesis. *J. Clin. Investig.* **2013**, *123*, 1694–1704. [CrossRef]

16. Xiang, Y.; Yao, X.; Chen, K.; Wang, X.; Zhou, J.; Gong, W.; Yoshimura, T.; Huang, J.; Wang, R.; Wu, Y.; et al. The G-protein coupled chemoattractant receptor FPR2 promotes malignant phenotype of human colon cancer cells. *Am. J. Cancer Res.* **2016**, *6*, 2599–2610.

17. Snapkov, I.; Oqvist, C.; Figenschau, Y.; Kogner, P.; Johnsen, J.; Sveinbjornsson, B. The role of formyl peptide receptor 1 (FPR1) in neuroblastoma tumorigenesis. *BMC Cancer* **2016**, *16*, 490. [CrossRef]

18. Prevete, N.; Liotti, F.; Visciano, C.; Marone, G.; Melillo, R.; de Paulis, A. The formyl peptide receptor 1 exerts a tumor suppressor function in human gastric cancer by inhibiting angiogenesis. *Oncogene* **2005**, *34*, 3826–3838. [CrossRef]

19. Sedlmayer, F.; Hell, D.; Muller, M.; Auslander, D.; Fussenegger, M. Designer cells programming quorum-sensing interference with microbes. *Nat. Commun.* **2019**, *9*. [CrossRef]

20. Gobbetti, T.; Coldewey, S.; Chen, J.; McArthur, S.; le Faouder, P.; Cenac, N.; Flower, R.; Thiemermann, C.; Perretti, M. Nonredundant protective properties of FPR2/ALX in polymicrobial murine sepsis. *Proc. Natl. Acad. Sci. USA* **2014**, *111*, 18685–18690. [CrossRef]

21. Gallo, I.; Rattazzi, L.; Piras, G.; Gobbetti, T.; Panza, E.; Perretti, M.; Dalley, J.; D'acquisto, F. Formyl peptide receptor as a novel therapeutic target for anxiety-related disorders. *PLoS ONE* **2014**, *9*, e114626. [CrossRef]

22. Wenceslau, C.; McCarthy, C.; Szasz, T.; Calmasini, F.; Mamenko, M.; Webb, R. Formyl peptide receptor-1 activation exerts a critical role for the dynamic plasticity of arteries via actin polymerization. *Pharm. Res.* **2019**, *141*, 276–290. [CrossRef]

23. He, H.; Ye, R. The formyl peptide receptors: Diversity of ligands and mechanism for recognition. *Molecules* **2017**, *22*, 455. [CrossRef]

24. Stama, M.; Oelusarczyk, J.; Lacivita, E.; Kirpotina, L.; Schepetkin, I.; Chamera, K.; Riganti, C.; Perrone, R.; Quinn, M.; Basta-Kaim, A.; et al. Novel ureidopropanamide based N-formyl peptide receptor 2 (FPR2) agonists with potential application for central nervous system disorders characterized by neuroinflammation. *Eur. J. Med. Chem.* **2017**, *141*, 703–720. [CrossRef]

25. Cooray, S.; Gobbetti, T.; Montero-Melendez, T.; McArthur, S.; Thompson, D.; Clark, A.; Flower, R.; Perretti, M. Ligand-specific conformational change of the G-protein-coupled receptor ALX/FPR2 determines proresolving functional responses. *Proc. Natl. Acad. Sci. USA* **2013**, *110*, 18232–18237. [CrossRef]

26. Winther, M.; Holdfeldt, A.; Gabl, M.; Wang, J.; Forsman, H.; Dahlgren, C. Formylated MHC Class Ib binding peptides activate both human and mouse neutrophils primarily through formyl peptide receptor 1. *PLoS ONE* **2016**, *11*, e0167529. [CrossRef]

27. Bufe, B.; Schuman, T.; Kappl, R.; Bogeski, I.; Kummerow, C.; Podgorska, M.; Smola, S.; Hoth, M.; Zufall, F. Recognition of bacterial signal peptides by mammalian formyl peptide receptors: A new mechanism for sensing pathogens. *J. Biol. Chem.* **2015**, *290*, 7369–7387. [CrossRef]

28. Kretschmer, D.; Rautenberg, M.; Linke, D.; Peschel, A. Peptide length and folding state govern the capacity of staphylococcal B-type phenol-soluble modulins to activate human formyl-peptide receptors 1 or 2. *J. Leukoc. Biol.* **2015**, *97*, 689–697. [CrossRef]

29. Marasco, W.; Phan, S.; Krutzsch, H.; Showell, H.; Feltner, D.; Nairn, R.; Becker, E.; Ward, P. Purification and identification of formyl-methionyl-leucyl-phenylalanine as the major peptide neutrophil chemotactic factor produced by Escherichia coli. *J. Biol. Chem.* **1984**, *259*, 5430–5439.

30. Piatkov, K.; Vu, T.; Hwang, C.; Varshavsky, A. Formyl-methionine as a degradation signal at the N-termini of bacterial proteins. *Microb. Cell* **2015**, *2*, 376–393. [CrossRef]

31. He, H.; Troksa, E.; Caltabioano, G.; Pardo, L.; Ye, R. Structural determinants for the interaction of formyl peptide receptor 2 with peptide ligands. *J. Biol. Chem.* **2014**, *289*, 2295–2306. [CrossRef]

32. Liu, Y.; Chen, K.; Wang, J. Chapter 91: FPR Ligands. In *Handbook of Biologically Active Peptides*, 2nd ed.; Kastin, A., Ed.; Elsevier Inc.: San Diego, CA, USA, 2013; pp. 671–680.

33. Towle, K.; Lohans, C.; Miskolzie, M.; Acedo, J.; van Belkum, M.; Vedaras, J. Solution structures of phenol-soluble modulins a1, a2, B2, virulence factors from Staphylococcus aureus. *Biochemistry* **2016**, *55*, 4798–4806. [CrossRef]

34. Zhang, Q.; Raoof, M.; Chen, Y.; Sumi, Y.; Sursal, T.; Wolfgang, J.; Broki, K.; Itagaki, K.; Hauser, C. Circulating mitochondrial DAMPs cause inflammatory responses to injury. *Nature* **2010**, *464*, 104–107. [CrossRef]

35. Wenceslau, C.; Szasz, T.; McCarthy, C.; Baban, B.; NeSmith, E.; Webb, R. Mitochondrial N-formyl peptides cause airway contraction and lung neutrophil infiltration via formyl peptide receptor activation. *Pulm. Pharmacol.* **2016**, *37*, 49–56. [CrossRef]

36. Zhang, X.; Wang, T.; Yuan, Z.; Dai, L.; Zeng, N.; Wang, H.; Liu, L.; Wen, F. Mitochondrial peptides cause proinflammatory responses in the alveolar epithelium via FPR-1, MAPKs, and AKT: A potential mechanism involved in acute lung injury. *Am. J. Physiol. Lung Cell Mol. Physiol.* **2018**, *315*, L775–L786. [CrossRef]

37. Dorward, D.; Lucas, C.; Doherty, M.; Chapman, G.; Scholefield, E.; Morris, A.; Felton, J.; Kipari, T.; Humphries, D.; Robb, C.; et al. Novel role for endogenous mitochondrial formylated peptide-driven formyl peptide receptor 1 signaling in acute respiratory distress syndrome. *Respir. Res.* **2017**, *72*, 928–936.

38. Lind, S.; Gabl, M.; Holdfeldt, A.; Martensson, J.; Sundqvist, M.; Nishino, K.; Dahlgren, C.; Mukai, H.; Forsman, H. Identification of residues critical for FPR2 activation by the cryptic peptide mitocryptide-2 originating from the mitochondrial DNA-encoded cytochrome b. *J. Immunol.* **2019**, ji1900060. [CrossRef]

39. Shawar, S.; Rich, R.; Becker, E. Peptides from the amino-terminus of mouse mitochondrially encoded NADH dehydrogenase subunit 1 are potent chemoattractants. *Biochem. Biophys. Res. Commun.* **1995**, *211*, 812–818. [CrossRef]

40. Bufe, B.; Zufall, F. The sensing of bacteria: Emergin principles for the detection of signal sequences by formyl peptide receptors. *Biomol. Concepts* **2016**, *7*, 205–214. [CrossRef]

41. Little, J.; Simtchouk, S.; Schindler, S.; Villanueva, E.; Gill, N.; Walker, D.; Wolthers, K.; Klergis, A. Mitochondrial transcription factor A (Tfam) is a pro-inflammatory extracellular signaling molecule recognized by brain microglia. *Mol. Cell Neurosci.* **2014**, *60*, 88–96. [CrossRef]

42. Wencesclau, C.; McCarthy, C.; Szasz, T.; Spitler, T.; Goulopoulou, S.; Webb, R.; DAMPs in Cardiovascular Disease Working Group. Mitochondrial damage-associated molecular patterns and vascular function. *Eur. Heart J.* **2014**, *35*, 1172–1177. [CrossRef]

43. Oie, C.; Snapkov, I.; Elvevold, K.; Sveinbjornsson, B.; Smedsrod, B. FITC conjugation markedly enhances hepatic clearance of N-formyl peptides. *PLoS ONE* **2016**, *11*, e0160602. [CrossRef]

44. Wang, J.; Chen, K. Chemotactic Peptide Ligands for Formylpeptide Receptors Influencing Inflammation. In *Handbook of Biologically Active Peptides*; Kastin, A., Ed.; Elsevier Inc.: San Diego, CA, USA, 2006; pp. 547–552.

45. Braun, M.; Wang, J.; Lahey, E.; Rabin, R.; Kelsall, B. Activation of the formyl peptide receptor by the HIV-derived peptide T-20 suppresses interleukin-12 p70 production by human monocytes. *Blood* **2001**, *97*, 3531–3536. [CrossRef] [PubMed]

46. Wood, M.; Cole, A.; Eade, C.; Chen, L.; Chai, K.; Cole, A. The HIV-1 gp41 ectodomain is cleaved by matriptase to produce a chemotactic peptide that acts through FPR2. *Immunology* **2014**, *142*, 474–483. [CrossRef]

47. Li, B.; Wetzel, M.; Mikovits, J.; Henderson, E.; Rogers, T.; Gong, W.; Le, Y.; Ruscetti, F.; Wang, J. The synthetic peptide WKYMVm attenuates the function of the chemokine receptors CCR5 and CXCR4 through activation of formyl peptide receptor-like 1. *Blood* **2001**, *97*, 2941–2947. [CrossRef] [PubMed]

48. Lin, Q.; Fang, D.; Hou, X.; Le, Y.; Fang, J.; Wen, F.; Gong, W.; Chen, K.; Wang, J.; Su, S. HCV peptide(C5a), an amphipathic a-helical peptide of hepatitis Virus C, is an activator of N-formyl receptor in human phagocytes. *J. Immunol.* **2011**, *186*, 2087–2094. [CrossRef]

49. Mills, J.S. Peptides derived from HIV-1, HIV-2, Ebola virus, SARS coronavirus and coronavirus 229E exhibit high affinity binding to the formyl peptide receptor. *Biochim. Biophys. Acta* **2006**, *1762*, 693–703. [CrossRef]

50. Bellner, L.; Thoren, F.; Nygren, E.; Liljeqvist, J.; Karlsson, A.; Eriksson, K. A proinflammatory peptide from herpes simplex virus type 2 glycoprotein G affects neutrophil, monocyte, and NK cell functions. *J. Immunol.* **2005**, *174*, 2235–2241. [CrossRef] [PubMed]

51. Bellner, L.; Karlsson, J.; Fu, H.; Boulay, F.; Dahlgren, C.; Eriksson, K.; Karlsson, A. A monocyte-specific peptide from herpes simplex virus type 2 glycoprotein G activates the NADPH-oxidase but not chemotaxis through a G-protein-coupled receptor distinct from the members of the formyl peptide receptor family. *J. Immunol.* **2007**, *179*, 6080–6087. [CrossRef]

52. Bloes, D.; Otto, M.; Peschel, A.; Kretschmer, D. Enterococcus faecium stimulates human neutrophils via the formyl-peptide receptor 2. *PLoS ONE* **2012**, *7*, e39910. [CrossRef]

53. Betten, A.; Bylund, J.; Christophe, T.; Boulay, F.; Romero, A.; Hellstrand, K.; Dahlgren, C. A proinflammatory peptide from Helicobacter pylori activates monocytes to induce lymphocyte dysfuntion and apoptosis. *J. Clin. Investig.* **2001**, *108*, 1221–1228. [CrossRef]

54. Su, S.; Gao, J.; Gong, W.; Dunlop, N.; Murphy, P.; Oppenheim, J.; Wang, J. T21/DP107, a synthetic leucine zipper-like domain of the HIV-1 envelope gp41, attracts and activates human phagocytes by using G-protein-coupled formyl peptide receptors. *J. Immunol.* **1999**, *162*, 5924–5930.

55. Shen, W.; Proost, P.; Li, B.; Gong, W.; Le, Y.; Sargeant, R.; Murphy, P.; Van Damme, J.; Wang, J. Activation of the chemotactic peptide receptor FPRL1 in monocytes phosphorylates the chemokine receptor CCR5 and attenuates cell responses to selected chemokines. *Biochem. Biophys. Res. Commun.* **2000**, *272*, 276–283. [CrossRef] [PubMed]

56. Le, Y.; Jiang, S.; Hu, J.; Gong, W.; Su, S.; Dunlop, N.; Shen, W.; Li, B.; Wang, J. N36, a synthetic N-terminal heptad repeat domain of the HIV-1 envelope protein gp41, is an activator of human phagocytes. *Clin. Immunol.* **2000**, *96*, 236–242. [CrossRef] [PubMed]

57. Deng, X.; Ueda, H.; Su, S.; Gong, W.; Dunlop, N.; Gao, J.; Ruscetti, F.; Murphy, P.; Wang, J. A synthetic peptide derived from human immunodeficiency virus type 1 gp120 downregulates the expression and function of chemokine receptors CCR5 and CXCR4 in monocytes by activating the 7-transmembrane G-protein-coupled receptor FPRL1/LXA4R. *Blood* **1999**, *94*, 1165–1173. [PubMed]

58. Le, Y.; Yazawa, H.; Gong, W.; Yu, Z.; Ferrans, V.; Murphy, P.; Wang, J. The neurotoxic prion peptide fragment PrP(106-126) is a chemotactic agonist for the G protein-coupled receptor formyl peptide receptor-like1. *J. Immunol.* **2001**, *166*, 1448–1451. [CrossRef] [PubMed]

59. Tcherniuk, S.; Cenac, N.; Comte, M.; Frouard, J.; Errazuriz-Cerda, E.; Galabov, A.; Morange, P.; Vergnolle, N.; Si-Tahar, M.; Alessi, M.; et al. Formyl peptide receptor 2 plays a deleterious role during influenza A virus infection. *J. Infect. Dis.* **2016**, *214*, 237–247. [CrossRef] [PubMed]

60. Brandenburg, L.; Konrad, M.; Wruck, C.; Koch, T.; Lucius, R.; Pufe, T. Functional and physical interactions between formyl-peptide-receptors and the scavenger MARCO and their involvement in amyloid beta 1-42-induced signal transduction in glial cells. *J. Neurochem.* **2010**, *113*, 749–760. [CrossRef] [PubMed]

61. Jarosz-Griffiths, H.; Noble, E.; Rushworth, J.; Hooper, N. Amyloid-B receptors: The good, the bad, and the prion protein. *J. Biol. Chem.* **2016**, *291*, 3174–3183. [CrossRef]

62. Brandenburg, L.; Koch, T.; Sievers, J.; Ralph, L. Internalization of PrP(106-126) by the formyl-peptide-receptor-like-1 in glial cells. *J. Neurochem.* **2006**, *101*, 718–728. [CrossRef]

63. Kim, Y.; Lee, B.; Kim, O.; Bae, Y.; Lee, T.; Suh, P.; Ryu, S. Pituitary adenylate cyclase-activating polypeptide 27 is a functional ligand for formyl peptide receptor-like 1. *J. Immunol.* **2006**, *176*, 2669–2675. [CrossRef]

64. Chedid, P.; Boussetta, T.; Dang, P.; Belambri, S.; Marzaioli, V.; Fasseau, M.; Walker, F.; Couvineau, A.; El-Benna, J.; Marie, J. Vasoactive intestinal peptide dampens formyl-peptide-induced ROS production and inflammation by targeting a MAPK-p47^{phox} phosphorylation pathway in monocytes. *Mucosal Immunol.* **2017**, *10*, 332–340. [CrossRef]

65. Carion, T.; Kracht, D.; Strand, E.; David, E.; McWhirter, C.; Ebrahim, A.; Berger, E. VIP modulates the ALX/FPR2 receptor axis toward inflammation resolution in a mouse model of bacterial keratitis. *Prostag. Other Lipid Mediat.* **2019**, *140*, 18–25. [CrossRef]

66. Elagoz, A.; Henderson, D.; Babu, P.; Salter, S.; Grahames, C.; Bowers, L.; Roy, M.; Laplante, P.; Grazzini, E.; Ahmad, S.; et al. A truncated form of CKβ8-1 is a potent agonist for human formyl peptide-receptor-like 1 receptor. *Br. J. Pharmacol.* **2004**, *141*, 37–46. [CrossRef] [PubMed]

67. Ye, R.; Boulay, F.; Wang, J.; Dahlgren, C.; Gerard, C.; Parmentier, M.; Serhan, C.; Murphy, P. International union of basic and clinical pharmacology. LXXIII. Nomenclature for the formyl peptide receptor family. *Pharm. Rev.* **2009**, *61*, 119–161. [CrossRef]

68. Rossi, F.; Napolitano, F.; Pesapane, A.; Mascolo, M.; Staibano, S.; Matucci-Cericinic, M.; Guiducci, S.; Ragno, P.; di Spigna, G.; Pistiglione, L.; et al. Upregulation of the N-formyl peptide receptors in Scleroderma Fibroblasts fosters the switch to myofibroblasts. *J. Immunol.* **2015**, *194*, 5161–5173. [CrossRef]

69. De Paulis, A.; Montuori, N.; Prevete, N.; Fiorentino, I.; Rossi, F.; Visconte, V.; Rossi, G.; Marone, G.; Ragno, P. Urokinase induces basophil chemotaxis through a urokinase receptor epitope that is an endogenous ligand for formyl peptide receptor-like 1 and -like 2. *J. Immunol.* **2004**, *173*, 5739–5748. [CrossRef]

70. Gargiulo, L.; Longanesi-Cattani, I.; Bifulco, K.; Franco, P.; Raiola, R.; Campiglia, P.; Grieco, P.; Peluso, G.; Stoppelli, P.; Carriero, M. Cross-talk between fMLP and vitronectin receptors triggered by urokinase receptor-derived SRSRY peptide. *J. Biol. Chem.* **2005**, *280*, 25225–25232. [CrossRef]

71. Devosse, T.; Dutoit, R.; Migeotte, I.; De Nadai, P.; Imbault, V.; Communi, D.; Salmon, I.; Parmentier, M. Processing of HEBP1 by chathepsin D gives rise to F2L, the agonist of formyl peptide Receptor 3. *J. Immunol.* **2011**, *187*, 1475–1485. [CrossRef] [PubMed]

72. Gao, J.; Guillabert, A.; Hu, J.; Le, Y.; Urizar, E.; Seligman, E.; Fang, K.; Yuan, X.; Imbault, V.; Communi, D.; et al. F2L, a peptide derived from heme-binding protein, chemoattracts mouse neutrophils by specifically activating Fpr2, the low-affinity N-formylpeptide receptor. *J. Immunol.* **2007**, *178*, 1450–1456. [CrossRef]

73. Lee, S.; Lee, M.; Lee, H.; Kim, S.; Shim, J.; Jo, S.; Lee, J.; Kim, J.; Choi, Y.; Baek, S.; et al. F2L, a peptide derived from heme-binding protein, inhibits LL-37-induced cell proliferation and tube formation in human umbilical vein endothelial cells. *FEBS Lett.* **2008**, *582*, 273–278. [CrossRef] [PubMed]

74. V18.1.proteinatlas.org. FAM3D Tissue Expression. Available online: https://www.proteinatlas.org/ENSG00000198643-FAM3D/tissue (accessed on 28 May 2019).

75. Peng, X.; Xu, E.; Liang, W.; Pei, X.; Chen, D.; Zheng, D.; Zhang, Y.; Zheng, C.; Wang, P.; She, S.; et al. Identification of FAM3D as a new endogenous chemotaxis agonist for the formyl peptide receptors. *J. Cell Sci.* **2016**, *129*, 1831–1842. [CrossRef]

76. Cao, T.; Yang, D.; Zhang, W.; Wang, Y.; Qiao, Z.; Gao, L.; Liang, Y.; Yu, B.; Zhang, P. FAM3D inhibits glucagon secretion via MKP1-dependent suppression of ERK1/2 signaling. *Cell Biol. Toxicol.* **2015**, *33*, 457–466. [CrossRef]

77. He, L.; Fu, Y.; Deng, J.; Shen, Y.; Wang, Y.; Yu, F.; Xie, N.; Chen, Z.; Hong, T.; Peng, X.; et al. Deficiency of FAM3D (Family with sequence similarity 3, member D), a novel chemokine, attenuates neutrophil recruitment and ameliorates abdominal aortic aneurysm development. *Arter. Thromb. Vas.* **2018**, *38*, 1616–1631. [CrossRef] [PubMed]

78. Buckingham, J.; Flower, R. Annexin A1. In *Stress: Neuroendocrinology and Neurobiology: Handbook of Stress Series*; Fink, G., Ed.; Elsevier Inc.: Parkville, Australia, 2013; Volume 2, pp. 257–263.

79. Odobasic, D.; Jia, Y.; Kao, W.; Fan, H.; Wei, X.; Gu, R.; Ngo, D.; Kitching, A.; Hodsworth, S.; Morand, E.; et al. Formyl peptide receptor activation inhibits the expansion of effector T cells and synovial fibroblasts and attenuates joint injury in models of rheumatoid arthritis. *Int. Immunopharmacol.* **2018**, *61*, 140–149. [CrossRef] [PubMed]

80. Vecchi, L.; Zoia, M.; Santos, T.; Beserra, A.; Ramos, C.; Colombo, B.; Maia, Y.; Andrade, V.; Mota, S.; Araujo, T.; et al. Inhibition of the AnxA1/FPR1 autocrine axis reduces MDA-MB-231 breast cancer cell growth and aggressiveness in vitro and in vivo. *BBA-Mol. Cell Res.* **2018**, *1865*, 1368–1382. [CrossRef]

81. Kao, W.; Gu, R.; Jia, Y.; Wei, X.; Fan, H.; Harris, J.; Zhang, Z.; Quinn, J.; Morand, E.; Yang, Y. A formyl peptide receptor agonist suppresses inflammation and bone damage in arthritis. *Br. J. Pharmacol.* **2014**, *171*, 4087–4096. [CrossRef] [PubMed]

82. Chen, Y.; Lin, M.; Lee, C.; Liu, S.; Wang, C.; Fang, W.; Chao, T.; Wu, C.; Wei, Y.; Chang, H.; et al. Defective formyl peptide receptor 2/3 and annexin A1 expression associated with M2a polarization of blood immune cells in patients with chronic obstructive pulmonary disease. *J. Transl. Med.* **2018**, *16*, 69. [CrossRef]

83. Lange, C.; Starrett, D.; Goetsch, J.; Gerke, V.; Rescher, U. Transcriptional profiling of human monocytes reveals complex changes in the expression pattern of inflammation-related genes in response to the annexin A1-derived peptide Ac1-25. *J. Leukoc. Biol.* **2007**, *82*, 1592–1604. [CrossRef]

84. Liao, W.; Wu, S.; Wu, G.; Pao, H.; Tang, S.; Huang, K.; Chu, S. Ac2-26, an Annexin A1 peptide, attenuates ischemia-repurfusion-induced acute lung injury. *Int. J. Mol. Sci.* **2017**, *18*, 1771. [CrossRef]

85. Karlsson, J.; Fu, H.; Boulay, F.; Dahlgren, C.; Hellstrand, C.; Hellstrand, K.; Movitz, C. Neutrophil NADPH-oxidase activation by an annexin A1 peptide is transduced by the formyl peptide receptor (FPR), whereas an inhibitory signal is generated independently of the FPR family receptors. *J. Leukoc. Biol.* **2005**, *78*, 761–762. [CrossRef]

86. Fredman, G.; Kamaly, N.; Spolitu, S.; Milton, J.; Ghorpade, D.; Chiasson, R.; Kuriakose, G.; Perretti, M.; Farokzhad, O.; Tabas, I. Targeted nanoparticles containing the proresolving peptide Ac2-26 protect against advanced atherosclerosis in hypercholesterolemic mice. *Sci. Transl. Med.* **2015**, *7*, 275ra. [CrossRef]

87. Geary, L.; Nash, K.; Adisetiyo, H.; Liang, M.; Liao, C.; Jeong, J.; Zandi Em Roy-Burman, P. CAF-secreted annexin A1 induces prostate cancer cells to gain stem cell-like features. *Mol. Cancer Res.* **2014**, *12*, 607–612. [CrossRef] [PubMed]

88. Zhao, Y.; Zhou, S.; Heng, C. Impact of serum amyloid A on tissue factor and tissue factor pathway inhibitor expression and activity in endothelial cells. *Arterioscler. Thromb. Vasc. Biol.* **2007**, *27*, 1645–1650. [CrossRef]

89. Lee, H.; Kim, S.; Shim, J.; Kim, H.; Yun, J.; Baek, S.; Kim, K.; Bae, Y. A pertussis toxin sensitive G-protein-independent pathway is involved in serum amyloid A-induced formyl peptide receptor 2-mediated CCL2 production. *Exp. Mol. Med.* **2010**, *42*, 302–309. [CrossRef]

90. Hinrichs, B.; Matthews, J.; Siuda, D.; O'Leary, M.; Wolfarth, A.; Saeedi, B.; Nusrat, A.; Neish, A. Serum Amyloid A1 is an epithelial prorestitutive factor. *Am. J. Pathol.* **2018**, *188*, 937–949. [CrossRef] [PubMed]

91. Ren, S.; Qi, X.; Jia, C.; Wang, Y. Serum amyloid A and pairing formyl peptide receptor 2 are expressed in corneas and involved in inflammation-mediated neovascularization. *Int. J. Opthalmol.* **2014**, *7*, 187–193.

92. Kim, S.; Kim, Y.; Jang, Y. Cutting edge: LL-37-mediated formyl peptide receptor-2 signaling in follicular dendritic cells contributes to B cell activation in Peyer's Patch germinal centers. *J. Immunol.* **2017**, *198*, 629–633. [CrossRef]

93. Pan, X.; Quan, W.; Wu, J.; Xiao, W.; Sun, Z.; Li, D. Antimicrobial peptide LL-37 in macrophages promotes colorectal cancer growth. *Zhonghua Zhong Liu Za Zhi* **2018**, *40*, 412–417. [PubMed]

94. Chen, X.; Zhou, X.; Qi, G.; Tang, Y.; Guo, Y.; Si, J.; Liang, L. Roles and mechanisms of human cathelicidin LL-37 in cancer. *Cell Physiol. Biochem.* **2018**, *47*, 1060–1073. [CrossRef]

95. Doring, Y.; Drechsler, M.; Wantha, S.; Kemmerich, L.; Lievens, D.; Vijayan, S.; Gallo, R.; Weber, C.; Soehnlein, O. Lack of neutrophil-derived CRAMP reduces atherosclerosis in mice. *Circ. Res.* **2012**, *110*, 1052–1056. [CrossRef]

96. Chen, K.; Xiang, Y.; Huang, J.; Gong, W.; Yoshimura, T.; Jiang, Q.; Tessarollo, L.; Le, Y.; Wang, J. The formylpeptide receptor 2 (Fpr2) and its endogenous ligand cathelin-related antimicrobial peptide (CRAMP) promote dendritic cell maturation. *J. Biol. Chem.* **2014**, *289*, 17553–17563. [CrossRef] [PubMed]

97. Yoshimura, T.; McLean, M.; Dzutsev, A.K.; Yao, X.; Chen, K.; Huang, J.; Gong, W.; Zhou, J.; Ziang, Y.; Badger, J.; et al. The antimicrobial peptide CRAMP is essential for colon homeostasis by maintaining microbiota balance. *J. Immunol.* **2018**, *200*, 2174–2185. [CrossRef] [PubMed]

98. Zhao, H.; Sonada, S.; Yoshikawa, A.; Ohinata, K.; Yoshikawa, M. Rubimetide, humanin, and MMK1 exert anxiolytic-like activities via the formyl peptide receptor 2 in mice followed by the successive activation of DP_1, A_{2A}, and $GABA_A$ receptors. *Peptides* **2016**, *83*, 16–20. [CrossRef] [PubMed]

99. Kamal, A.; Hayhoe, R.; Paramasivam, A.; Cooper, D.; Flower, R.; Solito, E.; Perretti, M. Antiflammin-2 activates the human formyl-peptide receptor like 1. *Sci. World J.* **2006**, *6*, 1375–1384. [CrossRef] [PubMed]

100. Sun, R.; Iribarren, P.; Zhang, N.; Zhou, Y.; Gong, W.; Cho, E.; Lockett, S.; Chertov, O.; Bednar, F.; Rogers, T.; et al. Identification of neutrophil granule protein cathepsin G as a novel chemotactic agonist for the G protein-coupled formyl peptide receptor. *J. Immunol.* **2004**, *173*, 428–436. [CrossRef] [PubMed]

101. Corminboeuf, O.; Leroy, X. FPR2/ALXR agonists and the resolution of inflammation. *J. Med. Chem.* **2015**, *58*, 5689–5701. [CrossRef] [PubMed]

102. Dean, S.; Wang, C.; Nam, K.; Maruyama, C.; Trump, B.; Baker, O. Aspirin triggered resolvin D1 reduces inflammation and restores saliva secretion in a Sjogrens syndrome mouse model. *Rheumatology* **2019**, *58*, 1285–1292. [CrossRef] [PubMed]

103. Guo, Z.; Hu, Q.; Xu, L.; Guo, Z.; Ou, Y.; He, Y.; Yin, C.; Sun, X.; Tang, J.; Zhang, J. Lipoxin A4 reduces inflammation through formyl peptide receptor 2/p38MAPK signaling pathway in subarachnoid hermorrhage rats. *Stroke* **2016**, *47*, 490–497. [CrossRef]

104. Tsuruki, T.; Yoshikawa, M. Orally administered FPRL1 receptor agonist peptide MMK-1 inhibits etoposide-induced alopecia by a mechanism different from intraperitoneally administered MMK-1. *Peptides* **2006**, *27*, 820–825. [CrossRef]

105. Tavano, R.; Segat, D.; Fedeli, C.; Malachin, G.; Lubian, E.; Mancin, F.; Papini, E. Formyl-peptide receptor agonists and amorphous SiO2-NPs synergistically and selectively increase the inflammatory responses of human monocytes and PMNs. *Nanobiomedicine* **2016**, *3*. [CrossRef]

106. Chen, J.; Bernstein, H.; Chen, M.; Wang, L.; Ishii, M.; Turck, C.; Coughlin, S. Tethered ligand library for discovery of peptide agonists. *J. Biol. Chem.* **1995**, *270*, 23398–23401. [CrossRef]

107. Hu, Y.; Cheng, N.; Wu, H.; Kang, S.; Ye, R.; Cai, J. Design, synthesis and characterization of fMLF-mimicking AApeptides. *ChemBioChem* **2014**, *15*, 2420–2426. [CrossRef] [PubMed]

108. Hecht, I.; Rong, J.; Sampaio, A.; Hermesh, C.; Rutledge, C.; Shemesh, R.; Toporik, A.; Beiman, M.; Dassa, L.; Niv, H.; et al. A novel peptide agonist of Formyl-peptide receptor-like 1 (ALX) displays anti-inflammatory and cardioprotective effects. *J. Pharmacol. Exp.* **2009**, *328*, 426–434. [CrossRef] [PubMed]

109. Gabl, M.; Holdfeldts, A.; Winther, M.; Oprea, T.; Bylund, J.; Dahlgren, C.; Forsman, H. A pepducin designed to modulate P2Y2R function interacts with FPR2 in human neutrophils and transfers ATP to an NADPH-oxidase-activating ligand through a receptor cross-talk mechanism. *BBA Cell Biol.* **2016**, *1863*, 1228–1237.

110. Winther, M.; Dahlgren, C.; Forsman, H. Formyl peptide receptors in mice and men: Similarities and differences in recognition of conventional ligands and modulating lipopeptides. *Basic Clin. Pharmacol. Toxicol.* **2017**, *122*, 191–198. [CrossRef] [PubMed]

111. Bae, G.; Lee, H.; Jung, Y.; Shim, J.; Kim, S.; Baek, S.; Kwon, J.; Park, J.; Bae, Y. Identification of novel peptides that stimulate human neutrophils. *Exp. Mol. Med.* **2012**, *44*, 130–137. [CrossRef] [PubMed]

112. Holdfeldt, A.; Svovbakke, S.; Winther, M.; Gabl, M.; Nielsen, C.; Perez-Gassol, I.; Larsen, C.; Wang, J.; Karlsson, A.; Dahlgren, C.; et al. The lipidated peptidomimetic Lau-((S)-Aoc)-(Lys-Bnphe)6-NH2 is a novel formyl peptide receptor 2 agonist that activates both human and mouse neutrophil NADPH oxidase. *J. Biol. Chem.* **2016**, *291*, 19888–19899. [CrossRef] [PubMed]

113. Madenspacher, J.; Azzam, K.; Gong, W.; Gowdy, K.; Vitek, M.; Laskowitz, D.; Remaley, A.; Wang, J.; Fessler, M. Apolipoproteins and apolipoprotein mimetic peptides modulate phagocyte trafficking through chemotactic activity. *J. Biol. Chem.* **2012**, *287*, 43730–43740. [CrossRef] [PubMed]

114. He, M.; Cheng, N.; Gao, W.; Zhang, M.; Zhang, Y.; Ye, R.; Wang, M. Characterization of Quin-C1 for its anti-inflammatory property in a mouse model of bleomycin-induced lung injury. *Acta Pharm. Sin.* **2011**, *32*, 601–610. [CrossRef] [PubMed]

115. Schepetkin, I.; Kirpotina, L.; Khlebnikov, A.; Jutila, M.; Quinn, M. Gastrin-releasing peptide-neuromedin B receptor antagonist PD176252, PD168368, and related analogs are potent agonists of human formyl-peptide receptors. *Mol. Pharm.* **2011**, *79*, 77–90. [CrossRef]

116. Kirpotina, L.; Khlebnikov, A.; Schepetkin, I.; Ye, R.; Rabiet, M.; Jutila, M.; Quinn, M. Identification of novel small-molecule agonists for human formyl peptide receptors and pharmacophore models of their recognition. *Mol. Pharm.* **2010**, *77*, 159–170. [CrossRef]

117. Crocetti, L.; Vergelli, C.; Cilibrizzi, A.; Graziano, A.; Khlebnikov, A.; Kirpotina, L.; Schepetkin, I.; Quinn, M.; Giovannoni, M. Synthesis and pharmacological evaluation of new pyridazine-based thioderivatives as formyl peptide receptor (FPR) agonists. *Drug Dev. Res.* **2013**, *74*. [CrossRef]

118. Frohn, M.; Xu, H.; Zou, X.; Chang, C.; McElvaine, M.; Plant, M.; Wong, M.; Tagari, P.; Hungate, R.; Burli, R. New 'chemical probes' to examine the role of the hFPRL1 (or ALXR) receptor in inflammation. *Bioorg. Med. Chem. Lett.* **2007**, *17*, 6633–6637. [CrossRef] [PubMed]

119. Cilibrizzi, A.; Schepetkin, I.; Bartolucci, G.; Crocetti, L.; Dal Piaz, V.; Giovannoni, M.; Graziano, A.; Kirpotina, L.; Quinn, M.; Vergelli, C. Synthesis, enantioresolution, and activity profile of chiral

6-methyl-2,4-disubstituted pyridazine-3(2H)-ones as potent N-formyl peptide receptor agonists. *Bioorg. Med. Chem.* **2012**, *20*, 3781–3792. [CrossRef] [PubMed]

120. Pinilla, C.; Edwards, B.; Appel, J.; Yates-Gibbins, T.; Giulanotti, M.; Medina-Franco, J.; Young, S.; Santos, R.; Sklar, L.; Houghten, R. Selective agonists and antagonists of formylpeptide receptors: Duplex flow cytometry and mixture-based positional scanning libraries. *Mol. Pharmacol.* **2013**, *84*, 314–324. [CrossRef] [PubMed]

121. Lacivita, E.; Schepetkin, I.; Stama, M.; Kirpotina, L.; Colabufo, N.; Perrone, R.; Khlebnikov, A.; Quinn, M.; Leopoldo, M. Novel 3-(1H-Indol-3-yl)-2-[3-(4-nitrophenyl)ureido]propanamides as selective agonists of human formyl-peptide receptor 2. *Bioorg. Med. Chem.* **2014**, *23*, 3913–3924. [CrossRef] [PubMed]

122. Lee, H.; Bae, Y. The anti-infective peptide, innate defense-regulator peptide, stimulates neutrophil chemotaxis via a formyl peptide receptor. *Biochem. Biophys. Res. Commun.* **2008**, *369*, 573–578. [CrossRef] [PubMed]

123. Nijnik, A.; Madera, L.; Ma, S.; Waldbrook, M.; Elliott, M.; Easton, D.; Mayer, M.; Mullaly, S.; Kindrachuk, J.; Jenssen, H.; et al. Synthetic cationic peptide IDR-1002 provides protection against bacterial infections through chemokine induction and enhanced leukocyte recruitment. *J. Immunol.* **2010**, *184*, 2539–2550. [CrossRef] [PubMed]

124. Qin, C.; May, L.; Li, R.; Cao, N.; Rosli, S.; Deo, M.; Alexander, A.; Horlock, D.; Bourke, J.; Yang, Y.; et al. Small-molecule-biased formyl peptide receptor agonist compound 17b protects against myocardial ischaemia-reperfusion injury in mice. *Nat. Commun.* **2017**, *8*, 14232. [CrossRef] [PubMed]

125. Cilibrizzi, A.; Quinn, M.; Kirpotina, L.; Schepetkin, I.; Holderness, J.; Ye, R.; Rabiet, M.; Biancalani, C.; Cesari, N.; Graziano, A.; et al. 6-methyl-2,4-disubstituted pyridazin-3(2H)-ones: A novel class of small-molecule agonists for formyl peptide receptors. *J. Med. Chem.* **2009**, *52*, 5044–5057.

126. Hakim, M.; Yang, S.; Lai, R. Centipede venoms and their components: Resources for potential therapeutic applications. *Toxins* **2015**, *7*, 4832–4851. [CrossRef] [PubMed]

127. Park, Y.; Park, B.; Lee, M.; Jeong, S.; Lee, H.; Sohn, D.; Song, J.; Lee, J.; Hwang, J.; Bae, Y. A novel antimicrobial peptide acting via formyl peptide receptor 2 shows therapeutic effects against rheumatoid arthritis. *Sci. Rep.* **2018**, *8*, 14664. [CrossRef] [PubMed]

128. Chen, Q.; Wade, D.; Kurosaka, K.; Wang, Z.; Oppenheim, J.; Yang, D. Temporin A and related frog antimicrobial peptides use formyl peptide receptor-like 1 as a receptor to chemoattract phagocytes. *J. Immunol.* **2004**, *173*, 2652–2659. [CrossRef] [PubMed]

129. Tsuruki, T.; Kishi, K.; Takahashi, M.; Tanaka, M.; Matsukawa, T.; Yoshikawa, M. Soymetide, an immunostimulating peptide derived from soybean β-conglycinin, is an fMLP agonist. *FEBS Lett.* **2003**, *540*, 206–210. [CrossRef]

130. Kim, S.; Zhang, F.; Gong, W.; Chen, K.; Xia, K.; Liu, F.; Fross, R.; Wang, J.; Linhardt, R.; Cotton, M. Copper regulates the interactions of antimicrobial piscidin peptides from fish mast cells with formyl peptide receptors and heparin. *J. Biol. Chem.* **2018**, *293*, 15381–15396. [CrossRef] [PubMed]

131. Park, Y.; Lee, S.; Jung, Y.; Lee, M.; Lee, H.; Lee, H.; Park, J.; Koo, J.; Koo, J.; Bae, Y. Promotion of formyl peptide receptor 1-mediated neutrophil chemotactic migration by antimicrobial peptides isolated from the centipede Scolopendra subspinipes mutilans. *BMB Rep.* **2016**, *49*, 520–525. [CrossRef] [PubMed]

 International Journal of
Molecular Sciences

Review

Humanized Mice as an Effective Evaluation System for Peptide Vaccines and Immune Checkpoint Inhibitors

Yoshie Kametani [1,2,*], Yusuke Ohno [1], Shino Ohshima [1], Banri Tsuda [3], Atsushi Yasuda [4], Toshiro Seki [4], Ryoji Ito [5] and Yutaka Tokuda [3]

[1] Department of Molecular Life Science, Division of Basic Medical Science, Tokai University School of Medicine; 143 Shimokasuya, Isehara-shi, Kanagawa 259-1193, Japan; y-ohno@tsc.u-tokai.ac.jp (Y.O.); shino-w@tokai-u.jp (S.O.)

[2] Institute of Advanced Biosciences, Tokai University, 4-1-1 Kitakaname, Hiratsuka-shi, Kanagawa 259-1292, Japan

[3] Department of Breast and Endocrine Surgery, Tokai University School of Medicine, 143 Shimokasuya, Isehara-shi, Kanagawa 259-1193, Japan; banri@is.icc.u-tokai.ac.jp (B.T.); tokuda@is.icc.u-tokai.ac.jp (Y.T.)

[4] Department of Internal Medicine, Division of Nephrology, Endocrinology and Metabolism, Tokai University School of Medicine, 143 Shimokasuya, Isehara-shi, Kanagawa 259-1193, Japan; yasuda1633@yahoo.co.jp (A.Y.); tsekimdpdd@tokai-u.jp (T.S.)

[5] Central Institute for Experimental Animals, 3-25-12 Tonomachi, Kawasaki-ku, Kawasaki-shi, Kanagawa 210-0821, Japan; rito@ciea.or.jp

[*] Correspondence: y-kametn@is.icc.u-tokai.ac.jp; Tel.: +81-463-93-1121 (ext. 2589)

Received: 5 November 2019; Accepted: 12 December 2019; Published: 16 December 2019

Abstract: Peptide vaccination was developed for the prevention and therapy of acute and chronic infectious diseases and cancer. However, vaccine development is challenging, because the patient immune system requires the appropriate human leukocyte antigen (HLA) recognition with the peptide. Moreover, antigens sometimes induce a low response, even if the peptide is presented by antigen-presenting cells and T cells recognize it. This is because the patient immunity is dampened or restricted by environmental factors. Even if the immune system responds appropriately, newly-developed immune checkpoint inhibitors (ICIs), which are used to increase the immune response against cancer, make the immune environment more complex. The ICIs may activate T cells, although the ratio of responsive patients is not high. However, the vaccine may induce some immune adverse effects in the presence of ICIs. Therefore, a system is needed to predict such risks. Humanized mouse systems possessing human immune cells have been developed to examine human immunity in vivo. One of the systems which uses transplanted human peripheral blood mononuclear cells (PBMCs) may become a new diagnosis strategy. Various humanized mouse systems are being developed and will become good tools for the prediction of antibody response and immune adverse effects.

Keywords: peptide vaccine; immune checkpoint inhibitor; humanized mouse; cancer antigen; immune suppression

---

1. Introduction

Peptide vaccines are widely accepted as a promising strategy to fight infectious disease and cancer. However, the efficacy of a peptide vaccine depends not only on the antigen presentation through antigen-presenting cells but also on the immune environment of each patient, since the immunity of patients with chronic infectious disease and/or cancers tend to be dampened. Therefore, to achieve a more personalized medicine, we need a more detailed diagnosis before treatment. We propose the use of the humanized mouse system established through transplanting human peripheral blood mononuclear cells (PBMCs) from a patient into an immunodeficient mouse, for the evaluation of the response to peptide vaccines and other reagents which influence patient immunity. We also describe the immune condition artificially induced by immune checkpoint inhibitors (ICIs) [1] and the reagents against immune-related adverse events (irAEs), followed by the current state-of-the-art advances of humanized mouse systems and the issues to overcome. Moreover, we will discuss whether it is possible to evaluate the patient immunity by using second-generation humanized mice.

2. Difficulties in the Development of Peptide Vaccines

The design of peptide vaccines relies on the potential of peptides to bind to the major histocompatibility complex (MHC) in order to be presented by antigen-presenting cells (APCs), such as dendritic cells (DC) and macrophages. However, the MHC binding affinity is not enough to predict the activation of immunity, because the immune condition is different among different patients. Therefore, the decrease in the immune competence should be evaluated when the vaccine is adopted for patients with cancer and/or affected by a chronic infection. The vaccine is not restricted to be used as an anticancer agent; it also includes the influenza vaccine, to be administered to cancer patients [2–4]. Moreover, if the immune checkpoint inhibitors (ICIs) are used for the purpose of immune activation, the situation becomes more complex. We discuss the factors in detail below.

2.1. Selection of the Adequate Peptide for Vaccination

Vaccines are categorized as preventive or therapeutic based on their function and are further classified into virus, peptide, DNA, or DC vaccines, depending on the antigen source. Various types of antigens and adjuvants have been developed and evaluated for vaccination against infectious diseases. The design of the peptide antigen is important for inducing the most effective output with each type of vaccine, as each pathogen has a unique strategy for infection and proliferation. However, for long-lasting memory production, protein/peptide-based antigens are essential because the memory requires the activation of T cells through antigen-presenting cells, such as DC and macrophages. On the other hand, antigens need to activate B cells by crosslinking B-cell receptors (surface Igs). Therefore, the antigen epitope should be exposed to the hydrophilic surface by protruding into the aqueous solution and, thus, being recognized by B cells in vivo.

For the vaccine components to activate T cells, the antigens should at least contain a highly immunogenic peptide with more than 8, and up to 30, residues which can be further presented by the patient MHC (class I for cytotoxic T-cell activation and class II for antibody production). Moreover, as the peptide sequence mutates easily within the virus, it should be selected to maintain the peptide primary structure. The peptide presentation is predicted for HLA and mouse MHC by using available algorithms [5–7]. However, the prediction is incomplete because more new HLA types have been reported [8–10], and even if a peptide is successfully presented by mouse MHC in an experimental design, it does not imply that the same peptide will be presented on HLA. Therefore, larger peptide antigens are typically used in order to include as many epitopes as possible to be presented by major HLAs.

The evaluation of adjuvants is also very important. The induction of inflammation by the adjuvant is effective for the enhancement of the immune response. However, inflammation induction may pose a risk and result in adverse effects for patients. Therefore, self-adjuvanting techniques have been developed for clinical use [11–14]. Among them, the conjugation of molecules related to the ligands of tool-like receptors (TLR) to target peptides may be a safe and effective vaccine adjuvant. The DNA vaccines now, on translational research, use genes of TLR-related molecules.

2.2. Antigens which Enable Activation of the Patient Immune System

While vaccination is the most effective strategy to prevent acute infectious diseases caused by bacteria and viruses, it is not easy to develop effective vaccines against cancer and chronic infectious diseases. Similarly, to antigens present in pathogenic bacteria and viruses, patients with cancer present tumor-associated antigens (TAA) with high antigenicity and immunogenicity. TAAs are classified into differentiation, tissue-specific, mutated, and overexpressed antigens [15]. The U.S. Food and Drug Administration (FDA) has already approved for clinical use several cancer vaccines based on TAAs [16,17]. Hepatitis B virus (HBV) and human papilloma virus (HPV) are examples of TAA-based vaccines [18]. There are also unique classic vaccines like sipuleucel-T, the first therapeutic cancer vaccine approved by the FDA [19]. Moreover, many cancer vaccine candidates are currently under investigation in clinical trials, including nucleic acid-containing liposomes and nanoparticles (DNA vaccines) and gp100 peptide (peptide vaccines) [20–22].

On the other hand, especially for tumor-associated peptide vaccines, even if the antigen presentation is satisfied, it is difficult to activate the patient immune system. In spite of the extensive development of vaccines which may induce an anticancer immune response in patients, this response may vary among patients, making the vaccine not always effective. The immune-reactive tumors are called hot tumors, whereas nonimmune reactive tumors are referred to as cold tumors. Hot tumors are thought to have much more cell mutations compared to cold tumors, suggesting that hot tumors have many more TAAs [23]. Accordingly, the hot tumor, which is immune-reactive for the patient, may become the target of peptide vaccines, whereas, in the case of nonreactive cold tumors, the peptide vaccine might be ineffective. In hot tumors, there are some antigens that are highly expressed because of their overexpression on cancer cells. Human epidermal growth factor receptor 2 (HER2) is an example of a TAA molecule, as HER2 is overexpressed on the tumors of patients with breast cancer, and the specific antibody Herceptin is very effective for suppressing cancer progression. Due to the success of antibody reagents, many other human antitumor IgGs have been developed, and their mechanism of action has been investigated [24]. However, the antitumor effect does not last long enough, and the mechanism underlying this effect has not been fully elucidated.

Another problem in the development of cancer vaccines is the incomplete prediction resulting from the algorithms used. Our immune system rejects self-antigen-reactive clones, which may contain cancer-specific clones. Therefore, many of the predicted peptides cannot induce the desired immune response, even though the peptide leads to an immune response in experimental animals. Even if the peptide functions as an antigen, cancer cells have heterogenous mutations in the tumor mass, and, thus, the reactivity of each cancer cell is predicted to be diverse. Therefore, a complete rejection of the cancer cells within the tumor mass is difficult if simply one TAA is selected as peptide antigen.

2.3. Immune Suppression in Patients Prevents the Effectiveness of Vaccines

The most important challenge in the design of a vaccine is the immune suppression caused by the patient. The levels of cytotoxic T-cell activation, antibody production, and productive inflammation are different among patients with cancer. Therefore, we cannot predict the patient immune response, even if the peptide vaccine induces an immune response that is similar to the one produced by a viral infection in a healthy individual.

Therefore, although peptide vaccines have been extensively developed, the effect of the anticancer peptide vaccine is very limited, even if the peptide is presented by class I HLA on the patient DCs and the beneficial effect remains, as reviewed by Wong et al. [25]. One of the reasons for this limited effect is that cancer cells are originally "self", and the immune response is basically suppressed by clonal deletion or regulatory immune cell reactions, even though the peptide-reactive CD8 T cells are often detected in the patient PBMC. Even if mutations occur, most of them are limited to a very small region, and the peptides recognized as "non-self" might be very few or suppressed. This mechanism is present in cold tumors.

Meanwhile, an autoimmune disease might be induced by the suppression of peripheral tolerance. The neutrophil extracellular traps (NETs) play a role in the development of autoimmunity [26,27]. NETs are networks of extracellular fibers that are primarily composed of DNA from neutrophils, which suppress the movement of pathogens. Neutrophils release granule proteins, together with chromatins, and form an extracellular fibril matrix of NETs. The autoantigens involved in neutrophil granular proteins contain very common proteins, such as actin and histones. The proteins vary with the stimulation, and they occasionally induce an autoimmune response. It is important to understand which condition determines if the immune system will or will not induce an autoimmune disease. Moreover, not only cancers, but also some pathogens, induce tolerance. Actually, immature DCs, which induce only an MHC-TCR signal, may induce anergy to self-reactive and non-self-reactive T cells [28].

3. Immune Checkpoint Inhibitors and Reagents for Side-Effect Regulation

Recently, adaptive immune-resistant tumor cells which express the programmed-death-L1 (PD-L1) antigen were reported in melanomas by Abiko et al. [29] and Taube et al. [30]. According to their reports, PD-L1 is largely induced on the local tumor cells by tumor-infiltrating lymphocytes (TILs)-derived IFN-γ because IFN-γ is the most potent inducer of PD-L1in inflammatory cytokines. Upregulation of PD-L1 by IFN-γ has been extensively described in various cell types [31–37]. Similarly, TNF-α, another pro-inflammatory cytokine, also upregulates PD-L1 expression via TNF-α-NF-κB pathway [38–40]. TNF-α is reported to synergistically act with IFN-γ to induce PD-L1 expression at both mRNA and protein levels. IFN-γ enhances the resistance of the adaptive immune response by PD-L1 induction in hepatocellular carcinoma cells which upregulate the expression of IFN-γ receptors [41]. PD-L1 is expressed not only in all hematopoietic cells but also in many non-hematopoietic cell types, such as endothelial and epithelial cells [42,43]. In contrast, PD-L2 expression is more restricted to professional antigen-presenting cells, such as DCs, B cells, and monocytes/macrophages. Besides PD-1, there are other known interacting partners for PD-L1 and PD-L2. PD-L1 also binds to CD80, whereas PD-L2 uses repulsive guidance molecule (RGM) domain family member B (RGMB) as an alternative binding partner. Both types of interaction also inhibit immune responses [44,45].

3.1. Patients with Cancer

Recently, the anticancer effect of various immune checkpoint antibodies was elucidated [46]. The "immune checkpoint antibody" induces the blockage of continuous T-cell activation in the periphery. PD-1 antigen is expressed on the long-lived activated T cells, exhausted T cells, and the follicular helper T cells (Tfh) [47,48]. Normally, PD-L1 is expressed on antigen-presenting cells and germinal center B cells [49,50]. Apoptosis is induced when the PD-1-expressing T cells encounter the PD-L1-expressing APCs [49]. When the PD-1/PD-L1 interaction is inhibited by the anti-PD-1 antibody, T cells survive, and the anticancer effect is prolonged. Other immune checkpoint molecules, such as CTLA-4, PD-1, TIM-3, and LAG-3, have been reported, and the ability of the antibodies against such immune checkpoint molecules is being evaluated as anticancer products [51,52]. The effect is remarkable, but the response is still limited to a fraction of patients with cancer. The effect is ordinary, not long-lasting, and the combination of these inhibitors and other anticancer drugs are under investigation.

Moreover, antibodies are so expensive that, before using them as therapeutic agents, a strategy is needed to distinguish among patients that are responsive to the treatment from those that are not. Many biomarkers have been reported to predict the efficacy of the treatment. However, the heterogeneity of tumor masses and the variety of antibodies available make it difficult to find such predictive biomarkers, and even PD-L1 expression might not be a promising marker. Collectively, many studies have suggested that PD-L1 expression on melanoma cells can represent a biomarker to test for the efficacy of anti-PD1 and related antibodies, such as Nivolumab, Ipilimumab, and Pembrolizumab [53–55], and other immune checkpoint inhibitors; however the PD-L1 expression is not always an effective marker for patients with cancer in other clinical trials [56,57]. For example, PD-L1 expression on melanoma cells in pretreatment tumor biopsy samples is reported to correlate with response rate, progression free survival, and overall survival in patients with advanced melanoma treated with anti-PD1 antibodies [55], but these antibodies are also effective for PD-L1-negative patients [57].

While the benefits of assessing PD-L1 expression on melanoma cells to predict the clinical outcomes of ICI.

It is already defined. treatment have been suggested, as above, there are still no common criteria of diagnosis. This fact limits the clinical usefulness of the diagnosis of PD-L1 expression, because the low sensitivity of immunohistochemical (IHC) assays using different antibody clones makes it difficult to establish staining platforms and scoring systems [54,55,57–59]. To avoid misprediction by IHC staining, Conroy et al. assessed the expression of PD-L1, using next-generation RNA sequencing, but the sensitivity of their system resembles that of IHC assay systems and is, in addition, more expensive [58]. Additional assays or completely different assay systems will be needed in the future to diagnose PD-L1 expression of patient cancer tissues, for the prediction of clinical outcomes for the ICI treatment of melanoma [60].

3.2. Patients with Infectious Diseases

Viral infections do not always enhance PD-L1 expression, because similar PD-L1 levels are detected in individuals not infected with viruses [61–64]. Increased PD-L1 levels are related to specific viruses, such as the following: Epstein–Barr virus (EBV) [65–68], hepatitis B virus (HBV) [69–71], hepatitis C virus (HCV) [72–75], human immunodeficiency virus (HIV) [63,76–79], human papilloma virus (HPV) [68,80–83], Merkel cell polyomavirus (MCPyV) [84], bovine leukemia virus (BLV) [85], and Kaposi sarcoma-associated herpes virus (KSHV) [86]. The pathobiological mechanisms by which viruses trigger the expression of PD-L1 have been elucidated. Pathogen-associated molecular patterns (PAMPs) such as lipopolysaccharides (LPS), double-stranded RNA, and non-methylated CpG, from virus, bacteria, and fungi, activate toll-like receptors (TLRs) to induce the immune response and protect the host against the infection. Therefore, the effect of PD-1/PD-L1 blockage by ICI might not be limited to blocking cancer-T-cell interaction. Other hematopoietic lineage cells expressing PD-1 and/or PD-L1 might also be affected. For example, a fraction of plasmablasts and regulatory B (Breg) cells also express PD-L1 [87,88]. Therefore, the blockage of the axis may affect the humoral immunity or Breg cells. However, the antigen-specific reaction in such a systemic immunity is difficult to analyze in vivo.

3.3. Steroid Hormones and ICI Side Effects

Glucocorticoids are a class of steroid hormones that are powerful immune-suppressants that produce an effect on the systemic immune response. Conditions such as pregnancy and chronic inflammation may induce glucocorticoid secretion. Glucocorticoids [89] secreted by the stimulation of chronic inflammation are widely used as anti-inflammatory drugs. While they induce various signals related to cytokine and Fc receptors that modify metabolism and immune responses, it was recently reported that glucocorticoids impair upstream B-cell-receptor and Toll-like-receptor 7 signaling, reduce transcriptional output from the immunoglobulin loci, and promote significant upregulation of genes encoding the immunomodulatory cytokine IL-10 and the terminal-differentiation factor BLI MP-1 [90]. Expression of κ light chain and the two variable regions are especially suppressed. If patients affected with cancer or severe infectious diseases increase their glucocorticoid levels in order to overcome the disease-induced inflammation, or if they are treated with glucocorticoid because of the regulation of anticancer drug-induced side effects, the anticancer Ig expression might be suppressed. If the inflammatory, glucocorticoid-abundant condition continues, the potential for antibody production in the patient may be dampened. Therefore, if the PBMC of patients is examined for the antibody-production response, we may be able to predict if the patient is exposed to such steroid-based immune suppression. Glucocorticoids have also been reported to enhance metastasis in breast cancer [91]; therefore, their effect on patients with cancer needs to be examined in more detail.

On the other hand, it has been reported that ICI treatments occasionally induce a typical side effect related to pituitary dysfunction. Notably, hypophysitis, a previously very rare disease, has emerged as a distinctive side effect of ipilimumab and occasionally of nivolumab [92]. These side effects are not limited only to the pituitary; they also affect the thyroid, adrenal glands, and other downstream-target organs [93].

4. Humanized Mouse Models for the Evaluation of the Human Immune Environment

As we mentioned above, the prediction of the protective immunity development by vaccination is difficult because the immune condition is diverse in each patient, and the appropriate ICIs and induced irAEs may not be predicted. In order to determine the protocol reflecting the immune condition of each patient, the so-called personalized medicine, a humanized mouse system reconstituted with the patient immunity, may be useful [94]. The immunization with vaccines may reveal not only the effect of a specific vaccine, but it may also provide information regarding the patient immune response to mimic the anticancer/pathogen response. The current status of the humanized mouse system involving next generation humanized mice and its limitations is shown in Figure 1 (cellular immunity) and discussed below [95,96].

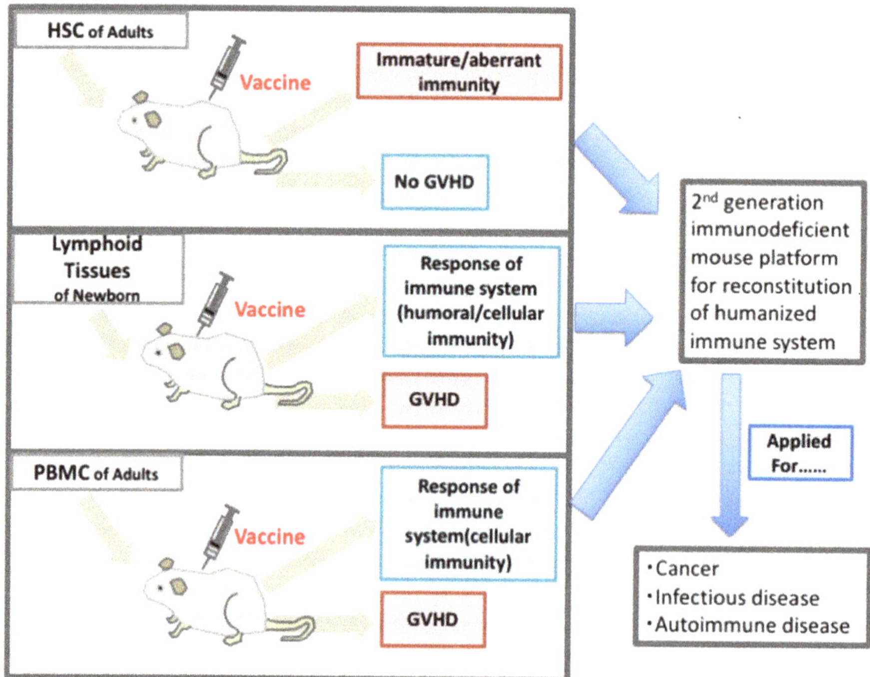

Figure 1. Three strategies for the reconstitution of human immunity in the immunodeficient mouse. The transplanted tissues are HSC, Lymphoid tissues or the fragments of mnewborn, and PBMCs. Many kinds of antigens and pathogens were used for the analysis.

4.1. Humanized Mice for Reconstitution of the Human Immune System with Hematopoietic Cells

The humanized mouse system was originally developed to evaluate the multipotency of human HSCs or progenitors. Severely immunodeficient mouse strains, as well as the transplantation techniques, have recently been developed [97–100]. After the discovery of the nonobese diabetic severe combined immunodeficient mouse (NOD-scid) model and its derivatives, transplantation of human hematopoietic stem cells (HSCs) into these mice led to the development of human lymphocytes and myeloid cells which, are localized in the primary and secondary lymphoid tissues of the mouse [101–103]. These mouse models have been used to analyze the differentiation of human hematopoietic and leukemic stem cells [104]. On the other hand, because of the success of humanized monoclonal antibody reagents such as trastuzumab and rituximab, completely human-type antibody production has also been attempted, using these mouse models transplanted with various types of human hematopoietic cells [105].

NOD/Shi-scid-IL2Rγ^{null} (NOG), developed at the Central Institute for Experimental Animals, and NOD scid gamma (NSG), developed at the Jackson Laboratory, are two representatives of severely immunodeficient mouse strains. Both mouse strains have a deficiency in IL-2rgc [97,106–108]. NOG mice possess a truncated IL-2rgc, and NSG mice have a complete deletion of the gene coding for IL-2rgc; the efficiency of the engraftment and the differentiation efficiency are comparable in the two strains. Both of them enabled the development of human T and B cells from human HSCs in a xenogenic environment. However, most of the human B cells differentiated in the mice expressed CD5, a marker of B1 cells, and the specific IgG antibody is not produced [109–113] (Table 1). We reconstructed human immunity in NOG mice transplanted with HSC and immunized with CH401MAP, a specific HER2 peptide antigen for patients with breast cancer, and keyhole limpet hemocyanin (KLH), or toxic shock syndrome toxin-1 (TSST-1), with Freund's complete adjuvant and measured the specific antibody titer by ELISA. As a result, although antigen-specific IgM and nonspecific IgG were detected in the sera, antigen-specific IgG was not detected in mice (Table 1) [114–116]. These mice did not develop a germinal center, which has a structure composed of T, B, and follicular DCs and plays a crucial role in highly specific crass-switched IgG antibody production. The results indicated that human T cells and B cells developed in the mouse environment could not induce cognate interaction, because the T cells are selected for mouse MHC in the thymus.

Table 1. Humanized mice with antigen-specific antibody production.

	Mouse Strain.	Transplanted Tissues	Antigen	Isotype	Reference
SCID-Hu	SCID	human fetal liver and thymic fragments under kidney capsule	pneumococcal vaccine	IgG	McCune JM 1988 [117], Aaberge IS et al., 1992 [118]
Hu-HSC	NOG	HSC(CB/MPB/BM) i.v.	DNP-KLH/CH401MAP/TSST-1	IgM	Matsumura T et al., 2003 [109], Kametani Y et al., 2006 [114]
	NSG; Balb/c-Rag1(-/-) gammac(-/-); C.B-17-scid/bg	HSC(CB/MPB/HFL) i.v.	KLH/inactivated H5N1 influenza virus	IgM, IgG	Lepus CM et al., 2009 [111]
	NOG	CD34 + HSC i.v.	OVA	Igs	Yajima M et al., 2008 [98], Watanabe Y et al., 2009 [112]
	NSG	human CD34 + HSC i.v.	OVA, HIV	IgM, IgG	Wtanabe S et al., 2007 [110], Singh M et al., 2012 [102]
	NOG-HLA-DR4/Ab KO	human CD34 + HSC i.v.	OVA	IgM, IgG	Suzuki M et al., 2012 [119]
	NSG-HIS-CD4/B	human CD34 + HSC i.v.	Plasmodium falciparum, circumsporo-zoite (PFCS) protein	IgG	Huang J et al., 2015 [120]
Hu-PBL	SCID	human PBMC i.v.	xenograft	IgM, IgG	Williams S et al., 1992 [121]
	NOG-IL-4-Tg	human PBMC i.v.	KLH/CH401MAP	IgG	Kametani Y et al., 2017 [116]
	DKO-NOG	human PBMC i.v.	human Liver xenograft,	Igs	Aono S et al. 2018 [122]
BLT	SCID	human fetal liver and thymic fragments under kidney capsule with autologous CD34 + HSC		IgG	McCune JM et al., 1988 [123], Aaberge IS et al., 1992 [118]
	NOD-SCID	human fetal liver and thymic fragments under kidney capsule with autologous CD34 + HSC	HIV-1, WNV envelope protein	IgM, IgG	Biswas et al., 2011 [113]
	NSG	human fetal liver and thymic fragments under kidney capsule with autologous CD34 + HSC	pneumococcal vaccine, Dengue virus infection, Zika virus, HIV -1 gp120	IgM, IgG, IgA	Jaiswal S et al., 2015 [124], Jangalwe S et al., 2016 [125], Schmitt K et al., 2018 [126], Gawron MA et al., 2019 [127]

Representative immune-humanized mouse systems which induced antibody production are shown. The data are based on PubMed, published from 1988 to 2019.

After the first trial with NOG and NSG mouse models, the animals with mouse MHC knockout and HLA transgenic antigen were developed to induce cognate interaction of T cells and B cells. Among them, HLA class I transgenic mice evoked antigen-specific cytotoxic T-cell response against HSV virion protein peptide [128] or WT1 peptide [129]. The success of the reconstitution of human cellular immune response was followed by an adoptive transfer therapy model using the humanized-mouse system [130]. Consequently, the established patient-derived xenograft (PDX) system, which transplants a patient's cancer tissues (minimal standard was reported by Meehan et al. [131]), combined with a patient's T cells, is widely accepted. The detail was intensively reviewed by other researchers [132–134].

On the other hand, the response of HLA class II transgenic mice did not completely mimic the human humoral immunity [119,120,125]. Moreover, mice need to be transplanted with the same HLA-bearing human HSCs, which restrict the samples to be examined. Among them, Ashizawa et al. reported that class I and class II MHC KO NOG mice (NOG dKO) transplanted with human PBMC and tumor cell lines showed higher anticancer effects after PD-1 antibody treatment [135]. In these mouse strains, transplanted tumor cells and immune cells can be engrafted, and the anticancer effect of human immune cells can be observed (reviewed by Chen et al. [95]). The mouse system had an advantage, which is that the restriction of HLA type could be avoided by using PBMC, which contain the same patient's T cells and antigen-presenting cells. However, they did not detect anticancer antibody production in this study.

Currently, various transgenic mouse strains expressing human cytokines and surface antigens, along with more severely immunodeficient mouse strains, are being developed to transplant human hematopoietic cells (HSC or PBMC). The category of newly-established mouse system includes myeloid cell development, cancer immunotherapy model, allergy model, and graft-versus-host disease (GVHD) model.

Another humanized mouse model, called BLT mice, has been reported. In this mouse model, immunodeficient mice are co-transplanted with human fetal liver and thymus tissues, along with autologous CD34 + HSCs. This mouse system is a modification of the SCID-Hu mice developed by MacCune [113,117,123]. In these mice, antigen-specific antibody production was partially achieved, and experiments on infection with bacteria or viruses were conducted [118]. Severely immunodeficient NSG mice are used to establish NSG–BLT mice [136]. A modified NSG mice, in which Human *SCF*, *GM-CSF*, and *IL-3* genes were transduced, was used to establish an improved BLT mouse strain. Based on the NSG mouse strain, human HSCs, fetal liver, and fetal thymus were transplanted, and mice were inoculated with dengue and/or Zika virus. As a result, these mice induced a higher immune response than that of conventional NSG mice, although graft-versus-host disease (GVHD) could not be avoided [124,126,127]. However, because of a serious ethical problem, Japanese researchers are unable to establish the BLT mouse system. The BLT model system succeeded in the induction of the cytotoxic immune response with no mature humoral immunity, maybe because the cytotoxicity is too high to maintain the antibody production (discussed in [94]).

Collectively, many of the strains support the differentiation of various hematopoietic cell lineages from human HSCs. Moreover, PBMC engrafts in the mice and can reconstitute human cellular immunity. However, human humoral immune response in a mouse model still needs further improvement: it is impossible, so far, to reconstruct the immune condition involving humoral immunity of various patients.

4.2. Humanized Mouse System to Evaluate Antigen-Specific Antibody Production

It is difficult to completely develop humoral immunity in humanized mice because of the reasons exposed above. While T cell–B cell interaction needs cognate interaction, humans have a large variety of HLA types, and it is difficult to cover all the HLA types present in a patient blood. Immunodeficient mice transplanted with PBMCs are promising tools to evaluate human immune responses to vaccines, compared to the HSC-transplanting mouse system. However, these mice usually develop severe GVHD [137]. With GVHD, mice develop a large amount of activated T cells, while B cells are decreased in parallel, and there is no humoral immune response. Therefore, it is difficult to evaluate the production of antigen-specific IgG production after antigen immunization in those mice. To evaluate antigen-specific IgG responses in PBMC-transplanted immunodeficient mice, we developed a novel NOD/Shi-scid-IL2rg^{null} (NOG) mouse strain that systemically expresses the human IL-4 gene (NOG-hIL-4-Tg) [116]. After human PBMC transplantation, GVHD symptoms were significantly suppressed in the Tg NOG, as compared to conventional NOG mice. In the kinetic analyses of human leukocytes, long-term engraftment of human T cells has been observed in peripheral blood of NOG-hIL-4-Tg, and then CD4+ T cells dominantly proliferated rather than CD8+ T cells. Furthermore, these CD4+ T cells produced large amounts of IL-4 but suppressed IFN-g expression, resulting in long-term suppression of GVHD. Most of the human B cells detected in the transplanted mice showed a plasmablast phenotype. Vaccination with HER2 multiple antigen peptide (CH401MAP) or keyhole limpet hemocyanin (KLH) successfully induced antigen-specific IgG production in PBMC-transplanted NOG-hIL-4-Tg. The HLA haplotype of donor PBMC might not be relevant to the ability of an antibody secretion after immunization. The reason why NOG-hIL-4-Tg retain B cells and succeeded in the specific antibody production was examined, and we found that the engrafted human lymphocytes decreased glucocorticoid receptor expression, which dampens the humoral immunity [138].

This evidence suggests that the PBMC-transplanted NOG-hIL-4-Tg mouse system is an effective tool to evaluate the production of antigen-specific IgG antibodies, following vaccination in individual cancer patients [116]. The mouse system can be used for the evaluation of the effect of ICIs on antibody production in the presence of human PBMCs, as well.

Of course, the vaccination is not limited to cancer vaccines. As plasmablasts are efficiently developed, the evaluation of vaccines against highly deleterious pathogens, such as Ebola virus, may become possible. Moreover, the donors recovered from such serious infectious disease may keep their memory B cells against the pathogen. Therefore, the transplantation of the PBMCs may develop plasma cells that secrete effective antipathogen antibodies. If we establish the technology for monoclonal antibody preparation, we may obtain the monoclonal antibody reagents for the treatment of such deleterious infectious diseases.

The humanized mouse systems discussed are summarized in Table 1.

5. Future Perspectives

Because the efficacy of the peptide vaccine is influenced by the immune-cell environment and the patient's body fluid content, we need to evaluate vaccines by constructing patient-mimicking conditions. If we can establish patient-PBMC-based check systems using the humanized mouse model for vaccination and additional reagents, we may check the vaccination efficiency, ICI, and IAEs at the same time. If those goals are achieved, they may enable a promising personalized medicine, such as in the case of the use of the mixed lymphocyte reaction for blood-type examination before transplantation. Therefore, it is urgent to develop humanized mice which reconstitute not only human immune cells but the environment of the actual patient. By using the PBMC-based humanized mouse system, various vaccines can be evaluated for their efficacy. We need to improve the humanized mouse system to fine-tune the peptide design for vaccine development.

Author Contributions: Conceptualization, Y.K., T.S., Y.T. and R.I.; analysis, Y.O., S.O., B.T., A.Y. and Y.K. drafted the work.

Funding: The NOG researchs were supported by Japan Society for the Promotion of Science by a Grant-in-Aid for Scientific Research (ITO) (S) [grant number 22220007] to MI; a Grant-in-Aid for Scientific Research (Kametani) (B) [grant number 17H03571] to YK; a Tokai University Grant-in-Aid to YK (2013–2014); and the MEXT-Supported Program for the Strategic Research Foundation at Private Universities (2012–2016).

Acknowledgments: We thank Yumiko Nakagawa for her excellent animal care skills. We thank the members of the Teaching and Research Support Center in the Tokai University School of Medicine for their technical skills.

Abbreviations

PBMC	peripheral blood mononuclear cell
MHC	major histocompatibility complex
DC	dendritic cell
irAE	immune-related adverse events
ICI	immune checkpoint inhibitors
HER2	Human epidermal growth factor receptor 2
IHC	immunohistochemistry
PD-L1	programmed-death-L1
NOD scid	nonobese diabetic severe combined immunodeficient
HSC	hematopoietic stem cells
NOG	NOD/Shi-scid-IL2Rγ^{null}
NSG	NOD scid gamma
GVHD	graft-versus-host disease

References

1. Marshall, H.; Djamgoz, M. Immuno-Oncology: Emerging Targets and Combination Therapies. *Front. Oncol.* **2018**, *8*, 315. [CrossRef] [PubMed]
2. Wilkinson, K.; Wei, Y.; Szwajcer, A.; Rabbani, R.; Zarychanski, R.; Abou-Setta, A.; Mahmud, S. Efficacy and safety of high-dose influenza vaccine in elderly adults: A systematic review and meta-analysis. *Vaccine* **2017**, *35*, 2775–2780. [CrossRef] [PubMed]
3. Nakashima, K.; Aoshima, M.; Ohfuji, S.; Suzuki, K.; Katsurada, M.; Katsurada, N.; Misawa, M.; Otsuka, Y.; Kondo, K.; Hirota, Y. Immunogenicity of trivalent influenza vaccine in patients with lung cancer undergoing anticancer chemotherapy. *Hum. Vaccin Immunother.* **2017**, *13*, 543–550. [CrossRef] [PubMed]
4. Waqar, S.; Boehmer, L.; Morgensztern, D.; Wang-Gillam, A.; Sorscher, S.; Lawrence, S.; Gao, F.; Guebert, K.; Williams, K.; Govindan, R. Immunogenicity of Influenza Vaccination in Patients With Cancer. *Am. J. Clin. Oncol.* **2018**, *41*, 248–253. [CrossRef]
5. Andreatta, M.; Karosiene, E.; Rasmussen, M.; Stryhn, A.; Buus, S.; Nielsen, M. Accurate pan-specific prediction of peptide-MHC class II binding affinity with improved binding core identification. *Immunogenetics* **2015**, *67*, 641–650. [CrossRef]
6. Andreatta, M.; Nielsen, M. Gapped sequence alignment using artificial neural networks: Application to the MHC class I system. *Bioinformatics* **2016**, *32*, 511–517. [CrossRef]
7. Rasmussen, M.; Fenoy, E.; Harndahl, M.; Kristensen, A.; Nielsen, I.; Nielsen, M.; Buus, S. Pan-Specific Prediction of Peptide-MHC Class I Complex Stability, a Correlate of T Cell Immunogenicity. *J. Immunol.* **2016**, *197*, 1517–1524. [CrossRef]
8. Wang, J.; Ji, X.; Ou, G.; Su, P.; Liu, Z. LA-B*46:40:02, a novel HLA-B*46 allele identified in a Chinese individual by sequence-based typing. *Hla* **2016**, *87*, 462–464. [CrossRef]
9. Li, J.; Zhang, X.; Lin, F.; Zhang, K.; Li, X.F. Identification of the novel HLA allele, HLA-B*40:06:07, by sequence-based typing. *Hla* **2018**, *92*, 326–327. [CrossRef]
10. Proust, B.; Masson, D.; Proust, B.; Baudron, M.; Dehaut, F. Identification of a novel HLA allele, HLA-B*08:163, in a platelet donor. *Hla* **2016**, *88*, 263–264. [CrossRef]
11. Moyle, P.; Toth, I. Self-adjuvanting lipopeptide vaccines. *Curr. Med. Chem.* **2008**, *15*, 506–516. [CrossRef] [PubMed]
12. Pasquevich, K.; García Samartino, C.; Coria, L.; Estein, S.; Zwerdling, A.; Ibañez, A.; Barrionuevo, P.; Oliveira, F.; Carvalho, N.; Borkowski, J.; et al. The protein moiety of Brucella abortus outer membrane protein 16 is a new bacterial pathogen-associated molecular pattern that activates dendritic cells in vivo, induces a Th1 immune response, and is a promising self-adjuvanting vaccine against systemic and oral acquired brucellosis. *J. Immunol.* **2010**, *184*, 5200–5212.
13. Shen, K.; Chang, L.; Leng, C.; Liu, S. Self-adjuvanting lipoimmunogens for therapeutic HPV vaccine development: Potential clinical impact. *Expert Rev. Vaccines* **2015**, *14*, 383–394. [CrossRef] [PubMed]

14. Zeng, B.; Middelberg, A.; Gemiarto, A.; MacDonald, K.; Baxter, A.; Talekar, M.; Moi, D.; Tullett, K.; Caminschi, I.; Lahoud, M.; et al. Self-adjuvanting nanoemulsion targeting dendritic cell receptor Clec9A enables antigen-specific immunotherapy. *J. Clin. Investig.* **2018**, *128*, 1971–1984. [CrossRef] [PubMed]

15. Wang, R. Human tumor antigens: Implications for cancer vaccine development. *J. Mol. Med.* **1999**, *77*, 640–655. [CrossRef]

16. Bezu, L.; Kepp, O.; Cerrato, G.; Pol, J.; Fucikova, J.; Spisek, R.; Zitvogel, L.; Kroemer, G.; Galluzzi, L. Trial watch: Peptide-based vaccines in anticancer therapy. *Oncoimmunology* **2018**, *7*, e1511506. [CrossRef]

17. Wang, Y.; Yang, L.; Zuo, J. Recent developments in antivirals against hepatitis B virus. *Virus Res.* **2016**, *213*, 205–213. [CrossRef]

18. Kanduc, D.; Shoenfeld, Y. From HBV to HPV: Designing vaccines for extensive and intensive vaccination campaigns worldwide. *Autoimmun. Rev.* **2016**, *15*, 1054–1061. [CrossRef]

19. Cheever, M.; Higano, C. PROVENGE (Sipuleucel-T) in prostate cancer: The first FDA-approved therapeutic cancer vaccine. *Clin. Cancer Res.* **2011**, *17*, 3520–3526. [CrossRef]

20. Guevara, M.; Persano, S.; Persano, F. Lipid-Based Vectors for Therapeutic mRNA-Based Anti-Cancer Vaccines. *Curr. Pharm. Des.* **2019**, *25*, 1443–1454. [CrossRef]

21. Park, S.; Kim, D.; Wu, G.; Jung, H.; Park, J.; Kwon, H.; Lee, Y. A peptide-CpG-DNA-liposome complex vaccine targeting TM4SF5 suppresses growth of pancreatic cancer in a mouse allograft model. *Onco Targets Ther.* **2018**, *11*, 8655–8672. [CrossRef] [PubMed]

22. Schwartzentruber, D.; Lawson, D.; Richards, J.; Conry, R.; Miller, D.; Treisman, J.; Gailani, F.; Riley, L.; Conlon, K.; Pockaj, B.; et al. gp100 peptide vaccine and interleukin-2 in patients with advanced melanoma. *N. Engl. J. Med.* **2011**, *364*, 2119–2127. [CrossRef] [PubMed]

23. Wargo, J.; Reddy, S.; Reuben, A.; Sharma, P. Monitoring immune responses in the tumor microenvironment. *Curr. Opin. Immunol.* **2016**, *41*, 23–31. [CrossRef] [PubMed]

24. Desjarlais, J.; Lazar, G.; Zhukovsky, E.; Chu, S. Optimizing engagement of the immune system by anti-tumor antibodies: An engineer's perspective. *Drug Discov. Today* **2007**, *12*, 898–910. [CrossRef] [PubMed]

25. Wong, K.; Li, W.; Mooney, D.; Dranoff, G. Advances in Therapeutic Cancer Vaccines. *Adv. Immunol.* **2016**, *130*, 191–249.

26. Papayannopoulos, V. Neutrophil extracellular traps in immunity and disease. *Nat. Rev. Immunol.* **2018**, *18*, 134–147. [CrossRef]

27. Skopelja-Gardner, S.; Jones, J.; Rigby, W. "NETtling" the host: Breaking of tolerance in chronic inflammation and chronic infection. *J. Autoimmun.* **2018**, *88*, 1–10. [CrossRef]

28. Steinman, R.; Nussenzweig, M. Avoiding horror autotoxicus: The importance of dendritic cells in peripheral T cell tolerance. *Proc. Natl. Acad. Sci. USA* **2002**, *99*, 351–358. [CrossRef]

29. Abiko, K.; Matsumura, N.; Hamanishi, J.; Horikawa, N.; Murakami, R.; Yamaguchi, K.; Yoshioka, Y.; Baba, T.; Konishi, I.; Mandai, M. IFN-γ from lymphocytes induces PD-L1 expression and promotes progression of ovarian cancer. *Br. J. Cancer* **2015**, *112*, 150–159. [CrossRef]

30. Taube, J.; Anders, R.; Young, G.; Xu, H.; Sharma, R.; McMiller, T.; Chen, S.; Klein, A.; Pardoll, D.; Topalian, S.; et al. Colocalization of inflammatory response with B7-h1 expression in human melanocytic lesions supports an adaptive resistance mechanism of immune escape. *Sci. Transl. Med.* **2012**, *4*, 127ra37. [CrossRef]

31. Mazanet, M.; Hughes, C. B7-H1 is expressed by human endothelial cells and suppresses T cell cytokine synthesis. *J. Immunol.* **2002**, *169*, 3581–3588. [CrossRef] [PubMed]

32. Brown, J.; Dorfman, D.; Ma, F.; Sullivan, E.; Munoz, O.; Wood, C.; Greenfield, E.; Freeman, G. Blockade of programmed death-1 ligands on dendritic cells enhances T cell activation and cytokine production. *J. Immunol.* **2003**, *170*, 1257–1266. [CrossRef] [PubMed]

33. Wintterle, S.; Schreiner, B.; Mitsdoerffer, M.; Schneider, D.; Chen, L.; Meyermann, R.; Weller, M.; Wiendl, H. Expression of the B7-related molecule B7-H1 by glioma cells: A potential mechanism of immune paralysis. *Cancer Res.* **2003**, *63*, 7462–7467. [PubMed]

34. Schoop, R.; Wahl, P.; Le Hir, M.; Heemann, U.; Wang, M.; Wüthrich, R. Suppressed T-cell activation by IFN-gamma-induced expression of PD-L1 on renal tubular epithelial cells. *Nephrol. Dial. Transplant.* **2004**, *19*, 2713–2720. [CrossRef] [PubMed]

35. de Kleijn, S.; Langereis, J.; Leentjens, J.; Kox, M.; Netea, M.; Koenderman, L.; Ferwerda, G.; Pickkers, P.; Hermans, P. IFN-γ-stimulated neutrophils suppress lymphocyte proliferation through expression of PD-L1. *PLoS ONE* **2013**, *8*, e72249. [CrossRef]

36. Garcia-Diaz, A.; Shin, D.; Moreno, B.; Saco, J.; Escuin-Ordinas, H.; Rodriguez, G.; Zaretsky, J.; Sun, L.; Hugo, W.; Wang, X.; et al. Interferon Receptor Signaling Pathways Regulating PD-L1 and PD-L2 Expression. *Cell Rep.* **2017**, *19*, 1189–1201. [CrossRef]

37. Lim, S.; Li, C.; Xia, W.; Cha, J.; Chan, L.; Wu, Y.; Chang, S.; Lin, W.; Hsu, J.; Hsu, Y.; et al. Deubiquitination and Stabilization of PD-L1 by CSN5. *Cancer Cell* **2016**, *30*, 925–939. [CrossRef]

38. Grinberg-Bleyer, Y.; Ghosh, S. A Novel Link between Inflammation and Cancer. *Cancer Cell* **2016**, *30*, 829–830. [CrossRef]

39. Kondo, A.; Yamashita, T.; Tamura, H.; Zhao, W.; Tsuji, T.; Shimizu, M.; Shinya, E.; Takahashi, H.; Tamada, K.; Chen, L.; et al. Interferon-gamma and tumor necrosis factor-alpha induce an immunoinhibitory molecule, B7-H1, via nuclear factor-kappaB activation in blasts in myelodysplastic syndromes. *Blood* **2010**, *116*, 1124–1131. [CrossRef]

40. Quandt, D.; Jasinski-Bergner, S.; Müller, U.; Schulze, B.; Seliger, B. Synergistic effects of IL-4 and TNFα on the induction of B7-H1 in renal cell carcinoma cells inhibiting allogeneic T cell proliferation. *J. Transl. Med.* **2014**, *12*, 151. [CrossRef]

41. Li, N.; Wang, J.; Zhang, N.; Zhuang, M.; Zong, Z.; Zou, J.; Li, G.; Wang, X.; Zhou, H.; Zhang, L.; et al. Cross-talk between TNF-α and IFN-γ signaling in induction of B7-H1 expression in hepatocellular carcinoma cells. *Cancer Immunol. Immunother.* **2018**, *67*, 271–283. [CrossRef] [PubMed]

42. Weyand, C.; Berry, G.; Goronzy, J. The immunoinhibitory PD-1/PD-L1 pathway in inflammatory blood vessel disease. *J. Leukoc. Biol.* **2018**, *103*, 565–575. [CrossRef] [PubMed]

43. Kythreotou, A.; Siddique, A.; Mauri, F.; Bower, M.; Pinato, D. PD-L1. *J. Clin. Pathol.* **2018**, *71*, 189–194. [CrossRef]

44. Butte, M.; Peña-Cruz, V.; Kim, M.; Freeman, G.; Sharpe, A. Interaction of human PD-L1 and B7-1. *Mol. Immunol.* **2008**, *45*, 3567–3572. [CrossRef] [PubMed]

45. Xiao, Y.; Yu, S.; Zhu, B.; Bedoret, D.; Bu, X.; Francisco, L.; Hua, P.; Duke-Cohan, J.; Umetsu, D.; Sharpe, A.; et al. RGMb is a novel binding partner for PD-L2 and its engagement with PD-L2 promotes respiratory tolerance. *J. Exp. Med.* **2014**, *211*, 943–959. [CrossRef]

46. Balar, A.; Weber, J. PD-1 and PD-L1 antibodies in cancer: Current status and future directions. *Cancer Immunol. Immunother.* **2017**, *66*, 551–564. [CrossRef]

47. Agata, Y.; Kawasaki, A.; Nishimura, H.; Ishida, Y.; Tsubata, T.; Yagita, H.; Honjo, T. Expression of the PD-1 antigen on the surface of stimulated mouse T and B lymphocytes. *Int. Immunol.* **1996**, *8*, 765–772. [CrossRef]

48. Haynes, N.; Allen, C.; Lesley, R.; Ansel, K.; Killeen, N.; Cyster, J. Role of CXCR5 and CCR7 in follicular Th cell positioning and appearance of a programmed cell death gene-1high germinal center-associated subpopulation. *J. Immunol.* **2007**, *179*, 5099–5108. [CrossRef]

49. Freeman, G.; Long, A.; Iwai, Y.; Bourque, K.; Chernova, T.; Nishimura, H.; Fitz, L.; Malenkovich, N.; Okazaki, T.; Byrne, M.; et al. Engagement of the PD-1 immunoinhibitory receptor by a novel B7 family member leads to negative regulation of lymphocyte activation. *J. Exp. Med.* **2000**, *192*, 1027–1034. [CrossRef]

50. Yamazaki, T.; Akiba, H.; Iwai, H.; Matsuda, H.; Aoki, M.; Tanno, Y.; Shin, T.; Tsuchiya, H.; Pardoll, D.; Okumura, K.; et al. Expression of programmed death 1 ligands by murine T cells and APC. *J. Immunol.* **2002**, *169*, 5538–5545. [CrossRef]

51. D'Arrigo, P.; Tufano, M.; Rea, A.; Vigorito, V.; Novizio, N.; Russo, S.; Romano, M.; Romano, S. Manipulation of the immune system for cancer defeat: A focus on the T cell inhibitory checkpoint molecules. *Curr. Med. Chem.* **2018**. [CrossRef] [PubMed]

52. De Sousa Linhares, A.; Leitner, J.; Grabmeier-Pfistershammer, K.; Steinberger, P. Not All Immune Checkpoints Are Created Equal. *Front. Immunol.* **2018**, *9*, 1909. [CrossRef] [PubMed]

53. Wolchok, J.; Kluger, H.; Callahan, M.; Postow, M.; Rizvi, N.; Lesokhin, A.; Segal, N.; Ariyan, C.; Gordon, R.; Reed, K.; et al. Nivolumab plus ipilimumab in advanced melanoma. *N. Engl. J. Med.* **2013**, *369*, 122–133. [CrossRef] [PubMed]

54. Carlino, M.; Long, G.; Schadendorf, D.; Robert, C.; Ribas, A.; Richtig, E.; Nyakas, M.; Caglevic, C.; Tarhini, A.; Blank, C.; et al. Outcomes by line of therapy and programmed death ligand 1 expression in patients with advanced melanoma treated with pembrolizumab or ipilimumab in KEYNOTE-006: A randomised clinical trial. *Eur. J. Cancer* **2018**, *101*, 236–243. [CrossRef]

55. Daud, A.; Wolchok, J.; Robert, C.; Hwu, W.; Weber, J.; Ribas, A.; Hodi, F.; Joshua, A.; Kefford, R.; Hersey, P.; et al. Programmed Death-Ligand 1 Expression and Response to the Anti-Programmed Death 1 Antibody Pembrolizumab in Melanoma. *J. Clin. Oncol.* **2016**, *34*, 4012–4109. [CrossRef]

56. Festino, L.; Botti, G.; Lorigan, P.; Masucci, G.; Hipp, J.; Horak, C.; Melero, I.; Ascierto, P. Cancer Treatment with Anti-PD-1/PD-L1 Agents: Is PD-L1 Expression a Biomarker for Patient Selection? *Drugs* **2016**, *76*, 925–945. [CrossRef]

57. Hodi, F.; Chiarion-Sileni, V.; Gonzalez, R.; Grob, J.; Rutkowski, P.; Cowey, C.; Lao, C.; Schadendorf, D.; Wagstaff, J.; Dummer, R.; et al. Nivolumab plus ipilimumab or nivolumab alone versus ipilimumab alone in advanced melanoma (CheckMate 067): 4-year outcomes of a multicentre, randomised, phase 3 trial. *Lancet Oncol.* **2018**, *19*, 1480–1492. [CrossRef]

58. Conroy, J.; Pabla, S.; Nesline, M.; Glenn, S.; Papanicolau-Sengos, A.; Burgher, B.; Andreas, J.; Giamo, V.; Wang, Y.; Lenzo, F.; et al. Next generation sequencing of PD-L1 for predicting response to immune checkpoint inhibitors. *J. Immunother. Cancer* **2019**, *7*, 18. [CrossRef]

59. Robert, C.; Long, G.; Brady, B.; Dutriaux, C.; Maio, M.; Mortier, L.; Hassel, J.; Rutkowski, P.; McNeil, C.; Kalinka-Warzocha, E.; et al. Nivolumab in previously untreated melanoma without BRAF mutation. *N. Engl. J. Med.* **2015**, *372*, 320–330. [CrossRef]

60. Kambayashi, Y.; Fujimura, T.; Hidaka, T.; Aiba, S. Biomarkers for Predicting Efficacies of Anti-PD1 Antibodies. *Front. Med.* **2019**, *6*, 174. [CrossRef]

61. Franzen, A.; Vogt, T.; Müller, T.; Dietrich, J.; Schröck, A.; Golletz, C.; Brossart, P.; Bootz, F.; Landsberg, J.; Kristiansen, G.; et al. PD-L1 (CD274) and PD-L2 (PDCD1LG2) promoter methylation is associated with HPV infection and transcriptional repression in head and neck squamous cell carcinomas. *Oncotarget* **2017**, *9*, 641–650. [CrossRef] [PubMed]

62. Said, E.; Al-Reesi, I.; Al-Riyami, M.; Al-Naamani, K.; Al-Sinawi, S.; Al-Balushi, M.; Koh, C.; Al-Busaidi, J.; Idris, M.; Al-Jabri, A. A Potential Inhibitory Profile of Liver CD68+ Cells during HCV Infection as Observed by an Increased CD80 and PD-L1 but Not CD86 Expression. *PLoS ONE* **2016**, *11*, e0153191. [CrossRef] [PubMed]

63. Okuma, Y.; Hishima, T.; Kashima, J.; Homma, S. High PD-L1 expression indicates poor prognosis of HIV-infected patients with non-small cell lung cancer. *Cancer Immunol. Immunother.* **2018**, *67*, 495–505. [CrossRef] [PubMed]

64. Choschzick, M.; Gut, A.; Fink, D. PD-L1 receptor expression in vulvar carcinomas is HPV-independent. *Virchows Arch.* **2018**, *473*, 513–516. [CrossRef] [PubMed]

65. Green, M.; Rodig, S.; Juszczynski, P.; Ouyang, J.; Sinha, P.; O'Donnell, E.; Neuberg, D.; Shipp, M. Constitutive AP-1 activity and EBV infection induce PD-L1 in Hodgkin lymphomas and posttransplant lymphoproliferative disorders: Implications for targeted therapy. *Clin. Cancer Res.* **2012**, *18*, 1611–1618. [CrossRef] [PubMed]

66. Severa, M.; Giacomini, E.; Gafa, V.; Anastasiadou, E.; Rizzo, F.; Corazzari, M.; Romagnoli, A.; Trivedi, P.; Fimia, G.; Coccia, E. EBV stimulates TLR- and autophagy-dependent pathways and impairs maturation in plasmacytoid dendritic cells: Implications for viral immune escape. *Eur. J. Immunol.* **2013**, *43*, 147–158. [CrossRef] [PubMed]

67. Fang, W.; Zhang, J.; Hong, S.; Zhan, J.; Chen, N.; Qin, T.; Tang, Y.; Zhang, Y.; Kang, S.; Zhou, T.; et al. EBV-driven LMP1 and IFN-γ up-regulate PD-L1 in nasopharyngeal carcinoma: Implications for oncotargeted therapy. *Oncotarget* **2014**, *5*, 12189–12202. [CrossRef]

68. Outh-Gauer, S.; Alt, M.; Le Tourneau, C.; Augustin, J.; Broudin, C.; Gasne, C.; Denize, T.; Mirghani, H.; Fabre, E.; Scotte, F.; et al. Immunotherapy in head and neck cancers: A new challenge for immunologists, pathologists and clinicians. *Cancer Treat. Rev.* **2018**, *665*, 54–64. [CrossRef]

69. Sun, C.; Lan, P.; Han, Q.; Huang, M.; Zhang, Z.; Xu, G.; Song, J.; Wang, J.; Wei, H.; Zhang, J.; et al. Oncofetal gene SALL4 reactivation by hepatitis B virus counteracts miR-200c in PD-L1-induced T cell exhaustion. *Nat. Commun.* **2018**, *9*, 1241. [CrossRef]

70. Balsitis, S.; Gali, V.; Mason, P.; Chaniewski, S.; Levine, S.; Wichroski, M.; Feulner, M.; Song, Y.; Granaldi, K.; Loy, J.; et al. Safety and efficacy of anti-PD-L1 therapy in the woodchuck model of HBV infection. *PLoS ONE* **2018**, *13*, e0190058. [CrossRef]

71. Park, C.; Cho, J.; Lee, J.; Kang, S.; An, J.; Choi, M.; Lee, J.; Sohn, T.; Bae, J.; Kim, S.; et al. Host immune response index in gastric cancer identified by comprehensive analyses of tumor immunity. *Oncoimmunology* **2017**, *6*, e1356150. [CrossRef] [PubMed]

72. Fouad, H.; Raziky, M.; Aziz, R.; Sabry, D.; Aziz, G.; Ewais, M.; Sayed, A. Dendritic cell co-stimulatory and co-inhibitory markers in chronic HCV: An Egyptian study. *World J. Gastroenterol.* **2013**, *19*, 7711–7718. [CrossRef] [PubMed]

73. Choi, Y.; Jin, N.; Kelly, F.; Sakthivel, S.; Yu, T. Elevation of Alanine Aminotransferase Activity Occurs after Activation of the Cell-Death Signaling Initiated by Pattern-Recognition Receptors but before Activation of Cytolytic Effectors in NK or CD8+ T Cells in the Liver During Acute HCV Infection. *PLoS ONE* **2016**, *11*, e0165533. [CrossRef] [PubMed]

74. Ojiro, K.; Qu, X.; Cho, H.; Park, J.; Vuidepot, A.; Lissin, N.; Molloy, P.; Bennett, A.; Jakobsen, B.; Kaplan, D.; et al. Modulation of Hepatitis C Virus-Specific CD8 Effector T-Cell Function with Antiviral Effect in Infectious Hepatitis C Virus Coculture Model. *J. Virol.* **2017**, *91*, e02129-16. [CrossRef]

75. Abdellatif, H.; Shiha, G. PD-L1 Expression on Circulating CD34 + Hematopoietic Stem Cells Closely Correlated with T-cell Apoptosis in Chronic Hepatitis C Infected Patients. *Int. J. Stem. Cells* **2018**, *11*, 78–86. [CrossRef]

76. Domblides, C.; Antoine, M.; Hamard, C.; Rabbe, N.; Rodenas, A.; Vieira, T.; Crequit, P.; Cadranel, J.; Lavolé, A.; Wislez, M. Nonsmall cell lung cancer from HIV-infected patients expressed programmed cell death-ligand 1 with marked inflammatory infiltrates. *AIDS* **2018**, *32*, 461–468.

77. Muthumani, K.; Shedlock, D.; Choo, D.; Fagone, P.; Kawalekar, O.; Goodman, J.; Bian, C.; Ramanathan, A.; Atman, P.; Tebas, P.; et al. HIV-mediated phosphatidylinositol 3-kinase/serine-threonine kinase activation in APCs leads to programmed death-1 ligand upregulation and suppression of HIV-specific CD8 T cells. *J. Immunol.* **2011**, *187*, 29322–29343. [CrossRef]

78. Meier, A.; Bagchi, A.; Sidhu, H.; Alter, G.; Suscovich, T.; Kavanagh, D.; Streeck, H.; Brockman, M.; LeGall, S.; Hellman, J.; et al. Upregulation of PD-L1 on monocytes and dendritic cells by HIV-1 derived TLR ligands. *AIDS* **2008**, *22*, 655–658. [CrossRef]

79. Planès, R.; BenMohamed, L.; Leghmari, K.; Delobel, P.; Izopet, J.; Bahraoui, E. HIV-1 Tat protein induces PD-L1 (B7-H1) expression on dendritic cells through tumor necrosis factor alpha- and toll-like receptor 4-mediated mechanisms. *J. Virol.* **2014**, *88*, 6672–6689. [CrossRef]

80. Hong, A.; Vilain, R.; Romanes, S.; Yang, J.; Smith, E.; Jones, D.; Scolyer, R.; Lee, C.; Zhang, M.; Rose, B. PD-L1 expression in tonsillar cancer is associated with human papillomavirus positivity and improved survival: Implications for anti-PD1 clinical trials. *Oncotarget* **2016**, *7*, 77010–77020. [CrossRef]

81. Lyford-Pike, S.; Peng, S.; Young, G.; Taube, J.; Westra, W.; Akpeng, B.; Bruno, T.; Richmon, J.; Wang, H.; Bishop, J.; et al. Evidence for a role of the PD-1:PD-L1 pathway in immune resistance of HPV-associated head and neck squamous cell carcinoma. *Cancer Res.* **2013**, *73*, 1733–1741. [CrossRef] [PubMed]

82. Yang, W.; Song, Y.; Lu, Y.; Sun, J.; Wang, H. Increased expression of programmed death (PD)-1 and its ligand PD-L1 correlates with impaired cell-mediated immunity in high-risk human papillomavirus-related cervical intraepithelial neoplasia. *Immunology* **2013**, *139*, 513–522. [CrossRef] [PubMed]

83. Lin, P.; Cheng, Y.; Wu, D.; Huang, Y.; Lin, H.; Chen, C.; Lee, H. A combination of anti-PD-L1 mAb plus Lm-LLO-E6 vaccine efficiently suppresses tumor growth and metastasis in HPV-infected cancers. *Cancer Med.* **2017**, *6*, 2052–2062. [CrossRef] [PubMed]

84. Lipson, E.; Vincent, J.; Loyo, M.; Kagohara, L.; Luber, B.; Wang, H.; Xu, H.; Nayar, S.; Wang, T.; Sidransky, D.; et al. PD-L1 expression in the Merkel cell carcinoma microenvironment: Association with inflammation, Merkel cell polyomavirus and overall survival. *Cancer Immunol. Res.* **2013**, *1*, 54–63. [CrossRef] [PubMed]

85. Ikebuchi, R.; Konnai, S.; Shirai, T.; Sunden, Y.; Murata, S.; Onuma, M.; Ohashi, K. Increase of cells expressing PD-L1 in bovine leukemia virus infection and enhancement of anti-viral immune responses in vitro via PD-L1 blockade. *Vet. Res.* **2011**, *42*, 103. [CrossRef]

86. Host, K.; Jacobs, S.; West, J.; Zhang, Z.; Costantini, L.; Stopford, C.; Dittmer, D.; Damania, B. Kaposi's Sarcoma-Associated Herpesvirus Increases PD-L1 and Proinflammatory Cytokine Expression in Human Monocytes. *MBio* **2017**, *8*, e00917-17. [CrossRef]

87. Good-Jacobson, K.; Szumilas, C.; Chen, L.; Sharpe, A.; Tomayko, M.; Shlomchik, M. PD-1 regulates germinal center B cell survival and the formation and affinity of long-lived plasma cells. *Nat. Immunol.* **2010**, *11*, 535–542. [CrossRef]

88. Lino, A.; Dang, V.; Lampropoulou, V.; Welle, A.; Joedicke, J.; Pohar, J.; Simon, Q.; Thalmensi, J.; Baures, A.; Flühler, V.; et al. LAG-3 Inhibitory Receptor Expression Identifies Immunosuppressive Natural Regulatory Plasma Cells. *Immunity* **2018**, *49*, 1. [CrossRef]

89. Cain, D.; Cidlowski, J. Immune regulation by glucocorticoids. *Nat. Rev. Immunol.* **2017**, *17*, 233–247. [CrossRef]

90. Franco, L.; Gadkari, M.; Howe, K.; Sun, J.; Kardava, L.; Kumar, P.; Kumari, S.; Hu, Z.; Fraser, I.; Moir, S.; et al. Immune regulation by glucocorticoids can be linked to cell type-dependent transcriptional responses. *J. Exp. Med.* **2019**, *216*, 384–406. [CrossRef]

91. Obradović, M.; Hamelin, B.; Manevski, N.; Couto, J.; Sethi, A.; Coissieux, M.; Münst, S.; Okamoto, R.; Kohler, H.; Schmidt, A.; et al. Glucocorticoids promote breast cancer metastasis. *Nature* **2019**, *567*, 540–544. [CrossRef] [PubMed]

92. Torino, F.; Barnabei, A.; Paragliola, R.; Marchetti, P.; Salvatori, R.; Corsello, S. Endocrine side-effects of anti-cancer drugs: mAbs and pituitary dysfunction: Clinical evidence and pathogenic hypotheses. *Eur. J. Endocrinol.* **2013**, *169*, R153–R164. [CrossRef] [PubMed]

93. Sznol, M.; Postow, M.; Davies, M.; Pavlick, A.; Plimack, E.; Shaheen, M.; Veloski, C.; Robert, C. Endocrine-related adverse events associated with immune checkpoint blockade and expert insights on their management. *Cancer Treat. Rev.* **2017**, *58*, 70–76. [CrossRef] [PubMed]

94. Kametani, Y.; Miyamoto, A.; Seki, T.; Ito, R.; Habu, S.; Tokuda, Y. The significance of humanized mouse models for the evaluation of the humoral immune response against cancer vaccines. *Pers. Med. Univ.* **2018**, *7*, 13–18. [CrossRef]

95. Chen, Q.; Wang, J.; Liu, W.; Zhao, Y. Cancer Immunotherapies and Humanized Mouse Drug Testing Platforms. *Transl. Oncol.* **2019**, *12*, 987–995. [CrossRef]

96. Ukai, H.; Sumiyama, K.; Ueda, H. Next-generation human genetics for organism-level systems biology. *Curr. Opin. Biotechnol.* **2019**, *58*, 137–145. [CrossRef]

97. Ito, M.; Kobayashi, K.; Nakahata, T. NOD/Shi-scid IL2rgamma(null) (NOG) mice more appropriate for humanized mouse models. *Curr. Top. Microbiol. Immunol.* **2008**, *324*, 53–76.

98. Yajima, M.I.K. Nakagawa A, Watanabe S, Terashima K, Nakamura H, Ito M, Shimizu N, Honda M, Yamamoto N, Fujiwara S, A new humanized mouse model of Epstein-Barr virus infection that reproduces persistent infection, lymphoproliferative disorder, and cell-mediated and humoral immune responses. *J. Infect. Dis.* **2008**, *198*, 673–682.

99. Brehm, M.; Wiles, M.; Greiner, D.; Shultz, L. Generation of improved humanized mouse models for human infectious diseases. *J. Immunol. Methods* **2014**, *410*, 3–17. [CrossRef]

100. Hasgur, S.; Aryee, K.; Shultz, L.; Greiner, D.; Brehm, M. Generation of Immunodeficient Mice Bearing Human Immune Systems by the Engraftment of Hematopoietic Stem Cells. *Methods Mol. Biol.* **2016**, *1438*, 67–78.

101. Willinger, T.; Rongvaux, A.; Takizawa, H.; Yancopoulos, G.; Valenzuela, D.; Murphy, A.; Auerbach, W.; Eynon, E.; Stevens, S.; Manz, M.; et al. Human IL-3/GM-CSF knock-in mice support human alveolar macrophage development and human immune responses in the lung. *Proc. Natl. Acad. Sci. USA* **2011**, *108*, 2390–2395. [CrossRef] [PubMed]

102. Singh, M.; Singh, P.; Gaudray, G.; Musumeci, L.; Thielen, C.; Vaira, D.; Vandergeeten, C.; Delacroix, L.; Van Gulck, E.; Vanham, G.; et al. An improved protocol for efficient engraftment in NOD/LTSZ-SCIDIL-2Rγnull mice allows HIV replication and development of anti-HIV immune responses. *PLoS ONE* **2012**, *7*, e38491. [CrossRef] [PubMed]

103. Ito, R.; Takahashi, T.; Katano, I.; Kawai, K.; Kamisako, T.; Ogura, T.; Ida-Tanaka, M.; Suemizu, H.; Nunomura, S.; Ra, C.; et al. Establishment of a human allergy model using human IL-3/GM-CSF-transgenic NOG mice. *J. Immunol.* **2013**, *191*, 2890–2899. [CrossRef] [PubMed]

104. Goyama, S.; Wunderlich, M.; Mulloy, J. Xenograft models for normal and malignant stem cells. *Blood* **2015**, *125*, 2630–2640. [CrossRef]

105. Villaudy, J.; Schotte, R.; Legrand, N.; Spits, H. Critical assessment of human antibody generation in humanized mouse models. *J. Immunol. Methods* **2014**, *410*, 18–27. [CrossRef]

106. Ito, M.; Hiramatsu, H.; Kobayashi, K.; Suzue, K.; Kawahata, M.; Hioki, K.; Ueyama, Y.; Koyanagi, Y.; Sugamura, K.; Tsuji, K.; et al. NOD/SCID/gamma(c)(null) mouse: An excellent recipient mouse model for engagement of human cells. *Blood* **2002**, *100*, 3175–3182. [CrossRef]

107. Shultz, L.; Ishikawa, F.; Greiner, D. Humanized mice in translational biomedical research. *Nat. Rev. Immunol.* **2007**, *7*, 118–130. [CrossRef]

108. Shultz, L.D.; Lang, P.A.; Christianson, S.W.; Gott, B.; Lyons, B.; Umeda, S.; Leiter, E.; Hesselton, R.; Wagar, E.J.; Leif, J.H.; et al. NOD/LtSz-Rag1null mice: An immunodeficient and radioresistant model for engraftment of human hematolymphoid cells, HIV infection, and adoptive transfer of NOD mouse diabetogenic T cells. *J. Immunol.* **2000**, *164*, 2496–2507. [CrossRef]

109. Matumura, T.; Kametani, Y.; Ando, K.; Hirano, Y.; Katano, I.; Ito, R.; Shiina, M.; Tsukmoto, H.; Saito, Y.; Tokuda, Y.; et al. Functional CD5+ B cells develop predominantly in the spleen of NOD/SCID/gc^{null} (NOG) mice transplanted either with human umbilical cord blood, bone marrow, or mobilized peripheral blood CD34+ cells. *Exp. Hematol.* **2003**, *31*, 789–797. [CrossRef]

110. Watanabe, S.; Terashima, K.; Ohta, S.; Horibata, S.; Yajima, M.; Shiozawa, Y.; Dewan, M.; Yu, Z.; Ito, M.; Morio, T.; et al. Hematopoietic stem cell-engrafted NOD/SCID/IL2Rgamma null mice develop human lymphoid systems and induce long-lasting HIV-1 infection with specific humoral immune responses. *Blood* **2007**, *109*, 212–218. [CrossRef]

111. Lepus, C.; Gibson, T.; Gerber, S.; Kawikova, I.; Szczepanik, M.; Hossain, J.; Ablamunits, V.; Kirkiles-Smith, N.; Herold, K.; Donis, R.; et al. Comparison of human fetal liver, umbilical cord blood, and adult blood hematopoietic stem cell engraftment in NOD-scid/gammac-/-, Balb/c-Rag1-/-gammac-/-, and C.B-17-scid/bg immunodeficient mice. *Hum. Immunol.* **2009**, *70*, 790–802. [CrossRef] [PubMed]

112. Watanabe, Y.; Takahashi, T.; Okajima, A.; Shiokawa, M.; Ishii, N.; Katano, I.; Ito, R.; Ito, M.; Minegishi, M.; Minegishi, N.; et al. The analysis of the functions of human B and T cells in humanized NOD/shi-scid/gammac(null) (NOG) mice (hu-HSC NOG mice). *Int Immunol.* **2009**, *21*, 843–858. [CrossRef] [PubMed]

113. Biswas, S.; Chan, G.H.; Sarkis, P.; Fikrig, E.; Zhu, Q.; Marasco, W. Humoral immune responses in humanized BLT mice immunized with West Nile virus and HIV-1 envelope proteins are largely mediated via human CD5+ B cells. *Immunology* **2011**, *134*, 419–433. [CrossRef] [PubMed]

114. Kametani, Y.; Shiina, M.; Katano, I.; Ito, R.; Ando, K.; Toyama, K.; Tsukamoto, H.; Matsumura, T.; Saito, Y.; Ishikawa, D.; et al. Development of human-human hybridoma from anti-Her-2 peptide-producing B cells in immunized NOG mouse. *Exp. Hematol.* **2006**, *34*, 1240–1248. [CrossRef]

115. Kametani, Y.; Shimada, S.; Mori, S.; Kojima, M.; Ohshima, S.; Kitaura, K.; Matsutani, T.; Okada, Y.; Yahata, T.; Ito, R.; et al. Antibody-secreting plasma cells with unique CD5+IgG+CD21lo phenotype developed in humanized NOG mice. *Clin. Res. Trials* **2016**, *2*, 164–173. [CrossRef]

116. Kametani, Y.; Katano, I.; Miyamoto, A.; Kikuchi, Y.; Ito, R.; Muguruma, Y.; Tsuda, B.; Habu, S.; Tokuda, Y.; Ando, K.; et al. NOG-hIL-4-Tg, a new humanized mouse model for producing tumor antigen-specific IgG antibody by peptide vaccination. *PLoS ONE* **2017**, *12*, e0179239. [CrossRef]

117. McCune, J. Development and applications of the SCID-hu mouse model. *Semin. Immunol.* **1996**, *8*, 187–196. [CrossRef]

118. Aaberge, I.; Michaelsen, T.; Rolstad, A.; Groeng, E.; Solberg, P.; Løvik, M. SCID-Hu mice immunized with a pneumococcal vaccine produce specific human antibodies and show increased resistance to infection. *Infect. Immunol.* **1992**, *60*, 4146–4153.

119. Suzuki, M.; Takahashi, T.; Katano, I.; Ito, R.; Ito, M.; Harigae, H.; Ishii, N.; Sugamura, K. Induction of human humoral immune responses in a novel HLA-DR-expressing transgenic NOD/Shi-scid/γcnull mouse. *Int. Immunol.* **2012**, *24*, 243–252. [CrossRef]

120. Huang, J.; Li, X.; Coelho-dos-Reis, J.; Zhang, M.; Mitchell, R.; Nogueira, R.; Tsao, T.; Noe, A.; Ayala, R.; Sahi, V.; et al. Human immune system mice immunized with Plasmodium falciparum circumsporozoite protein induce protective human humoral immunity against malaria. *J. Immunol. Methods* **2015**, *427*, 42–50. [CrossRef]

121. Williams, S.; Umemoto, T.; Kida, H.; Repasky, E.; Bankert, R. Engraftment of human peripheral blood leukocytes into severe combined immunodeficient mice results in the long term and dynamic production of human xenoreactive antibodies. *J. Immunol.* **1992**, *149*, 2830–2836. [PubMed]

122. Aono, S.; Tatsumi, T.; Yoshioka, T.; Tawara, S.; Nishio, A.; Onishi, Y.; Fukutomi, K.; Nakabori, T.; Kodama, T.; Shigekawa, M.; et al. Immunological responses against hepatitis B virus in human peripheral blood mononuclear cell-engrafted mice. *Biochem. Biophys Res. Commun.* **2018**, *503*, 1457–1464. [CrossRef] [PubMed]

123. McCune, J.; Namikawa, R.; Kaneshima, H.; Shultz, L.; Lieberman, M.; Weissman, I. The SCID-hu mouse: Murine model for the analysis of human hematolymphoid differentiaiton and function. *Schience* **1988**, *241*, 1632–1639. [CrossRef] [PubMed]

124. Jaiswal, S.; Smith, K.; Ramirez, A.; Woda, M.; Pazoles, P.; Shultz, L.; Greiner, D.; Brehm, M.; Mathew, A. Dengue virus infection induces broadly cross-reactive human IgM antibodies that recognize intact virions in humanized BLT-NSG mice. *Exp. Biol. Med.* **2015**, *240*, 67–78. [CrossRef] [PubMed]

125. Jangalwe, S.; Shultz, L.; Mathew, A.; MA, B. Improved B cell development in humanized NOD-*scid IL2Rγ^{null}* mice transgenically expressing human stem cell factor, granulocyte-macrophage colony-stimulating factor and interleukin-3. *Immunol. Inflamm. Dis.* **2016**, *4*, 427–440. [CrossRef]

126. Schmitt, K.; Charlins, P.; Veselinovic, M.; Kinner-Bibeau, L.; Hu, S.; Curlin, J.; Remling-Mulder, L.; Olson, K.; Aboellail, T.; Akkina, R. Zika viral infection and neutralizing human antibody response in a BLT humanized mouse model. *Virology* **2018**, *515*, 235–242. [CrossRef]

127. Gawron, M.; Duval, M.; Carbone, C.; Jaiswal, S.; Wallace, A.; Martin, J.C.; Dauphin, A.; Brehm, M.; Greiner, D.; Shultz, L.; et al. Human Anti-HIV-1 gp120 Monoclonal Antibodies with Neutralizing Activity Cloned from Humanized Mice Infected with HIV-1. *J. Immunol.* **2019**, *202*, 799–804. [CrossRef]

128. Srivastava, R.; Khan, A.; Spencer, D.; Vahed, H.; Lopes, P.; Thai, N.; Wang, C.; Pham, T.; Huang, J.; Scarfone, V.; et al. HLA-A02:01-restricted epitopes identified from the herpes simplex virus tegument protein VP11/12 preferentially recall polyfunctional effector memory CD8+ T cells from seropositive asymptomatic individuals and protect humanized HLA-A*02:01 transgenic mice against ocular herpes. *J. Immunol.* **2015**, *194*, 2232–2248.

129. Najima, Y.; Tomizawa-Murasawa, M.; Saito, Y.; Watanabe, T.; Ono, R.; Ochi, T.; Suzuki, N.; Fujiwara, H.; Ohara, O.; Shultz, L.; et al. Induction of WT1-specific human CD8+ T cells from human HSCs in HLA class I Tg NOD/SCID/IL2rgKO mice. *Blood* **2016**, *127*, 722–734. [CrossRef]

130. Jespersen, H.; Lindberg, M.; Donia, M.; Söderberg, E.; Andersen, R.; Keller, U.; Ny, L.; Svane, I.; Nilsson, L.; Nilsson, J. Clinical responses to adoptive T-cell transfer can be modeled in an autologous immune-humanized mouse model. *Nat. Commun.* **2017**, *8*, 707. [CrossRef]

131. Meehan, T.; Conte, N.; Goldstein, T.; Inghirami, G.; Murakami, M.; Brabetz, S.; Gu, Z.; Wiser, J.; Dunn, P.; Begley, D.; et al. PDX-MI: Minimal Information for Patient-Derived Tumor Xenograft Models. *Cancer Res.* **2017**, *77*, e62–e66. [CrossRef] [PubMed]

132. Okada, S.; Vaeteewoottacharn, K.; Kariya, R. Application of Highly Immunocompromised Mice for the Establishment of Patient-Derived Xenograft (PDX) Models. *Cells* **2019**, *8*, 889. [CrossRef] [PubMed]

133. Shi, J.; Li, Y.; Jia, R.; Fan, X. The fidelity of cancer cells in PDX models: Characteristics, mechanism and clinical significance. *Int. J. Cancer* **2019**. [CrossRef] [PubMed]

134. Marangoni, E.; Poupon, M. Patient-derived tumour xenografts as models for breast cancer drug development. *Curr. Opin. Oncol.* **2014**, *26*, 556–561. [CrossRef]

135. Ashizawa, T.; Iizuka, A.; Nonomura, C.; Kondou, R.; Maeda, C.; Miyata, H.; Sugino, T.; Mitsuya, K.; Hayashi, N.; Nakasu, Y.; et al. Antitumor Effect of Programmed Death-1 (PD-1) Blockade in Humanized the NOG-MHC Double Knockout Mouse. *Clin. Cancer Res.* **2017**, *23*, 149–158. [CrossRef]

136. Joo, S.Y.; Chung, Y.S.; Choi, B.; Kim, M.; Kim, J.H.; Jun, T.G.; Chang, J.; Sprent, J.; Surh, C.D.; Joh, J.W.; et al. Systemic human T cell developmental processes in humanized mice cotransplanted with human fetal thymus/liver tissue and hematopoietic stem cells. *Transplantation* **2012**, *94*, 1095–1102. [CrossRef]

137. Ito, R.; Katano, I.; Kawai, K.; Yagoto, M.; Takahashi, T.; Ka, Y.; Ogura, T.; Takahashi, R.; Ito, M. A Novel Xenogeneic Graft-Versus-Host Disease Model for Investigating the Pathological Role of Human CD4+ or CD8+ T Cells Using Immunodeficient NOG Mice. *Am. J. Transplant.* **2017**, *17*, 1216–1228. [CrossRef]

138. Seki, T.; Miyamoto, A.; Ohshima, S.; Ohno, Y.; Yasuda, A.; Tokuda, Y.; Ando, K.; Kametani, Y. Expression of glucocorticoid receptor shows negative correlation with human B-cell engraftment in PBMC-transplanted NOGhIL-4-Tg mice. *Biosci. Trends* **2018**, *12*, 247–256. [CrossRef]

International Journal of
Molecular Sciences

Review

Detection of Antigen-Specific T Cells Using In Situ MHC Tetramer Staining

Hadia M. Abdelaal [1,2], Emily K. Cartwright [1] and Pamela J. Skinner [1,3,*]

[1] Department of Veterinary and Biomedical Sciences, University of Minnesota, St. Paul, MN 55108, USA; moham698@umn.edu (H.M.A.); cartw082@umn.edu (E.K.C.)
[2] Department of Microbiology and Immunology, Zagazig University, Zagazig 44519, Egypt
[3] Microbiology Research Facility, 689 23rd Avenue SE, University of Minnesota, Twin Cities, MN 55455, USA
* Correspondence: skinn002@umn.edu; Tel.: +1-612-624-2644; Fax: +1-612-625-5203

Received: 1 October 2019; Accepted: 16 October 2019; Published: 18 October 2019

Abstract: The development of in situ major histocompatibility complex (MHC) tetramer (IST) staining to detect antigen (Ag)-specific T cells in tissues has radically revolutionized our knowledge of the local cellular immune response to viral and bacterial infections, cancers, and autoimmunity. IST combined with immunohistochemistry (IHC) enables determination of the location, abundance, and phenotype of T cells, as well as the characterization of Ag-specific T cells in a 3-dimensional space with respect to neighboring cells and specific tissue locations. In this review, we discuss the history of the development of IST combined with IHC. We describe various methods used for IST staining, including direct and indirect IST and IST performed on fresh, lightly fixed, frozen, and fresh then frozen tissue. We also describe current applications for IST in viral and bacterial infections, cancer, and autoimmunity. IST combined with IHC provides a valuable tool for studying and tracking the Ag-specific T cell immune response in tissues.

Keywords: T cells; In situ tetramer staining; MHC tetramer; immune response; antigen-specific; confocal microscopy; fresh tissue

1. Introduction

T cells play a pivotal role in the adaptive immune response. They perform a wide range of immune functions, including, but not limited to, providing help for B cells, protecting against intracellular and extracellular pathogens, detecting and killing cancer cells, and preventing autoimmunity [1]. T cells recognize the antigen (Ag) via a T cell receptor-cluster of differentiation 3(TCR-CD3) complex in the context of peptide-major histocompatibility complex (p-MHC) on the surface of antigen-presenting cells (APCs). Cluster of differentiation 4 (CD4)[+] T cells recognize antigens processed by APCs placed into a groove of MHC class II molecules (MHCII), whereas Cluster of differentiation 8 (CD8)[+] T cells recognize antigens presented by MHC class I molecules (MHCI). Regardless of the class of MHC, TCR: p-MHC interaction is required for initiating the T cell signaling cascade, leading to T cell activation [2].

The development of flow cytometric analysis of Ag-specific CD8[+] and CD4[+] T cells using fluorochrome-conjugated p-MHCI and p-MHCII tetramer staining, respectively, has dramatically increased our understanding of the cellular immune response [3,4]. Using flow cytometry, we are able to determine the quantity, function, and phenotype of Ag-specific T cells [5–7] and identify associations between the human leukocyte antigen (HLA) haplotype and disease progression [6–8]. Despite these important contributions, a major limitation of flow cytometry is the inability to visualize the localization of Ag-specific T cells, both with regards to their interactions with other cells, as well as their distribution within the tissue compartment. In addition, the dissociation of tissues into a single cell suspension for flow cytometry tends to underestimate the quantity of Ag-specific T cells within non-lymphoid tissues, such as the female reproductive tract (FRT), lung, and liver [9]. Thus, while flow cytometric analysis of Ag-specific T cells is extremely valuable, it fails to determine the spatial relationships between Ag-specific T cells and target cells and underestimates their total numbers in tissue, which are crucial for a complete understanding of the cellular adaptive immune response.

We developed a method for the in situ detection of Ag-specific CD8$^+$ T cells using MHCI tetramers. Using in situ MHCI tetramer (IST) staining combined with IHC, we were able to directly visualize and quantify Ag-specific CD8$^+$ T cells and their specific location within the tissue compartment [10]. We used fresh, unfixed, lightly fixed, and frozen spleens from TCR transgenic mice. Ag-specific CD8$^+$ T cells were readily detected in the spleens from the transgenic mice, and we found that fresh tissues by far produced the best quality staining [10]. At the same time as we were performing these studies, Haanen et al. developed and used similar IST staining methods combined with IHC to detect virus-specific CD8$^+$ T cells in TCR-transgenic and wild-type virus-infected mice, as well as to detect endogenous CD8$^+$ T cells directed against epitope tagged tumor cells in mice [11].

Since this time, we and others have also developed IST methods that use p-MHCII multimers to detect Ag-specific CD4$^+$ cells in situ [12–16]. With MHC class I and class II IST technologies, we are able to determine the spatial and temporal location and abundance of Ag-specific T cell responses in tissues. By combining IST with IHC, we are able to stain Ag-specific T cells, as well as cellular markers. Additional cellular markers allow phenotypic characterization of Ag-specific T cells and the surrounding cells in the tissue, which can include target cells. We have recently produced a video demonstrating IST staining [17], and these IST staining methods have previously been reviewed [2,18–20]. This review article builds on previous reviews, incorporates new methodologies, and describes more recently developed applications.

2. In Situ Tetramer Staining

Tetramers designed for IST are the same as those used in conventional flow cytometry. They both consist of four MHC monomers loaded with a specific peptide to interact with the T cell specific to that peptide [3].

3. Direct vs. Indirect IST

The two common methods used to detect Ag-specific T cells in situ are direct and indirect IST. Direct IST requires the use of MHCI or MHCII tetramer conjugated directly to a bright fluorophore, like allophycocyanin (APC) or phycoerythrin (PE) to directly label Ag-specific T cells [11,21,22]. Direct staining can also be done by using MHC-dextran multimers (dextramers) [16,23]. These multimers have more p-MHC complexes and more fluorochromes, allowing for brighter signal than standard tetramers with only one fluorophore. In addition, Tjernlund et al. used Qdot 655 multimers to directly detected SIV-specific CD8$^+$ T cells [7,24].

Indirect IST uses antibody staining directed against the fluorophore on the tetramer to amplify the signal [10,14,17,25–36]. For example, as described in Figure 1, the four biotinylated monomers are bound to an FITC-labeled ExtrAvidin molecule (a fluorescently labeled avidin). Conjugation to this FITC-avidin molecule allows amplification of the tetramer signal using an anti-FITC antibody. In this case, tissue is labeled with FITC-conjugated MHCI tetramers, followed by incubation with rabbit- α-FITC antibodies for signal amplification. Then, a secondary antibody, such as Cy3 labeled α-rabbit IgG, further amplifies the signal. Figure 1 also shows concurrent staining with CD3 antibodies to label T cells (in blue) and CD20 antibodies to label B cells (in green). Figure 2 shows a representative image of a spleen tissue section stained indirectly with FITC-conjugated MHCI tetramers to detect virus-specific CD8$^+$ T cells (in red) and counterstained with antibodies against CD3 to label T cells (in blue) and CD20 to detect B cells (in green).

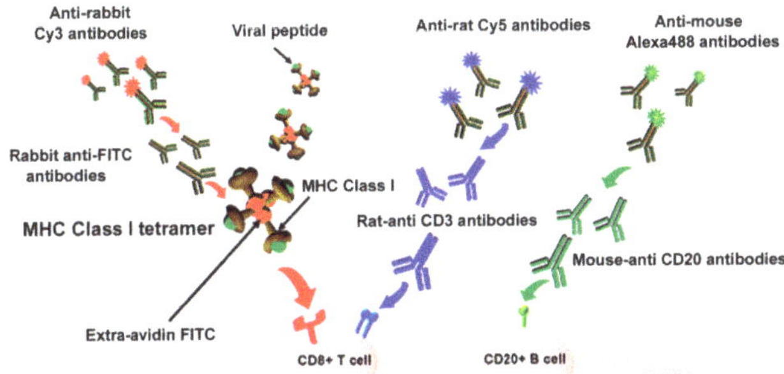

Figure 1. In situ major histocompatibility complex (MHC) class I (MHCI) tetramer staining combined with immunohistochemistry (IHC) to detect virus-specific CD8$^+$ T cells. Schematic diagram of in situ MHC tetramer (IST) combined with IHC to detect virus-specific CD8$^+$ T cells in fresh, unfixed tissue sections. An MHCI tetramer consists of four biotinylated MHC-class I monomers loaded with a viral peptide (or another antigenic peptide) bound to a fluorescently labeled avidin molecule. After primary incubation with MHCI tetramers, sections are fixed and then anti-FITC antibodies are used to amplify the tetramer signal. This signal is then further amplified using Cy3-tagged anti-Rabbit IgG antibodies. Sections can be counterstained with CD3 antibodies to label T cells (blue), and CD20 antibodies to label B cells (green).

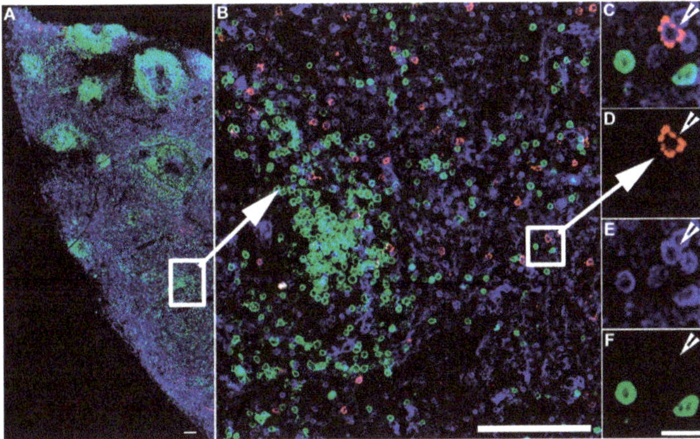

Figure 2. IST detection of virus-specific CD8$^+$ T cells. IST Combined with IHC in spleen sections from an SIV infected rhesus macaque. Fresh unfixed spleen section was stained with *Mamu-A*01* tetramers loaded with SIV Gag/CM9 peptides detect SIV-specific CD8$^+$ T cells (Red color), and counterstained with CD3 antibodies to label T cells blue, and CD20 antibodies to label B cells green and delineate B cell follicles. Confocal images were collected using a 20 X objective and 3 μm z-steps. (**A**) shows a montage of several projected confocal z-series fields. The scale bar = 100 μm. (**B**) shows an enlargement of the selected area in panel (**A**), which is a confocal Z-scan showing the distribution of tetramer$^+$ T cells within the spleen. The scale bar = 100 μm. (**C–F**) are enlargements for the selected area in panel B and shows that an SIV-specific CD8$^+$ T cell is tetramer$^+$ (**C,D**), CD3$^+$ (**E**), and CD20$^-$ (**F**), scale bars = 10 μm. Arrowheads point to a virus-specific CD8$^+$ T cell.

MHC tetramers conjugated to PE and APC can similarly be used for indirect staining [21,22,37–41]. In addition, antibodies directed against streptavidin can be used. For example, Vries et al. used indirect MHCI IST to detect melonoma-specific CD8$^+$ T cell following dendritic cell vaccination of melanoma patient, where they used a rabbit anti-streptavidin that recognizes MHCI tetramer-associated streptavidin molecules. They amplified the signal from the anti-streptavidin antibodies using goat-anti-rabbit Alexa594 [42]. Another application of indirect tetramer staining involves the use of the horseradish peroxidase (HRP)-conjugated tetramer. Instead of a fluorochrome, Yang et al. used tetramers conjugated to HRP–streptavidin and amplified the signal with the addition of biotin-conjugated tyramide [21,43].

Both methods have their advantages and drawbacks. Direct staining is a simpler procedure, can result in lower background staining, and provides more options to co-label other proteins since no secondary antibody is involved in labeling TCRs. However, direct staining provides a weaker signal intensity and is relatively more expensive because it requires as much as 40 times the tetramer of the indirect staining method [18]. In contrast, indirect labeling is a multi-step procedure that is more time consuming. Indirect staining, however, yields a more intense signal, resulting in a much higher signal to noise ratio and is relatively less expensive because it requires smaller amounts of the tetramer reagents.

4. IST Staining on Fresh and Frozen Tissue

IST staining can be done on fresh tissue sections, fresh then frozen tissue, or frozen tissue sections. In situ tetramer staining is ideally performed using unfixed, fresh tissue sections to maintain the structure and mobility of TCRs to interact with p-MHC tetramers [10,11]. To generate fresh 200 µm tissue sections, either a Vibratome or Compresstome can be used. However, a Compresstome is much more efficient in generating sections and accommodates larger section sizes [25]. While fresh tissue sections are ideal, there are some circumstances where fresh samples are not feasible. For example, some studies require that samples be shipped overnight. Some studies have limited tissue sampling, size availability, or their tissue was already frozen and archived. To determine if these conditions were feasible to perform IST, we performed IST on tissue samples that were stored at 4 °C overnight in PBS, lightly pre-fixed or frozen [10]. We found that there was no difference in the quality of the staining that was done on either spleen sections directly after dissection or spleen sections that were stored overnight in PBS at 4 °C. Moreover, we found that the IST also worked on lightly fixed spleen tissue from TCR transgenic mice (defined as 2% formaldehyde or 50% methanol and 50% acetone). While the IST worked on lightly fixed tissues, it yielded a higher background and less intense signal than the fresh, unfixed tissue. Additionally, IST worked on 10 µm-thick frozen sections but also resulted in weaker signal intensity compared to that from fresh tissue section [10]. Vyth-Dreese et al. compared direct tetramer staining on fresh viable tissue sections versus cryopreserved tissue sections and was able to detect Ag-specific T cells only in viable tissue sections. However, they were able to detect Ag-specific T cells in fresh skin tissue sections that were pre-stained with tetramers and then cryopreserved [22]. Similarly, others have now successfully performed indirect IST on fresh tissue pre-incubated with tetramers, and then fixed and snap-frozen. Later, frozen sectioning was done followed by IHC and tetramer amplification [38,44].

For staining frozen sections, IST has been described on unfixed and fixed tissue samples. Fixation was done before tetramer incubation, after tetramer incubation, or fixed both before and after tetramer incubation. We performed IST on unfixed spleen tissue stored in OCT freezing medium, and fixation was done post-tetramer incubation [10]. Similarly, Oerke et al. and Tully et al. used indirect IST using PE tetramers on frozen sections, and fixation was done after tetramer incubation [40,41]. Tjernlund et al. used Qdot 655 multimers to directly stain frozen section where fixation was done post-tetramer incubation [7,24]. In addition, Yuhong et al. used indirect IST to detect *Mycobacterium tuberculosis* (*M. tb*)-specific CD4$^+$ T cells in lymph node and lung from untreated *M. tb* patients. For this study, IST was done on frozen sections that were first fixed in 4% PFA then incubated overnight with tetramer [39]. Similarly, Vries et al. lightly fixed frozen tissue sections before starting IST and fixed them again after incubation with tetramers [42].

In summary, performing IST on fresh (not frozen) unfixed tissue has several advantages compared to pre-fixed or thin frozen sections. IST with fresh tissue sections results in the highest staining intensity over background fluorescence. In addition, the use of fresh 200 μm-thick sections provide more information about the tissue because it allows the examination of 20 times more tissue than a thin 10 μm-thick frozen sections. When trying to detect a rare population of Ag-specific T cells, the more tissue examined the greater chances are of detecting rare cells. Moreover, the thick fresh sections can be examined using confocal microscopy to provide a 3-D view of the location and the interaction of Ag-specific CD8$^+$ T cells with other cells and tissue structures. On the other hand, frozen sections offer some great advantages in that they enable the detection of Ag-specific T cells in archived samples. Additionally, with frozen sections, tissue samples can be stored and processed when needed, which makes it easier to answer future questions that might arise.

5. Specificity and Sensitivity of IST

A critical and remarkable property of the cellular adaptive immune response is specificity, where selective activation and expansion of a very small population of Ag-specific T cells is required. Therefore, the specificity and sensitivity of IST is the key to its success in detecting such small fractions of Ag-specific T cells [4]. Because high background autofluorescence is inherent when imaging whole tissues, ensuring the proper negative controls for IST staining is crucial [10]. There are several methods used to confirm the specificity of tetramer staining using IST. In the interaction between the p-MHC complex and TCR, both the amino acid sequence of the peptide and the haplotype of MHC are critical for determining specificity of the T cell. Both of these variables can be altered in an experiment to ensure the tetramer staining is specific. When changing the amino acid sequence of the peptide, the same fluorescently labeled MHC molecule is used, but the peptide loaded should not be present in the experimental system. For example, in studies of *Mamu-A*001:01* rhesus macaques, FITC-labeled *Mamu-A*001:01* tetramers loaded with an irrelevant peptide FV10 (FLPSDYFPSV), a peptide from the hepatitis B viral core protein served as a negative control for FITC-labeled MHCI *Mamu-A*001:01* SIV GagCM9 (181–189) (CTPYDINQM) tetramers and for FITC-labeled MHCI *Mamu-A*001:01* SIV Tat STPESANL (SL8) tetramers [17,29,30,35]. As another example, a study using tissues from human study participants used HLA- B*57 tetramers loaded with MART (ELAGIGILTV), a peptide from melanoma protein, to serve as a negative control for FITC-labeled HLA- B*57 HIV-1 Gag IW9 (ISPRTLNAW) or QW9 (QASQEVKNW) tetramers during detection of HIV-specific tissue-resident CD8$^+$ T cells within the gastrointestinal tract in a chronic infection [27].

Alternatively, studies have revealed specificity by using MHC-mismatched tetramers. These are tetramers loaded with the peptide of interest but not able to bind to T cells in the tissue due to cells in the tissue expressing different MHCI molecules [10,11]. A third type of negative control includes using a tissue that does not have Ag-specific cells of interest. In this case, the tissue and tetramer are the same haplotype, but the individual animal or study participant that was sampled was not infected with microbes [10,18,21].

We found that the specificity and sensitivity of IST staining is comparable to that of flow cytometry. Following the adoptive transfer of transgenic T cells into a wild-type mouse, the spleen of the recipient mouse was split in half. One half was used to determine the number of tetramer$^+$ CD8$^+$ T cells using flow cytometry, and the other half was used for IST staining. Both techniques showed that ~1% of the CD8$^+$ cells were tetramer$^+$ [10]. Haanen et al. found that the background staining in tissues permits detection limits of 0.1–1% of T cells, whereas the limit of detection of flow cytometry is less than 0.1% [13]. Nonetheless, the sensitivity of IST is sufficient to detect endogenous antigen-specific T cell responses.

6. Applications for In Situ Tetramer Staining

As mentioned previously, traditional tissue processing for MHC tetramer staining by flow cytometry requires dissociation of the tissue into a single cell suspension. Though flow cytometry is powerful in providing information about the phenotypes of Ag-specific cells, it does not show where specifically in the tissue they are located, what the phenotype of cells are in specific locations or what cells they are interacting within specific tissue compartments. This information can be critical for understanding T cell responses to infections, predicting vaccine efficacy, and investigating an immune response within the tumor microenvironment.

In HIV and SIV infection, CD4[+] T follicular helper (T_{FH}) cells are a major site of viral persistence during antiretroviral therapy and are a critical barrier to eradication [34,45–50]. Studies using IST have shown that low levels of virus-specific CD8[+] T cells in lymphoid follicles permit ongoing viral replication in T_{FH} [33,36]. This knowledge can inform future vaccine and cell therapy design for HIV infection by monitoring or inducing the accumulation of HIV-specific CD8[+] T cells in B cell follicles [17,28,29,34–36,51].

Another application combines IST with in situ hybridization (ISTH) to determine a direct spatial and temporal relationship between virus-specific CD8[+] T cells (effector cells) and virus-infected target cells. Looking at both the SIV infection of non-human primates (NHP) and LCMV infection in mice, it was determined that the location, timing, and abundance of antigen-specific T cells directly relates to the number of infected target cells [26,52].

In barrier tissues, like the lung, gut, and skin, CD4[+] and CD8[+] T cells take up residence following an infection. These tissue resident T cells (T_{RM}), are the first line of defense against secondary pathogen exposure. IST staining has been used to increase our understanding of the immune response at these sites, including cell types required for the generation and maintenance of T_{RM}. In a murine model of HSV-2 infection, IST staining was used to determine CD301b[+] dendritic cells (DCs) are critical for initiating and maintaining the CD8[+] T_{RM} population in the female reproductive tract (FRT) [53]. They show that activation by CD301b[+] DCs activates CD8[+] T_{RM} to produce interferon-gamma (IFN-γ), and this response is necessary for protection from HSV's challenge.

IST staining has also been used to visualize in situ dynamics of the immune response to *Listeria monocyotogenes* (LM) [54]. As early as day 3 post-infection, LM-specific CD8[+] T cells are detected in the spleen of infected mice and located at the border of the T and B cell zones. They also show an interaction of LM-specific CD8[+] T cells with CD11c[+] DCs in clustered foci within the T cell zones. Interestingly, after both influenza and LM infection, memory Ag-specific CD8[+] T cells can be found within B cell follicles. Upon a secondary challenge, there are many more foci of Ag-specific CD8[+] T cells within T cell zones, consistent with a robust CD8[+] memory T cell response to the secondary challenge.

Outside of infectious diseases, IST staining can be applied to other fields, including cancer biology and autoimmunity. In one study, IST was used to determine the feasibility of a dendritic cell-based vaccine for melanoma. In the three study participants examined, researchers detected tumor-specific CD8[+] T cells within the tumor following vaccination [42]. This has important implications for predicting the efficacy of cancer vaccines, as circulating T cell responses are not always a good indicator of protection. In another study, researchers investigated the CD8[+] T cell response in type 1 diabetes (T1D) [55]. Using IST, they described the first confirmation of Ag-specific, autoreactive CD8[+] T cells in the islet lesions from T1D patients. Interestingly, they also showed that recent onset patients (<1 year duration of the disease) had a more clonally restricted CD8[+] T cell response in the islets, whereas long standing patients (>1 year duration of the disease) had a more diverse CD8[+] T cell response [55]. Much like predicting vaccine efficacy, understanding the immune response at the site of autoimmunity, and not just in peripheral blood, is critical to improving future treatment of cancer and autoimmune diseases.

While the bulk of IST staining has been done using MHCI tetramers to study the CD8[+] T cell response, MHCII tetramers are available to investigate the CD4[+] T cell response to infections and autoimmunity. In one study examining patients with active *Mycobacteriaum tuberculosis* (*M. tb*) infection, IST staining showed Ag-specific CD4[+] T cells producing IFN-γ and TNF-α in the lymph nodes, lung granulomas, and cavernous tissue [12]. In experimental autoimmune encephalomyelitis (EAE), antigen-specific CD4$^+$ T cells were detected in the lymph nodes and central nervous system (CNS) of diseased animals, but not in uninfected animals [13].

7. Limitations of IST Staining

While there are many important applications for IST staining, this technique has significant limitations. Many of these limitations are a factor of tetramer technology. Unlike a peptide pool, used to broadly probe an antigen-specific T cell response, a tetramer can only have one peptide presented and, therefore, will only interact with T cells specific for that peptide. This can potentially cause researchers to underestimate the total immune response to a pathogen or vaccine, and similarly, can lead to over interpretation of results for that epitope. Most often, tetramers are made against the immunodominant epitope. The immunodominant epitope is the peptide that the majority of the CD4$^+$ or CD8$^+$ T cell response is generated against. However, there are often subdominant responses that might be overlooked with tetramer staining.

Because tetramer technology takes advantage of the specific interaction between the p-MHC complex and the TCR, use of the technology requires determining the MHC genotype of an individual prior to examining the Ag-specific response. There may be limitations in the availability of tetramers for MHC molecules encoded by a particular allele. Additionally, there is the possibility that the immune response you are visualizing cannot be generalized to other MHC molecules. For example, people expressing *HLA-B27* and *-B57* MHC molecules are more often elite controllers of HIV infection [56,57]. In rhesus macaques, *Mamu-A*001:01* [58], *-B*008:01 molecules* [59], and *-B*017:01* [60] alleles are also associated with enhanced control of SIV infection. While understanding the immune response in elite controllers of HIV/SIV infection is valuable information, it is not generalizable to all CD8$^+$ T cell responses during HIV/SIV infection.

Another limitation that can be a problem for the in situ visualization of Ag-specific CD4$^+$ T cells is the affinity threshold. The affinity threshold required for staining of the p-MHC with tetramer is higher than that required for activation of TCR, which biases tetramer technology to detect primarily high affinity T cells [61]. Work in recent years has shed light on low affinity T cells contribute significantly to the immune response [62,63]. Though this has been described primarily for Ag-specific CD4$^+$ T cells, it can be found in Ag-specific CD8$^+$ T cells, as well [64]. Researchers have begun to address this limitation by increasing the number of p-MHC complexed in multimers. They have changed the scaffold from biotin to dextran which allows more p-MHC and more fluorescent molecules to bind, both of which help increase the detection of antigen-specific T cells [65].

One of the challenges unique to IST staining is that the best results are obtained using fresh tissue samples [18,21]. We showed that while fixed and frozen tissue can be used in IST, the best results are gathered from fresh, unfixed samples [10]. Nonetheless, sometimes experimental conditions do not allow for the use of fresh, unfixed samples.

We have not been successful at getting IST staining to work well and consistently in different experimental systems with fixed or frozen tissues. However, as mentioned above, others have demonstrated success in their experimental systems. Certainly, having a robust, reliable method to track and phenotype Ag-specific T cells in fixed and frozen tissues would be a great advantage to the scientific community. Advancements in existing IST staining methodologies, or the development of new methods, are warranted to achieve this goal. Future methods to detect Ag-specific T cells in fixed and frozen tissues may be on the horizon and may not rely on MHC-tetramers or multimers. For example, in situ hybridization methods that detect the unique hypervariable regions of TCR genes, termed complementarity-determining regions (CDRs), might be an effective means to track Ag-specific T cells in fixed and frozen sections. Indeed, Advanced Cell Diagnostics have recently developed an in situ hybridization method called BaseScope that may allow detection of CDRs.

In summary, IST combined with IHC has radically enhanced our understanding of the Ag-specific T cell response. Not only does it enable the determination of the magnitude and phenotype of Ag-specific CD4$^+$ and CD8$^+$ T-cell responses in situ, but it also is a critical tool in tracking their location within tissue compartments and cell–cell interactions. IST staining has been, and continues to be, used to enhance our understanding of the local cellular immune response in many areas of research, including cancer biology, vaccinology, viral pathogenesis, bacterial infection, and autoimmune diseases.

Author Contributions: H.M.A wrote the manuscript and made the figures; E.K.C. assisted with drafting the manuscript; P.J.S. obtained funding, helped in drafting the manuscript, and provided oversight.

Funding: This work was funded by National Institutes of Health, grant numbers 1R01 AI143380-01 and 1UM1AI26617 and the APC was funded by 1R01 AI143380-01 and 1UM1AI26617.

Conflicts of Interest: The authors declare no conflict of interest.

References

1. Kumar, B.V.; Connors, T.J.; Farber, D.L. Human T Cell Development, Localization, and Function throughout Life. *Immunity* **2018**, *48*, 202–213. [CrossRef] [PubMed]
2. Constantin, C.M.; Bonney, E.E.; Altman, J.D.; Strickland, O.L. Major Histocompatibility Complex (MHC) Tetramer Technology: An Evaluation. *Biol. Res. Nurs.* **2002**, *4*, 115–127. [CrossRef] [PubMed]
3. Altman, J.D.; Moss, P.A.H.; Goulder, P.J.R.; Barouch, D.H.; McHeyzer-Williams, M.G.; Bell, J.I.; McMichael, A.J.; Davis, M.M. Phenotypic Analysis of Antigen-Specific T Lymphocytes. *Science* **1996**, *274*, 94–96. [CrossRef] [PubMed]
4. Novak, E.J.; Liu, A.W.; Nepom, G.T.; Kwok, W.W. MHC Class II Tetramers Identify Peptide-Specific Human CD4+ T Cells Proliferating in Response to Influenza A Antigen. *J. Clin. Investig.* **1999**, *104*, 63–67. [CrossRef] [PubMed]
5. Goulder, P.J.R.; Watkins, D.I. Impact of MHC Class I Diversity on Immune Control of Immunodeficiency Virus Replication. *Nat. Rev. Immunol.* **2008**, *8*, 619–630. [CrossRef] [PubMed]
6. Kuroda, M.J.; Schmitz, J.E.; Charini, W.A.; Nickerson, C.E.; Lifton, M.A.; Lord, C.I.; Forman, M.A.; Letvin, N.L. Emergence of CTL Coincides with Clearance of Virus during Primary Simian Immunodeficiency Virus Infection in Rhesus Monkeys. *J. Immunol.* **1999**, *162*, 5127–5133.
7. Tjernlund, A.; Zhu, J.; Laing, K.; Diem, K.; McDonald, D.; Vazquez, J.; Cao, J.; Ohlen, C.; McElrath, M.J.; Picker, L.J.; et al. In Situ Detection of Gag-Specific CD8+cells in the GI Tract of SIV Infected Rhesus Macaques. *Retrovirology* **2010**, *7*, 15–17. [CrossRef]
8. Loffredo, J.; Valentine, L.; Watkins, D. Beyond Mamu-A* 01+ Indian Rhesus Macaques: Continued Discovery of New MHC Class I Molecules That Bind Epitopes from the Simian AIDS Viruses. *HIV Mol. Immunol.* **2006**, 29–51.
9. Steinert, E.M.; Schenkel, J.M.; Fraser, K.A.; Beura, L.K.; Manlove, L.S.; Igyártó, B.Z.; Southern, P.J.; Masopust, D. Quantifying Memory CD8 T Cells Reveals Regionalization of Immunosurveillance HHS Public Access. *Cell* **2015**, *161*, 737–749. [CrossRef]
10. Skinner, P.J.; Daniels, M.A.; Schmidt, C.S.; Jameson, S.C.; Haase, A.T. Cutting Edge: In Situ Tetramer Staining of Antigen-Specific T Cells in Tissues. *J. Immunol.* **2000**, *165*, 613–617. [CrossRef]
11. Haanen, J.B.; van Oijen, M.G.; Tirion, F.; Oomen, L.C.; Kruisbeek, A.M.; Vyth-Dreese, F.A.; Schumacher, T.N. In Situ Detection of Virus- and Tumor-Specific T-Cell Immunity. *Nat. Med.* **2000**, *6*, 1056–1060. [CrossRef]
12. Li, Y.; Zhu, Y.; Zhou, L.; Fang, Y.; Huang, L.; Ren, L.; Peng, Y.; Li, Y.; Yang, F.; Xie, D.; et al. Use of HLA-DR*08032/E7 and HLA-DR*0818/E7 Tetramers in Tracking of Epitope-Specific CD4 + T Cells in Active and Convalescent Tuberculosis Patients Compared with Control Donors. *Immunobiology* **2011**, *216*, 947–960. [CrossRef] [PubMed]
13. Bischof, F.; Hofmann, M.; Schumacher, T.N.M.; Vyth-Dreese, F.A.; Weissert, R.; Schild, H.; Kruisbeek, A.M.; Melms, A. Analysis of Autoreactive CD4 T Cells in Experimental Autoimmune Encephalomyelitis after Primary and Secondary Challenge Using MHC Class II Tetramers. *J. Immunol.* **2004**, *172*, 2878–2884. [CrossRef] [PubMed]
14. Dileepan, T.; Kim, H.O.; Cleary, P.P.; Skinner, P.J. In Situ Peptide-MHC-II Tetramer Staining of Antigen-Specific CD4+ T Cells in Tissues. *PLoS ONE* **2015**, *10*, e0128862. [CrossRef] [PubMed]
15. Massilamany, C.; Gangaplara, A.; Jia, T.; Elowsky, C.; Li, Q.; Zhou, Y.; Reddy, J. In Situ Detection of Autoreactive CD4 T Cells in Brain and Heart Using Major Histocompatibility Complex Class II Dextramers. *J. Vis. Exp.* **2014**, *90*, 1–7. [CrossRef]

16. Massilamany, C.; Gangaplara, A.; Jia, T.; Elowsky, C.; Kang, G.; Riethoven, J.J.; Li, Q.; Zhou, Y.; Reddy, J. Direct Staining with Major Histocompatibility Complex Class II Dextramers Permits Detection of Antigen-Specific, Autoreactive CD4 T Cells in Situ. *PLoS ONE* **2014**, *9*, e87519. [CrossRef]
17. Li, S.; Mwakalundwa, G.; Skinner, P.J. In Situ MHC-Tetramer Staining and Quantitative Analysis to Determine the Location, Abundance, and Phenotype of Antigen-Specific CD8 T Cells in Tissues. *J. Vis. Exp.* **2017**, *2017*, 1–8. [CrossRef]
18. Skinner, P.J.; Haase, A.T. Skinner In Situ Tetramer Staining. *J. Immunol. Methods* **2002**, *268*, 29–34. [CrossRef]
19. Benechet, A.P.; Menon, M.; Khanna, K.M. Visualizing T Cell Migration in Situ. *Front. Immunol.* **2014**, *5*, 363. [CrossRef]
20. Skinner, P.J. In Situ MHC Tetramer Staining: In Situ Tetramers. In *Analyzing T Cell Responses: How to Analyze Cellular Immune Responses against Tumor Associated Antigens*; Springer: Amsterdam, The Netherlands, 2005; pp. 219–225. [CrossRef]
21. Skinner, P.J.; Haase, A.T. In Situ Staining Using MHC Class I Tetramers. *Curr. Protoc. Immunol.* **2004**, *64*, 17. [CrossRef]
22. Vyth-Dreese, F.A.; Kim, Y.H.; Dellemijn, T.A.M.; Schrama, E.; Haanen, J.B.A.G.; Spierings, E.; Goulmy, E. In Situ Visualization of Antigen-Specific T Cells in Cryopreserved Human Tissues. *J. Immunol. Methods* **2006**, *310*, 78–85. [CrossRef] [PubMed]
23. Andersen, M.H.; Pedersen, L.O.; Capeller, B.; Bröcker, E.B.; Becker, J.C.; thor Straten, P. Spontaneous Cytotoxic T-Cell Responses against Survivin-Derived MHC Class I-Restricted T-Cell Epitopes in Situ as Well as Ex Vivo in Cancer Patients. *Cancer Res.* **2001**, *61*, 5964–5968. [PubMed]
24. Tjernlund, A.; Burgener, A.; Lindvall, J.M.; Peng, T.; Zhu, J.; Öhrmalm, L.; Picker, L.J.; Broliden, K.; Mcelrath, M.J.; Corey, L. In Situ Staining and Laser Capture Microdissection of Lymph Node Residing SIV Gag-Specific CD8 + T Cells-A Tool to Interrogate a Functional Immune Response Ex Vivo. *PLoS ONE* **2016**, *11*, e0149907. [CrossRef] [PubMed]
25. Abdelaal, H.M.; Kim, H.O.; Wagstaff, R.; Sawahata, R.; Southern, P.J.; Skinner, P.J. Comparison of Vibratome and Compresstome Sectioning of Fresh Primate Lymphoid and Genital Tissues for in Situ MHC-Tetramer and Immunofluorescence Staining. *Biol. Proced. Online* **2015**, *17*, 2. [CrossRef] [PubMed]
26. Li, Q.; Skinner, P.J.; Ha, S.-J.; Duan, L.; Mattila, T.L.; Hage, A.; White, C.; Barber, D.L.; O'Mara, L.; Southern, P.J.; et al. Visualizing Antigen-Specific and Infected Cells in Situ Predicts Outcomes in Early Viral Infection. *Science* **2009**, *323*, 1726–1729. [CrossRef] [PubMed]
27. Kiniry, B.E.; Li, S.; Ganesh, A.; Hunt, P.W.; Somsouk, M.; Skinner, P.J.; Deeks, S.G.; Shacklett, B.L. Detection of HIV-1-Specific Gastrointestinal Tissue Resident CD8 + T-Cells in Chronic Infection. *Mucosal Immunol.* **2018**, *11*, 909–920. [CrossRef]
28. Webb, G.M.; Li, S.; Mwakalundwa, G.; Folkvord, J.M.; Greene, J.M.; Reed, J.S.; Stanton, J.J.; Legasse, A.W.; Hobbs, T.; Martin, L.D.; et al. The Human IL-15 Superagonist ALT-803 Directs SIV-Specific CD8+ T Cells into B-Cell Follicles. *Blood Adv.* **2018**, *2*, 76–84. [CrossRef]
29. Mylvaganam, G.H.; Rios, D.; Abdelaal, H.M.; Iyer, S.; Tharp, G.; Mavinger, M.; Hicks, S.; Chahroudi, A.; Ahmed, R.; Bosinger, S.E.; et al. Dynamics of SIV-Specific CXCR5+ CD8 T Cells during Chronic SIV Infection. *Proc. Natl. Acad. Sci. USA* **2017**, *114*, 1976–1981. [CrossRef]
30. Reynolds, M.R.; Rakasz, E.; Skinner, P.J.; White, C.; Abel, K.; Ma, Z.-M.; Compton, L.; Napoe, G.; Wilson, N.; Miller, C.J.; et al. CD8+ T-Lymphocyte Response to Major Immunodominant Epitopes after Vaginal Exposure to Simian Immunodeficiency Virus: Too Late and Too Little. *J. Virol.* **2005**, *79*, 9228–9235. [CrossRef]
31. Li, S.; Folkvord, J.M.; Rakasz, E.G.; Abdelaal, H.M.; Wagstaff, R.K.; Kovacs, K.J.; Kim, H.O.; Sawahata, R.; MaWhinney, S.; Masopust, D.; et al. Simian Immunodeficiency Virus-Producing Cells in Follicles Are Partially Suppressed by CD8 + Cells In Vivo. *J. Virol.* **2016**, *90*, 11168–11180. [CrossRef]
32. Hong, J.J.; Reynolds, M.R.; Mattila, T.L.; Hage, A.; Watkins, D.I.; Miller, C.J.; Skinner, P.J. Localized Populations of CD8low/-MHC Class I Tetramer+ SIV-Specific T Cells in Lymphoid Follicles and Genital Epithelium. *PLoS ONE* **2009**, *4*, e4131. [CrossRef] [PubMed]
33. Connick, E.; Folkvord, J.M.; Lind, K.T.; Rakasz, E.G.; Miles, B.; Wilson, N.A.; Santiago, M.L.; Schmitt, K.; Stephens, E.B.; Kim, H.O.; et al. Compartmentalization of Simian Immunodeficiency Virus Replication within Secondary Lymphoid Tissues of Rhesus Macaques Is Linked to Disease Stage and Inversely Related to Localization of Virus-Specific CTL. *J. Immunol.* **2014**, *193*, 5613–5625. [CrossRef] [PubMed]

34. Connick, E.; Mattila, T.; Folkvord, J.M.; Schlichtemeier, R.; Meditz, A.L.; Ray, M.G.; McCarter, M.D.; MaWhinney, S.; Hage, A.; White, C.; et al. CTL Fail to Accumulate at Sites of HIV-1 Replication in Lymphoid Tissue. *J. Immunol.* **2007**, *178*, 6975–6983. [CrossRef] [PubMed]

35. Sasikala-Appukuttan, A.K.; Kim, H.O.; Kinzel, N.J.; Hong, J.J.; Smith, A.J.; Wagstaff, R.; Reilly, C.; Piatak, M.; Lifson, J.D.; Reeves, R.K.; et al. Location and Dynamics of the Immunodominant CD8 T Cell Response to SIVΔnef Immunization and SIVmac251 Vaginal Challenge. *PLoS ONE* **2013**, *8*, e81623. [CrossRef] [PubMed]

36. Li, S.; Folkvord, J.M.; Kovacs, K.J.; Wagstaff, R.K.; Mwakalundwa, G.; Rendahl, A.K.; Rakasz, E.G.; Connick, E.; Skinner, P.J. Low Levels of Siv-Specific CD8+ T Cells in Germinal Centers Characterizes Acute SIV Infection. *PLoS Pathog.* **2019**, *15*, e1007311. [CrossRef] [PubMed]

37. Stratmann, T.; Martin-Orozco, N.; Mallet-Designe, V.; Poirot, L.; McGavern, D.; Losyev, G.; Dobbs, C.M.; Oldstone, M.B.A.; Yoshida, K.; Kikutani, H.; et al. Susceptible MHC Alleles, Not Background Genes, Select an Autoimmune T Cell Reactivity. *J. Clin. Investig.* **2003**, *112*, 902–914. [CrossRef] [PubMed]

38. Tan, H.-X.; Wheatley, A.K.; Esterbauer, R.; Jegaskanda, S.; Glass, J.J.; Masopust, D.; De Rose, R.; Kent, S.J. Induction of Vaginal-Resident HIV-Specific CD8 T Cells with Mucosal Prime-Boost Immunization. *Mucosal Immunol.* **2017**, *13*, 994. [CrossRef]

39. Huang, Y.; Huang, Y.; Fang, Y.; Wang, J.; Li, Y.; Wang, N.; Zhang, J.; Gao, M.; Huang, L.; Yang, F.; et al. Relatively Low Level of Antigen-Specific Monocytes Detected in Blood from Untreated Tuberculosis Patients Using CD4+ T-Cell Receptor Tetramers. *PLoS Pathog.* **2012**, *8*, e1001036. [CrossRef]

40. Oerke, S.; Höhn, H.; Zehbe, I.; Pilch, H.; Schicketanz, K.H.; Hitzler, W.E.; Neukirch, C.; Freitag, K.; Maeurer, M.J. Naturally Processed and HLA-B8-Presented HPV16 E7 Epitope Recognized by T Cells from Patients with Cervical Cancer. *Int. J. Cancer* **2005**, *114*, 766–778. [CrossRef]

41. Tully, G.; Kortsik, C.; Höhn, H.; Zehbe, I.; Hitzler, W.E.; Neukirch, C.; Freitag, K.; Kayser, K.; Maeurer, M.J. Highly Focused T Cell Responses in Latent Human Pulmonary Mycobacterium Tuberculosis Infection. *J. Immunol.* **2005**, *174*, 2174–2184. [CrossRef]

42. De Vries, I.J.M.; Bernsen, M.R.; Van Geloof, W.L.; Scharenborg, N.M.; Lesterhuis, W.J.; Rombout, P.D.M.; Van Muijen, G.N.P.; Figdor, C.G.; Punt, C.J.A.; Ruiter, D.J.; et al. In Situ Detection of Antigen-Specific T Cells in Cryo-Sections Using MHC Class I Tetramers after Dendritic Cell Vaccination of Melanoma Patients. *Cancer Immunol. Immunother.* **2007**, *56*, 1667–1676. [CrossRef] [PubMed]

43. Yang, J.; Jaramillo, A.; Liu, W.; Olack, B.; Yoshimura, Y.; Joyce, S.; Kaleem, Z.; Mohanakumar, T. Chronic Rejection of Murine Cardiac Allografts Discordant at the H13 Minor Histocompatibility Antigen Correlates with the Generation of the H13-Specific CD8+ Cytotoxic T Cells. *Transplantation* **2003**, *76*, 84–91. [CrossRef] [PubMed]

44. Lee, Y.J.; Wang, H.; Starrett, G.J.; Phuong, V.; Jameson, S.C.; Hogquist, K.A. Tissue-Specific Distribution of INKT Cells Impacts Their Cytokine Response. *Immunity* **2015**, *43*, 566–578. [CrossRef] [PubMed]

45. Fukazawa, Y.; Lum, R.; Okoye, A.A.; Park, H.; Matsuda, K.; Bae, J.Y.; Hagen, S.I.; Shoemaker, R.; Deleage, C.; Lucero, C.; et al. B Cell Follicle Sanctuary Permits Persistent Productive Simian Immunodeficiency Virus Infection in Elite Controllers. *Nat. Med.* **2015**, *21*, 132–139. [CrossRef]

46. Gratton, S.; Cheynier, R.; Dumaurier, M.J.; Oksenhendler, E.; Wain-Hobson, S. Highly Restricted Spread of HIV-1 and Multiply Infected Cells within Splenic Germinal Centers. *Proc. Natl. Acad. Sci. USA* **2000**, *97*, 14566–14571. [CrossRef]

47. Perreau, M.; Savoye, A.-L.; De Crignis, E.; Corpataux, J.-M.; Cubas, R.; Haddad, E.K.; De Leval, L.; Graziosi, C.; Pantaleo, G. Follicular Helper T Cells Serve as the Major CD4 T Cell Compartment for HIV-1 Infection, Replication, and Production. *J. Exp. Med.* **2013**, *210*, 143–156. [CrossRef]

48. Embretson, J.; Zupancic, M.; Ribas, J.L.; Burke, A.; Racz, P.; Tenner-Racz, K.; Haase, A.T. Massive Covert Infection of Helper T Lymphocytes and Macrophages by HIV during the Incubation Period of AIDS. *Nature* **1993**, *362*, 359–362. [CrossRef]

49. Hufert, F.T.; van Lunzen, J.; Janossy, G.; Bertram, S.; Schmitz, J.; Haller, O.; Racz, P.; von Laer, D. Germinal Centre CD4+ T Cells Are an Important Site of HIV Replication in Vivo. *AIDS* **1997**, *11*, 849–857. [CrossRef]

50. Folkvord, J.M.; Armon, C.; Connick, E. Lymphoid Follicles Are Sites of Heightened Human Immunodeficiency Virus Type 1 (HIV-1) Replication and Reduced Antiretroviral Effector Mechanisms. *AIDS Res. Hum. Retrovir.* **2005**, *21*, 363–370. [CrossRef]

51. Haran, K.P.; Hajduczki, A.; Pampusch, M.S.; Mwakalundwa, G.; Vargas-Inchaustegui, D.A.; Rakasz, E.G.; Connick, E.; Berger, E.A.; Skinner, P.J. Simian Immunodeficiency Virus (SIV)-Specific Chimeric Antigen Receptor-T Cells Engineered to Target B Cell Follicles and Suppress SIV Replication. *Front. Immunol.* **2018**, *9*, 492. [CrossRef]

52. Li, Q.; Skinner, P.J.; Duan, L.; Haase, A.T. A Technique to Simultaneously Visualize Virus-Specific CD8+ T Cells and Virus-Infected Cells in Situ. *J. Vis. Exp.* **2009**, *30*, 11–13. [CrossRef] [PubMed]

53. Shin, H.; Kumamoto, Y.; Gopinath, S.; Iwasaki, A. CD301b+ Dendritic Cells Stimulate Tissue-Resident Memory CD8+ T Cells to Protect against Genital HSV-2. *Nat. Commun.* **2016**, *7*, 13346. [CrossRef] [PubMed]

54. Khanna, K.M.; McNamara, J.T.; Lefrançois, L. In Situ Imaging of the Endogenous CD8 T Cell Response to Infection. *Science* **2007**, *318*, 116–120. [CrossRef] [PubMed]

55. Coppieters, K.T.; Dotta, F.; Amirian, N.; Campbell, P.D.; Kay, T.W.H.; Atkinson, M.A.; Roep, B.O.; von Herrath, M.G. Demonstration of Islet-Autoreactive CD8 T Cells in Insulitic Lesions from Recent Onset and Long-Term Type 1 Diabetes Patients. *J. Exp. Med.* **2012**, *209*, 51–60. [CrossRef]

56. Kaslow, R.A.; Carrington, M.; Apple, R.; Park, L.; Muñoz, A.; Saah, A.J.; Goedert, J.J.; Winkler, C.; O'Brien, S.J.; Rinaldo, C.; et al. Influence of Combinations of Human Major Histocompatibility Complex Genes on the Course of HIV-1 Infection. *Nat. Med.* **1996**, *2*, 405–411. [CrossRef]

57. Migueles, S.A.; Sabbaghian, M.S.; Shupert, W.L.; Bettinotti, M.P.; Marincola, F.M.; Martino, L.; Hallahan, C.W.; Selig, S.M.; Schwartz, D.; Sullivan, J.; et al. HLA B*5701 Is Highly Associated with Restriction of Virus Replication in a Subgroup of HIV-Infected Long Term Nonprogressors. *Proc. Natl. Acad. Sci. USA* **2000**, *97*, 2709–2714. [CrossRef]

58. Mothe, B.R.; Weinfurter, J.; Wang, C.; Rehrauer, W.; Wilson, N.; Allen, T.M.; Allison, D.B.; Watkins, D.I. Expression of the Major Histocompatibility Complex Class I Molecule Mamu-A*01 Is Associated with Control of Simian Immunodeficiency Virus SIVmac239 Replication. *J. Virol.* **2003**, *77*, 2736–2740. [CrossRef]

59. Loffredo, J.T.; Maxwell, J.; Qi, Y.; Glidden, C.E.; Borchardt, G.J.; Soma, T.; Bean, A.T.; Beal, D.R.; Wilson, N.A.; Rehrauer, W.M.; et al. Mamu-B*08-Positive Macaques Control Simian Immunodeficiency Virus Replication. *J. Virol.* **2007**, *81*, 8827–8832. [CrossRef]

60. Yant, L.J.; Friedrich, T.C.; Johnson, R.C.; May, G.E.; Maness, N.J.; Enz, A.M.; Lifson, J.D.; O'Connor, D.H.; Carrington, M.; Watkins, D.I. The High-Frequency Major Histocompatibility Complex Class I Allele Mamu-B*17 Is Associated with Control of Simian Immunodeficiency Virus SIVmac239 Replication. *J. Virol.* **2006**, *80*, 5074–5077. [CrossRef]

61. Laugel, B.; Van Den Berg, H.A.; Gostick, E.; Cole, D.K.; Wooldridge, L.; Boulter, J.; Milicic, A.; Price, D.A.; Sewell, A.K. Different T Cell Receptor Affinity Thresholds and CD8 Coreceptor Dependence Govern Cytotoxic T Lymphocyte Activation and Tetramer Binding Properties. *J. Biol. Chem.* **2007**, *282*, 23799–23810. [CrossRef]

62. Martinez, R.J.; Andargachew, R.; Martinez, H.A.; Evavold, B.D. Low-Affinity CD4+ T Cells Are Major Responders in the Primary Immune Response. *Nat. Commun.* **2016**, *7*, 13848. [CrossRef] [PubMed]

63. Martinez, R.J.; Evavold, B.D. Lower Affinity T Cells Are Critical Components and Active Participants of the Immune Response. *Front. Immunol.* **2015**, *6*, 468. [CrossRef] [PubMed]

64. Rius, C.; Attaf, M.; Tungatt, K.; Bianchi, V.; Legut, M.; Bovay, A.; Donia, M.; thor Straten, P.; Peakman, M.; Svane, I.M.; et al. Peptide–MHC Class I Tetramers Can Fail to Detect Relevant Functional T Cell Clonotypes and Underestimate Antigen-Reactive T Cell Populations. *J. Immunol.* **2018**, *200*, 2263–2279. [CrossRef] [PubMed]

65. Dolton, G.; Lissina, A.; Skowera, A.; Ladell, K.; Tungatt, K.; Jones, E.; Kronenberg-Versteeg, D.; Akpovwa, H.; Pentier, J.M.; Holland, C.J.; et al. Comparison of Peptide-Major Histocompatibility Complex Tetramers and Dextramers for the Identification of Antigen-Specific T Cells. *Clin. Exp. Immunol.* **2014**, *177*, 47–63. [CrossRef] [PubMed]

MDPI

St. Alban-Anlage 66

4052 Basel

Switzerland

Tel. +41 61 683 77 34

Fax +41 61 302 89 18

www.mdpi.com

International Journal of Molecular Sciences Editorial Office

E-mail: ijms@mdpi.com

www.mdpi.com/journal/ijms